Cephalopod Behaviour SECOND EDITION

With their large brains, elaborate sense organs and complex behaviour, cephalopods are among the world's most highly evolved invertebrates. This second edition summarises the wealth of exciting new research data stemming from over 500 papers published since the first volume appeared. It adopts a comparative approach to causation, function, development and evolution as it explores cephalopod behaviour in natural habitats and the laboratory. Extensive colour and black-and-white photography illustrates various aspects of cephalopod behaviour to complement the scientific analysis. Covering the major octopus, squid and cuttlefish species as well as the shelled *Nautilus*, this is an essential resource for undergraduate and advanced students of animal behaviour, as well as researchers new to cephalopods, in fields such as neuroscience and conservation biology. By highlighting the gaps in current knowledge, the text looks to inform and to stimulate further study of these beautiful animals.

Roger T. Hanlon is a Senior Scientist at the Marine Biological Laboratory in Woods Hole, Massachusetts, and Professor of Ecology and Evolutionary Biology at Brown University. An expert SCUBA diver, he studies the behaviour of cephalopods across the globe and has showcased his research in over 40 television programmes, including for the BBC, NOVA, Discovery Channel and National Geographic.

John B. Messenger is a zoologist interested in sensory physiology and the neural bases of animal behaviour. He has taught at the universities of Cambridge, Naples and Sheffield, and has studied living cephalopods in several marine stations, including Banyuls-sur-Mer, Ine (Japan), Naples, Plymouth and Woods Hole.

Cephalopod
Behaviour

SECOND EDITION

ROGER T. HANLON

Marine Biological Laboratory, Woods Hole

JOHN B. MESSENGER

Department of Zoology, University of Cambridge

CAMBRIDGE
UNIVERSITY PRESS

CAMBRIDGE
UNIVERSITY PRESS

University Printing House, Cambridge CB2 8BS, United Kingdom

One Liberty Plaza, 20th Floor, New York, NY 10006, USA

477 Williamstown Road, Port Melbourne, VIC 3207, Australia

314–321, 3rd Floor, Plot 3, Splendor Forum, Jasola District Centre,
New Delhi – 110025, India

79 Anson Road, #06–04/06, Singapore 079906

Cambridge University Press is part of the University of Cambridge.

It furthers the University's mission by disseminating knowledge in the pursuit of
education, learning and research at the highest international levels of excellence.

www.cambridge.org
Information on this title: www.cambridge.org/9780521897853
DOI: 10.1017/9780511843600

Second edition © Roger T. Hanlon and John B. Messenger 2018

First published 1996
First paperback edition 1998
Sixth printing 2008
Second edition 2018

Printed in the United States of America by Sheridan Books, Inc

A catalogue record for this publication is available from the British Library.

Library of Congress Cataloging-in-Publication Data
Names: Hanlon, Roger T., author. | Messenger, J. B., author.
Title: Cephalopod behaviour / Roger T. Hanlon, Marine Biological Laboratory, Woods Hole,
John B. Messenger, University of Cambridge.
Description: Cambridge, United Kingdom ; New York : Cambridge University Press, 2017. |
Includes bibliographical references and index.
Identifiers: LCCN 2017005649 | ISBN 9780521897853 (hardback : alk. paper) |
ISBN 9780521723701 (paperback : alk. paper)
Subjects: LCSH: Cephalopoda–Behavior.
Classification: LCC QL430.2 .H37 2017 | DDC 594/.5–dc23 LC record available at
https://lccn.loc.gov/2017005649

ISBN 978-0-521-89785-3 Hardback
ISBN 978-0-521-72370-1 Paperback

Additional resources for this publication at www.cambridge.org/cephalopods

Dedication

RTH dedicates this book to Arlene, Erin and Grayson; and to many mentors and colleagues along the way, but especially to Patrick Bateson, Robert Hinde and Martin Wells who taught him the principles and values of studying the complex subject of behaviour.

JBM dedicates this second edition to the memory of his parents, Maggie and Frank, who encouraged his childhood fascination with animals.

Contents

Online Videos That Complement This Book

Animal behaviour often involves sequences of action, and many of the data presented in this book were extracted from field and lab video. Therefore, we provide additional resources to enable interested readers to witness the dynamics of many aspects of cephalopod behaviour. These are available at www.cambridge.org/cephalopods.

Preface to the Second Edition

As many readers will know, this new edition is long overdue, partly because of personal problems experienced by both authors, but also because we have had to examine a far larger number of new papers than we ever imagined when, many years ago, we cheerfully agreed to Cambridge University Press's invitation to write a second edition. It would be gratifying to think that our first edition helped stimulate new research into cephalopod behaviour, but there is no doubt that in the past 20 years, there has been a welcome surge of fascinating investigations into many aspects of the lives of these animals. As a result, we have come to realise, for example, that squid can detect 'infra-sound' and that polarisation sensitivity is far more important in the lives of cephalopods than previously thought. Some of this new work has been done in the laboratory, but there

have also been some beautiful field observations – of, for example, the horizontal and vertical movements of the jumbo squid (revealed by special tagging); of the extraordinary reproductive behaviour of the giant Australian cuttlefish in shallow waters; and, in the depths, a glimpse of the complex flashing behaviour of the squid *Taningia*, as well as the first ever observations of living giant squid from submersibles.

In this edition, we have attempted to do justice to all this new work, with the inevitable result that the book has become very long, for which we apologise. The subject matter of the new edition is organised in the same way as in the first. The majority of the chapters reflect the combined inputs of RTH and JBM; but Chapters 5, 6 and 11 are the work of RTH alone. As before, the order of the authors is alphabetical.

Preface to the First Edition

Cephalopods are a very ancient and specialised group within the Mollusca. They are highly evolved, and in many features of their behaviour they have more in common with fish than with other invertebrates. In this book, we attempt to summarise what is known about the behaviour of modern coleoids – the cuttlefish, squid and octopuses – and their ancient relative, *Nautilus*. Curiously, given the great interest in cephalopod behaviour for more than half a century, it is the first book devoted exclusively to this theme. We have brought together field and laboratory data to present as complete a picture as we can of the life of cephalopods. In each chapter, we have drawn attention to those aspects of behaviour that will, we think, merit further study, not only to advance knowledge of cephalopods, but to add comparative data to behavioural biology. Where information is available, we have drawn on current theory in our analysis, but generally we have tried to avoid interpretations in terms of theory that may be out of date in a few years. We prefer to let the data speak for themselves. Our conceptual approach is primarily ethological, although we include functional morphology, physiology and neurobiology as needed. Where possible, we have considered questions of the function, causation, development and evolution of cephalopod behaviour. By the end, we hope to have shed some light on how and why these short-lived creatures with large brains behave the way they do.

The book is aimed at three audiences. First, our fellow 'teuthologists', biologists who study cephalopods and are experts in their respective fields, but who might find that knowledge of cephalopod behaviour could help them understand the biology of the group better. Second, students of animal behaviour (be they ethologists, behavioural ecologists, sociobiologists or comparative psychologists; undergraduates, graduates or faculty members), biologists who are more familiar with their own various kinds of animals but who might

be interested in looking over the hedge into someone else's garden, albeit briefly. To these latter we would say at the outset: don't expect too much from us, nor judge us too harshly. Our garden has not been cultivated for very long, and our animals are not easily accessible. Third, fisheries and conservation biologists, who might find that information about behaviour, especially reproductive behaviour and ecology, could help them manage their economically important resources better. The book is not aimed specifically at neurobiologists, neuroethologists or biomedical scientists, although, of course, we very much hope they will find some things of interest to them. This does not mean that we regard the nervous system as unimportant, nor that there are no new problems on the neural basis of cephalopod behaviour. Far from it! But much of this work has been before the public eye for some time, and most of it is easily accessible in books or reviews (see Chapter 1); moreover, M. Nixon and J.Z. Young are about to publish their definitive account of cephalopod neurobiology: *The Brains and the Lives of Cephalopods*.

We apologise to those authors whose work does not appear here; in a book of this sort, not every reference could be cited, and inevitably we may have overlooked a paper by an especially productive author. We are particularly aware that we may not have done justice to our Japanese and Russian colleagues. Neither of us reads those languages and having to rely on inadequate English summaries must mean that we have missed some points that would have added to our general theme. On the other hand, there is not, to the best of our belief, a substantial body of behavioural data in either language.

Both the authors are zoologists by training and by inclination, but their backgrounds and expertise differ. RTH has wide experience of cephalopods underwater and has logged many hundreds of dives in the Caribbean, the Mediterranean, the Red Sea and throughout the Indo-Pacific. He is also proficient in mariculture and for the past two decades has led a team that has cultured nearly 20 shallow-water cephalopod species through their life cycles. Trained in marine biology and oceanography, his interests are catholic but have revolved around the behaviour of cephalopods. JBM is a laboratory-based scientist with interests in the broad area of neurobiology and sensory physiology; he has carried out learning experiments with octopuses and cuttlefish in the Naples Zoological Station, and has also done experimental work with a variety of cephalopods in Plymouth, Banyuls and Galveston. The order of the authors is alphabetical.

Acknowledgements

We are indebted to many colleagues who kindly accepted our requests to review text and figures for earlier drafts of this second edition. Their expert input has been of enormous help in preventing mistakes, although, of course, we accept full responsibility for any remaining errors.

We particularly thank: Shelley Adamo, Jennifer Basil, Jean Boal, Ulli Budelmann, C.-C. Chiao, Nicki Clayton, Robyn Crook, Ludovic Dickel, Andy Dunstan, Karina Hall, Binyamin Hochner, Crissy Huffard, Yoko Iwata, Gilles Laurent, Anne Lindgren, Justin Marshall, Lydia Mäthger, Aran Mooney, Alexandra Schnell, David Sinn, Michelle Staudinger, Martin Stevens, Joan Stevenson-Hinde, Jan Strugnell, Mike Vecchione, Janet Voight and Richard Young. Many colleagues provided images for the book, which we acknowledge in the figure captions. We thank Basia Goszczynska especially, for her expertise in assembling the many plates, for her skill in modifying the dozens of graphs and line drawings, and her patience in liaising with Cambridge University Press to produce the figures.

We are hugely indebted to our editor, Katrina Halliday, for providing continual encouragement and professional judgement in the later stages of the book's preparation. Her diplomacy and gentle persuasion are enormously appreciated. We also thank Ilaria Tassistro for her unfailing support as we worked on successive versions of the numerous figures, and last but not least, we thank our superb copy editor, Lindsay Nightingale.

We thank the numerous authors and journals who kindly allowed us to use their published figures. Finally, we thank Kate Messenger for her sterling work on the Index.

RTH writes: this edition was undertaken while concurrently directing a sizeable research laboratory at the Marine Biological Laboratory and conducting field work worldwide. Thanks are bestowed liberally on many diving partners, colleagues, collaborators, students and postdocs who have contributed to the research discoveries. His family has graciously endured and supported many of the trials and tribulations of the peripatetic life style of a field biologist. Numerous funding agencies have made this research possible, and their support is appreciated greatly. He is fortunate to reside in the Marine Resources Center facility of MBL, which has a sophisticated seawater system and ample floor space and tank designs that have uniquely enabled laboratory testing of cephalopods for diverse behavioural research projects. He thanks the Directors and staff of the MBL for their support of our research over the past two decades.

JBM writes: this edition was completed after his retirement from the University of Sheffield, while he was a guest of the Zoology Department, Downing Street, Cambridge, UK. He would like to express his immense gratitude to Professor Malcolm Burrows for the hospitality and intellectual stimulus that he received in this wonderful department.

He would also like to place on record his heartfelt thanks to the many people — teachers, students, colleagues and friends — who through their intellectual example and stimulus have helped him throughout his life: at school, Eric Dickins, Basil Harvey, Geoffrey Matthews, Gordon Reed, William Tallents; at university, Carl Pantin (those wonderful lectures!), Laurence Picken, Bill Thorpe, Martin Wells, John (JZ) Young; and later, Eric Denton, Ernst Florey, Francesco Ghiretti and Michael Land. For the encouragement and support given him at various stages of his career he also thanks his parents, as well as Sydney Butterworth, Ian Chester Jones, William Hecker, John Morrison, Nino Salvatore and James Smeall. Finally, his very special thanks go to Mary Jane Drummond, partner for 50 years, for everything — and much, much more.

CHAPTER ONE

Introduction

Cephalopods are among the most beautiful of all animals, and their behaviour is complex and fascinating. There are about 750 species living today, but their biomass is considerable, and their body form and life style are highly varied (Fig. 1.1). The smallest adult cephalopod (*Idiosepius*) is less than 10 mm long; the giant squid (*Architeuthis*) and the colossal squid (*Mesonychoteuthis*) have body lengths of over 5 m and are easily the largest of all the invertebrates. *Octopus micropyrsus* rarely exceeds 50 g, but the giant octopus of the Pacific, *Enteroctopus dofleini*, grows to over 250 kg. Although there are cephalopods in the tropics and in the polar regions, they barely penetrate into the Baltic and are absent from the Black Sea and from fresh water. Several octopuses live between tide levels, but the finned octopods and the vampire squid *Vampyroteuthis* live in the deep ocean, usually below 1000 m (Fig. 1.2), and some octopods are known to exploit the productivity associated with thermal vents (Rocha *et al.*, 2002; Voight, 2008). Many cephalopods, for example the ommastrephid squid, abound in the upper layers of the seas, where they compete with fish for food and fall prey to albatrosses, fulmars and penguins as well as marine mammals. Such squid, which are slender and streamlined, are fast, active predators; but in the mid-waters of the oceans there are many millions of 'ammoniacal' squid, delicate, slow-moving animals that retain ammonium chloride in spaces in the body for neutral buoyancy. These occur in such large numbers in some oceans that they can support an enormous sperm whale population (Denton, 1974). In many parts of the world, cephalopods are sufficiently numerous to form the basis of economically important fisheries.

Cephalopods can change colour and pattern with breath-taking rapidity. They are the chameleons of the sea, although far superior to them in the speed and diversity of their changes, and many can generate complex patterns on the skin for camouflage and for communication. They can swim, pounce, walk, burrow, and soar on up-currents, even 'fly' (Muramatsu *et al.*, 2013). Most of them can eject ink at a pursuing predator, and many deep-sea forms have complex light organs all over their bodies. They have large eyes, and anyone who has seen an octopus in an aquarium will have had the uncanny impression of being carefully watched. Laboratory experiments have shown that octopuses can learn very quickly, even by vertebrate standards, and show flexible and adaptive behaviour in many situations. These findings agree with the fact that the cephalopod brain is relatively enormous by invertebrate standards, and large compared with fish and reptiles (Fig. 1.3).

All the evidence is that these are advanced invertebrates (Wells, 1978), and it is not surprising that they can show complex behaviour more reminiscent of fish than of their molluscan relatives (Packard, 1972). Yet unlike fish, cephalopods grow surprisingly quickly and, *Nautilus* apart, few of them seem to live more than a few years (Chapter 6).

1.1 Aims and Perspective

This book explores the behaviour of living cephalopods in their natural habitats: how they find food, escape from their enemies, migrate, signal to one another and reproduce. The emphasis is on life in the sea, but because much of what is known about cephalopod behaviour is based upon laboratory observations and experiments, we have not hesitated to draw on these data, nor to touch upon the extensive literature on cephalopod anatomy and physiology where this is relevant. Our approach throughout is essentially

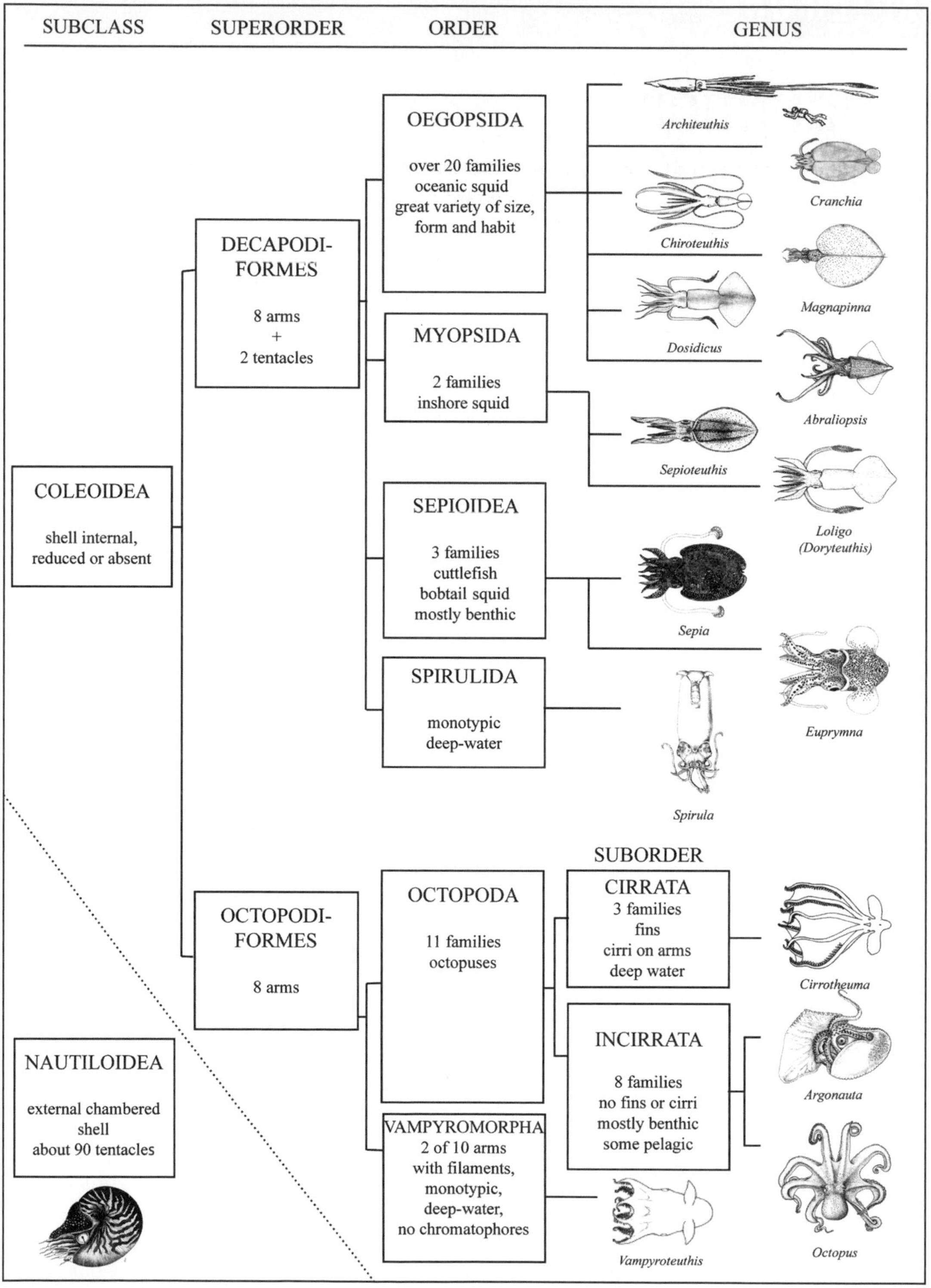

Figure 1.1 The major groups of cephalopods, showing phylogenetic relationships and diversity. Based on Young, Vecchione and Mangold (2008). See Table 1.1.

Table 1.1 Classification of living genera mentioned in text

The classification in this table (and in Fig. 1.1) is based on that of the Tree of Life Project (Young, Vecchione & Mangold, 2008). Families and genera are listed alphabetically. N.B. In cephalopods, as in other groups, nomenclature changes: in this table, and throughout the book, we use the generic name as it was cited in the original published article. Hence the frequent reference in the text to *Loligo plei*, even though this squid has become (in 2011) *Doryteuthis plei*.

CLASS CEPHALOPODA
SUBCLASS NAUTILOIDEA
Family Nautilidae
Allonautilus
Nautilus
SUBCLASS COLEOIDEA

I SUPERORDER DECAPODIFORMES (= DECABRACHIA)		II SUPERORDER OCTOPODIFORMES (= OCTOBRACHIA)
Order Oegopsida	**Order Myopsida**	**Order Vampyromorpha**
Family Architeuthidae	Family Loliginidae	Family Vampyroteuthidae
Architeuthis	*Alloteuthis*	*Vampyroteuthis*
Family Bathyteuthidae	*Doryteuthis*	
Bathyteuthis	*Loligo*	**Order Octopoda**
Family Brachioteuthidae	*Lolliguncula*	
Brachioteuthis	*Photololigo*	SUBORDER CIRRATA
[=*Benthoteuthis*]	*Pickfordiateuthis*	Family Cirroteuthidae
Family Chiroteuthidae	*Sepioteuthis*	*Cirroteuthis*
Chiroteuthis		*Cirrothauma*
Grimalditeuthis	**Order Sepioidea**	Family Opisthoteuthidae
Family Chtenopterygidae		*Opisthoteuthis*
Chtenopteryx	SUBORDER SEPIIDA	Family Stauroteuthidae
Family Cranchiidae	Family Sepiidae	*Stauroteuthis*
Bathothauma	*Metasepia*	
Cranchia	*Sepia*	SUBORDER INCIRRATA
Leachia	*Sepiadarium*	Family Alloposidae
Megalocranchia	*Sepiella*	*Alloposus* [=*Haliphron*]
Mesonychoteuthis	*Sepioloidea*	Family Amphitretidae
Sandalops		*Amphitretus*
Taonius	SUBORDER SEPIOLIDA	Family Argonautidae
Teuthowenia	Family Sepiolidae	*Argonauta*
Family Enoploteuthidae	*Euprymna*	Family Bolitaenidae
Abralia	*Heteroteuthis*	*Bolitaena* [=*Eledonella*]
Abraliopsis	*Neorossia*	*Japetella*
Watasenia	*Rossia*	Family Octopodidae
Family Gonatidae	*Sepietta*	*Abdopus*
Gonatus	*Sepiola*	*Ameloctopus*
Family Histioteuthidae	Family Idiosepiidae	*Amphioctopus*
Histioteuthis	*Idiosepius*	*Bathypolypus*
Family Joubiniteuthidae	Family Sepiadariidae	*Bentheledone*
Joubiniteuthis	*Sepioloidea*	*Benthoctopus*
Family Lepidoteuthidae		*Eledone*
Pholidoteuthis	**Order Spirulida**	*Enteroctopus*
	Family Spirulidae	*Euaxoctopus*
	Spirula	*Graneledone*

table continues

Table 1.1 (cont.)

Family Lycoteuthidae
    Lycoteuthis
    [=*Oregoniateuthis*
    =*Thaumatolalampas*]
    Nematolampas
    Selenoteuthis
Family Magnapinnidae
    Magnapinna
Family Mastigoteuthidae
    Idioteuthis
    Mastigoteuthis
Family Neoteuthidae
    Neoteuthis
Family Octopoteuthidae
    Octopoteuthis
    Taningia
Family Ommastrephidae
    Dosidicus
    Eucleoteuthis
    Illex
    Nototodarus
    Ommastrephes
    Sthenoteuthis
    Symplectoteuthis
    Todarodes
Family Onychoteuthidae
    Moroteuthis [=*Onykia*]
    Onychoteuthis
Family Pyroteuthidae
    Pterygioteuthis
    Pyroteuthis
Family Thysanoteuthidae
    Thysanoteuthis
	Hapalochlaena
	Macrotritopus
	Octopus
	Scaeurgus
	Thaumeledone
	Thaumoctopus
	Vulcanoctopus
	Family Ocythoidae
	    Ocythoe
	Family Tremoctopodidae
	    Tremoctopus
	Family Vitreledonellidae
	    Vitreledonella

zoological: that is, we always try to consider the whole animal and its relation to the environment in which it must survive and leave offspring.

Many previous authors have written about cephalopod behaviour, beginning with Aristotle, who wrote about the common shallow-water forms in the Mediterranean – *Sepia, Loligo* and *Octopus* – as long ago as 330 BC (see translation, 1910). Over a century ago, Henry Lee (1875) gave an excellent and accurate account of cephalopod biology, including their behaviour. F.W. Lane wrote a popular, well-illustrated book, *Kingdom of the Octopus* (1957), which conveys much of the excitement and glamour of octopuses and squid, including those of the deep sea. Another popular 'behaviour' book with excellent photographs, many of them taken underwater, is *Octopus and Squid: The Soft Intelligence*, by Cousteau and Diolé (1973). *Cephalopods: A World Guide* by Norman (2000) has useful illustrations, as has *Octopus – The Ocean's Intelligent Invertebrate*, by Mather, Anderson and Wood (2010). The comprehensively illustrated book on body patterning in cephalopods by Borrelli, Gherardi and Fiorito (2006) also touches on behaviour.

Cephalopods obviously exert a great attraction for novelists, and several have included episodes about giant cephalopods or 'kraken' in their novels. Victor Hugo (*Toilers of the Sea*, 1866), Jules Verne (*Twenty Thousand Leagues under the Sea*, 1869), Ian Fleming

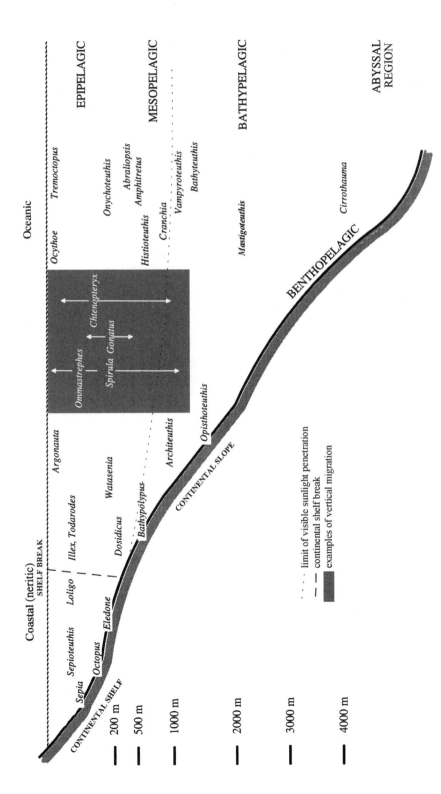

Figure 1.2 A simplified diagram showing the approximate distribution of some coleoids in the different zones of the sea. The depth axis is not to scale; arrows indicate approximate limits of vertical migration of some genera (heavily modified from Packard, 1972, with kind permission from Wiley).

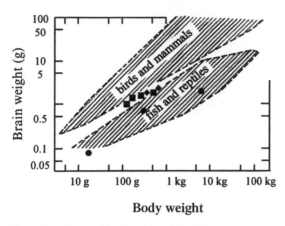

Figure 1.3 Brain and body weight relationships in vertebrates and some cephalopods (three *Octopus* species [●], four squid species [■] and two *Sepia* [◆]; from Packard, 1972, with kind permission from Wiley).

(*Dr No*, 1958) and Peter Benchley (*Beast*, 1991) all give highly entertaining fictitious accounts of the behaviour of malevolent giant cephalopods.

Among serious texts touching upon behaviour are the books by Wells (1962a), *Brain and Behaviour in Cephalopods*, and Young (1964a), *A Model of the Brain*; but the emphasis in these is physiological and neurobiological. The same is true of many of the papers in *The Biology of Cephalopods* (Nixon & Messenger, 1977) and of Wells' second book *Octopus: Physiology and Behaviour of an Advanced Invertebrate* (1978). There are many behavioural data in *Cephalopod Life Cycles* (Boyle, 1983b, 1987); in the several volumes of *The Mollusca* (Wilbur, 1983–1988); in the *Céphalopodes* volume of the *Traité de Zoologie* by Mangold (1989); in *The Brains and Lives of Cephalopods* (Nixon & Young, 2003); and in *Cephalopods. Ecology and Fisheries* (Boyle & Rodhouse, 2005), but in none of these is behaviour a major theme. Until recently there was only one monograph and one book devoted entirely to behaviour: these are the fascinating, although perhaps over-interpreted, ethological accounts 'The behavior and natural history of the Caribbean reef squid, *Sepioteuthis sepioidea*', by Moynihan and Rodaniche (1982), and *Communication and Noncommunication by Cephalopods*, by Moynihan (1985a). While we were preparing this second edition, however, a valuable new review of *Cephalopod Cognition* (Darmaillacq, Dickel & Mather, 2014) was published.

Apart from all this, there is a surprisingly large body of data on cephalopod behaviour scattered throughout the scientific literature, as well as anecdotal material in such unlikely sources as fishery reports, diving magazines and on the Internet. In this book, we attempt to review all the scientific literature in the light of our own first-hand experience and knowledge of living cephalopods, in order to present some general features of cephalopods from a behavioural rather than a physiological perspective.

Because cephalopods are likely to be unfamiliar animals to many readers, we begin with a brief introduction to their phylogeny and evolution, before going on to outline the ways in which their behaviour has been studied.

1.2 Classification and Phylogeny

There are two major divisions within present-day cephalopods: the shelled nautiloids, dealt with separately in Chapter 10 because they are so different, and the coleoids, containing all other living cephalopods (Fig. 1.1). The coleoids comprise the Decapodiformes and the Octopodiformes. The Decapodiformes is a morphologically and biogeographically diverse group that includes the inshore and oceanic squid, as well as the cuttlefish. The Octopodiformes comprises the deep-water finned octopods, the more familiar octopuses and the vampire squid. The approximately 750 species of living cephalopods are distributed among more than 150 genera and nearly 50 families. Table 1.1 lists the genera mentioned in this book.

Cephalopods are an ancient group of unusual molluscs that evolved from monoplacophoran-like ancestors in the Palaeozoic. They are thus older than the vertebrates, dating from at least the late Cambrian (Fig. 1.4). The earliest fossil cephalopods were small creatures (20 mm), but subsequently they increased greatly in size. Some of the shelled ammonites that were a dominant component of the marine fauna during the Mesozoic were over 3 m in diameter, and some of the Ordovician endocerids were 10 m long (House, 1988; Teichert, 1988).

Yet a key feature in the evolution of the coleoids was the reduction, internalisation or loss of the shell so that, with so few hard parts being preserved, there has inevitably been considerable speculation about the phylogeny of this group. Recently, however, molecular studies have helped to clarify relationships and to

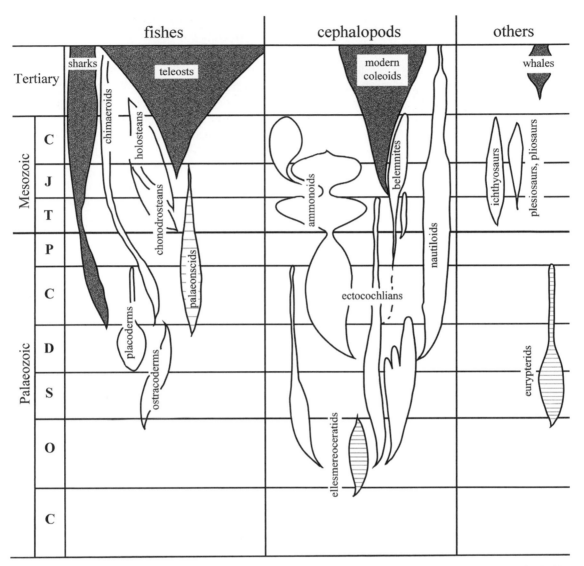

Figure 1.4 The co-evolution of coleoid cephalopods and marine vertebrates, showing that the major adaptive radiation of coleoids began in the Mesozoic, when fish and marine reptiles abounded. Major modern groups are shaded; groups believed to have been freshwater are striped (modified from Packard, 1972, with kind permission from Wiley).

establish approximate times of divergences of the major coleoid lines. The recent review of cephalopod evolution by Kröger, Vinther and Fuchs (2011), which emphasises the broad agreement among the fossil, embryological and molecular evidence, is a useful reference for the general reader, and the following account is based on it (Fig. 1.5).

The earliest cephalopods evolved about 530 million years ago, during the Cambrian, from molluscs thought to be like present-day monoplacophorans, which also gave

rise to the bivalves and gastropods. The nautiloids and coleoids diverged about 416 million years ago (Kröger, Vinther & Fuchs, 2011). The latter internalised their shells and diverged in the late Palaeozoic about 276 million years ago into the Decapodiformes (Decabrachia) and Octopodiformes (Octobrachia) (Table 1.1) (Strugnell et al., 2005, 2006). The radiation of the Decapodiformes may have occurred in the middle Mesozoic (Kröger, Vinther & Fuchs, 2011), but the relationships within this order,

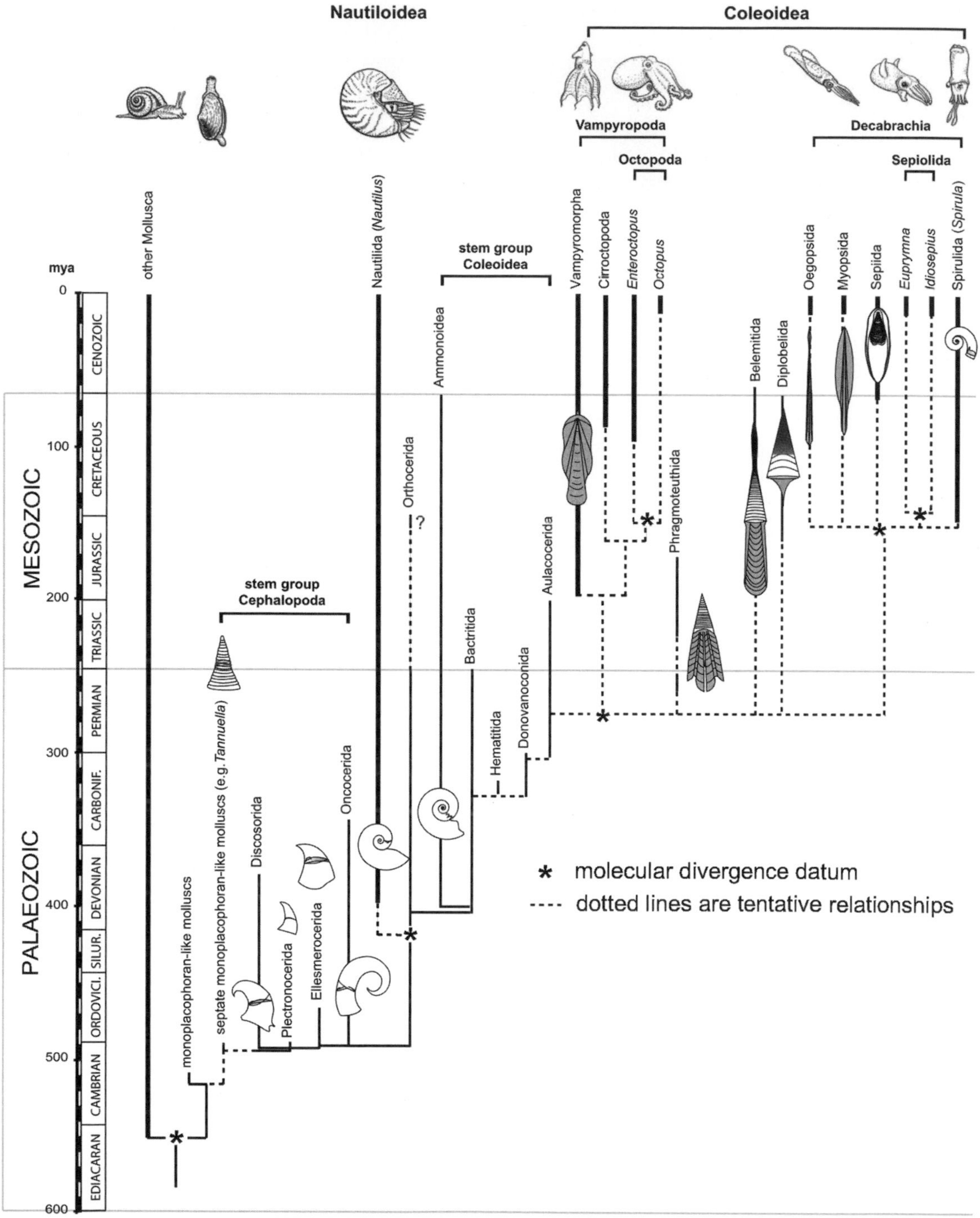

Figure 1.5 A molecularly calibrated time-tree of cephalopod evolution, modified from Kröger, Vinther and Fuchs (2011), with kind permission from Wiley. Nodes marked with an asterisk are molecular and based on the analyses of Kröger *et al.*, except for the divergence of *Spirula* from other Decabrachia, which is based on Warnke *et al.* (2011). Bold lines indicate the fossil record of extant lineages; dotted lines indicate tentative relationships among modern coleoids (see text). N.B. 'Vampyropoda' is equivalent to our 'Octopodiformes' or 'Octobrachia' (Table 1.1).

which is thought to be monophyletic (Lindgren *et al.*, 2004; Lindgren, 2010), are poorly understood (Strugnell *et al.*, 2005, 2006; Lindgren *et al.*, 2012). Some molecular evidence suggests that the myopsids (which include the widely distributed genera *Loligo* and *Doryteuthis*) may be closely related to the benthic sepioids, including the cuttlefish (Lindgren, 2010), whereas other studies support a closer relationship between the myopsids and oegopsids (Allcock *et al.*, 2011).

Whatever the evolutionary relationships between major coleoid taxa, there can be no doubt that the diversity of modern coleoids is remarkable. Some zoologists believe that the wide structural variation in modern coleoids compared with the relatively small variation in fossil shells suggests that they may be at the peak of their development today (Clarke, 1988b). This is highly speculative because about 10 500 species of shelled cephalopods have been identified from fossils (Lehmann, 1981; Ward & Bandel, 1987) and at least 28 fossil cephalopod orders have been described (Teichert, 1988), compared with only four extant coleoid orders. However, Clarke (1988b) has suggested that this may simply reflect the fact that more palaeontologists than biologists have investigated cephalopod evolution.

1.3 Evolution and Behaviour

Although cephalopods are molluscs (Chapter 2), they resemble modern teleosts to an extraordinary extent in their morphology, physiology and ecology and, as we shall see, in their behaviour. As Packard (1972) put it, 'cephalopods functionally are fish', and most students of behaviour who have seen a squid or an octopus would probably agree that they are more fish-like than molluscan.

Packard (*op. cit.*) not only first drew attention to the many close similarities between fish and coleoids, but also went on to argue persuasively that the evolution of the coleoids has been influenced strongly by competition and predation pressures from fish and marine reptiles from the Mesozoic onwards (Fig. 1.4). In particular, Packard suggested that the adaptive radiation of fish and reptiles in inshore waters forced the shelled (ectocochleate) cephalopods into deeper, offshore waters. One consequence of this was the loss of the chambered shell, because of depth limits imposed by hydrostatic pressure acting on the low-pressure gas

spaces in the shell. Subsequently, some of the new shell-less coleoids returned to shallow waters to compete again with the teleosts. It is certainly a remarkable feature of cephalopod evolution that all the externally shelled cephalopods except *Nautilus* have become extinct, and Packard's hypothesis has been widely accepted.

There seems little doubt that the evolution of the cephalopods is associated with the development of low-pressure buoyancy mechanisms (Fig. 1.6); indeed, it is the enclosure of an air space within the shell that distinguishes the early cephalopods from other molluscs (Donovan, 1964: Denton & Gilpin-Brown, 1973; Denton, 1974; Teichert, 1988). The behavioural consequences of this are set out clearly in the review by Packard (1972): the ability to regulate buoyancy, followed by reduction and internalisation of the shell, led to the development of the mantle musculature for fast jet-propelled locomotion (Vermeij, 1987; Wells, 1990; Voight, Portner & O'Dor, 1994; Finn & Norman, 2010). Associated with this were changes in the skin as it became important for camouflage and communication, the development of more effective sense organs, an increase in brain size and the emergence of the more sophisticated behaviour that forms the subject of this book.

However, the report of coleoids from the early Devonian (Bandel *et al.*, 1983; Stürmer, 1985) suggests that any predation pressure on the early coleoids may have been exerted by such early vertebrates as the chondrosteans or placoderms (Fig. 1.4). On the basis of this and other evidence, Aronson (1991) has modified what he terms the 'Packard scenario' and proposed that both ectocochleates and the early coleoids lived in shallow as well as deep water during the Palaeozoic and early Mesozoic, and that subsequently the ectocochleates were eliminated from shallow water (Fig. 1.7). In a stimulating account, Grasso and Basil (2009), relying on recent evidence that cephalopods evolved far earlier than previously thought, stress that the emerging coleoids may have had to contend with cartilaginous fish, crustacean and other cephalopod predators as well as with teleosts.

Even if the timing of the predatory influence in the Packard scenario is wrong, there is good evidence in modern forms that predation can be a critical factor in

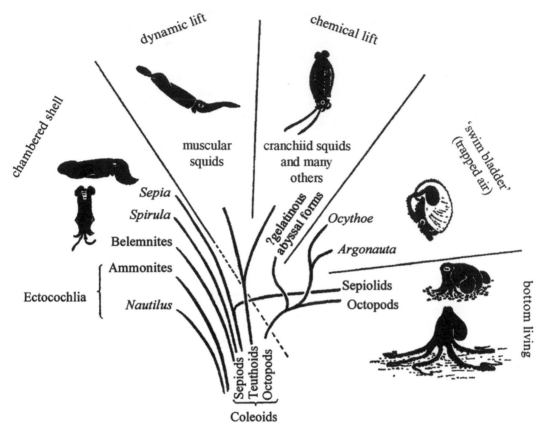

Figure 1.6 Adaptive radiation of cephalopod buoyancy mechanisms. In all but a few living forms, the low-pressure gas space in the chambered shell has been replaced by other means of achieving lift. Chemical lift may involve retention of ammonia, the replacement of SO_4^{2-} by Cl^-, or storage of oil (modified from Packard, 1972, with kind permission from Wiley).

the distribution of octopuses, for Aronson (1991) has shown that there is an inverse relationship between the number of predatory teleosts and the octopus population density (Chapter 9). He points out that predatory pressures in the remote past could have shaped behaviours that would be well suited to present-day environments where predation is high.

The essential point is that behaviour and morphology must both be taken into account when considering the interrelationships and evolution of animals and their predators. To take one more example, the skin of octopuses has not only evolved elaborate devices for cryptic and warning coloration, but these devices may be especially suited to the characteristics of the visual system of vertebrates, which are their chief predators in the sea and with

which they have been interacting behaviourally for many millions of years (Packard, 1988a; Grasso & Basil, 2009; see also Chapters 3 and 7).

1.4 Studying Cephalopod Behaviour

The account of cephalopod behaviour that we give in the following chapters draws upon several quite different sources of information, which are considered in this section.

Meanwhile, it may be worth reminding new investigators of techniques widely used by ethologists but not yet adopted by students of cephalopod behaviour, in the search for quantitative data. Perhaps the key point is that, if at all possible, we need to obtain quantitative data as a basis for our analyses. Useful reviews of behavioural

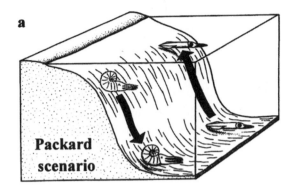

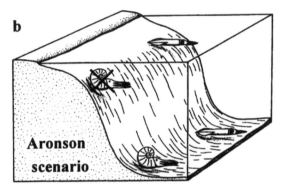

Figure 1.7 Two schemes for the origin of the coleoids. (**a**) According to Packard, competition during the Mesozoic pushed the ectocochleates into deep water, where they lost the shell prior to re-invading shallow water. (**b**) According to Aronson, ectocochleates and coleoids co-existed in shallow and deep water during the Palaeozoic. Subsequently, the ectocochleates were eliminated from shallow water (reproduced with permission from Aronson, 1991: *Bulletin of Marine Science* **49**, 245–255, Ecology, paleobiology and evolutionary constraint in the octopus, Aronson, R. B., © (1991).

methodology include those by Hazlett (1977), Colgan (1978), Drummond (1981), Ellis (1985), Noakes and Baylis (1990), Harvey and Pagel (1991), Lehner (1998), Ploger and Yasukawa (2003), Dawkins (2007), and especially Martin and Bateson's book *Measuring Behaviour* (2007).

1.4.1 Laboratory Studies

With the development of marine biological stations in the late nineteenth century at places such as Naples (Italy), Woods Hole (USA) and Plymouth (UK), the opportunity arose for biologists to see living cephalopods at close quarters. Many researchers visiting these laboratories by the sea were primarily physiologists, but there are several papers touching on cephalopod senses and behaviour dating from the beginning of this century. After the Second World War, J.Z. Young began his comparative study of learning at the Naples Zoological Station, and from the late 1940s to the mid 1960s attracted all kinds of biologists as well as experimental psychologists to Naples to study various aspects of the behaviour of *Octopus vulgaris* and *Sepia officinalis*, although it is curious that no ethologists were involved in this work. Laboratory studies continue to be important, and currently there are several institutes worldwide where cephalopod behaviour is the major focus of research, for example Luc-sur-Mer (France), Naples and Woods Hole.

Recent developments in the design of seawater systems and husbandry methods are making live cephalopods available in many more facilities around the world, generating new observations on different species. Several laboratories worldwide have built small-scale facilities for studying cephalopod behaviour (e.g. Boal, 2011) and the construction of larger laboratory tanks with observation windows has enabled researchers to study intraspecific relationships, foraging and predator–prey interactions (e.g. Yang *et al.*, 1989; Shashar *et al.*, 2000; Adamo *et al.*, 2006; Pronk, Wilson & Harcourt, 2010; Staudinger, Hanlon & Juanes, 2011).

1.4.2 Studies at Sea

Finding out about life under the sea is notoriously difficult, so that our understanding of cephalopod behaviour may seem rudimentary to an ornithologist or entomologist. What little we do know derives from snorkelling and SCUBA diving and from various types of diving vessels, manned and unmanned, some of whose potential is only now being realised.

SCUBA Diving

With the development of SCUBA diving in the 1950s, a trickle of field observations on shallow-water cephalopods began to add to our laboratory-based knowledge. The data were at first fragmentary, but studies like those of Altman (1967) and Kayes (1974) using SCUBA and snorkelling were important landmarks

of their kind. Yarnall (1969) produced one of the best
early studies of octopuses in the field by using a hide
positioned over shallow clear water. Studies such as
those by Moynihan and Rodaniche (1982), Hartwick,
Ambrose and Robinson (1984), Ambrose (1986), Aronson
(1986), Mather (1991b), Sauer et al. (1997), Norman, Finn
and Tragenza (2001), Hall and Hanlon (2002), Hanlon
et al. (2005), Huffard, Boneka and Full (2005), and
Shashar and Hanlon (2013), among others, have shown
how powerful field studies can be and, more
importantly, have shown that it is possible to test
sensory and behavioural hypotheses in the field. Use of
teams of observers to aid with longer-term observational
work in shallow water has helped us to understand daily
rhythms, time budgets and spawning activities (e.g.
Kayes, 1974; Mather, 1988, 1991b; Forsythe & Hanlon,
1997; Hanlon, Forsythe & Joneschild, 1999; Naud et al.,
2004; Hanlon, Conroy & Forsythe, 2008). Saturation
diving from underwater habitats and the use of closed-
circuit re-breathers have also extended observation
times and have proved useful in some studies (Hochberg
& Couch, 1971; Hanlon et al., 1979; Hanlon & Hixon,
1980).

Oceanographic Research Vessels

Ocean-going research ships have traditionally
provided much information about many aspects of the
habits and life style of mid- and deep-water animals,
and this is true for cephalopods too: for example,
depth distribution, vertical migration and
bioluminescence, all of which are considered in
Chapter 9. Some researchers have been able to make
direct behavioural observations and experiments on
living deep-sea animals, because of improved capture
and holding facilities on board research vessels. An
example of such a study is the direct demonstration of
ventral counter-illumination by the squid *Abralia* and
Abraliopsis (R.E. Young, Roper & Walters, 1979; see
Chapter 9). Until methods of net-capture improve
considerably, however, such studies are likely to be
short-term. Meanwhile, many oceanic and deep-sea
cephalopods are so difficult to approach that, for the
foreseeable future, we may still have to rely on
evidence from specimens that survive for only a few
hours after being caught by research vessels (e.g.
Zylinski & Johnsen, 2011).

Submersible Vehicles

Exciting new data about mid- and deep-water
cephalopods have been gathered by submersibles,
manned or otherwise, with video-recording equipment.
For example, rare cirrate octopods have been observed
swimming horizontally and hanging like an umbrella
above the bottom (Roper & Brundage, 1972). We have
learned of the strange postures that some squid adopt in
mid-water (Vecchione & Roper, 1991) and of the curious
'ballooning' behaviour of a cirrate octopus when
touched by the grab-arm of a submersible at nearly
3000 m (Boletzky, Rio & Roux, 1992). Inking and various
body patterns have also been observed (e.g. Bush &
Robison, 2007; Bush, Robison & Caldwell, 2009) as well
as brooding octopus aggregations (Drazen et al., 2003).
Perhaps most spectacular of all, a previously unknown
squid with enormous fins and extremely long and
unusual arms was video-taped in the Atlantic, Indian
and Pacific oceans, reminding us how poorly
documented the deep ocean is (Vecchione et al., 2001).
This squid, for which a large adult specimen has not yet
been captured, is a member of the Family
Magnapinnidae (Vecchione & Young, 2006).

Manned submersibles have also provided some
direct observations of behaviour in deep-sea
cephalopods, although their bright lights, loud sounds
and restricted bottom time generally preclude any
detailed behavioural observations (Moiseev, 1991;
Vecchione & Roper, 1991; Guerra et al., 2002b;
Vecchione et al., 2002).

Remotely operated vehicles (ROVs) have also been
useful in many shallow-water studies, commencing with
the squid studies of Waller and Wicklund (1968) and
Vecchione (1988). Their potential has increased as ROVs
became smaller and more manageable, and in shallow
water their use has augmented several SCUBA diving
studies. For example, the spawning behaviour of *Loligo
reynaudii* in South Africa (Sauer, Smale & Lipinski,
1992), and the spawning rhythms and behavioural
interactions of *Loligo opalescens* in California, were
studied at depths beyond safe SCUBA diving limits
(Hunt et al., 2000; Forsythe, Kangas & Hanlon, 2004;
Hanlon, Kangas & Forsythe, 2004). In southern
Australia, Hanlon et al. (2007) used a very small ROV
with a sensitive camera to discover that cuttlefish
camouflage themselves at night. And in a remarkable use
of ROVs, Dunstan and his collaborators (e.g. Dunstan,

Ward & Marshall, 2011) used vehicles and various telemetry devices to study the behaviour of *Nautilus* on the Great Barrier Reef (see Chapter 10).

In the past decade, autonomous underwater vehicles (AUVs), which unlike ROVs are untethered, unmanned vehicles that were developed originally to map the seabed, have been used by Williams *et al.* (2009) to map the distribution and camouflage diversity of cuttlefish at night.

Acoustic Sensing of Squid

For some time commercial fishermen have been developing acoustic methods of finding squid, and scientists have now joined this effort (Goss, Middleton & Rodhouse, 2001) to help determine biomass levels and schooling dynamics at inshore spawning grounds, where a great deal of fishing occurs, especially for loliginids. Identification by so-called 'fish finders' has now advanced to a point at which squid can be identified with certainty and differentiated from schooling fish or invertebrates (Roberts *et al.*, 2002; Benoit-Bird *et al.*, 2008; Matteson, Benoit-Bird & Gilly, 2009). Side-scan sonar has been adapted to map the distribution and abundance of loliginid egg masses laid on the sea floor (Foote *et al.*, 2006), thus providing some clues about the timing of spawning as well as aberrant egg laying (Young *et al.*, 2011).

Telemetry

Telemetry devices have become smaller and more advanced both in the quality and quantity of information that they can gather from cephalopods behaving naturally in the sea. For example, the pioneering studies of Carlson, McKibben and DeGruy (1984) and Ward *et al.* (1984) using simple tags or sonic transmitters with *Nautilus* led to devices that can record other physiological data either in the laboratory (O'Dor, Wells & Wells, 1990) or at sea (O'Dor *et al.* 1993; Dunstan, Ward & Marshall, 2011). Accelerometry, which quantifies energy expenditure via measurement of body acceleration, has been applied recently to field studies of energy expenditure by cuttlefish (Payne *et al.*, 2010a). Pop-up satellite tags logging temperature and depth have increased knowledge of squid movements in open-ocean and mesopelagic habitats (Gilly *et al.*, 2006).

Radio-acoustic positioning telemetry (RAPT) records pressure-flow (for energy consumption) and location (for animal movements) with accuracy down to 1 metre in the sea (O'Dor *et al.*, 1998; O'Dor, 2002) and has led to new behavioural and physiological insights in octopuses, cuttlefish and squid (Sauer *et al.*, 1997; Aitken, O'Dor & Jackson, 2005; Rigby & Sakurai, 2005). The recent introduction of affordable data-logging acoustic receivers (Heupel, Semmens & Hobday, 2006) has provided opportunities for tracking mobile squid over hundreds of kilometres (Stark, Jackson & Lyle, 2005). Acoustic telemetry has emerged as a leading technique for field studies of cuttlefish, although caution is necessary when inferring behaviour in the absence of adequate controls (Payne *et al.*, 2010a,b).

Recording Behaviour Visually

Much of cephalopod behaviour involves rapid complex changes in body patterning and appearance so that recording the precise details of this is essential. Over the past two decades, digital filming has replaced conventional photography and videography, and because these newer techniques offer so many advantages, we outline some of them in Box 1.1.

1.4.3 Indirect Methods

Comparative morphology alone, in any animal group, can give some indication about habit and life style. In a cephalopod, the length of the arms, for example, or the thickness of the mantle muscle, the orientation and shape of the eye, or the nature of the photophores all provide some clues about its life. Two particularly useful indirect sources of information have proved to be the study of beaks and brains. To these we can now add DNA fingerprinting.

The Study of Beaks

Other animals are far better than humans at catching cephalopods, and, as we shall see in Chapter 5, whales, seals, birds, a variety of fish and other cephalopods consume great quantities of cephalopods. One of the parts of a cephalopod that is resistant to digestion is the chitinous beak (Chapter 2), and the important discovery of Clarke (1962, 1977, 1986) that cephalopods can be identified to family, genus or even species from features of their beaks has provided a powerful way of studying the distribution of these elusive animals in the sea. As a

---

Box 1.1 Digital imaging, video playback, spectrometry and hyperspectral imaging

High-definition digital still cameras and video have become readily available in the past five years and have transformed the way that imagery can be used as a data source for behavioural studies. Digital images require no processing, and the visual information becomes immediately available to the researcher. High-definition television (HDTV) video cameras provide very crisp imagery of behavioural sequences, and have the further advantage that each video frame (shooting, for example, at 30 frames per second) is the equivalent of a 2-megapixel still image. Such an image has sufficient resolution to be used as a photograph for publication, or to be processed for measurement of colour, contrast or pattern for the particular behavioural investigation being carried out.

Video playback of behaviours to animals for experimental studies has also progressed rapidly with digital video (Ord *et al.*, 2002). For cephalopods, the issue is complicated by their ability to see polarised light (Chapter 2), but several researchers have overcome this by choosing the appropriate viewing devices (e.g. Pronk, Wilson & Harcourt, 2010), and this promises to increase the types of experiments that can be conducted on cephalopods,

especially for studying communication and predator–prey interactions.

Spectrometers for field and laboratory use have also become readily available, and small portable units have found use in vision research, for example in measuring the light environment and sampling the colour and contrast match of camouflaged cephalopods on different backgrounds (Chapter 3). Hyper-spectral imagers combine a photographic image with many spectra in each pixel. Such equipment promises to revolutionise the way nature is imaged, and will open new avenues of research by enabling us to view the world through the eyes of other animals, not solely through those of humans (Chiao *et al.*, 2011).

Many other techniques and equipment are available for cephalopod ethologists, and we should not forget some of the older ones that have hardly been applied, such as semi-permanent underwater television (Myrberg, 1973) and video cameras with remotely controlled zoom lenses and servo-motors for scanning (DiMarco & Hanlon, 1997); such devices are now digital and more readily adaptable to laboratory and field experimentation.

complement to Clarke's pioneering work, mainly on the lower beak, recent research has developed the upper beak as an informative character (Xavier & Cherel, 2009). Beaks from the stomach contents of a whale, penguin or shark can tell us about where and at what depths particular cephalopods live, their abundance and their diel rhythm (Chapter 9). Conversely, cephalopod diets can be determined by various molecular techniques (Boyle & Rodhouse, 2005), and the use of stable isotopes of nitrogen and carbon can help to trace ontogenetic changes in feeding habits and trophic position of cephalopods in marine food webs (e.g. Cherel & Hobson, 2005).

The Study of the Brain

The detailed studies of cephalopod brains by J.Z. Young and his colleagues (Chapter 2) have gone a long way towards attributing function to each of the numerous

lobes. It is therefore possible to infer something about the habits of a cephalopod by examining its brain. For example, the presence of a giant fibre system involving the arms, as occurs in the small deep-water squid *Pterygioteuthis*, suggests some special form of prey capture (Young, 1977a). The structure of the statocysts, too, can be correlated with the type and speed of locomotion (Young, 1988). Thus, studying serial sections of the brains of preserved cephalopods can, in the hands of a good anatomist, provide a surprising amount of information about the life, and therefore behaviour, of animals that most of us will never see alive. As Young (1977a) puts it, 'Such knowledge, gained inductively, can be used, with caution, for deductions about the probable habits of species not yet investigated at sea. At least it can give the field observer some suggestions of what to look for when he is fortunate enough to observe these oceanic and deep-sea cephalopods alive.' Nixon and Young have continued with this approach in their exhaustive treatment of *The Brains and Lives of Cephalopods* (Nixon & Young, 2003).

DNA Fingerprinting

This newer methodology, involving polymorphic microsatellite loci, has been used effectively in two ways: for paternity assessment in sexual selection studies (Chapter 6) and for population genetics studies that provide indirect evidence of migrations and reproductive isolation. For the former, studies such as those by Naud and her colleagues (e.g. Naud *et al.*, 2004; Chapter 6) have effectively pinpointed paternity and discovered details of sperm competition and cryptic female choice. For the latter, details of population structure have been determined (e.g. Shaw *et al.*, 2004; Buresch, Gerlach & Hanlon, 2006; Doubleday *et al.*, 2009), although these results are still somewhat controversial (Shaw, 2002). For example, the meaning and interpretation of null alleles, which occur in microsatellite loci, for population studies is being questioned (Shaw *et al.*, 2010; Gerlach, Buresch & Hanlon, 2012; Shaw *et al.*, 2012) as research into the molecular evolution of microsatellites continues. Despite some limitations, DNA fingerprinting is developing into a valuable new technique.

Anecdote

Finally, we should recognise that fishermen's tales will be with us until the end of time, and the legend of the 'kraken' that devours ships will not die easily (Lane, 1957; Ellis, 1999). Interestingly, the widespread availability of digital cameras has increased the flow of anecdotal evidence but has also been responsible for a number of hoaxes. Nevertheless, fishermen, as well as recreational and professional SCUBA divers and photographers, can often give valuable advice on where and how to find a particular cephalopod. We know so little about most cephalopods that such information is usually worth listening to.

All these approaches to studying cephalopod behaviour have their advantages and disadvantages, but it should be obvious that there is no best or 'right' way to study these, or any other, animals. However, it is important that observations and records of what living cephalopods actually do in the sea are constantly borne in mind by laboratory investigators, however refined their analyses and experiments.

1.5 Summary

The behaviour of cephalopods has interested humans since classical times, and there is an extensive literature on many aspects of their behaviour, which we try to bring together in this book. Much of our understanding of these animals derives from laboratory studies, but this has been augmented, especially in the past few years, by research at sea. Snorkellers and SCUBA divers in shallow waters, and biologists on ocean-going vessels and submersibles, are now providing important new behavioural data on these – for the most part – extremely inaccessible animals.

Cephalopods are transformed molluscs, more like fish than limpets. They evolved from slow-moving, shelled animals that also gave rise to the present-day snails and bivalves. The earliest cephalopods date from the Cambrian, some 530 million years ago, so that they are older than the vertebrates. Shelled cephalopods (belemnites, ammonites) were abundant throughout the late Palaeozoic and Mesozoic but gradually died out, with the exception of two genera of nautiloids (Chapter 10). The coleoids, whose behaviour forms the substance of this book, diverged from the nautiloids about 420 million years ago, internalising or losing the shell; and molecular dating suggests that the octopodiformes and decapodiformes diverged about 280 million years ago. Coleoids now number about 750 species, and some authorities regard the present-day coleoids as being at the peak of their flowering. Their evolution can be understood in terms of the acquisition of neutral buoyancy and movement away from the seabed, so that they came to occupy the 'behaviour space' occupied by fish and the early marine reptiles, with which they competed. As a result, most of them are agile, active animals: all have a large brain and well-developed sense organs, which are the subject of the next chapter.

CHAPTER TWO

Senses, Effectors and the Brain

The behaviour of an animal depends, among other things, on the kinds of information collected by its sense organs, on the nature of its 'effectors', or motor apparatus, and on the organisation of the brain (Messenger, 1979b). These topics are considered in this chapter, beginning with the sense organs, whose organisation can provide useful clues about what kinds of stimuli might be influencing an animal. However, one can also get an idea about the life style of an animal by examining its effectors, even at the gross level: what are the limbs like? Has it a sticky tongue, or poisonous fangs, or light-generating organs? Because molluscs are likely to be unfamiliar animals to many readers and because cephalopods are so different from other invertebrates and from vertebrates, it is necessary to describe their various effectors in some detail. Obviously, the brain is also fundamentally important in determining behaviour, selecting as it does an appropriate course of action from among its many 'programmes' of action (Young, 1978). However, this is dealt with only briefly here because the information is available elsewhere (e.g. Young, 1971; Wells, 1978; Budelmann, 1994, 1995; Budelmann, Schipp & Boletzky, 1997; Nixon & Young, 2003; Grasso & Basil, 2009).

Cephalopods belong to the phylum Mollusca, an ancient and successful group that also contains the chitons, various kinds of marine and terrestrial snails, and the bivalves. All these forms have a similar body plan, fundamentally different from that of vertebrates: essentially there is a head, a foot and a visceral mass covered by a mantle (Fig. 2.1). The mantle secretes the protective shell, so familiar to us in limpets, snails, oysters and clams, but in most living cephalopods the shell has been internalised, reduced or lost entirely.

The cephalopod head is well developed, with sense organs and the largest brain of any invertebrate. The cephalopod foot has become modified, partly into the funnel, partly into a set of grasping, manipulative appendages that extend from the head. Following common usage, the eight appendages found in all coleoids are termed the *arms* throughout this book; the term *tentacle* is reserved for the two longer appendages that cuttlefish and squid (the Decapodiformes, Table 1.1) use for prey capture. (It is also used for the totally different appendages of *Nautilus* (Chapter 10).) The scientific name of the group (Gr. *kephale*, head; *pous, podos*, foot) derives from the close association of the head with the ancestral foot.

The mantle is so organised around the viscera that it forms a sheltered space, the mantle cavity, in which the gills lie and into which the gut, 'kidneys' and gonads open. In cephalopods, this mantle cavity constitutes a system for locomotion ('jetting') as well as for respiration. The viscera include elaborate circulatory, excretory, digestive and reproductive systems. More details about the morphology and physiology of cephalopods can be found in the texts of Wells (1978), Wilbur (1983–88) and Mangold (1989). Several useful reviews of sensory mechanisms and functions of cephalopods are available as well (Budelmann, 1996; Hanlon & Shashar, 2003; Gleadall & Shashar, 2004).

2.1 Sense Organs

All the evidence to date suggests that cephalopods rely on photoreceptors, mechanoreceptors and chemoreceptors to guide their lives.

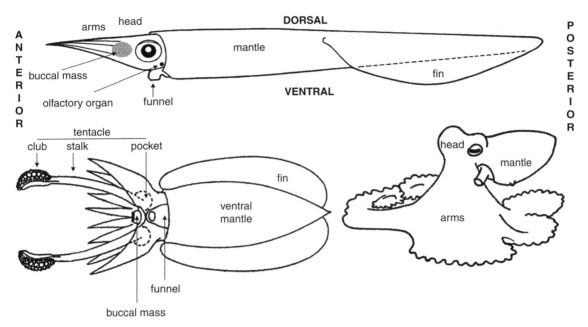

Figure 2.1 Diagrams of a squid, octopus and cuttlefish (seen from below) to define the terms used throughout the text and to show the main effectors and sense organs.

2.1.1 Photoreceptors

Eyes

The eyes are one of the most conspicuous features of all coleoid cephalopods. With one or two exceptions, they are large, and sometimes enormous. In a 250-g *Octopus vulgaris* (the size used in the Naples Zoological Station for training experiments; see below), the eye is about 20 mm in diameter, but it can be far larger in some squid. The largest eyes recorded in any animal have been found in the giant squid, *Architeuthis*, and the colossal squid, *Mesonychoteuthis*, for which diameters of about 270 mm have been reliably reported (Nilsson *et al.*, 2012).

Superficially, the coleoid eye resembles that of a vertebrate (Fig. 2.2a): there is a large posterior chamber, lens, iris, retina, choroid, sclera and argentea (Young, 1971; Messenger, 1981, 1991). The lens has a graded refractive index and a focal length approximately 2.5 times the radius (Matthiessen's ratio), which enables a high-quality image to be produced of objects at distances from infinity to a few centimetres, thus minimising the need for accommodation. In those cephalopods so far examined, the lens is generally transparent down to wavelengths of 400 nm (Shashar *et al.*, 1998) but in *Spirula* and *Chiroteuthis* the lenses are transparent to the near ultraviolet (Denton & Warren, 1968; see also §9.7). The argentea is generally iridescent and may be a component of body patterning in transparent cephalopods.

The eyes are positioned laterally on the head and, as in vertebrates, there are differences in the degree of binocularity in different cephalopods. In the cuttlefish *Sepia*, there is an anterior binocular visual field used for prey capture (Muntz, 1977), and there is experimental evidence for depth perception in this species (Messenger, 1977b). However, for the most part, cuttlefish and squid view the world monocularly except when approaching prey. In *Octopus*, however, vision is essentially monocular (with much reduced binocular overlap), and there has recently been a claim that individual octopuses have a preferred eye for attacking prey (Byrne, Kuba & Griebel, 2002; Byrne, Kuba & Meisel, 2004), although this proposition has been questioned on methodological grounds by Boal and Fenwick (2007). In *Sepia*, too, there is evidence for visual lateralisation, which may be the result of eye-use preference (Jozet-Alves *et al.*, 2012).

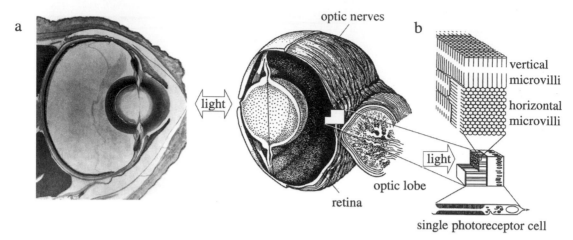

Figure 2.2 The coleoid eye. (**a**) Section through the eye of *Octopus vulgaris* showing considerable gross similarity to a mammalian eye (modified from Boycott & Young, 1956, with kind permission). (**b**) Drawing of a dissection of the eye and optic lobe of an octopus, to show the pupil, lens and optic nerves leaving the back of the retina and entering the outer region of the optic lobe (Young, 1971: *The Anatomy of the Nervous System of* Octopus vulgaris, Young, J. Z., Oxford: Clarendon Press, by permission of Oxford University Press). Diagram of a section of retina (centre) with a single receptor cell shown below and the horizontal/vertical arrays of microvilli shown above (republished with permission of Rockefeller University Press, from Saibil & Hewat, 1987: *Journal of Cell Biology* **105**, 19–28, Saibil, H. R. & Hewat, E., © (1987); permission conveyed through Copyright Clearance Center, Inc.).

Attached to the eye are extraocular eye muscles that enable the eye to be moved in the head as a result of movements of the body (compensatory movements) or as a result of visual influence (Budelmann & Young, 1984). In some cephalopods, there are convergent eye movements that are probably used for depth perception (Messenger, 1977b), and there may be special muscles associated with convergence (Budelmann & Young, 1993).

Despite the similarity to the vertebrate eye, the retina contains no rods or cones: the eye is of the rhabdomeric type found in other molluscs and also in arthropods. The receptor cells are long (about 400 μm) and thin (about 5 μm diameter) and bear microvilli (containing the visual pigment) at right angles to the long axis (Fig. 2.2b). Half the receptor cells have their microvilli oriented vertically and half horizontally as the eye is held in the orbit. This orthogonal arrangement provides the basis for polarised light sensitivity, which has been demonstrated physiologically (Sugawara, Katagiri & Tomita, 1971) and behaviourally (see below).

Several workers have estimated the acuity of octopuses: the smallest estimate is about 5 minutes of arc (Muntz & Gwyther, 1988b), which is a value comparable to that found in fish and many other vertebrates. Packard (1969b) showed that in *O. vulgaris* the size of the receptors does not change as the animal grows, so that, theoretically at least, large cephalopods should have excellent acuity. Perhaps this could be an advantage in foraging as it could increase the distance at which prey or predators could be seen (Gibson, 1983). In *Sepia*, recent behavioural experiments have yielded slightly higher acuity values (of about 34 minutes of arc), but interestingly confirmed that acuity does increase with age (size) and with increasing light intensity (Groeger, Cotton & Williamson, 2005). Sweeney, Haddock and Johnsen (2007) examined the lenses of eight species of mesopelagic cephalopods and calculated their 'modulation transfer function' as a measure of resolving ability. On the basis of the optics, they found that squid have vision as good as that of fish or birds; but they unexpectedly found that *Vampyroteuthis*, a deep-water form thought to be very ancient and close to the line leading to octopods (Chapter 1), had the best acuity of all. Of course, retinal photoreceptor maps, or better, behavioural measures of acuity, are also needed before we can make firm claims about acuity.

In a very interesting recent paper, Nilsson and his colleagues have argued that the extremely large eyes found in giant squid may have evolved specifically to detect large predators such as sperm whales at distances of over 120 m, which would allow them to take avoiding action before they are detected by the whales' sonar system (Nilsson *et al.*, 2012).

Although there is only one type of receptor, the density of receptors is often higher in certain areas of the retina. This can be related to habit and life style (Young, 1963; see also Talbot & Marshall, 2011). In the bottom-living *Sepia* and *Octopus*, for example, there is a horizontal equatorial strip of longer, thinner receptor cells that are presumed to be a kind of 'fovea' or sensitive area (Young, 1963). In the posterior part of the *Sepia* strip, the density of receptor cells is especially high, and it is here that the image of the prey is brought during prey capture (Young, 1963; see also Chapter 4). In squid, there are marked differences between the retinas of coastal and oceanic forms (Makino & Miyazaki, 2010). The posterior retina always has the highest density of photoreceptor cells, but such regions are situated dorsally in inshore genera so that the principal visual axis is directed forward and downwards, whereas in mid-water genera the region of highest density lies ventrally so that the axis is directed forward and upwards (Fig. 2.3). It has long been known that in the deep-water *Bathyteuthis* there is a ventral 'fovea' for gazing upwards (Chun, 1910), and the firefly squid, *Watasenia scintillans*, also has a specialised ventral zone, with a double-banked retina, the only known example in cephalopods (see below).

Many cephalopods are active at night or in the deep sea, and Hanlon *et al.* (2007) found that the giant Australian cuttlefish, *Sepia apama*, shows camouflage patterns appropriate to the surrounding substrate on moonless nights. Subsequent laboratory experiments on the European cuttlefish *S. officinalis* showed that it could actively change its camouflage pattern at levels of light approaching that of starlight (0.003 lux in photopic units) (Allen *et al.*, 2010b). It is well established that during dark-adaptation, the pupil dilates and the screening pigment in the retinal support cells retracts to allow more light to reach the photoreceptors (Young,

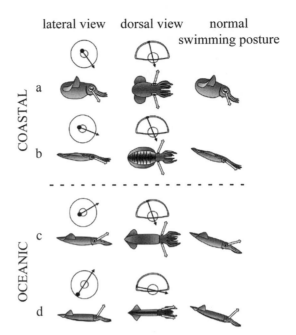

Figure 2.3 Coastal and oceanic squid have marked differences in their principal visual axes, based on the location of highest photoreceptor density in the retina (black dot). In the coastal species *Euprymna morsei* (**a**) and *Sepioteuthis lessoniana* (**b**), the highest density lies dorsally so that the axis of keenest vision is directed downward. In the oceanic species *Todarodes pacificus* (**c**) and *Eucleoteuthis luminosa* (**d**), the highest density lies ventrally so that keenest vision is directed upwards. Note also differences in the horizontal plane: the gaze of *Eucleoteuthis* is directed straight ahead. Modified from Makino and Miyazaki (2010) by permission of Oxford University Press.

1963; Daw & Pearlman, 1974; Muntz, 1977; and Gleadall *et al.*, 1993). This has important implications for camouflage (Chapter 3), for feeding (Chapter 4) and for activity patterns (Chapter 9). Recently, it has been shown that the pupillary response is extremely fast: full contraction in *Sepia* takes less than 1 second, making it among the fastest in the animal kingdom (Douglas, Williamson & Wagner, 2005). Interestingly the pupils act independently, in this and other coleoids, in contrast to *Nautilus*, in which there is a consensual response (Hurley, Lange & Hartline, 1978).

The visual pigments of more than 20 cephalopods have been characterised (for summaries see Messenger, 1981; Kito *et al.*, 1987; Seidou *et al.*, 1990). In all but five of those species, there is only a single visual pigment, whose peak absorbance varies considerably but in a way that is for the most part consistent with the ecology of the animal (Munz, 1958, 1964). Thus, the oegopsid squid have a 'deep sea' pigment (λ_{max} 480 nm), the epipelagic *Loligo pealeii* has a 'surface water' pigment (λ_{max} 493 nm) and *Sepiella* has a 'rocky shore' pigment (λ_{max} 500 nm). The octopod pigments absorb maximally at shorter wavelengths (470–475 nm) (Muntz & Johnson, 1978). The fact that all these animals have only one visual pigment is in agreement with other evidence that most cephalopods probably lack colour vision, which usually requires at least two pigments and generally involves more (Tansley, 1965).

The first evidence for multiple visual pigments in a cephalopod was found in the firefly squid, *Watasenia scintillans* (Matsui *et al.*, 1988; Seidou *et al.*, 1990; Michinomae *et al.*, 1994). This species has three pigments: a 'conventional' pigment based on the substance retinal (peak absorbance 484 nm), which is found in most parts of the retina; a pigment in the outer layer of the ventral double bank absorbing maximally at 470 nm; and a pigment in the inner layer absorbing maximally at 500 nm. The latter two pigments are highly unusual chemically, and they may have evolved comparatively recently. The fact that there are three visual pigments as well as a double-banked retina (Denton, 1990) suggests strongly that *Watasenia* may have colour vision. It is a bioluminescent species, so that perhaps these adaptations have evolved to enable it to discriminate between the light produced by a conspecific and the natural, downwelling light (Chapter 9). Subsequently three other squid (teuthoids) and one octopod have been found to have multiple visual pigments (Table 2.1); all of these forms are mesopelagic or

nocturnal, suggesting again that colour vision, if present, may have evolved to detect bioluminescence (Kito *et al.*, 1992).

All accounts of the living coleoids, however anecdotal, agree that vision must be well developed. However, in one species we have direct evidence of visual ability: *Octopus vulgaris* can readily be trained to make a variety of visual discriminations. Boycott and Young (1955b) were the first to show this, and their training technique was later perfected by Sutherland, who, using careful controls and elegantly designed experiments, went on to demonstrate unequivocally that octopuses have visual abilities comparable to those of vertebrates (Sutherland, 1957a, b, 1963, 1968, 1969). It would be inappropriate here to analyse these results in detail, and there are several comprehensive reviews of this work (Sanders, 1975; Wells, 1978; Messenger, 1981). In summary, we may note that octopuses can discriminate between two shapes on the basis of (1) brightness, (2) size, (3) horizontal/vertical orientation, (4) form and (5) plane of polarisation (Fig. 2.4).

Black and white shapes, whether they be rectangles, squares or circles, are readily discriminable (as are grey rectangles of different intensity), and large shapes are easily discriminable from small. Vertical and horizontal rectangles are discriminable, provided the eye is held horizontally in the orbit, as it normally is in life. Removal of the statocysts (organs of balance, §2.1.2) abolishes this orientation, and the discrimination becomes impossible (Wells, 1960; Budelmann, 1970, 1976). Training octopuses on this discrimination has revealed the curious fact that naive octopuses have a strong preference for a vertical rectangle over a horizontal one, if the shapes are

Table 2.1 Cephalopods with multiple visual pigments

Species	Visual pigments		
	A_1-based	A_2-based	A_4-based
Watasenia	✓	✓	✓
Bathyteuthis	✓	✓	✓
Pyroteuthis	✓		✓
Pterygioteuthis	✓		✓
Japetella	✓		✓

Based on data in Seidou *et al.* (1990); Kito *et al.* (1992).

Brightness

Size

Orientation

Form

Plane of polarization

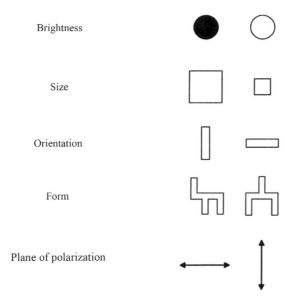

Figure 2.4 Discrimination training experiments have revealed five classes of information that octopuses can extract from their visual world (reprinted from Messenger, 1979a: *Endeavour* **3**, 92–99, The eyes and skin of *Octopus*: compensating for sensory deficiencies, Messenger, J. B., © (1979), with permission from Elsevier).

moved up and down; if they are moved from side to side, the horizontal is preferred (Sutherland & Muntz, 1959). Such preferences could perhaps be related to the habits of their prey in the sea: fish tend to move along their long axis. Young cuttlefish, too, show a marked preference for line stimuli moving along the long axis, like the *Mysis* on which they feed (Wells, 1962b; see Chapter 8). Experimentally naive octopuses also show a marked preference for pointed shapes moved in the direction of the point (Sutherland, 1959) and seem to prefer large shapes to small ones (Sutherland & Carr, 1963). The significance of these preferences is not clear.

There is no reason to suppose that *Octopus vulgaris* is an unusual cephalopod in its visual capabilities. The eyes and optic lobes of most cephalopods are well developed (Nixon & Young, 2003); indeed, in most decapods the optic lobes are larger relative to the central brain than in *Octopus vulgaris* (Box 8.2; Young, 1988). Moreover, Allen, Michels and Young (1985) have shown that the squid *Lolliguncula brevis* can discriminate between a white horizontal rectangle

and a white sphere, and Messenger (1977b) found that *Sepia officinalis* could easily discriminate between small and large squares irrespective of distance.

Polarisation Vision

Polarisation vision (sometimes termed polarisation light sensitivity) was first shown in cephalopods by discrimination training. Octopuses were found to readily learn to attack (or not) a torch covered with a Polaroid sheet according to whether the electric- (or e-) vector was oriented vertically or horizontally; they were less convincingly able to discriminate between obliquely oriented e-vectors (Moody & Parriss, 1960, 1961). At the same time, these authors provided the first ultrastructural evidence that the photoreceptors in the *Octopus* retina are microvillar and constitute a two-plane rhabdomeric system (*ibid.*). Subsequently, physiological evidence for differential sensitivity to the plane of polarisation in *Octopus* was provided by the elegant experiments of Tasaki and Karita (1966), and Sugawara, Katagiri and Tomita (1971); Saidel, Lettvin and MacNichol (1983) and Saidel *et al.* (2005) obtained similar results in *Loligo*. Using the optomotor response of cephalopods, Talbot and Marshall (2010a, b) demonstrated polarisation vision in *Sepia plangon* and *S. mestus* as well as the striped pyjama squid, *Sepioloidea lineolata*. Curiously, Darmaillacq and Shashar (2008) reported that *Sepia elongata* did not respond to polarised stripes in a typical optomotor test despite having microvilli arranged orthogonally, which allows polarisation sensitivity in other cephalopods; the authors suggest future studies on this species to understand this anomaly. Recent evidence suggests that *Sepia plangon* is capable of extremely high-resolution of e-vector angle discrimination, which may be important in 'breaking' intensity-based background matching by potential prey (Temple *et al.*, 2012).

The retinas of *Sepia* (Wolken, 1958), *Loligo* (Zonana, 1961), *Sepiella* (Yamamoto, 1985), *Histioteuthis* and *Octopoteuthis* (Shashar, Milbury & Hanlon, 2002), and *Watasenia* (Michinomae *et al.*, 1994) also have microvilli arranged in two planes orthogonally, and the presumption is that all coleoids are potentially sensitive to polarised light (for references see Messenger, 1981, 1991).

Recently, a number of fascinating behavioural studies have added to the morphological and physiological evidence and begun to reveal how important polarised light sensitivity must be in the life of many cephalopods. The most significant of these findings are highlighted in Box 2.1. For a stimulating discussion of whether polarisation is an independent component of the visual input of cephalopods, or essentially only a contrast-enhancing device, the reader is referred to the thoughtful discussion of Gleadall and Shashar (2004).

Colour blindness

There is one category of visual discrimination that octopuses, cuttlefish and most squid seem unable to make: colour discriminations (Box 2.2). Until 1973, there were conflicting reports about whether cephalopods had colour vision. All these accounts were flawed in experimental design (or lack of it) and because no account had been taken of possible brightness differences that could have contributed to a positive result. Aware of these problems, Messenger and his colleagues attempted to train over

---

Box 2.1 Behavioural evidence that polarisation vision is important in the life of cephalopods

OCTOPUS

1. Moody and Parriss (1960, 1961) showed that *O. vulgaris* could discriminate between two polarised light sources whose e-vectors are oriented at 0° and 90°, although not, apparently, at 45° and 135°. Rowell and Wells (1961) confirmed that the discrimination was made intra-ocularly.

2. Shashar and Cronin (1996) demonstrated that *O. vulgaris* and *O. briareus* could discriminate between targets with and without a pattern produced by 90° polarisation contrast, whether the patterns were oriented at 0°/ 90° or 45°/135°. This strongly suggests that polarisation vision could be an important aid to object recognition in their natural environment.

SEPIA

1. Shashar, Rutledge and Cronin (1996) viewed living cuttlefish through a polarising filter and discovered a prominent polarisation pattern (on the arms, around the eyes and on the front of the head) that could be expressed or not according to behavioural circumstances (see further proof in Chiou *et al.*, 2007). The responses of cuttlefish to their own reflected image changes if the polarised structure of the image is altered. The implication that cuttlefish may use polarisation patterns (invisible to most vertebrate predators) for intraspecific communication has yet to be fully substantiated (Boal *et al.*, 2004; Mäthger, Shashar & Hanlon, 2009; see Chapter 7).

2. Shashar *et al.* (2000) showed clearly that cuttlefish preferentially attack fish with normal polarisation

patterns reflecting from their scales rather than fish whose polarisation reflection has been experimentally reduced by a filter. Reflection-based radiance matching by silvery fish scales has a downside in that the reflected light appears to be partially polarised (Denton, 1970), and this experiment suggests that cuttlefish could use polarisation vision to 'break' the camouflage of their fish prey.

3. Cartron *et al.* (2012) demonstrated that cuttlefish can use polarised light cues for orientation. The experimenters trained animals to find shelter in a Y-maze with black and white landmarks and the e-vector of polarised light as relevant cues. Subjects not only learned the maze quickly, but could use either type of cue when the other was missing. Moreover, when presented with conflicting spatial information, the majority followed the e-vector rather than the landmarks. This striking evidence that the sun's position in the sky could potentially be used by cuttlefish for orientation suggests that field studies into this area would be rewarding.

***LOLIGO* AND OTHER GENERA**

1. Jander, Daumer and Waterman (1963) found that young *Euprymna scolopes* (a sepiolid) and *Sepioteuthis lessoniana* (a squid) could orient themselves to polarised light whose e-vector lies at 0°, 45°, 90° and 135°.

2. Shashar and Hanlon (1997) showed that *Loligo pealeii* and *Euprymna scolopes* can exhibit polarised light patterns on different parts of the body.

3. Shashar, Hanlon and Petz (1998) demonstrated that hatchling *Loligo pealeii* attack living zooplankton at 70% greater distances in a linearly polarised light field than in a depolarised light field. They also showed that adult squid preferred to attack small glass beads that had been heat-stressed to produce a polarisation pattern rather than beads that were not polarisation-active. These experiments suggest that squid could use polarisation vision to 'break' camouflage by transparency, which is extremely common in the sea.

4. Pignatelli *et al.* (2011) showed that three species of cephalopod (*Sepia plangon*, *Sepioloidea lineola* and *Sepioteuthis lessoniana*) gave escape responses to looming polarised signals, suggesting that polarisation vision may assist cephalopods in predator evasion. This study, and that of Temple *et al.* (2012), demonstrated surprising acuity in polarisation discrimination in cephalopods, which could assist in prey detection.

SUMMARY

There is now a general consensus that polarisation vision in cephalopods can be important for detecting silvery fish and transparent organisms, and possibly for predator avoidance. However, it is important to bear in mind that polarisation vision is only effective at relatively short distances; Shashar *et al.* (2011) estimate that in the clearest oceanic waters there is an upper limit of 15 m that will decrease markedly with turbidity. Some cephalopods can exhibit polarised light patterns on specific parts of the body, but the evidence that any cephalopod uses such patterns for communication is still equivocal (Chapter 7).

70 experimentally naive *Octopus vulgaris* to make discriminations between vertical rectangles painted different hues that were matched for brightness by human subjects (Messenger, Wilson & Hedge, 1973; Messenger, 1977a). They used a variety of training techniques, but octopuses never learned to discriminate on the basis of hue, although in the course of the same experiments they did discriminate on the basis of brightness or orientation. The same authors also demonstrated that *O. vulgaris* gave no optomotor response in a nystagmus apparatus to stripes of equal brightness but different hue.

All these results, together with the morphological, biochemical and physiological data, suggest strongly that *O. vulgaris* is colour blind. And in the cuttlefish, *Sepia officinalis*, the behavioural experiments of Marshall and Messenger (1996) and Mäthger *et al.* (2006) provide strong evidence for colour blindness (Box 2.2). To what extent other cephalopods are colour blind is not known, although there is some experimental evidence that *O. apollyon* (Roffe, 1975) and also *Todarodes pacificus* (Flores, Igarashi & Mikami, 1978; Flores, 1983) may be colour blind. So, apart from the five genera cited in Table 2.1, current evidence suggests that most cephalopods are colour blind. This seems curious in view of the remarkable ability of many shallow-water cephalopods to match the colour of their background (Mäthger *et al.*, 2008), and we shall return to this point in Chapter 3 (see also Messenger, 1979a).

Photosensitive Vesicles

Cephalopods, like many other molluscs, possess photoreceptors other than the eyes. In octopods, there are photosensitive vesicles (PSV) attached to the stellate ganglion in the mantle, and in decapods there are PSV lying close to the olfactory lobe in the brain. These organs have the ultrastructure and biochemistry of photoreceptors, and there is good physiological evidence that they respond to light (Mauro, 1977). Their function is not well understood. They may be especially well developed in mid- and deep-water cephalopods (Young, 1977, 1978; Aldred, Nixon & Young, 1978); but they are also conspicuous in epipelagic genera such as *Illex* and *Todarodes*. R.E. Young (1977) has suggested that the PSV may serve to monitor the downwelling light to regulate the light that they emit from their own photophores, and in *Abralia* there is good evidence to support this idea (Young, Roper & Walters, 1979). Indeed, in some genera it is possible to differentiate between the dorsal PSV, which monitor downwelling light, and the ventral PSV, which monitor the emission of the ventral photophores (Young, 1973). There has also been speculation that the PSV could be involved in the diel vertical migration undertaken by so many cephalopods (Chapter 9). Whatever their function, we need to remember that in cephalopods, as in other molluscs (Messenger, 1981), there may be photic influences on behaviour that are not processed via the eye. A claim that there are putative photoreceptor organs

Box 2.2 Evidence that most cephalopods may be colour blind

BIOCHEMISTRY, MORPHOLOGY, PHYSIOLOGY
Visual pigment
The cephalopod retina contains two photopigments. Rhodopsin, the visual pigment, is found in the outer segments, and retinochrome, an accessory pigment involved in the regeneration of rhodopsin, in the inner segment (Hara & Hara, 1972). The rhodopsin is similar to rhodopsins found elsewhere in the animal kingdom (Hubbard & St George, 1958). In 23 of the 28 cephalopods whose retina has been examined, there is only one visual pigment, strong prima facie evidence that most cephalopods, certainly shallow-water species, may be unable to discriminate wavelength (see text).

Eye morphology
With a single visual pigment, it may still be possible to discriminate wavelength if the retina contains multiple banks of receptors, such as the double layers of rods known in some deep-sea teleosts (Denton & Locket, 1989; Denton, 1990). Light reaching the deeper layer will be different from that absorbed by the superficial layer, so that the two banks will have different spectral sensitivities. Such an arrangement of banked receptors has only been found in one species, the squid *Watasenia scintillans* (see text).

Physiology of the retina
The gross electrical response of the retina to a flash of light is termed the electroretinogram (ERG). The ERG has been measured in the squid *Ommastrephes sloanii pacificus* and in three species of *Octopus* (*O. briareus, O. dofleini, O. vulgaris*) (Orlov & Byzov, 1961, 1962; Byzov, Orlov & Utina, 1962; Hamasaki, 1968a, b). In the isolated retina, a given ERG can be produced by light of different wavelengths if compensations are made for brightness; changing from light of one wavelength to another produces no change in response (no Purkinje shift) such as would be expected if there were receptors with different spectral sensitivities. During dark-adaptation, the size of the ERG increases smoothly: there is no sharp discontinuity such as occurs in vertebrate eyes that have a duplex retina. In these species, therefore, colour vision seems improbable.

BEHAVIOURAL EXPERIMENTS WITH *OCTOPUS VULGARIS*
Optomotor responses
Octopuses and cuttlefish make visually induced compensatory eye, head and body movements when surrounded by a revolving black-and-white striped background (Packard, 1969b; Messenger, 1970). Octopuses also make these responses to stripes of different intensity grey; to dark and light red stripes; and to dark and light cyan stripes. They do not make them with stripes of different hue when the stripes have been matched for brightness by a dark-adapted human subject (Messenger, Wilson & Hedge, 1973).

Learning experiments
Octopuses were trained in three different types of experiment to discriminate between vertical rectangles differing in brightness or in hue with: (1) successive presentation; (2) successive presentation using two cues; or (3) simultaneous presentation. In total, 79 experimentally naive octopuses were used. The experimental design was such that each individual animal was tested on brightness and hue during the course of the experiment. In all the experiments, octopuses that could learn to discriminate on the basis of brightness failed to learn on the basis of hue. Blue, green and red were not discriminable from each other under the conditions of the experiments; nor were yellow and violet from matching greys (Messenger, Wilson & Hedge, 1973; Messenger, 1977a). Additional details and references are given by Messenger (1981, 1991).

BEHAVIOURAL EXPERIMENTS WITH *SEPIA OFFICINALIS*
Marshall and Messenger (1996) devised another way of investigating the visual capability of cephalopods. Taking advantage of the well-known ability to 'match' the background using the neurally controlled chromatophores, they placed cuttlefish on a variety of specially designed backgrounds, containing white, blue, green, yellow or red gravel in various combinations. These experiments demonstrated that in these circumstances, cuttlefish do not

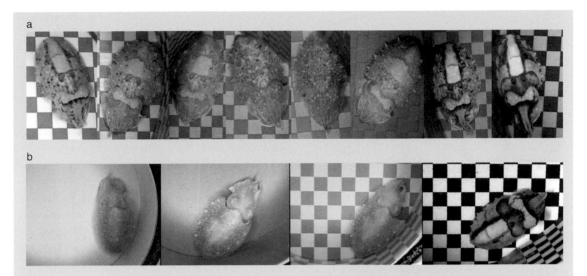

adjust the appearance of their chromatophores in response to different wavelengths in the substrate, only to different brightnesses.

In 2006, Mäthger and her colleagues also used the body patterns of cuttlefish to investigate this problem. They tested cuttlefish on backgrounds comprising specially designed chequerboards that elicit disruptive patterns if there is sufficient contrast (panel **a** & panel **b** far right). However, when cuttlefish were placed on a blue–yellow chequerboard whose

colours were matched in intensity to an eye with a single visual pigment (λmax 492 nm, Brown & Brown, 1958), they did not show disruptive patterning, suggesting they viewed this as a uniform background, like the plain yellow and plain blue backgrounds (panel **b**).

(Images from Mäthger *et al.*, 2006: *Vision Research* **46**, 1746–1753, Color blindness and contrast perception in cuttlefish (*Sepia officinalis*) determined by a visual sensorimotor assay, Mäthger, L. M., Barbosa, A., Miner, S. & Hanlon, R. T. © (2006), with permission from Elsevier.)

on the head of hatchling cuttlefish and squid (Sundermann, 1990) has never been followed up. Nor has the suggestion that the nuchal organs, which lie in the mantle just behind the brain, are also photoreceptors (Parry, 2000).

Much more intriguing is the recent discovery that the same opsins that are found in the eye occur in the skin of the cuttlefish *Sepia*, raising the possibility of a distributed light sense that might mediate certain camouflage responses (Mäthger, Roberts & Hanlon, 2010) (see also Chapter 3). Previously it had been shown that the bioluminescent organ of *Euprymna* not only contains the same opsin as that in the retina but that it responds physiologically to light, presumably that produced by its symbiotic bacterial partners (Tong, *et al.*, 2009).

2.1.2 Mechanoreceptors

Statocysts

These are elaborate, paired organs situated in the cartilage below the brain (Fig. 2.5a), and they provide

the information about gravity and acceleration necessary for maintaining orientation and for making compensatory head, eye and body movements. They also control the countershading reflexes and are responsible for detecting 'infrasound' (see below). Surgical destruction of both statocysts causes the animal to spiral and zigzag in all three planes as it swims (Boycott, 1960; Dijkgraaf, 1961); when the statolith (see below) is prevented from developing normally, the statocyst becomes dysfunctional, and hatchling squid 'spin' and twist when swimming, unable to feed (Colmers *et al.*, 1984). Surgical destruction of statocysts also disturbs the orientation of the eyes (Budelmann, 1970) so that octopuses cannot learn a simple horizontal/vertical discrimination or recognise the plane of polarised light (Wells, 1960; Rowell & Wells, 1961).

There are major differences between decapod and octopod statocysts, but in both groups each statocyst contains two receptor systems: the *macula–statolith–statoconia* system and the *crista–cupula*

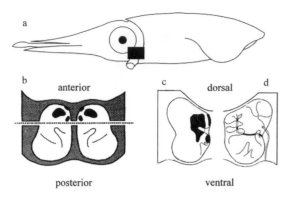

Figure 2.5 (**a**) Lateral view of a squid showing location of the statocysts (black rectangle). (**b**) Ventral view of statocysts embedded in cartilage (shaded). Solid black shapes represent the six maculae; the dotted line shows the position of a transverse cut made to reveal an anterior view (**c**) and a posterior view (**d**) of each statocyst. (**c**) Three maculae and two projections (anticristae). (**d**) Part of the crista system and eight projections (simplified from Budelmann, 1990: *Squid as Experimental Animals*, The statocysts of squid (1990), 421–439, Budelmann, B. U., © Plenum Press, with permission of Springer). The maculae detect gravity and linear acceleration; the cristae detect rotational acceleration.

system. Both systems rely on numerous polarised hair cells to detect mechanical disturbance (Budelmann, 1990).

The macula system utilises information about the direction of gravity and linear acceleration to regulate the orientation of the head, eyes and body (Budelmann, 1970); it also regulates the countershading reflexes, whereby the chromatophores on the ventral side of a cephalopod are expanded to maintain countershading when the animal moves out of normal orientation (see §5.1.2). In decapods, there are three maculae in each statocyst (Fig. 2.5b); in octopus, there is only one. The largest decapod macula and the octopod macula carry the statolith, a small structure made of aragonite with an organic matrix, which commonly shows rings. In many squid species, these are laid down with a definite periodicity, so that they can provide information about age (e.g. Jackson *et al.*, 1993; Jackson, 1994; Jackson *et al.*, 1997; Jackson & Moltschaniwskyj, 2002).

The crista system provides the brain with information about rotational acceleration to enable it to regulate the position of the head, funnel and especially the eyes (Budelmann, Sachse & Staudigl, 1987). In decapods, there are four cristae set at right angles to each other (Fig. 2.5d) so that the hairs on the cristae will be stimulated differentially when the animal rotates about the pitch, yaw or roll axis. In octopods, there are nine cristae. There are considerable differences in detailed organisation between the statocysts of different cephalopods, and these can be related to their locomotor habits (Young, 1984; Stephens & Young, 1982; Maddock & Young, 1984; Young, 1989; Williamson, 1991). In fast-moving loliginid and ommastrephid squid, the statocyst wall has been sculptured into a series of projections, or anticristae (Fig. 2.5c, d), that virtually form canals, not unlike the semi-circular canals of vertebrates. The most complete canals occur in *Dosidicus gigas*, the Pacific jumbo squid, an active, voracious predator (§9.7.2). In the neutrally buoyant squid, such as the cranchiids, which move much more slowly, canals are absent.

'Hearing'

The statocysts also provide the brain with another kind of information: low-frequency vibration, or 'infrasound'. The question of whether cephalopods 'hear' has been raised many times (Baglioni, 1910; Wells & Wells, 1956; Hubbard, 1960; Young, 1960b; Moynihan, 1985b; Taylor, 1986; Hanlon & Budelmann, 1987; Budelmann, 1992, 1996) and has only recently been resolved. Strictly, cephalopods cannot 'hear' since they have no sense organs with gas-filled cavities that can be compressed by the pressure waves of underwater sound: what they respond to is particle motion detection. Several early reports had noted that octopuses were sensitive to vibrations caused by tapping the tank, but later attempts to train octopuses to respond to frequencies of 100 Hz, 900 Hz and 2000 Hz were unsuccessful (Hubbard, 1960). Yet Dijkgraaf (1963) reported that *Sepia* respond to a sound stimulus of 180 Hz, and Maniwa (1976) claimed that not only are *Todarodes* attracted to an intermittent pure tone of 600 Hz, but that sound projectors emitting such a frequency have been employed successfully in commercial fisheries to enhance squid capture by day or under a full moon.

Packard, Karlsen and Sand (1990) trained *Sepia*, *Loligo* and *Octopus* to associate 'sound' stimuli with a small shock. They placed the animals in a special

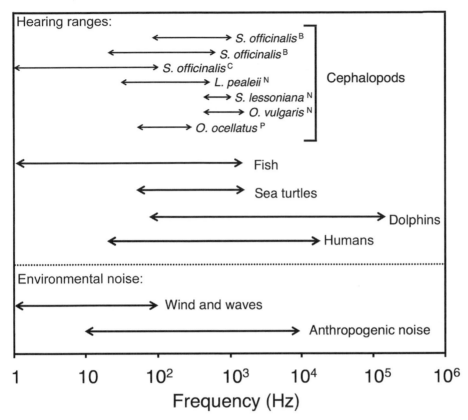

Figure 2.6 Hearing ranges in marine animals compared with ambient noise. The hearing ranges of various cephalopods were determined using behavioural (B), conditioned (C), neurological (N) or physiological (P) responses. References, from top to bottom: Samson *et al.* (2014), Komak *et al.* (2005), Packard, Karlsen and Sand (1990), Mooney *et al.* (2010), Hu *et al.* (2009), Kaifu, Akamatsu and Segawa (2008). See text for details.

acoustic tube that could deliver stimuli at frequencies of between 1 Hz and 100 Hz and recorded changes in their breathing and jetting movements. Their infrasound audiograms showed that the most sensitive part of the range was below 10 Hz. Unfortunately, Packard and his collaborators provided no evidence about the structures responsible for sound detection, but their suggestion that the statocysts are responsible was plausible since recordings from the crista and macula nerves have shown responses to low frequencies (Maturana & Sperling, 1963; Budelmann, 1976; Williamson, 1988).

Only recently has it been established unequivocally that it is the statocysts that mediate cephalopods' sensitivity to low-frequency particle motion. In a series of careful experiments, Mooney *et al.* (2010) succeeded in recording auditory evoked potentials (AEPs) from

electrodes implanted near the statocysts of sedated longfin squid (*Loligo pealeii*) and showed that the animals responded to low-frequency particle motion but not to pressure. AEPs were obtained for frequencies between 30 Hz and 500 Hz, with the lowest thresholds between 100 and 300 Hz (Fig. 2.6). Such responses were not obtained if the electrodes were placed far from the statocysts, and they disappeared after statocyst ablation, thus excluding the possibility that the 'lateral lines' or neck proprioceptors (below) were involved in the response. Mooney and his colleagues, considering the biological significance of their results, speculate that such low-frequency sound detection may have evolved to detect the movements of predators such as fish or toothed whales, although not the high-frequency whistles and clicks of odontocetes, as suggested by Hu *et al.* (2009). Separate behavioural experiments demonstrated that the longfin squid did not react to intense ultrasonic clicks of echolocating toothed whales (Wilson *et al.*, 2007). Yet the sense may also be employed to detect prey and the mantle movements of conspecifics in a shoal, and, in the near-shore environment, yield information useful for orientation or even navigation. Clearly this is an exciting area for future research.

A downside to this sensitivity to infrasound has recently been reported, however. André *et al.* (2011) have obtained ultrastructural evidence in four genera of cephalopods that controlled exposure to low-frequency sounds causes substantial damage to hair cells on the macula and crista. In the sea, this could be life-threatening, and the authors suggest that noise pollution associated with human activities may impact cephalopods as well as mammals.

There is a recent report claiming that spiny lobsters (*Panulirus*) produce sounds ('stridulations') to deter attacks by *Octopus briareus* (Bouwma & Herrnkind, 2009); but the authors provide no details of these sounds, nor did they experiment with octopuses whose statocysts had been removed.

Lateral Line Analogue

Although the senses and behaviour of cephalopods have been studied intensively in the laboratory for over 100 years, it was only recently that the epidermal head and arm lines of *Sepia officinalis* and *Lolliguncula brevis* were shown to respond to local water displacements with microphonic potentials like those recorded from fish lateral lines (Budelmann & Bleckmann, 1988). It is known that at least nine genera of cephalopods have epidermal lines (Budelmann *et al.*, pers. comm.).

The epidermal lines (Naef, 1928; Sundermann, 1983) are well defined in hatchling cuttlefish (Fig. 2.7), and in the adult their position can easily be seen because the lines coincide with the pink iridophore arm stripes. They contain many thousands of hair cells, and physiological studies have shown that these respond to sinusoidal water movements within the 0.5–400 Hz range (Bleckmann, Budelmann & Bullock, 1991). The most effective frequencies are between 75 and 100 Hz, with the minimal threshold at 0.06 μm peak-to-peak water displacement at the level of the receptor cells. This implies that the system ought to be sensitive enough to

allow a cuttlefish to detect a fish 1 m long at a distance of 30 m (Budelmann, Riese & Bleckmann, 1991).

There is also behavioural evidence that cuttlefish can detect local water movements in the range 20 Hz to 600 Hz (Komak *et al.*, 2005). Juvenile animals, stimulated by sinusoidal water movements of constant amplitude but varying frequency, gave a variety of non-habituating responses. Not all frequencies were responded to, however, suggesting that some kinds of water movements are more important than others for cuttlefish; such movements, which seem to be more important for younger, smaller animals, may perhaps alert them to predators. Whether or not the epidermal lines mediate these responses remains to be established.

However, there is already some evidence that this receptor system could be used by cuttlefish to locate prey: intact cuttlefish still capture 50% of prawns left in the tank in total darkness, whereas those with most of the epidermal lines destroyed caught only 30%. Presumably this receptor system could locate predators too; it might also be used by schooling squid to help maintain station during darkness (§7.2) and to transmit what Partridge and Pitcher (1980) term a 'fright response' in teleosts.

Neck Proprioceptive Hair Cells

After discovering proprioceptive hair cells on the neck of the squid *Lolliguncula*, Preuss and Budelmann (1995a) showed that these were involved in the control of head-to-body position. Moreover, these cells, together with the statocysts, also influence a *dorsal light reflex*, whereby squid illuminated from the side roll the head and body towards the light (Preuss & Budelmann, 1995b). Such a response, which is well known in fish, reveals that the processing of sensory information in the cephalopod central nervous system is complex: there is clearly multi-modal interaction of visual and at least two kinds of mechanical information in the regulation of posture.

Touch and Pressure Receptors

Little is known about general pressure sensitivity in cephalopods, although Knight-Jones and Morgan (1966) claim that juvenile *Loligo forbesi* move upwards in response to increased pressure, as does *Nautilus* (Jordan, Chamberlain & Chamberlain, 1988; Chapter 10).

All cephalopods seem to be highly sensitive to touch, and in *Octopus* several kinds of mechanoreceptors have

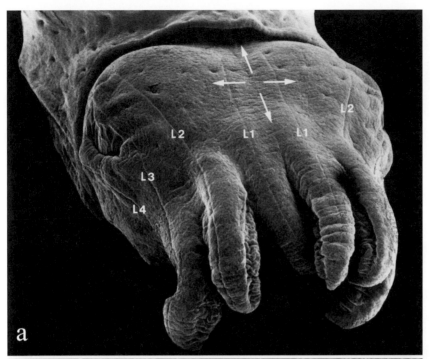

Figure 2.7 (**a**) Scanning electron micrograph of the 'lateral lines' on the head of a young cuttlefish (*Sepia officinalis*). The grooves on the head and along the dorsal surface of the arms (L1–L4) contain hair cells (**b**) that are polarised in the direction of the arrows in (**a**). Scale bar is 10 μm. Republished with permission of Wiley-Liss, from Budelmann, Schipp & Boletzky (1997): *Microscopic Anatomy of Invertebrates*, Vol. 6A: *Mollusca* II, 119–414, Budelmann, B. U., Schipp, R. & Boletzky, S. V., © (1997); permission conveyed through Copyright Clearance Center, Inc.

been described in the rim of the sucker (Graziadei, 1971; Graziadei & Gagne, 1976a, b; Fig. 2.8). In octopuses, tactile information from the suckers passes to the highest centres in the brain for learning, and many experiments have been made on touch learning in these animals. They can readily learn to discriminate rough objects from smooth, whether these are shells, cylinders or spheres; they can also distinguish between different degrees of roughness. Experiments suggest that it is the proportion of mechanoreceptors stimulated that is used by the brain in making these discriminations; Wells (1978) gives a useful summary of this work. It is interesting that although octopuses can easily discriminate visually between vertical and horizontal rectangles, they cannot distinguish by touch between

the orientation of grooves cut around or along a perspex cylinder. This is because they have no proprioceptive information about the position of the suckers in space, presumably because the arms are extremely flexible and lack joints (Wells & Wells, 1957a). Nor can octopuses be trained to discriminate between heavier and lighter objects (Wells, 1961). In fact, the nervous system of the arms, which contains more neurons than the whole of the central brain (Young, 1971), is in some ways curiously divorced from the rest of the brain, and many of the arms' actions are performed without reference to the brain (Sumbre *et al.*, 2001). It is noteworthy that octopuses do not readily learn to make skilled manipulative movements of objects that they can see, which could explain why attempts to train them in an

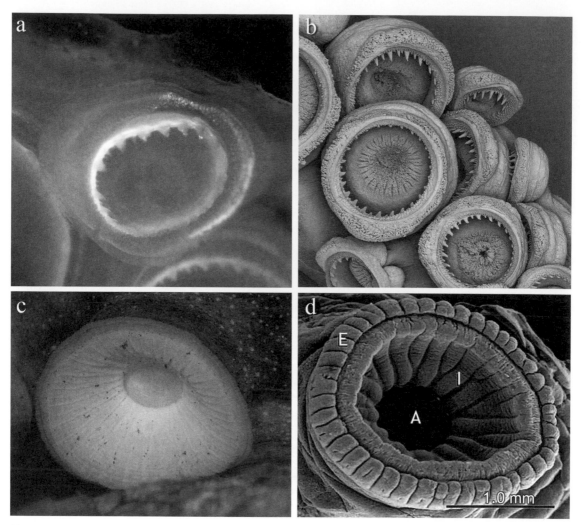

Figure 2.8 Suckers of cephalopods. The tentacular clubs of squid such as *Loligo pealeii* have semi-rigid proteinaceous rings with serrated teeth (**a**) to cut down the shear forces of wriggling prey such as fish and crustaceans. A scanning electron microscope (SEM) image (**b**) shows details of this complex organ. Octopus arm suckers are structured differently (**c, d**) for more versatile functions. A SEM image shows the acetabulum (A), the infundibulum (I) and the rim of loose epithelium (E). (Photograph **a** by R. Hanlon; **b**, from James Weaver; **c**, *Octopus cyanea* © Fred Bavendam; **d**, *Octopus bimaculoides* from Kier & Smith, 2002, with kind permission from Oxford University Press.)

operant conditioning situation have generally failed (Chapter 8 and Wells, 1978) and why they cannot easily remove the lids of jars containing crabs (Fiorito, von Planta & Scotto, 1990) or insert a stone into a gaping bivalve to prevent it closing, a myth that seems to date back to Pliny (AD 77).

The ability of octopuses to learn tactile discriminations is associated with two lobes in their brain that are lacking in decapods: the inferior frontal and subfrontal lobes, which are associated with the posterior buccal lobe (§2.3). By making a series of lesions, Wells (1959a) showed that interference with this part of the brain seriously affects touch memory, and more recently it has been shown that there are structural changes to cells in the subfrontal and posterior buccal lobes during tactile learning (Robertson & Lee, 1990). It is likely that these regions are involved in taste memories too.

2.1.3 Chemoreceptors

Chemoreception is widespread and important in all animals, and it is useful to recognise three increasingly sensitive chemoreceptor systems, although strict distinctions blur in aquatic animals: a general chemical sense, a chemotactile sense, and distance chemoreception, or olfaction (Messenger, 1979b).

Octopuses, and perhaps all cephalopods, are sensitive to chemical as well as tactile stimuli all over the body, and octopuses and cuttlefish are known to have chemoreceptors in the lips (Graziadei, 1965) and on the suckers. The sucker rims in particular contain large numbers of tapered, ciliated cells (Graziadei, 1964a). In *Octopus*, there are about 10 000 of these primary receptors on each sucker; since there are 200 suckers on each arm there are about 16 million of these cells in an adult animal. In *Sepia*, there are only about 100 cells per sucker and far fewer suckers on the arms (Graziadei, 1964b). This is the only estimate available for a decapod, and it emphasises the different life styles of the two groups: octopuses use their arms to detect food that is out of sight, while cuttlefish and squid use their arms mainly for holding prey caught after a visual attack.

The sensitivity of *Octopus vulgaris* suckers to 'taste-by-touch' has been investigated experimentally by Wells and his associates (Wells, 1963; Wells, Freeman & Ashburner, 1965), who found that animals can easily learn to discriminate between objects soaked in solutions of sucrose, hydrochloric acid and quinine. Moreover, they can detect these substances at 10 to 1000 times lower concentrations than humans can. They can also discriminate between seawater and seawater with 10^{-5} M KCl added. Experiments like these reveal that octopuses can probably detect quite small differences in the taste of objects that they handle.

In *Octopus vulgaris*, Chase and Wells (1986) have shown that animals move upstream towards a solution of adenosine 5'-monophosphate (AMP) and, to a lesser extent, glycine and glutamic acid. These substances, in particular AMP, are known to be attractants for other species at similar concentrations (10^{-5} M). This behaviour is not altered after removal of the olfactory organs (see below).

The ventilation rate of an isolated *Octopus* increases when it is exposed to water in which crabs have lived (Boyle, 1983a), and *Eledone* is also sensitive to chemicals such as proline, glycine and especially betaine, all of which are constituents of crustacean flesh (Boyle, 1986a). However, a more extensive study of chemotaxis in a Y-maze, where octopuses were given a clear choice between substances, shows that the matter may be more complicated (Lee, 1992). Although *O. maya* was attracted strongly to the maze compartment with crab extract, proline or ATP (rather than AMP), betaine appears to arrest movement, even though it does cause arousal. All these findings suggest that distance chemoreception may be more important in guiding octopuses to their food than has been assumed hitherto.

In the cuttlefish, *Sepia*, there is also evidence that olfactory cues can influence the animals' behaviour but unfortunately the results are inconsistent. In one series of experiments, it was found that, in sexual encounters, female cuttlefish consistently preferred, and mated with, recently mated males (Boal, 1997), and the author plausibly suggests that this must be because of chemical cues. However, subsequently Boal and Marsh (1998), using a simple Y-maze for choice, found that the test subjects were not significantly more responsive to water that had passed over conspecifics (of either sex) than they were to control water. Yet when Boal and Golden (1999) monitored ventilation rate in young cuttlefish, they found clear responses to ink from a conspecific, to water containing food, and to water that had contained sea turtles as well as to novel seawater, suggesting considerable sensitivity to water-borne chemical cues. More recently, increased ventilation in *Sepia* has been reported after exposure to odours from eggs and ovary extracts, suggesting that the eggs might be a source of pheromones (Boal *et al.*, 2010) perhaps not unlike those reported in *Loligo* by Cummins *et al.* (2011). There has even been a claim that exposure to certain odours in the first week of life can alter *visual* preference for prey by hatchling cuttlefish (Guibé, Boal & Dickel, 2010), showing that the influence of chemical cues on cuttlefish behaviour may be complex.

It should also be noted that, as with the *Octopus*, *Eledone*, *Sepia* and *Loligo* data above, the relative importance of suckers, lips and 'olfactory organs' (see below) in such responses remains to be established.

In squid, compelling evidence for chemical cues influencing behaviour comes from the observations and experiments by Hanlon and his colleagues on the squid *Loligo pealeii* (King, Adamo & Hanlon, 2003;

Buresch *et al.*, 2003, 2004). The females of these loliginids lay large masses of fertilised eggs (in clusters of egg capsules) on the sea floor. When males see these, they approach them, and if they touch the egg capsules, it leads to an immediate and dramatic escalation to the most extreme male–male agonistic behaviour, even in the absence of females (Cummins *et al.*, 2011) (Fig. 7.2). Experiments show that the outer tunic of the egg capsules contains a specific protein (a member of the β-microseminoprotein family found also in mammalian reproductive secretions) that acts as a contact pheromone. Recombinant β-microseminoprotein manufactured from female oviducal and accessory nidamental glands, and painted on a glass flask containing egg capsules, elicited similar extreme fighting behaviour among males. This is a multi-modal sensory process: visual attraction followed by contact chemoreception of the protein molecules on the egg capsule, resulting in a cascade of extreme aggression comparable to that elicited by contact with the eggs themselves. The receptors mediating this response are unknown but may lie on the arm suckers.

The function of the so-called olfactory organs, which have been known for about 150 years, is still far from clear: it is only recently that they have been shown unequivocally to be chemoreceptors, and we now know there are no fewer than five morphological types of olfactory receptor neurons with different odorant specificity (Mobley, Michel & Lucero, 2008). They are small structures located below and behind the eyes, where they would be situated ideally to monitor the quality of the inhalant respiratory stream. Yet M.J. Wells (pers. comm., 1993) found no difference in ventilation rates before and after olfactory organ removal in experiments that systematically varied oxygen tension. The olfactory organ has the ultrastructure of a chemoreceptor organ (Woodhams & Messenger, 1974; Emery, 1975, 1976), containing several types of ciliated cells, which in *Loligo opalescens* are kept clean by a strong flow of mucus from secretory cells (Gilly & Lucero, 1992). In restrained *L. opalescens*, squid ink, L-DOPA and dopamine (precursors of melanin) pipetted onto the olfactory organ elicited the jet escape response, suggesting that one of its functions may be to detect the presence of alarmed conspecifics (Gilly & Lucero,

op. cit.; Lucero, Horrigan & Gilly, 1992). They may well have other functions, of course, and the suggestion that the organs could be sensitive to sexual pheromones from conspecifics (Woodhams & Messenger, 1974) may have found some support from the recent experiments of Cummins *et al.* (2011) mentioned above. Gilly and Lucero tested the effects of crude extracts of gonads or accessory reproductive organs but, as they themselves admit, since their criterion for an effect was an escape response, an attractant substance could not have been detected by their experimental method. It is interesting that the olfactory organs are especially well developed in some mid-water cephalopods, projecting from the head on stalks so that they look like dipteran halteres.

2.1.4 Other Possible Senses

To date, there is no evidence to suggest that cephalopods have an electric sense, but the experiments necessary to exclude this possibility have not been done. It is well known that elasmobranchs (and some teleosts) can respond to the Earth's magnetic field, and the discovery that a gastropod mollusc, *Tritonia*, can show geomagnetic orientation (Lohmann & Willows, 1987) suggests that anyone with the suitable apparatus should at least consider testing a cephalopod.

We now come on to the difficult issue of nociception and pain, for which practically no data exist. The ability to sense and respond to noxious stimuli is an almost universal trait among animals, and it is likely that cephalopods possess 'nociceptors' to deal with such stimuli. There are many free nerve endings in the skin, although their function is uncertain (Bullock, 1965). In the laboratory, octopuses have been observed responding adversely to stinging sea anemones (Ross, 1971), and cuttlefish to nipping crabs. Moreover, deliberately inflicted noxious stimulation (noxious under the assumption that it directly activates nociceptors), in the form of a small electric shock (4–10 V, a.c.), has been consistently and successfully employed to shape octopuses' responses to objects during discrimination training (Chapter 8). In these circumstances, octopuses appear to react as a human does when receiving a mild nettle sting.

Crook *et al.* (2011) tested responsiveness in the squid *Loligo pealeii* to the approach and contact of an inert plastic filament applied to different parts of the body both before and after a discrete injury to the distal third of one arm, and showed that squid expressed behavioural hypersensitivity persisting for at least two days after injury. These alterations parallel forms of nociceptive plasticity in other animals. This long-lasting sensitisation of defensive responses parallels nociceptive sensitisation described in mammals and, thus far, in several other invertebrates. In contrast to what is observed in mammals, and possibly fish (Sneddon, 2009), injury failed to elicit overt wound-directed behaviour by squid (such as grooming).

Whether cephalopods are capable of experiencing the 'affective component' (see Machin, 2005; Carere, Wood & Mather, 2011) required for pain is a matter of ongoing debate (Crook & Walters, 2011) because pain is currently defined as an 'unpleasant sensory and emotional experience associated with actual or potential tissue damage, or described in terms of such damage' (Merskey & Bogduk, 1994). For mammals, long-term sensitisation of defensive responses is often used as an indicator of persisting pain, but this link has been questioned because of lack of evidence for the emotional (affective) component. Crook *et al.* (2011) and Crook, Hanlon and Walters (2013) did not observe spontaneous behaviours in squid suggestive of on-going pain or distress after arm injury. Nor did Alupay, Hadjisolomou and Crook (2014) observe pain-like states in the octopus *Abdopus aculeatus*, although, unlike squid, they did see directed wound attention, site-specific sensitisation, and long-term decreased thresholds for initiating escape responses. Since evidence for behaviours suggestive of pain-like sensations in any invertebrate is limited, we must be cautious in assuming that cephalopods experience pain. When considering this issue, it may be more helpful to compare cephalopods with fish rather than mammals (Sneddon, 2009; Braithwaite, 2010).

2.2 Effectors

There seems to be no better word than 'effectors' to describe the organs of the body that carry out, or 'effect', the response an animal makes to a stimulus. The commonest effectors are muscles, but in cephalopods there are several other kinds of effectors, some of them rather unusual: the chromatophores; the different types of reflecting cells; the photophores; the ink sac; the arm appendages; the suckers; and the buccal mass, which includes the beak, radula, salivary palp and salivary glands.

2.2.1 Muscles

The behaviour of most animals is expressed through muscular activity or inactivity, and cephalopods are no exception to this. Cephalopod muscles are typically molluscan in that they are obliquely striated. They contract quickly and powerfully. Their mitochondria tend to be organised into a central core within each sarcomere and, unusually by vertebrate standards, they rely heavily on carbohydrates and amino acids rather than fats for their metabolism (Hochachka, 1994). The muscles are arranged three-dimensionally in closely packed blocks of constant volume with a connective tissue framework (Bone, Pulsford & Chubb, 1981) so that the contraction of one particular set of muscles transmits forces to others in the block. This arrangement is termed a muscular hydrostat (Kier, 1988) and there is morphological and other evidence that the mantle muscles (Ward & Wainwright, 1972), fin muscles (Kier, Messenger & Miyan, 1985), arm and tentacle muscles (Kier, 1985, 1988) and suckers (Kier & Smith, 1990, 2002; Stella & Kier, 2004) are arranged in this way.

The muscles of cephalopods form six major structures (Fig. 2.1): the arms and tentacles; the head and eye muscles; the mantle; the funnel; the fins (Box 2.3); and the skin muscles. There are also the radial muscles of the chromatophores (see next section).

These various sets of muscles are used in all the behaviours described in this book, but locomotion is sufficiently varied in the group to merit a brief comment here. Cephalopods can swim, jet, walk or even 'fly'. Swimming is achieved by movements of the fins, as in *Sepia*, or by umbrella-like movements of the arms and interbrachial web, as in some octopods. In jetting, which is mainly used for escape, there is rapid expulsion of water through the mobile funnel. 'Flying', more properly gliding, is limited to a few neritic oceanic squid, such as *Ommastrephes*, *Onychoteuthis* and *Dosidicus*, which can accelerate out of the water and apparently travel for distances of up to 50 m (Lane, 1957; Cole & Gilbert, 1970; Packard, 1972). Walking implies use of the arms: it is mostly limited to the benthic

octopods, but some other genera also walk
(e.g. *Metasepia*; Roper & Hochberg, 1988). Describing
walking and arm movement in octopuses has proved
difficult because of the 'fluid' nature of the arms'
movements (Wells, 1978). Mather (1998) has provided an
inventory of arm movements in *Octopus*, and Hochner
and his collaborators have carried out an extended series
of experiments on arm control in this animal (Gutfreund
et al. 1996, 1998; Matzner, Gutfreund & Hochner, 2000;
Sumbre *et al.*, 2001, 2005, 2006) that are beyond the
scope of this book. Grasso (2008) has addressed the
interesting question of how the numerous suckers are

coordinated with each other, and with movements of the
arm that bears them. He found that the animals used
different suckers in certain tasks in combination with
a variety of arm movements; that when restricted to
using a single arm, octopuses tend to use the middle
portion of that arm; and that there is clear evidence
that information about the state and position of a given
sucker is available to local arm circuits as well to the
brain. Byrne *et al.* (2006a) have evidence for arm
preferences in *Octopus*: the anterior arms are strongly
preferred for reaching and exploration and some
individuals may have a lateral bias; there is also some

Box 2.3 Cephalopods: the main muscle groups

FINS

These are present in all decapods, in *Vampyroteuthis* and in
the cirrate octopods. Their shape and size vary greatly and
can be related to the way the animal moves. For example,
the epipelagic *Loligo* has triangular fins that give lift and
stability in the roll plane; the hovering, neutrally buoyant
Sepia has long thin fins that can beat in either direction so
that the animal can make precise turns about the yaw
plane; and the squid *Sepioteuthis* has fins somewhat
intermediate in form suitable for manoeuvring around
objects in its reef habitat (Boycott, 1965). Some of the
gently floating cranchiids and *Spirula* have small propeller-
like fins; the powerful ommastrephids have large fins
whose posterior margins are swept back to provide
streamlining. In benthic forms such as the sepiolids or
Sepia, the fins can be important in concealment, for they
can be rippled strongly to throw sand over the edge of the
mantle as the animal buries itself.

MANTLE

The muscles of the mantle are important for the
characteristic jet propulsion of these animals. The
musculature is complex, with the circular muscles
supplying the main force for the expiratory jet. In strong
jetters, the mantle muscle is very thick; in weaker ones it is
correspondingly thinner (Trueman & Packard, 1968). The
mantle is also important for respiration, although the
muscles of the collar may play the major role in moving
water into the mantle cavity (Bone, Brown & Travers, 1994).

FUNNEL

This is mobile and is moved to help steer the animal. It is
also used by some benthic forms to blow a cavity in the
substrate to facilitate burying, to direct a disturbing jet of
water at prey or predator, and by female octopuses to
aerate and keep clean their egg masses.

HEAD AND EYES

Head movements can be important in steering. Eye
movements are critically important for localising prey or
predator. There is a pupillary response, and in some
species the pupil can be widely dilated as part of a
communication display (Chapter 7).

ARMS AND TENTACLES

The arms are important for feeding (Chapter 4), for mating
(Chapter 6: the male passes spermatophores to the females
with a modified arm), for steering in decapods, and for
walking in most octopods. They may also be used for burying
(e.g. sepiolids, Boletzky & Boletzky, 1970) and signalling
(Chapters 5, 6 and 7). The arms are the functional equivalent
of the pharynx of a fish and enormously increase the potential
for taking large prey (Packard, 1972). The arms typically bear
suckers, principally for grasping and retaining prey, but with
several other functions (see text). The tentacles, present in
decapods only, are used for prey capture. They have a shaft
with no or very reduced suckers and a distal club where the
suckers are well developed (Fig. 4.3). The tentacles can be
quite short, as in *Joubiniteuthis*, or extremely long as in

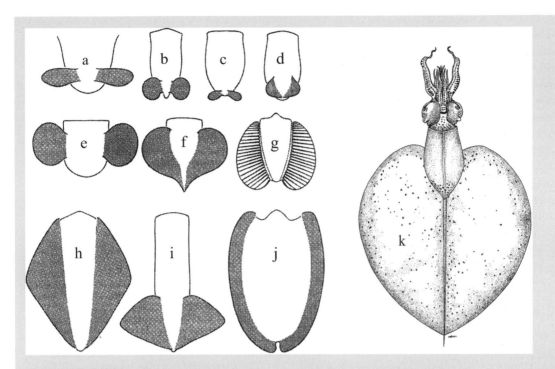

Figure B.2.3.1 Different fin types, not to scale (Mangold, 1989: *Traité de zoologie* T5 Vol 4 – *Métazoaïres (suite)* – *Céphalopodes* by Grassé/Mangold, © 1989, Masson, Paris. Reproduced by permission of Dunod Editeur, 11, rue Paul Bert, 92247 Malakoff). (**a**) *Vampyroteuthis*. (**b**) *Idiosepius*. (**c**) Very young ommastrephid. (**d**) Young *Thysanoteuthis*. (**e**) *Rossia*. (**f**) *Mastigoteuthis*. (**g**) *Chtenopteryx*. (**h**) Adult *Thysanoteuthis*. (**i**) *Ommastrephes*. (**j**) *Sepia*. (**k**) *Magnapinna*.

Chiroteuthis; in the giant squid *Architeuthis* they can exceed 10 m in a specimen whose total length is 15 m (Mangold, 1989). In *Sthenoteuthis*, the tentacles may be locked together at one point and used like tongs (Clarke, 1966).

SKIN PAPILLAE
In sepioids and octopods, there is a complex (and poorly described) musculature that changes the physical texture of the skin so that it can vary between being smooth and very spiky: the cylindrical or plate-like papillae can be 10 mm or more high (Fig. 3.4), and skin texture makes an important contribution to body patterning, especially for camouflage (Allen *et al.*, 2009). In larval octopods, the Koelliker's bristles used in hatching (and also, apparently, for protection) are extruded from a pit by the action of muscles (Boletzky, 1978; Packard, 1988b).

CHROMATOPHORES
These are complex organs, each with 15–25 radial muscles innervated directly from the brain. There may be hundreds, thousands or millions in the skin of an adult octopus or cuttlefish. They are considered at length in Chapter 3.

evidence that arm choice may be influenced by eye use (Byrne *et al.*, 2006b). An important recent paper has also shown that octopuses can visually guide their arm movements in an operant task requiring the location of food in a three-choice maze (Gutnick *et al.*, 2011). This is the first evidence that these animals can combine proprioceptive information about arm position with visual information.

Reports that some oceanic squid can take to the air – that they can 'fly' – have been known for some time

(Lane, 1957; Cole & Gilbert, 1970; Packard, 1972). Two important recent papers, however, not only present detailed evidence for rocket-propulsion, confirming that squid such as *Sthenoteuthis* and *Dosidicus* accelerate through air by expelling water out of their mantles, but indicate that velocities in air can exceed those in water by almost fourfold (Muramatsu *et al.*, 2013; O'Dor *et al.*, 2013). This raises the interesting possibility that flying by squid may be commoner than previously thought, because flight could reduce the energetic cost of migrating (Chapter 9) as well as risks from predation (Chapter 5).

Many cephalopods have independently evolved neutral buoyancy mechanisms that profoundly affect body design, fin shape and locomotor behaviour (Denton, 1974; Clarke, 1988a; see also Fig. 1.6).

2.2.2 Chromatophore Organs

The chromatophore organs are strictly part of the muscular system, but they constitute such an unusual system of effectors, and one that is so important in the life of cephalopods, that it is appropriate to treat them separately. They make cephalopods among the most beautiful of all animals, and they are responsible for their continually changing appearance, a feature that has attracted biologists' attention since the early nineteenth century.

It was not until the 1960s, however, that Florey and his collaborators (Florey, 1966, 1969; Cloney & Florey, 1968; Florey & Kriebel, 1969) combined electrophysiological and ultrastructural techniques to give us a firm basis for understanding the working of these complex organs, which are quite unlike the branched chromatophore cells with mobile pigment granules that are found in crustaceans, teleosts, amphibians and reptiles. This, and subsequent, work is conveniently summarised in the review by Messenger (2001) (see also Tublitz, Gaston & Loi, 2006; Suzuki *et al.*, 2011; and Bell *et al.*, 2013).

Each chromatophore organ comprises an elastic sacculus containing pigment granules, surrounded by a series of about 15–25 radial muscles. When the muscles contract, the sacculus is greatly expanded; when they relax, the elastic energy stored in the sacculus presumably causes it to retract (Fig. 2.9). Because the muscles are under nervous control, expansion and retraction can be very rapid and some chromatophores can be expanded while others remain retracted so that *patterns* can be generated in the skin in a way impossible in other animals: the significance of this, which can be so important in the life of shallow-water cephalopods, is explored extensively in Chapter 3. Chromatophores may be sparse or absent in deep-water forms.

The pigments contained in the sacculus are black, brown, red, orange or yellow. That is, pigmentary coloration in cephalopods is restricted to the longer wavelengths. As far as is known, all the greens, blues and violets in the skin of cephalopods are structural colours, produced by reflecting cells.

2.2.3 Iridophores and Leucophores

There are two other structures in the skin of cephalopods that are important for camouflage and signalling: the iridophores and leucophores. Uniquely in *Octopus dofleini*, Cloney and Brocco (1983) also described 'reflector cells', which have to date not been reported elsewhere. The iridophores and leucophores occur in specific regions of the dermis, around the eyes and ink sac, and sometimes in the photophores (next section).

The *iridophores* are multi-layer stacks of very thin electron-dense platelets alternating with layers of cytoplasm. The platelets were reported to be made of protein (Cooper, Hanlon & Budelmann, 1990) in the squid *Lolliguncula brevis*. In *Euprymna*, a sepiolid, the proteins have been characterised: they are very unusual, presumably independently evolved, and have been named *reflectins* (Crookes *et al.*, 2004). Their structure, their organisation and the way in which they produce colours by constructive interference have been thoroughly investigated (e.g. Denton & Land, 1971; Land, 1972; Cooper, Hanlon & Budelmann, 1990; Ghoshal *et al.*, 2013). In many places the platelets are arranged parallel to the surface of the skin, but sometimes they are long and serpentine and lie on their side, perpendicular to the skin surface (Mirow, 1972). The suggestion has been made that they act as diffraction gratings (Schäfer, 1936). Whether they produce colours by interference or diffraction, it is these structures that provide the blues and greens of cephalopod body coloration: for example in the ventral

a

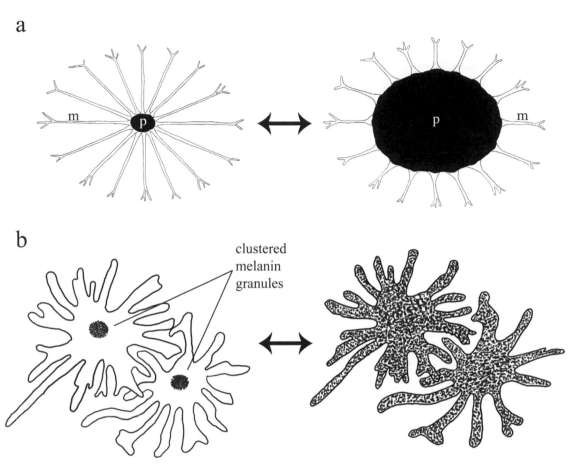

b

clustered
melanin
granules

Figure 2.9 Diagrammatic comparison between a cephalopod chromatophore organ and a vertebrate or crustacean chromatophore cell. (**a**) Cephalopod pigment sacculus (p) retracted (left) and expanded (right) by action of radial muscles (m; only 16 of which are illustrated). See text and colour Figs. 3.2, 3.3 and 3.5. (**b**) Fish chromatophore cell with the pigmented melanin granules clustered (left) and dispersed (right), always within the branched but fixed cell.

mantle skin of some cuttlefish, or on the ventral mantle of some squid (Figs. 3.2, 3.3 and 3.5). They are probably responsible for the blue rings in the ocelli of some octopuses, and are certainly responsible for the blue rings and lines of *Hapalochlaena* spp. (Mäthger *et al.*, 2009, 2012; see also Chapter 7). They also produce the silver used to conceal the eyes and the ink sac by reflection.

It had been assumed that the iridophores were 'passive' cells that did not change their appearance as a result of direct control by the animal. However, there is evidence in the squid *Lolliguncula brevis* that the iridophores are controlled actively by acetylcholine, ACh (Hanlon *et al.*, 1990). The thickness of the platelets and the state of the material within the platelets can change as

a result of treatment with ACh, and this is associated with a change in colour or a change from non-iridescence to iridescence (Cooper, Hanlon & Budelmann, 1990; Mäthger & Denton, 2001; Mäthger & Hanlon, 2007; Izumi *et al.*, 2010). Acetylcholine is present in the dermal iridophore layer of *Lolliguncula* and *Loligo*, and Wardill *et al.* (2012) and Gonzalez-Bellido *et al.* (2014) showed that physiological stimulation of peripheral nerves, separate from those of the chromatophores, produces iridescence. It is clear now that active (or physiological) change in the iridophores plays a significant part in the lives of some cephalopods, for example in signalling (Chapters 5, 6 and 7).

Leucophores, which occur in many cuttlefish, the squid *Sepioteuthis* and many octopuses, are responsible for the 'white spots' in these animals (Figs. 3.1, 3.2). Each leucophore is a branched, flattened cell covered with typically 1000–2000 proteinaceous spheres 250–1250 nm in diameter (Mäthger *et al.*, 2013). These spheres, termed leucosomes by Cloney and Brocco (1983), function as almost perfect diffusers, producing white light by scattered reflection. Yet although they appear white in white light, because they are broadband reflectors they will look blue in blue light or yellow in yellow light (Messenger, 1974). This ability to reflect the predominant wavelengths in their environment equally in all directions, and from all angles of viewing (Mäthger *et al.*, 2013), is thought to be important for achieving general colour resemblance to the background (Chapter 3). In *Octopus vulgaris*, they occur at the centre of the circular skin patches (Packard & Sanders, 1971; Froesch & Messenger, 1978), or they can occur across several skin patches as in *Octopus rubescens* (Fig. 3.2n, o, p), and in both species they can be elevated at the tip of the papillae, presumably to maximise reflectance.

2.2.4 Photophores

Photophores, or light organs, are found in all orders of coleoids, although they are absent from most common coastal forms; they are especially frequent among oegopsid squid (Herring, 1977, 1988) and they have probably evolved independently many times (Young & Bennett, 1988). The photophores are mostly on the ventral body surface and are especially common below the eyes, on the fins, arms and tentacles, and also on the suckers (Johnsen, Balser & Widder, 1999) (Figs. 6.3, 9.11, 9.13, 9.14, 9.15).

Little is known about how the photophores are controlled. There may be nervous control, either directly or via muscles associated with the extensive capillary supply to the organs (Young & Arnold, 1982). Another possibility is that the organs may be obscured by the expansion of overlying chromatophore, by ink movements, by intervention of a pigment screen or even by rotation of the whole organ (Herring, 1977, 1988). It is certain, however, that light organs can produce a steady glow or give intermittent flashes (e.g. Herring, 1977; Roper & Vecchione, 1993). The former is associated with ventral counter-illumination (Chapter 9), the latter occur during signalling (Chapter 7) or prey capture. Kubodera, Koyama and Mori (2007) report that the large mesopelagic squid *Taningia* emits short, bright flashes from the arm tip photophores as it attacks bait rigs.

2.2.5 Ink Sac

Another unusual effector organ characteristic of most cephalopods is the ink sac, which comprises the ink gland and a reservoir with sphincters (Mangold, 1989). The ink is a suspension of almost pure melanin, thought to derive from tyrosine via L-DOPA (Chedekel, Murr & Zeise, 1992), as well as a variety of free amino acids (Derby *et al.*, 2007). Cephalopods can produce dense clouds of ink that are used as a smoke screen, or 'pseudomorphs', small mucus-bound blobs that act as decoys (Schäfer, 1956; and Chapter 5). Although the ink must obviously disturb predators that hunt by sight, there is now good evidence that it may also act as a visual alarm substance for conspecifics (Wood, Pennoyer & Derby, 2008): in *Sepioteuthis sepioidea* shoals, the release of ink elicits a variety of warning ('deimatic') responses (Chapter 3), even when the ink is in a separate adjacent tank. However, transparent, melanin-free ink released into the same tank as the animals does not. Lucero, Farrington and Gilly (1994) have shown that the ink of *Loligo opalescens* contains L-DOPA and dopamine, both of which are active on olfactory neurons. The claim that octopus ink may also 'paralyse' the olfactory sense of its eel predators (MacGinitie & MacGinitie, 1968) seems never to have been tested rigorously, although there is some biochemical evidence to support the idea (Prota *et al.*, 1981; Derby, 2007).

The ink sac is absent from several coleoids that live at great depths, for example *Vampyroteuthis*, the cirrate octopods and some deep-water genera of the family Octopodidae (Mangold, 1989) including *Vulcanoctopus* (Gleadall *et al.*, 2010). It may be significant that in the poisonous blue-ringed octopus, *Hapalochlaena*, the ink sac is greatly reduced; although the young sometimes eject brown ink, they probably never do so after the fourth week of life (Tranter & Augustine, 1973).

2.2.6 Suckers

The arms typically bear another cephalopod feature: the suckers (Fig. 2.8). These are complex organs, usually organised into one or two rows, and there are differences in the morphology and innervation of decapod and octopod suckers (§2.1.2); they also vary greatly in size (Nixon & Dilly, 1977; Packard, 1988b; Mangold, 1989). They may rest directly on the arm or be on stalks, conferring greater mobility. The rim of the outer chamber of the sucker (the infundibulum) is stiffened with protein (not chitin; see Mizerez et al., 2009) and is often toothed in some genera. Inside, the infundibular surface is covered with minute pegs or denticles (Nixon & Dilly, 1977), which may improve the seal at the rim margin and enhance friction between sucker surface and substrate, making resistance to shear forces more effective (Kier & Smith, 1990). The musculature permits very strong adhesion (Kier & Smith, 2002; Grasso, 2008), and there is a small ganglion below each sucker that serves to coordinate motor and sensory information (Graziadei, 1971; Nesher et al., 2014). In a few teuthoid families, the arms and/or tentacles bear hooks (Engeser & Clarke, 1988); these can be quite large, but little seems to be known about their use (Fig. 4.3).

In squid such as Loligo or Ommastrephes, the suckers are probably only concerned with prey holding, but, as Packard (1988b) has pointed out, the suckers of benthic octopuses may have no fewer than six functions: (1) locomotion; (2) anchorage and prey holding; (3) sampling, collecting and manipulating small objects; (4) mediating chemotactile recognition and learning; (5) cleaning; and (6) visual display (Packard, 1961; see Chapters 6 and 7), which is also true for the modified, light-emitting suckers of the deep-sea finned octopod, Stauroteuthis (Johnsen, Balser & Widder, 1999). For Tremoctopus we could add offence or defence, because the suckers in each row on the dorsal arms of this epipelagic octopod are modified to hold pieces of the Portuguese Man-of-War jellyfish, Physalia, complete with their nematocysts (Jones, 1963).

The arms of one group of Octopoda bear small cirri, hence the name cirrate octopods; the cirri may sense the small creatures on which these animals feed (Villanueva & Guerra, 1991).

2.2.7 The Buccal Mass: Beak, Radula and the Salivary System

The cephalopod 'buccal mass' is a large and complex structure, comprising the beak with its associated muscles, the radula, the salivary papilla, the salivary glands with their ducts and the submandibular gland. It lies in front of the brain, in the centre of the arms. The beak is chitinous and its musculature generally massive (Kear, 1994) so that it can be a formidable weapon, as anyone who has been bitten by a cephalopod will know. The beaks of different cephalopods vary considerably, and features of the lower beak are sufficiently different to be of great taxonomic value and permit identification to generic or even species level (Clarke, 1966, 1980, 1986).

The radula is a ribbon bearing nine transverse teeth (13 in Nautilus) that is moved back and forth like a rasp to break up food and convey it to the pharynx: its mode of functioning is still not fully understood (Messenger & Young, 1999). At the front end, the teeth become worn in feeding and are replaced from behind by new ones formed in the radular sac. There are accessory supporting muscles or 'bolsters' whose function is to ensure that the erect radular teeth are spread to make a linear 'rake' at the bending plane to excavate and transport food with greater efficiency (ibid.). In some octopods it is known to be involved in the initial stages of drilling holes in the shells of molluscs and the exoskeleton of crustaceans, but the essential organ for drilling is the salivary papilla (Nixon, 1979, 1980), which carries the duct from the posterior salivary gland (PSG). The papilla and the eversible tip of the duct bear small teeth, and the secretions of the PSG are delivered precisely to the drilling site (Nixon & Maconnachie, 1988). In Octopus vulgaris, the saliva contains substances that are important for the breakdown of the mineral and organic matrix of mollusc shells and crustacean skeletons (Nixon, Maconnachie & Howell, 1980; Nixon & Maconnachie, 1988), but we do not know yet what these are.

In Sepia officinalis, Octopus vulgaris and Eledone cirrhosa, this PSG secretion contains a 'cocktail' of other substances, including dopamine and serotonin, toxins, proteolytic enzymes and chitinases. The principal toxin is cephalotoxin (Ghiretti, 1959, 1960; Ghiretti & Cariello, 1977; Cariello & Zanetti, 1977). In the blue-ringed octopus, Hapalochlaena, the PSG produces tetrodotoxin, the well-known poison found also in puffer fish and some amphibians. This is a potent neurotoxin that specifically blocks sodium channels, and the bite of this animal is known to be fatal to humans (Sheumack et al., 1978). In octopuses, the saliva also contains proteases

(Moroshita, 1974) and at least two salivary chitinases (Grisley, 1993) that probably aid in loosening crab muscle from the skeleton (Grisley & Boyle, 1987, 1990) or mollusc muscle from the shell (Nixon, 1984).

The anterior salivary glands secrete abundant mucus (Gennaro, Lorincz & Brewster, 1965) and hyaluronidase, to spread the toxin from the posterior salivary glands (Romanini, 1952); the submandibular gland may produce a mucus to lubricate movements of the salivary papilla (Young, 1971).

More details of feeding are given in Chapter 4. Its nervous control has been the subject of papers by several workers (for reviews, see Young, 1971; Boyle, 1986b).

2.2.8 Adhesive Devices

Several species of cephalopods belonging to the genera *Sepia*, *Euprymna* and *Idiosepius* (Fig. 5.13c) can adhere to hard surfaces, or fronds of algae, perhaps using adhesive secretions from specialised glands on the mantle, fin or arms, perhaps also by suction (von Byern & Klepal, 2006). In *Euprymna*, the secretion on the dorsal mantle is used to retain sand granules for camouflage (Moynihan, 1983a; Shears, 1988).

2.3 The Brain

Because cephalopods are highly evolved molluscs and have elaborate sense organs and effectors, their brains must routinely handle much more information than a limpet or a snail. It is not surprising, therefore, that the cephalopod CNS is substantially larger and the various ganglia more concentrated than in other molluscs.

The coleoid brain comprises a central part, lying round the oesophagus, and paired optic lobes laterally. These lie close to the central brain in decapods, but in octopods there is a distinct optic tract (Fig. 2.10). Centrally, there are sets of sub- and supra-oesophageal lobes, linked by the peri-oesophageal magnocellular lobes. In *Loligo*, no fewer than 30 different lobes can be recognised (Maddock & Young, 1987) so that it is obviously inappropriate here even to list these, much less describe them in detail. Fortunately, there are several useful reviews of the extensive literature on the cephalopod nervous system (Bullock, 1965; Boyle,

1986b; Budelmann, 1995; Budelmann, Schipp & Boletzky, 1997; Nixon & Young, 2003), and the reader is directed to these as well as to the key references for *Sepia* (Boycott, 1961), for *Loligo* (Young, 1974, 1976, 1977b, 1979; Messenger, 1979c) and for *Octopus* (Young, 1971). These are the three genera that have received the most attention from neurobiologists, who have used a combination of anatomical techniques, electrical stimulation and lesioning to gain an understanding of the functional organisation of the cephalopod brain.

Effectors are innervated by motoneurons originating in the lower and intermediate motor centres, which for the most part are situated in the sub-oesophageal lobes: these regions of the brain could be regarded as equivalent to the vertebrate spinal cord (Boycott & Young, 1950; Boycott, 1961). They are controlled by the higher motor centres, especially by the various basal lobes, which, in turn, are controlled by the optic lobes. Stimulation of the higher motor centres by implanted electrodes elicits discrete and increasingly complex motor responses as the stimulus strength rises, but curiously there is no evidence of a somatotopic motor representation in the brain of cephalopods, as there is in arthropods and vertebrates (Boycott, 1961; Budelmann & Young, 1984, 1993; Zullo et al., 2009). One of the best known of cephalopod motor systems is the giant fibre system of decapods (Young, 1939; Pumphrey & Young, 1938). Its more important features are summarised in Box 2.4.

Sense organs project to the brain along various sensory pathways, the major ones being in the arms, statocyst nerves and optic nerves. Some of the information they deliver is processed in the lower motor centres, but much of it is made available to the higher motor centres, and to the areas concerned with establishing memories (Young, 1965b, 1971, 1991). There is ample opportunity for convergence of visual, statocyst and proprioceptive information in the cerebellar-like peduncle and basal lobes, which act as regulators of motor programmes set up in the optic lobes (Young, 1977a; Messenger, 1983). Physiological proof of convergence onto oculomotor neurons has been provided by Williamson and Budelmann (1991).

The optic lobes, which in *Octopus* contain some 65 million nerve cells (Young, 1971), are part higher

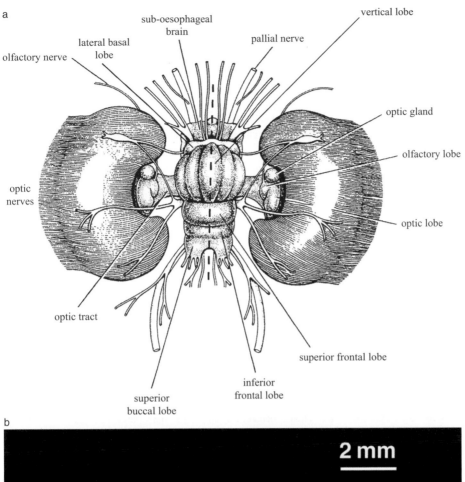

a

olfactory nerve

lateral basal lobe

sub-oesophageal brain

pallial nerve

vertical lobe

optic gland

olfactory lobe

optic lobe

optic nerves

optic tract

superior buccal lobe

inferior frontal lobe

superior frontal lobe

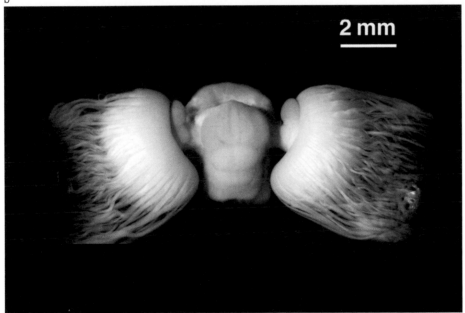

b

2 mm

Figure 2.10 (**a**) Dorsal view of the brain of *Octopus vulgaris*. Note the large optic lobes (visual processing areas and visual memory store); the inferior frontal lobe (which together with the subfrontal and superior buccal lobes forms the tactile memory store); and the vertical and superior frontal lobes (accessory to the visual memory store). Dashed line indicates plane of section in Fig. 2.11 (from Young, 1971: *The Anatomy of the Nervous System of* Octopus vulgaris, by permission of Oxford University Press). (**b**) Freshly dissected brain of *Octopus bimaculoides* (courtesy of Shuichi Shigeno).

motor centre and part visual analyser, for it is here (in the outer plexiform layers) that information from the eyes is processed as second-order visual cells synapse with incoming optic nerve fibres (Young, 1971). One detail to note is that the optic nerves usually pass through a dorso-ventral chiasm as they proceed from the eye to the optic lobe. This presumably 'rectifies' the inverted retinal image as it is mapped onto the optic lobe and brings the visual and gravity 'maps' into

correspondence (Young, 1962b). It may be significant that only in the deep-water *Cirrothauma* and in *Nautilus* (Chapter 10) is the chiasm lacking: neither of these animals has a lens in the eye and they presumably lack keen vision.

The optic lobes are also the site of the visual memory store and can be thought of as constituting the 'highest' of the brain centres, comparable to the vertebrate forebrain. The laying down of memories

---

Box 2.4 Jetting and giant fibres

It is well known that cephalopods are jet-propelled (although for much of the time most cephalopods do not move by jetting). It is also well known that they have nerves with very large axons, the so-called 'giant fibres' (but not all cephalopods have giant fibres: they are absent in octopods, and in *Vampyroteuthis*).

The 'classic' giant fibre system is that first described fully by J.Z. Young (1939) in the squid *Loligo*. It comprises a set of first-, second- and third-order fibres, and its function is to mediate an escape response as a result of visual, tactile or other stimuli. A pair of first-order giant cells in the brain (with wide synaptic input) gives rise to axons that fuse

Figure B.2.4.1 The giant axon in the first stellar nerve of the squid *Lolliguncula brevis*. Reprinted from Hulet, Hanlon & Hixon, 1980: *Trends in Neuroscience* **3**, iv–v, *Lolliguncula brevis* – a new squid species for the neuroscience laboratory, Hulet, W. H., Hanlon, R. T. & Hixon, R. F. © (1980), with permission from Elsevier.

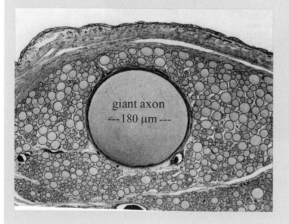

giant axon
---180 μm---

to form a bridge before synapsing with seven pairs of second-order axons. Six pairs of second-order axons pass directly to the retractor muscles of the head and funnel. The axons of the other pair run to the stellate ganglia on either side of the mantle. Here, the second-order fibres synapse with all the 10–12 third-order fibres whose cell bodies lie in the ganglion. Each of these giant fibres leaves the ganglion in one of the stellar nerves, which also contain many hundreds of medium-to-small fibres (see figure), and passes to the mantle circular muscle. The third-order giant fibres are graded in diameter and hence in conduction velocity (Pumphrey & Young, 1938): the thickest fibre (that used by neurobiologists) passes to the farthest part of the mantle, the thinnest to the region closest to the ganglion.

The significance of these morphological arrangements is (1) that a stimulus reaching either of the first-order giant cells will activate the second-order cells bilaterally; and (2) that the second-order fibres to the stellate ganglion will activate all the third-order fibres simultaneously. Thus the circular muscles of the mantle will contract simultaneously throughout its length to expel the largest possible mass of water through the funnel for the escape jet. The effectiveness of this device for jet propulsion has been measured directly by Packard (1969a), who showed that a 100-g squid reaches peak velocity in about 1/8 second.

Experiments have now shown that the escape response greatly improves during early development: see Chapter 8 (§8.1.1).

There is morphological evidence that the giant fibre system is organised differently in different squid (Martin, 1965, 1977) but no physiological evidence exists about giant fibre function in any other cephalopod. In the squid *Pterygioteuthis* there are giant fibres extending to the arms.

A comparative account of the giant fibre system and its development is given in Martin (1977): for a description of Hodgkin and Huxley's classic experiments on the electrophysiology of the giant axons see Hodgkin (1964). Wells and O'Dor (1991) describe the evolution of jetting in different cephalopods and the limitations of jet propulsion.

involves a series of other lobes, however, notably the vertical and superior frontal lobes (Fig. 2.10), which act somehow to balance the tendencies to attack or retreat when responses are followed by some kind of pain (Young, 1964b, 1965b). It was shown long ago that when these parts of the brain are stimulated electrically the animal makes no response (Bert, 1867). Surgical removal of the 'silent areas' has no apparent effect on the behaviour of an octopus until the animal is examined in a learning situation: then impairment of acquisition or retention becomes evident (Boycott & Young, 1950). Many experiments have since confirmed the importance of these accessory memory lobes (e.g. Fiorito & Chichery, 1995), although opinions differ as to the way they operate (Young, 1965b, 1991; Wells, 1978). However, in a very interesting paper, Hochner *et al.* (2003), who used brain slices to investigate these regions of the brain, demonstrated a long-term potentiation (LTP) in the vertical lobe similar to that found in vertebrates. LTP has long been associated with learning and memory (but see Martin, Grimwood & Morris, 2000) and the fact that it involves the transmitter L-glutamate (Messenger, 1996), in *Octopus* as in vertebrates, suggests that similar mechanisms of synaptic plasticity to subserve the laying down of memories may have arisen independently in molluscs and vertebrates. This system continues to be investigated electrophysiologically by Hochner and his associates (e.g. Hochner, Shomrat & Fiorito, 2006; Shomrat *et al.*, 2008, 2011), who have discovered important similarities to learning processes in mammals. The fact that large parts of the cephalopod brain are devoted entirely to the memory system forewarns us that learning is likely to play an important part in the life of these animals (Chapter 8).

In *Octopus*, the brain also contains a chemotactile memory system (§2.1.2 and 2.1.3) located in the subfrontal and inferior frontal lobes (Wells, 1978; Robertson, 1994; Robertson, Bonaventura & Kohm,

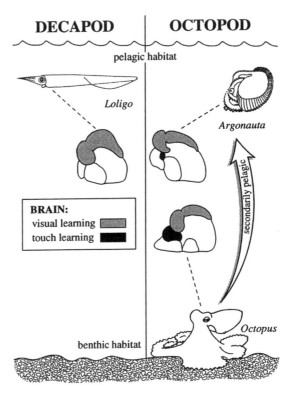

Figure 2.11 *Octopus* has a touch memory system that is lacking in squid and is substantially reduced in secondarily pelagic octopods such as *Argonauta*. All three forms have large eyes and optic lobes (not shown) and a well-developed vertical lobe system that is important for visual learning. Only the supra-oesophageal brain is shown, in midline sagittal section (see Fig. 2.10).

1994). These lobes are absent in decapods, however, and it is thought that they evolved as the octopods adopted a benthic mode of life, came to use the elongated arms as exploratory, sensing devices, and needed to process and store the information they collected. In support of this is the fact that in those octopods that have become secondarily pelagic, such as *Tremoctopus* or *Argonauta*, the subfrontal system is poorly developed (Young, 1964a) (Fig. 2.11).

2.4 Summary and Future Research Directions

The senses and effectors of cephalopods have been described at length for the benefit of readers unfamiliar with animals that are organised very differently from vertebrates. The brain has received absurdly short treatment in this book, partly because there are full accounts elsewhere and partly because the emphasis of this book is on behaviour.

The visual system of cephalopods continues to receive considerable attention, and recent work on polarised light sensitivity suggests that this aspect of vision definitely merits further study (see Marshall, Cronin & Wehling, 2011). However, given the overwhelming evidence that most cephalopods are probably colour blind (Section 2.1.1), it is perhaps not worth making major efforts to pursue this issue, although data on the visual pigments of individual species will always be welcome (Table 2.1). Much more exciting is the possibility that the skin may have light-sensing capabilities: the recent discovery of opsins in the skin of cuttlefish suggests that there may be other modes of light sensing for camouflage or communication (Mäthger, Roberts & Hanlon, 2010).

Our survey of the sense organs brings out the fact that we need to know more about the different kinds of mechanical and chemical stimuli that may be influencing cephalopod behaviour. How important is infrasound sensitivity in cephalopods? More information about distance chemoreception is also required and especially the role of the enigmatic olfactory organs. How might pheromones mediate sexual behaviours (see for example Cummins *et al.* 2011)? The search for other possible sensory modalities, for example a geomagnetic sense, might also be rewarding.

Among effector systems, the feeding apparatus of octopuses has been thoroughly investigated, but this only points up the fact that we know almost nothing about prey handling and processing in any other cephalopods (see also Chapter 4).

Studies of the brain have been extensive, but it must be emphasised that the lack of electrophysiological data of the kind available in vertebrates, arthropods and gastropod molluscs to complement behavioural data means that a 'neuroethology' of cephalopods is still a way off. The techniques that were emerging 20 years ago have unfortunately not yet yielded any significant advances (Bullock & Budelmann, 1991; Novicki *et al.*, 1992; Williamson & Budelmann, 1991; Budelmann, Bullock & Williamson, 1995) although the studies of Hochner *et al.* (2003) and Mooney *et al.* (2010) are exceptions. The application of non-invasive imaging techniques with these animals would seem to be especially worth investigating.

CHAPTER THREE

Body Patterning and Colour Change

The beautiful play of colour and pattern in the skin of a living cephalopod appears almost magical. The extraordinary patterns in the skin and the speed with which they change are due mainly to hundreds, thousands or millions of chromatophore organs under neuromuscular control directly from the brain. Cephalopod chromatophores are thus unique in the animal kingdom, and in many shallow-water forms, nearly all behaviours are inextricably bound up with chromatophore activity. Feeding, avoiding predators, mating and communication (Chapters 4–7) all involve the chromatophores, although, as this chapter will make clear, these act in conjunction with reflecting cells and other structures to produce the final appearance of the animal. This appearance is termed a *body pattern*, and in this chapter we shall show how it can be interpreted as being constructed in a hierarchical fashion. Another theme of this chapter is that body patterns are themselves behaviours.

Many authors write about the 'colours', 'colour changes', 'colour patterns' or 'chromatophore patterns' of cephalopods. Such terms can be misleading because they emphasise the chromatic components of body patterns, at the expense of three other classes of component, namely textural, postural and locomotor components, which together contribute to what, following Packard, we term the *body pattern*.

3.1 The Organisation of Body Patterns

Packard and his collaborators (Packard & Sanders, 1969, 1971; Packard & Hochberg, 1977) have developed a very useful hierarchical system to describe and analyse the body patterns of cephalopods. We follow this throughout this book and suggest that it might help if future workers think carefully before adopting a different system. The essential point is that the whole appearance of the animal, the *body pattern*, is a combination of chromatic, textural, postural and locomotor *components*. The components may be recombined at any time to create a different body pattern. The components are themselves constructed of *units*; these units too may be combined in different ways to make different components. Finally, the units are built out of diverse *elements*, the most obvious of which are the chromatophores. Box 3.1 shows these relationships, and Table 3.1 indicates that many authors have adopted this hierarchical system. To explain Packard's system, it may be helpful to start at the bottom of the hierarchy.

3.1.1 Elements

The term elements refers to the basic morphological entities that produce different colours, intensities or textures of the skin. There are five types of element in cephalopods: chromatophores, reflecting cells, internal organs, muscles and photophores. Figure 3.1 illustrates where the chromatophores and the various types of reflecting cells are positioned in the dermis and how they contribute to a cephalopod's appearance, and Fig. 3.2 shows the extraordinary diversity of appearances that they can produce.

Chromatophores are the most important skin elements. They provide pigmentary colours embracing the longer wavelengths: yellow, orange, red, brown and black (Fig. 3.2); most species have three colour classes of chromatophores (Fig. 3.3a). A few squid genera such as

---

Box 3.1 The hierarchy of body patterns: levels and definitions

Body pattern – the appearance of the whole animal

1. Chronic patterns	– lasting minutes or hours
2. Acute patterns	– lasting seconds or minutes

Components – the constituents of a pattern

1. Chromatic	– shapes of specific size and position in the skin:
a. dark	– resulting from expanded chromatophores or internal body parts seen through the skin
b. light	– light or white features often enhanced by reflecting cells
2. Textural	– smoothness/papillation of skin
3. Postural	– various positions of body parts
4. Locomotor	– typical movements

Units – groups of elements contributing to components

The static morphological array of elements in the skin; e.g. the patch and groove system of *Octopus vulgaris*.

Elements – the basic structures in the skin contributing to units

1. Chromatophore organs	– pigmentary colours
2. 'Reflecting cells' Iridophores Reflector cells Leucophores	– structural colours
3. Internal organs	– e.g. ink sac, testis, eyeball
4. Muscles	– for skin texture, posture, locomotion
5. Photophores	– important in many mid-water forms (Chapter 9); not discussed here

Lolliguncula and *Alloteuthis* have only two colours, and some oegopsids have only one colour. The chromatophores of cephalopods differ fundamentally from those of other animals: they are neuromuscular organs rather than cells and are not controlled hormonally. Our understanding of their structure and function relies heavily on the elegant studies of Florey and his colleagues on the squid *Loligo opalescens* (Florey, 1966, 1969; Cloney & Florey, 1968; Florey & Kriebel, 1969). In essence, these showed that each chromatophore organ comprises an elastic sacculus containing pigment granules, to which is attached a set of 10–30 radial muscles, each with its nerves running directly from the brain without synapse. The excitatory fibres stimulate the muscles to contract, expanding the chromatophore; in the absence of excitation, energy stored in the elastic sacculus retracts it. There is no inhibitory innervation, but serotonin (5-HT), released from other nerve fibres non-synaptically, relaxes the muscles by suppressing the release of calcium.

The control of the chromatophores is complex: the excitatory neuromuscular transmitter is L-glutamate (Florey, Dubas & Hanlon, 1985; Messenger, Cornwell & Reed, 1997) but in the cuttlefish, *Sepia*, the chromatophores are also expanded, on a slower time scale, by peptides (Loi *et al.*, 1996). Furthermore, Mattiello *et al.* (2010) showed in *Sepia* that L-glutamate acts via two pathways: rapid expansion is mediated by AMPA receptors, while slow, maintained expansion is mediated by NMDA receptors that are influenced by nitric oxide (NO); a somewhat similar dual functionality was reported earlier in squid (Lima *et al.*, 2003). It is not clear why there should be two systems involved in slow, long-term regulation of the chromatophores, but it is probably essential for camouflage that they can be kept expanded for long periods. For a fuller discussion of the physiology and pharmacology of the chromatophores, see the extensive review by Messenger (2001).

The chromatophores are especially important elements for patterning. The number and density of chromatophores in the skin vary greatly with species, the ontogenetic stage and the part of the body (Packard, 1985). The size of the chromatophores differs between species as well; in squid such as *Loligo*, the largest fully expanded chromatophore may be 1.5 mm in diameter (Hanlon, 1982) as compared with only about 0.3 mm in *Sepia* (Hanlon & Messenger, 1988). In the species examined, chromatophores of different colour classes also differ in size: yellows are always the smallest, oranges and reds are intermediate, and browns and blacks the largest in any given species. Chromatophore density can range from about 230 per mm^2 in *Octopus vulgaris* (Packard & Sanders, 1969), to 50 per mm^2 in

Table 3.1 Cephalopods Whose Body Patterns Have Been Analysed in Detail*

	Hierarchical system	Other
Order Sepioidea (cuttlefish)		
Sepia officinalis	Hanlon & Messenger, 1988	Holmes, 1940
Sepia papuensis	Roper & Hochberg, 1988	
Sepia latimanus	(this book)	Corner & Moore, 1980
Metasepia pfefferi	Roper & Hochberg, 1988	
Sepiola affinis	Mauris, 1989	
Idiosepius pygmaeus		Moynihan, 1983a
Euprymna scolopes		Moynihan, 1983b
Order Teuthoidea (squid)		
Sepioteuthis sepioidea	(this book)	Moynihan & Rodaniche, 1982
		Mather, Griebel & Byrne, 2010
Sepioteuthis lessoniana	(this book)	Moynihan & Rodaniche, 1982
Sepioteuthis australis	Jantzen & Havenhand, 2003b	Moynihan & Rodaniche, 1982
Pickfordiateuthis pulchella		
Lolliguncula brevis	Dubas *et al.*, 1986	
Lolliguncula panamensis		Moynihan & Rodaniche, 1982
Alloteuthis subulata	Lipinski, 1985;	
	Cornwell, Messenger & Hanlon, 1997	
Loligo forbesi	Hanlon, 1988;	
	Porteiro, Martins & Hanlon, 1990	
Loligo opalescens	Hanlon, 1982	Hurley, 1977
Loligo pealeii	Hanlon, Hixon & Hulet, 1983;	
	Hanlon, 1988; Hanlon *et al.*, 1999	
Loligo plei	Hanlon, 1978, 1982, 1988	
Loligo vulgaris	Hanlon, 1988	Tardent, 1962; Neill, 1971
Loligo vulgaris reynaudii	Hanlon, Smale & Sauer, 1994	
Octopoteuthis deletron	Bush, Robison & Caldwell, 2009	
Order Octopoda (octopuses)		
Abdopus aculeatus	Huffard, 2007	
Octopus bimaculoides	Forsythe & Hanlon, 1988;	
	Packard & Hochberg, 1977	
Octopus briareus	Hanlon & Wolterding, 1989	
Octopus burryi	Hanlon & Hixon, 1980;	
	Forsythe & Hanlon, 1985	
Octopus chierchiae		Moynihan & Rodaniche, 1982
Octopus cyanea	Roper & Hochberg, 1988	Van Heukelem, 1966
Octopus insularis	Leite & Mather, 2008	
Octopus ornatus	Roper & Hochberg, 1988	
Octopus rubescens	Packard & Hochberg, 1977	Dorsey, 1976;
		Warren, Scheier & Riley, 1974
Octopus spilotus		Moynihan & Rodaniche, 1982
Octopus vulgaris	Packard & Sanders, 1971;	Cowdry, 1911
	Packard & Hochberg, 1977	
Eledone cirrhosa	Boyle & Dubas, 1981	
Hapalochlaena cf. *maculosa*	Roper & Hochberg, 1988	
Hapalochlaena cf. *fasciata*	Roper & Hochberg, 1988	

* Many of these species are illustrated and compared in Borrelli, Gherardi and Fiorito (2006).

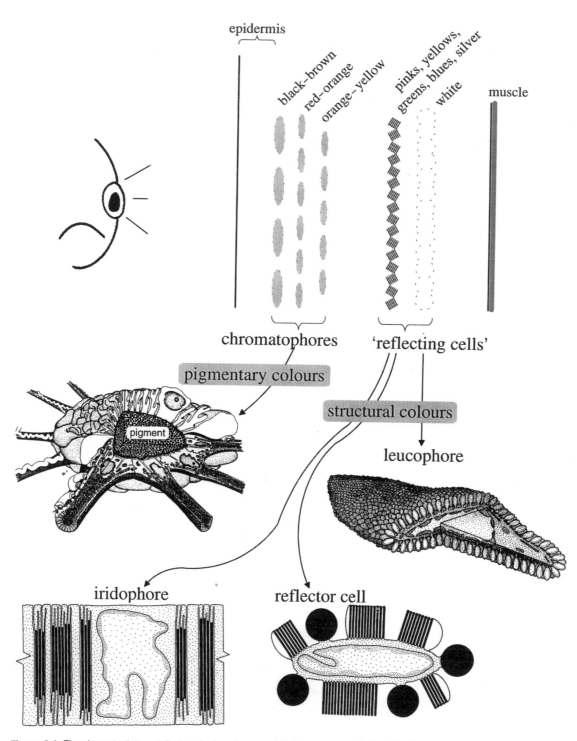

Figure 3.1 The elements that contribute to body patterns and their arrangement in the skin of an octopus being viewed by a fish. Deepest in the skin lie the leucophores (broadband reflectors); above them are the iridophores (or, in some other species, reflector cells); and more superficially the chromatophores (yellow, orange, red, black). It is important to note that the precise arrangement

adult *Sepia officinalis*, to only 8 per mm^2 in adult *Loligo plei* (Hanlon & Messenger, 1988). Since there is an inverse relationship between the size of chromatophores and the density, this means that squid like *Loligo*, with fewer and larger chromatophores, cannot achieve patterning as finely detailed as *Sepia* or *Octopus*. Cephalopods that have more complex patterning repertoires have higher densities of chromatophores.

There are at least two types of *reflecting cells* that produce iridescent structural colours or white: iridophores and leucophores. *Iridophores* are multi-layer stacks of thin protein plates alternating with layers of cytoplasm: in some species, the stacks have been shown to function as 'ideal' quarter-wavelength reflectors that reflect colours by constructive interference (Denton & Land, 1971; Land, 1972; Mäthger & Denton, 2001; Mäthger *et al.*, 2008; Ghoshal *et al.*, 2013). The wavelengths produced include reds, oranges, blues and greens, and the latter short-wave colours complement the yellow, red and brown pigments of the chromatophores for camouflage (Fig. 3.2a, g, k; Fig. 3.3). However, the colours from the iridophores can also be used for signalling, and some squid can produce red, green and blue stripes (Mäthger *et al.*, 2008); the silvery reflections of cephalopod eyes and ink sacs are also produced by iridophores. Above the eyes of loliginid squid are fluorescent reflectors that are very bright, which may be especially useful at depth because of their fluorescent properties (Mäthger & Denton, 2001).

It was thought previously that cephalopod iridophores were inert, passive reflectors (Messenger, 1979a), but there is now clear evidence that squid can actively turn some of their iridophores on and off, over a period of between a few seconds and two minutes (Cooper, Hanlon & Budelmann, 1990; Mäthger, Collins & Lima, 2004; Mäthger & Hanlon, 2007; Wardill *et al.*, 2012). Such changes are brought about by ACh,

released non-synaptically from nerves in the skin, acting on muscarinic receptors (Hanlon *et al.*, 1990; Mäthger, Collins & Lima, 2004). The change in wavelength is achieved when the platelets become thinner (with increased ACh) and the interplatelet distance thicker, thus turning the colour from red to blue (Cooper, Hanlon & Budelmann, 1990; Tao *et al.*, 2010; DeMartini, Krogstad & Morse, 2013; Ghoshal *et al.*, 2013). In *Octopus dofleini*, Cloney and Brocco (1983) have described what they term 'reflector cells'. Their platelets are arranged differently from squid iridophores (Fig. 3.1), but nothing is known about their optical properties, and they may function like squid iridophores; comparative studies in other cephalopod species will be informative. Iridescence produces a polarised signature and thus may be useful to cephalopods as intraspecific signals (Boal *et al.*, 2004; Mäthger & Hanlon, 2006; see §2.1.1).

Leucophores are broadband reflectors of ambient light so that they look white in white light, blue in blue light and so on. They contribute to the distinctive 'white markings' of cuttlefish and octopuses (Fig. 3.2c, d, f, g, h, k, n, o, p; Packard & Sanders, 1971; Froesch & Messenger, 1978; Messenger, 1979a; Cloney & Brocco, 1983). The leucophores bear numerous tiny protein spheres, 250–1250 nm in diameter, termed leucosomes. These scatter ambient light in all directions so that the leucophores function as near-perfect diffusers, appearing equally bright from all angles of view (Mäthger *et al.*, 2013). Unlike iridophores and chromatophores, there are no nerves or muscles associated with leucophores; nor do the leucophores polarise reflected light.

In some species, certain *internal organs* such as the ink sac, eyeball (sclera), the testis, or the red accessory nidamental gland in females, are sometimes made visible through the translucent mantle to contribute to body patterning.

Caption for Figure 3.1 (*cont.*) of the elements varies from species to species and sometimes in different parts of the body (based on Cloney & Florey, 1968: *Zeitschrift fur Zellforschung* **89**, 250–280, Ultrastructure of cephalopod chromatophore organs, Cloney, R. A. & Florey, E., © (1968), with permission of Springer; Packard & Hochberg, 1977: *Symposia of the Zoological Society of London* **38**, 191–231, Skin patterning in *Octopus* and other genera, Packard, A. & Hochberg, F. G., © (1977), with kind permission; Cloney & Brocco, 1983: *American Zoologist* **23**, 581–592, Chromatophore organs, reflector cells, iridocytes and leucophores in cephalopods, Cloney, R. A. & Brocco, S. L., © (1983), the Society for Integrative and Comparative Biology, by permission of Oxford University Press).

Cuttlefish – *Sepia officinalis*

Squid – *Sepioteuthis sepioidea*

Squid – *Doryteuthis plei*

Octopus – *Octopus rubescens*

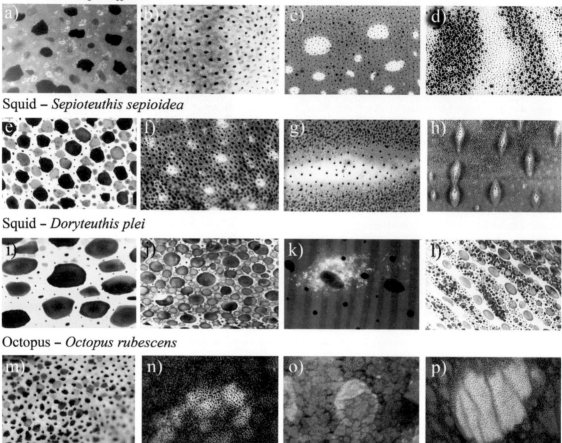

Figure 3.2 Colour production in the skin of cephalopods.

Sepia officinalis: (**a**) chromatophores with underlying iridophores visible, (**b**) chromatophores, (**c**) leucophores, (**d**) skin component dark Zebra band illustrating selective neural excitation of chromatophores.

Sepioteuthis sepioidea: (**e**) chromatophores, (**f**) chromatophores with leucophores, (**g**) large leucophore, (**h**) Zebra band component.

Doryteuthis plei: (**i**) chromatophores, (**j**) fully expanded chromatophores, (**k**) large iridophore splotch on fin, (**l**) selective excitation to produce Lateral flame component.

Octopus rubescens: (**m**) chromatophores mostly retracted, (**n**) expanded chromatophores except over leucophores, (**o**) outline of leucophore and 'patch and groove' arrangement of chromatophores, (**p**) close-up of leucophore with retracted chromatophores above it. All photos by R. Hanlon.

In cuttlefish and octopuses, there are *muscles* in the dermis devoted specifically to changing the physical texture[1] of the skin, altering it from smooth to rugose to highly papillate almost instantly. Figure 3.4 shows some examples of this quite extraordinary ability, at least by the standards of most animals. In *Sepia officinalis*, for example, there are no fewer than nine sets of skin papillae (Allen *et al.*, 2009).

---

[1] Throughout this book, we use the word 'texture' to describe the three-dimensional appearance of the skin surface of a cephalopod; this is not to be confused with the term 'visual texture' (sometimes simply 'texture') used by many students of visual psychophysics, which we avoid.

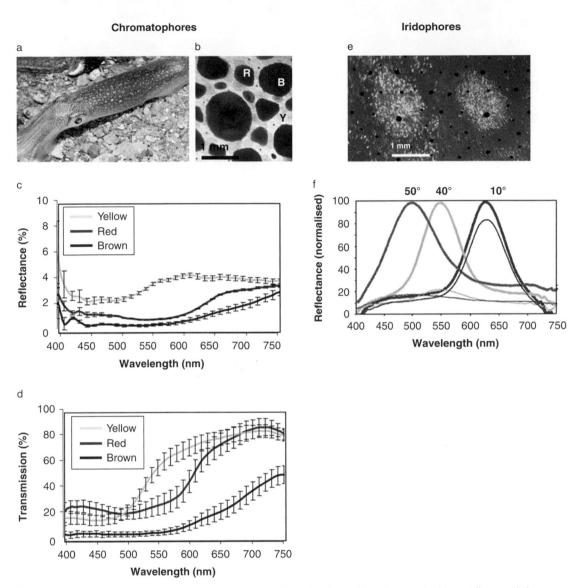

Figure 3.3 Squid chromatophore organs and reflective iridophores. (**a**) Live squid (*Loligo pealeii*) with partially expanded chromatophores and round gold iridophore splotches; (**b**) yellow (Y), red (R), brown (B) chromatophores in transmitted light; (**c**) spectral reflectance and (**d**) transmission of each colour class of chromatophore; (**e**) dorsal mantle iridophore splotch is blue when viewed at ca. 50° incidence and red when viewed at near-normal incidence; (**f**) reflectance and colour of a single iridophore at different angles; thick lines are parallel plane of polarisation, thin lines are perpendicular plane (from Mäthger & Hanlon, 2007: *Cell and Tissue Research* **329**, 179–186, Malleable skin coloration in cephalopods: selective reflectance, transmission and absorbance of light by chromatophores and iridophores, Mäthger, L. M. & Hanlon, R. T., © (2007), with permission of Springer).

Photophores are distributed in a highly ordered manner in the skin of some cephalopods and almost certainly contribute to body patterns. This has not yet been studied, however (see §9.7.1), and we do not discuss them further here.

It must be emphasised that the various elements in a particular patch of skin do not act independently. Expansion of overlying chromatophores obviously alters the quality and amount of light reflected from the iridophores or leucophores deeper in the skin (Fig. 3.1). Full

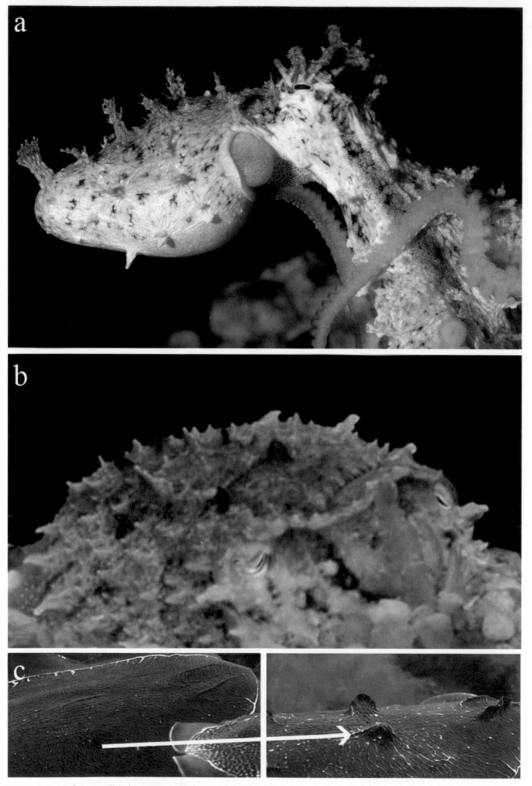

Figure 3.4 (**a**) Skin papillae in octopus *Abdopus aculeatus*, (**b**) oblique view of a cuttlefish *Sepia officinalis* with strongly expressed papillae and (**c**) still images taken from video footage of *S. apama* in South Australia showing smooth skin and strong expression of papillae (from Allen *et al.*, 2009: *Journal of Comparative Physiology A: Neuroethology Sensory Neural and Behavioral Physiology* **195**, 547–555, Cuttlefish use visual cues to control three-dimensional skin papillae for camouflage, Allen, J. J., Mäthger, L. M., Barbosa, A. & Hanlon, R. T., © Springer-Verlag 2009, with permission of Springer).

expansion of dark chromatophores could eliminate all light reflected from the iridophores, for example, but even partial expansion modulates iridescence, as can be seen in Fig. 3.5. There is, therefore, an extremely subtle regulation of skin coloration in cephalopods at the element level. This is considered more fully in §3.1.6.

3.1.2 Units

A unit is a particular morphological arrangement of elements in the skin. For example, some octopus species have their chromatophores and reflecting cells organised into circular patches surrounded by a groove (Fig. 3.2o; Packard & Sanders, 1971); and some cuttlefish and squid have an arrangement in which central large dark chromatophores are surrounded by smaller lighter ones (Fig. 3.2i, j) (Hanlon, 1982; Hanlon & Messenger, 1988). However, not all squid chromatophores are arranged like this. For example, in *Loligo plei* there are units that help produce the dramatic Lateral flame chromatic component used in agonistic contests (Fig. 3.2l; also Fig. 6.24 and Box 7.1). For the student of behaviour, the category of units is relatively unimportant because one needs to be so close to a stationary animal to recognise them. It is the components that we, and presumably other animals, see and respond to.

3.1.3 Components

Components are made up of groups of units. These are physiological entities that depend upon specific neural excitation for their appearance. *Chromatic components* are classified as either dark or light. They are highly diverse, and Table 3.2 shows how numerous the chromatic components can be in some cephalopod species (36, for example, in *Sepia officinalis*). Dark components appear when chromatophores are expanded or ocelli are expressed, when a dark internal organ is made visible through translucent skin, or even when eye pupils are dilated. Light components appear when the chromatophores retract to reveal pale whitish or translucent underlying skin and perhaps a pale internal organ, iridescence or bright white leucophores.

Textural components describe smoothness or papillation of the skin; papillae vary greatly in size and shape and sometimes form large flaps. Skin papillae are muscular hydrostats similar in basic morphology to an octopus arm, elephant trunk or human tongue (Kier & Smith, 1985; Allen *et al.*, 2013, 2014). *Postural components* describe the attitude of the arms, tentacles, head, eyes, mantle, interbrachial web or fins. *Locomotor components* include a variety of acts: resting, sitting, burying, bottom suction, standing, walking, ambulating, strutting, scuttling, jetting, hovering, free falling, crouching, pouncing, groping, puffing, inking, breath holding, hyperventilating and head bobbing. The authors cited in Table 3.1 have used all of these terms.

The terminology for naming components (and body patterns) has not been universally agreed upon. Because of the potential for confusion, we suggest that some convention be followed for the future. Drummond (1981) provided useful guidelines for understanding the differences between 'naming', 'describing' and 'defining' behaviours, and these will be helpful as they are applied to components and body patterns of cephalopods. In this book, we propose one guideline for chromatic components: 'bands' or 'bars' to name transverse components; 'stripes' or 'streaks' to name components aligned along the longitudinal body axis. 'Lines' may refer to components that lie in either orientation. This terminology agrees with that used by ichthyologists (e.g. Robins *et al.*, 1991). For convention, we capitalise the first letter of a component name and the first letter of each word of a body pattern or display (e.g. White head bar; Acute Disruptive).

3.1.4 Body Patterns

The body pattern is the total appearance of the animal at any given moment; it is made up of all the chromatic, textural, postural and locomotor components shown at that time. Packard and Sanders (1969, 1971) formalised the concept (first introduced by Cowdry, 1911) of referring to categories of patterns as *chronic patterns*, which last for hours or days and are mostly used for camouflage, and *acute patterns*, which last only seconds or minutes. Many acute patterns are highly ritualised and usually occur when the animal is actively signalling to another animal. In the context of communication, some acute body patterns are called displays; see Chapter 7.

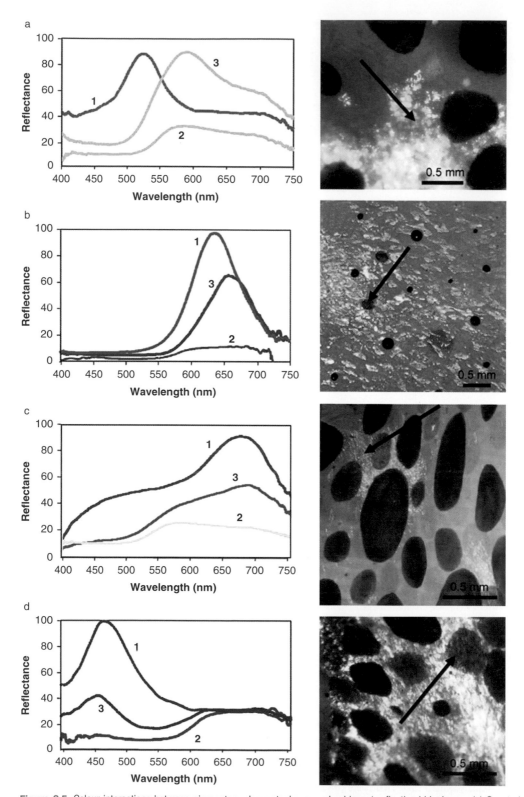

Figure 3.5 Colour interactions between pigmentary chromatophores and subjacent reflective iridophores. (**a**) Spectral shifts from green (iridophore only, curve 1) or yellow (chromatophore only, curve 2) to a mix of green and yellow (chromatophore expands over

Table 3.2 Numbers of chromatic components and body patterns in some cephalopods

	Chromatic components*			Body patterns*		
	Light	Dark	Total	Chronic	Acute	Total
Order Sepioidea						
Sepia officinalis	18	19	36	6	7	13
Sepia latimanus	9	14	23	5	7	12
Metasepia pfefferi	12	8+	20+	3	5	8
Sepia papuensis	7	5+	12+	4	5	9
Sepiola affinis	3	8	11	1	3	4
Order Teuthoidea						
Sepioteuthis sepioidea	13	17	30	3	13	16
Sepioteuthis lessoniana	10	10+	20+	1	5	6
Loligo pealeii	10	20	30	6	5	11
Sepioteuthis australis	5	12	17	–	–	7+
Loligo plei	7	13	20	4	9	13
Octopoteuthis deletron	10	10	20	–	–	4+
Loligo forbesi	6	11	17	3	3	6
Loligo vulgaris	6	10	16	1	1	2+
Loligo opalescens	6	7+	13+	1	2	3+
Lolliguncula brevis	4	9+	13+	1	5	6
Alloteuthis subulata	7	8+	15+	1	2	3+
Order Octopoda						
Octopus vulgaris	6	13	19	3	11	14
Octopus bimaculoides	7	13	20	2	6	8
Octopus cyanea	5	14	19	2	7	9
Octopus insularis	7	9	16	6	3	9
Octopus burryi	6	9	15	3	5	8
Octopus rubescens	5	10	15	3	6	9
Hapalochlaena maculosa	7	8	15	2	5	7
Hapalochlaena fasciata	7	7	14	2	4	6
Octopus briareus	5	9	14	3	10	13
Abdopus aculeatus	7	6	13	5	2	7
Octopus ornatus	6	4	10	1	3	4
Eledone cirrhosa	7	8	15	2	?	2+
Ranges:	3–18	4–20	10–36	1–6	1–13	2–16

* Re-calculated from various authors according to the hierarchical system.
+ Indicates that more are known but not yet named or described properly.

There are difficulties in naming body patterns, partly because of the problems inherent in pattern recognition and partly because patterning in cephalopods is dynamic and subject to immediate change and countless gradations (see §3.2.1). Conservative and highly stereotyped patterns may be named easily, but some of the patterns used for camouflage, for example the Mottles or Disruptives (Chapter 5), intergrade to such an extent that different authors are rarely going to agree on the same name.

Caption for Figure 3.5 (*cont.*) iridophores, curve 3). Similar shifts can be measured with red chromatophore over pink iridophores (**b**), yellow chromatophore over pink iridophores (**c**), and red chromatophore over blue iridophores (**d**) (from Mäthger & Hanlon, 2007: *Cell and Tissue Research*, **329**, 179–186, Malleable skin coloration in cephalopods: selective reflectance, transmission and absorbance of light by chromatophores and iridophores, Mäthger, L. M. & Hanlon, R. T., © Springer-Verlag 2007, with permission of Springer).

3.1.5 The Patterning Hierarchy

The hierarchical classification of body patterns as it relates to behaviour is shown in Fig. 3.6. To illustrate how this classification works, consider the young cuttlefish, *Sepia officinalis*, shown in Fig. 3.7. This animal is showing a *body pattern* that we term 'Strong Disruptive', which breaks up the animal's outline into mosaics that look like random samples of the background. The pattern comprises the four different types of *component*: conspicuous *chromatic* components include the White square and White head bar, and the dark Mantle spots and Anterior head bar; the chief *textural* component is Major lateral papillae; the main *postural* component is Flattened body; and the main *locomotor* component is partially Buried. Examining the chromatic component dark Anterior head bar in greater detail shows that it is made up of *units* of expanded chromatophores forming a transverse bar and that the individual *elements* are the

fully expanded brown chromatophores (see also Fig. 3.2d). The hierarchy depicted in Fig. 3.6 corresponds in a general way with the organisation of the chromatophore system in the brain; that is, body patterns are organised at the higher levels in the brain (optic lobes), while the motoneurons controlling the chromatophores lie in the lower motor centres of the brain (i.e. chromatophore lobes, Fig. 3.9).

To make this hierarchical scheme more meaningful, it is worth relating the appearance of the cuttlefish in Fig. 3.7 to the general principles of body patterning proposed by Packard and Hochberg (1977):

(1) '... components vary in the extent of their expression from barely perceptible to fully expressed'. This is highly useful for camouflage. For example, in Fig. 3.7, we draw attention to three chromatic components that are expressed weakly, averagely and strongly.

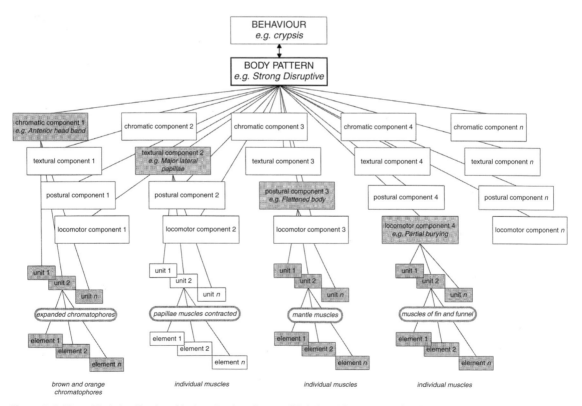

Figure 3.6 Hierarchical classification of body patterning. See text (§3.1.5) and Fig. 3.7 (modified from Hanlon & Messenger, 1988, by kind permission of the Royal Society).

(2) '... single components ... do not appear in isolation but appear along with other components of the same category...'. That is, there are usually several chromatic components shown in a single body

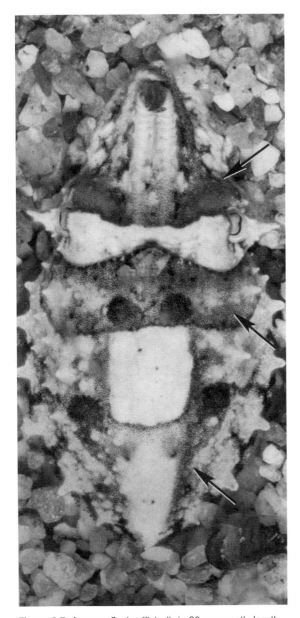

Figure 3.7 A young *Sepia officinalis* (<20 mm mantle length, ML) in a Disruptive Pattern for concealment. See Fig. 3.6. The arrows indicate three skin components that are expressed increasingly strongly: Median mantle stripe (bottom), dark Anterior mantle bar (middle) and dark Anterior head bar (top) (from Hanlon & Messenger, 1988, by kind permission of the Royal Society).

pattern. For example, the cuttlefish in Fig. 3.7 is simultaneously showing White posterior triangle, White square, and White head bar as well as the dark components Anterior mantle bar, Anterior head bar, four Mantle spots and many more. We shall encounter some exceptions to this principle, as in unilateral expression of some components, or certain components that are, by nature, single components; for example, All Dark in the squid *Loligo reynaudii* (Fig. 3.19a) is a single component that acts as a body pattern.

(3) '... different components combine together (i.e. in parallel) to give distinct patterns'. For example, in Fig. 3.7 besides the chromatic components already mentioned, the animal is showing the textural components Major lateral papillae, the postural component Flattened body, and the locomotor component partially Buried.

(4) '... patterns are reflections of the whole behaviour of the animal. Put another way, we can say that body patterns are themselves components of particular behavioural sequences.' This obviously applies to the individual animal in Fig. 3.7, but this is such an important point that we develop it more fully in §3.4 and indeed throughout the book.

It is essential to recall that body patterns, like behaviours, can grade into one another so that in the freely behaving cephalopod one can sometimes observe a flowing sequence of body patterns with what appears at first to be an indefinite number of intermediate stages. This effect is a corollary of the first principle that components vary in the extent of their expression.

3.1.6 Spectral Characteristics of Cephalopod Skin

Mäthger and her colleagues have, for the first time, measured the quality and quantity of light reflected from the skin of a squid, *Loligo pealeii*, and shown how the chromatophores and iridophores interact to produce colours covering the whole of the visible spectrum (Mäthger & Hanlon, 2007; Mäthger *et al.*, 2009).

This species has three classes of chromatophore, with brown, red or yellow pigments, and below them a single type of iridophores that can nevertheless produce red, orange, yellow, green or blue iridescence. Using a spectrometer, Mäthger first characterised the spectral characteristics of the chromatophores (Fig. 3.3a–d), then the light reflected by the iridophores at different angles of incidence (Fig. 3.3e, f), and also the change in wavelength under the influence of ACh (§3.1.1), which induces a colour shift towards the shorter wavelength end of the spectrum (blue, ultraviolet).

Importantly, Mäthger and Hanlon (2007) have shown how the iridophores and chromatophores interact, and how the quality of light reflected from the iridophores is filtered by the overlying chromatophores (Fig. 3.5). The authors have also shown that while the iridophores polarise reflected light at certain angles, the chromatophores do not; instead, they transmit polarised light reflected from iridophores (Mäthger & Hanlon, 2006). Moreover, in cuttlefish and octopus, there is a layer of leucophores lying beneath the iridophores, and these influence the colours produced by the skin. Thus, there is a rather complex mixture of *pigmentary* and *structural* coloration operating in the skin of cephalopods, and this is the anatomical basis of their optical diversity (see also Ghoshal *et al.*, 2013; Deravi *et al.*, 2014).

These findings provide, for the first time, a quantitative basis for understanding the subtle and dynamic changes continually at play in the skin of living cephalopods. The red and yellow chromatophores transmit more light and are important in modulating iridescence. The brown ones transmit less light, and their function may be to block iridescence or leucophore whiteness, as well as to generate the dark components of body patterning. It seems certain, however, that the three pigmentary colours combined with the multilayer reflectors and diffusers enable cephalopods to reflect any wavelength in the visible spectrum. Collectively, these findings reaffirm the complexity of cephalopod skin (Messenger, 2001). Most of these results have been obtained in a loliginid squid, whose skin elements and patterning are simpler than those shown by cuttlefish and octopus. Future comparative research is certain to reveal new mechanisms of colour production and contrast regulation by the skin (e.g. Mäthger *et al.*, 2013; Bell *et al.*, 2014).

3.2 Control of Body Patterning

Changes in body patterning in cephalopods are accomplished chiefly by the chromatophores acting under direct control of the brain. In this section, we first summarise what is known about the physiology of chromatophore control in cephalopods and then consider the important consequences of having neurally controlled colour change.

3.2.1 The Chromatophore Motor System in the Brain

Figure 3.9 shows in a diagrammatic and simplified way the main pathway between the eyes and the skin of a squid. This pathway can be summarised as:

eye → optic lobe → lateral basal lobe → anterior and posterior chromatophore lobes → skin

It needs emphasising that there is no evidence for any kind of feedback in this pathway, certainly no visual feedback, as Fig. 3.10a makes clear.

Not surprisingly, the chromatophores are driven mainly by the *eyes*, although Packard and Brancato (1993) reported that octopus chromatophores respond directly to light, and more recently a fascinating paper has established that some areas of cuttlefish skin contain the same opsins that occur in the retina (Mäthger, Roberts & Hanlon, 2010), although no functionality of skin opsins has been established thus far. Each eye projects to its large *optic lobe*, where visual information is processed and where motor commands are formulated and transmitted to the intermediate and lower motor centres in what is essentially a hierarchical system. In a lightly anaesthetised cuttlefish, or in a free-swimming animal with electrodes implanted in the optic lobe, direct electrical stimulation of the optic lobe produces recognisable components or even body patterns (Boycott, 1961; Chichery & Chanelet, 1976).

Each optic lobe projects to the ipsilateral *lateral basal lobe* (the entire system is bilateral). The lateral basal lobe sends tracts to the anterior and posterior chromatophore lobes. Stimulation in a lateral basal lobe produces darkening or paling from the chromatophores of the arms, head and mantle, but never patterning (Boycott, 1961); sometimes these responses are bilateral, sometimes ipsilateral only. Direct electrical stimulation of the *anterior chromatophore lobe* (ACL) produces only ipsilateral darkening (never paling) of chromatophores in the skin of the arms and head; stimulation of the *posterior chromatophore lobe* (PCL) produces only ipsilateral darkening of the mantle skin (Boycott, 1961). The motoneurons controlling the chromatophores thus lie in the ACL and PCL (Loi *et al.*, 1996; Messenger, 2001) and, in *Octopus*, stimulation here also causes erection of long skin papillae on the ipsilateral mantle (Miyan & Messenger, 1995).

These stimulation experiments confirm the anatomical organisation of these lobes and suggest that the motor system controlling the skin is organised hierarchically. On the basis of visual input, cells in the optic lobe select the appropriate body pattern; information about this pattern is sent to the lateral basal lobe, which plays an important part in ensuring that both sides of the animal as well as the anterior and posterior parts respond together. From the lateral basal lobe, there are tracts running to the anterior and posterior chromatophore lobes, which contain the motoneurons whose axons leave the brain and run without synapse to the chromatophore radial muscles in the skin (Sereni & Young, 1932; Dubas *et al.*, 1986). In *Sepia*, there is evidence that some cell bodies of chromatophore motoneurons lie outside the PCL, in the fin lobe and in the anterior suboesophageal mass (Gaston & Tublitz, 2004), and no doubt future studies will reveal that the system in the brain that controls the chromatophores is more complicated than is suggested here.

In *Sepia*, there is a countershading reflex (§5.1.1) driven by receptors in the statocyst, and there is evidence that the controlling pathway for this runs via the lateral basal lobes (Ferguson, Messenger & Budelmann, 1994).

Unfortunately, no further details are known about where in the brain the chromatic components are organised. Nor is it known whether there is a simple topographical map of the skin in the chromatophore lobes, although this seems unlikely (Dubas *et al.*, 1986; Gaston & Tublitz, 2004). Nevertheless, there is a precise organisation of chromatophore motor fields in the skin. This has been demonstrated in *Sepia officinalis* (Maynard, 1967; Hanlon & Messenger, 1988; Gaston & Tublitz, 2004), in *Lolliguncula brevis* (Ferguson, Martini & Pinsker, 1988), in *Octopus vulgaris* (Froesch, 1973;

Figure 3.8 Neurally controlled polyphenism in young *Sepia officinalis*. Uniform Pattern (top) and Disruptive Pattern (bottom) show two extremes of polyphenism for concealment that is tailored for the immediately adjacent microhabitat. Close inspection of the Disruptive Pattern shows how the nervous system can selectively grade the expansion of the chromatophores (from Hanlon & Messenger, 1988).

Packard, 1974) and in *Eledone cirrhosa* (Dubas & Boyle, 1985). Figure 3.11a and b illustrate the neural control of chromatophores in the Dark zebra bands and surrounding the White fin spots of *Sepia*.

Neural control also allows highly detailed chromatic components to be expressed with fine-tuning of brightness, contrast and colour (§3.1.6). Figure 3.7 illustrates *Sepia*'s ability to show either a mild gradation of contrast across a light/dark boundary or the sharpest of delineations (also see Fig. 3.8).

At the level of the skin, the chromatophores, reflecting cells and muscles are organised precisely with respect to one another, not only morphologically, but also physiologically. What actually appears on the skin at a particular moment is the result of the activity of a certain set of motoneurons in the chromatophore lobes. Such neurons fire specific 'motor fields' of chromatophores and iridophores (i.e. physiological units) that are selected from what Packard (1982) terms the 'static morphological array' of chromatophore organs in the skin. An extraordinary example of precise and swift neural control is the 'Passing Cloud' Display of

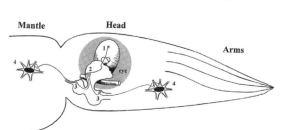

Figure 3.9 Diagrammatic representation of the system in the squid brain controlling the chromatophores (much simplified). Only one side of the brain is shown. (**1**) Optic lobe; (**2**) lateral basal lobe; (**3**) posterior and anterior chromatophore lobes; and (**4**) chromatophore organs. See text §3.2.1.

cuttlefish (e.g. Laan *et al.*, 2014; Fig. 4.6). It is important to recall that a chromatophore or a group of chromatophores and iridophores can participate in more than one component (Fig. 3.11). For a discussion of these ideas, see Florey (1966), Packard and Hochberg (1977), Packard (1982, 1995), Messenger (2001), Tublitz, Gaston and Loi (2006), and Wardill *et al.* (2012).

3.2.2 Visual Perception and Camouflage

How does an octopus or cuttlefish, finding itself on a new background in the sea, almost instantly select an appropriate body pattern – and disappear? And do so on such diverse backgrounds as sand, mud, seagrass, kelp bed or coral reef? There is abundant evidence that body patterning is visually driven, but what sort of information about the external world is being processed in the optic lobe to produce the camouflage?

In the past decade, two groups of investigators (at the Marine Biological Laboratory, Woods Hole, USA; and at the University of Sussex in the UK) have begun to

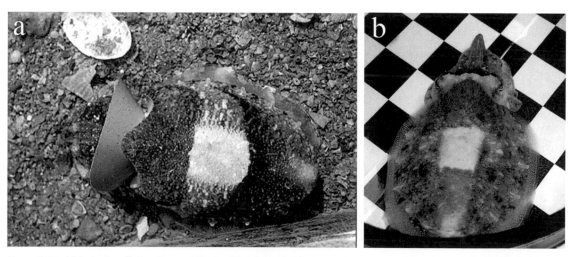

Figure 3.10 (**a**) A plastic ruff placed around the 'neck' of the cuttlefish *Sepia officinalis* (ML 120 mm) prevents it from seeing its own mantle but does not prevent it from showing an appropriate Disruptive body pattern by cueing on the white shell (from Messenger, 2001: *Biological Reviews* **76**, 473–528, Cephalopod chromatophores: neurobiology and natural history, Messenger, J. B., © (2001), Cambridge Philosophical Society, with kind permission from John Wiley and Sons). (**b**) When artificial white checks are approximately the size of the cuttlefish's White square component on its mantle, the cuttlefish expresses a Disruptive body pattern (reprinted from Chiao, Chubb & Hanlon, 2007: *Vision Research* **47**, 2223–2235, Interactive effects of size, contrast, intensity and configuration of background objects in evoking disruptive camouflage in cuttlefish, Chiao, C. C., Chubb, C. & Hanlon, R. T., © (2007), with permission from Elsevier).

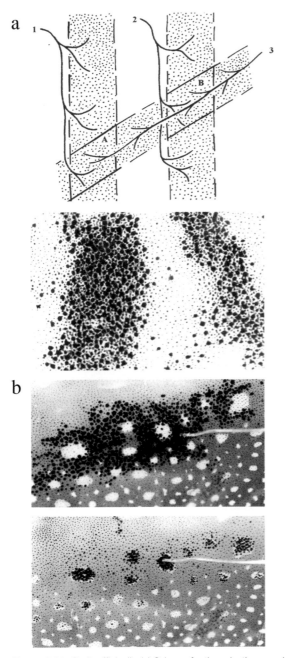

answer such questions by studying the body patterning of cuttlefish (*Sepia officinalis, S. pharaonis*) placed on carefully constructed natural or artificial backgrounds. Marshall and Messenger (1996) initiated this technique to demonstrate that cuttlefish show Mottle body patterns on a variegated substrate containing stones of different brightness, to match it closely. They do not do so, however, with stones of the same brightness but different wavelength, suggesting that cuttlefish perceive such backgrounds as uniform, i.e. that they are colour blind. This agrees with other data (see Chapter 2) and was subsequently confirmed in detailed experiments on *Sepia officinalis* by Mäthger et al. (2006).

Subsequently Chiao, Hanlon and their colleagues in Woods Hole refined and developed this approach in a series of papers that studied both the background input and animal pattern output quantitatively. They constructed natural backgrounds and computer-generated chequerboards of different size and brightness on which they could place cuttlefish and record the body pattern response. Hanlon and Messenger (1988) had already found that cuttlefish, on a wide variety of backgrounds, adopt one of only three basic body patterns for camouflage, Uniform, Mottle or Disruptive (Fig. 3.12), which can now be characterised quantitatively by image analysis according to the coarseness (spatial frequency) and contrast of the pattern components (the 'granularity program'; Barbosa et al., 2008; Chiao et al., 2010).

Chiao and Hanlon (2001a) specifically looked for the features eliciting Disruptive, with its conspicuous White square in the centre of the dorsal mantle. They found that to elicit a Disruptive pattern consistently, the size of the squares of the chequerboard had to be between 40% and 120% of that of the cuttlefish's own White square (Fig. 3.10b). Interestingly, it was subsequently found that the chequer size needed to evoke Disruptive in cuttlefish remains at this value throughout ontogeny (over a size range of 7 to 196 mm mantle length, ML), suggesting that a single 'visual sampling rule' suffices

Figure 3.11 *Sepia officinalis*. (**a**) Scheme for the selective neural excitation that produces the Dark zebra bands (based on Maynard, 1967; Packard & Hochberg, 1977). Note how the chromatophores in fields A and B receive dual innervation. (**b**) Physiological units that emphasise or conceal White fin spots are revealed by direct stimulation of the skin with a silver electrode. Reprinted from Hanlon & Messenger, 1988: *Philosophical Transactions of the Royal Society of London B* **320**, 437–487,

Caption for Figure 3.11 (*cont.*) Adaptive coloration in young cuttlefish (*Sepia officinalis* L.): The morphology and development of body patterns and their relation to behaviour, Hanlon, R. T. & Messenger, J. B., © (1988), with kind permission of the Royal Society.

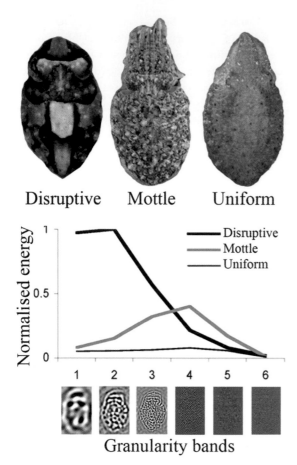

Figure 3.12 The three basic camouflage patterns of the cuttlefish, *Sepia officinalis*. Disruptive: coarse granularity (i.e. bands 1 and 2), high contrast, with skin components of varying shape and orientation. Mottle: medium granularity (bands 3 and 4) and medium contrast; Uniform/stipple: minimal contrast. The granularity bands (*x* axis) are derived from a Fourier transform of the cuttlefish photograph. Normalised energy (*y* axis) is a measure of contrast (from Barbosa *et al.*, 2008: Vision Research 48, 1242–1253, Cuttlefish camouflage: the effects of substrate contrast and size in evoking uniform, mottle or disruptive body patterns. Barbosa, A., Mäthger, L. M., Buresch, K. C. *et al.* (2008) with permission from Elsevier).

for life (Barbosa *et al.*, 2007). They also showed that, given appropriately sized chequers, the cuttlefish regulated the intensity of its Disruptive pattern according to the contrast of the black and white squares of the chequerboard (Fig. 3.13). Perhaps more surprisingly, they found that as few as four white squares placed on a black background (the white

contributing only 1.25% of the total substrate area) could elicit Disruptive. Chiao and Hanlon (2001b) also showed that cuttlefish produce the disruptive White square when other white shapes are presented on a black ground, suggesting that it is area – not shape or aspect ratio – that is the important cue here. In a separate experiment using natural substrates, Mäthger *et al.* (2007) showed that white rocks on a sandy substrate elicited the White square and White head bar, whereas dark rocks on the same background did not, even though the contrast between all rocks and substrate was measured to be equal. This confirms that light objects in the background guide most aspects of disruptive patterning in *Sepia officinalis*.

Edges in the visual background also influence Disruptive patterns, as noted by Chiao, Kelman and Hanlon (2005), who showed, by using low-pass filters to remove edges and high-pass filters to remove contrast, that cuttlefish use visual information about both the edges and contrast of objects to produce Disruptive on a natural substrate. Zylinski, Osorio and Shohet (2009b) explored how edges alone affect Disruptive pattern expression, and Chiao *et al.* (2013) followed up those experiments to demonstrate experimentally how various aspects of edges (including contrast polarity, contrast strength, and the presence or absence of line terminators) turn on light versus dark components of Disruptive patterns (Fig. 3.14). These and other experiments are beginning to reveal that there can be complex interactions among various features of even quite simple visual environments (Chiao, Chubb & Hanlon, 2007; Barbosa *et al.*, 2008).

There is abundant evidence that cryptic body patterning is highly dependent on the spatial scale of the background. The three basic cryptic body patterns of Uniform, Mottle and Disruptive can be elicited consistently on a simple set of artificial substrates of increasing scale: homogeneous light background, small chequerboards and large chequerboards, respectively (Fig. 3.15a; Chiao *et al.*, 2010). To address in more detail how spatial scales modulate the expression of camouflage, Chiao *et al.* (2010) created nine substrate textures that differed only in spatial frequency. Cuttlefish transitioned from Uniform and Mottle at very low frequencies, then produced a variety of Disruptive patterns as predicted for background matching (Fig. 3.15b). Curiously, they showed remnants of

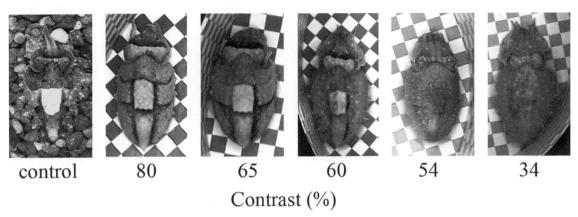

Figure 3.13 Contrast has a strong influence on body patterns of *Sepia officinalis*. On high-contrast natural backgrounds (left), Disruptive patterns are shown. On artificial backgrounds, as contrast is reduced, the disruptive body components drop out, and the cuttlefish takes on a uniform body pattern below 60% contrast of the chequers (modified from Barbosa *et al.*, 2008: *Vision Research* **48**, 1242–1253, Cuttlefish camouflage: the effects of substrate contrast and size in evoking uniform, mottle or disruptive body patterns. Barbosa, A., Mäthger, L. M., Buresch, K. C. *et al.* (2008), with permission from Elsevier).

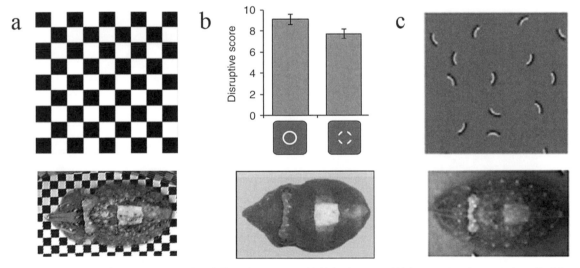

Figure 3.14 Edges influence the production of Disruptive patterns. (**a**) Light chequers of high contrast and correct size will elicit disruptive patterns, yet well-delineated circles with intact edges (**b**) and even scattered bits of quarter circles (**c**) elicit the same pattern, drawing attention to the fact that both size and contrast help to elicit disruptive patterning. In b, the cuttlefish is demonstrating its ability for 'contour completion' by filling in the gaps of the partial circle. (Image in b modified from Zylinski, Darmaillacq & Shashar, 2012: *Proceedings of the Royal Society B: Biological Sciences*, **279**, 2386–2390, Visual interpolation for contour completion by the European cuttlefish (*Sepia officinalis*) and its use in dynamic camouflage, Zylinski, S., Darmaillacq, A. S. & Shashar, N., © (2012), by permission of the Royal Society; images in a, c modified from Chiao *et al.*, 2013: *Vision Research* **83**, 40–47, How visual edge features influence cuttlefish camouflage patterning, Chiao, C. C., Ulmer, K. M., Siemann, L. A. *et al.*, © (2013), with permission from Elsevier).

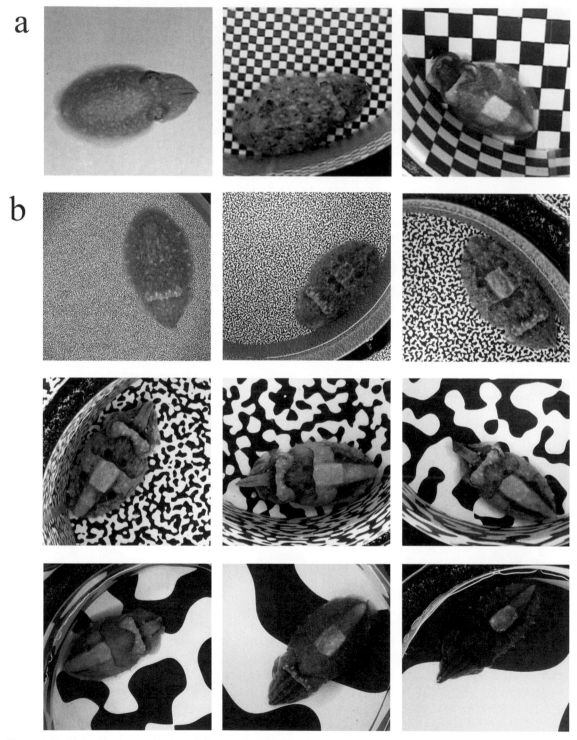

Figure 3.15 Scale of background features influences cuttlefish body patterns. (**a**) Cuttlefish show a Uniform pattern on a uniform light background, a Mottle pattern on small chequers and Disruptive on larger checks (modified from Chiao, Chubb & Hanlon, 2007: *Vision Research* **47**, 2223–2235, Interactive effects of size, contrast, intensity and configuration of background objects in evoking

a

b

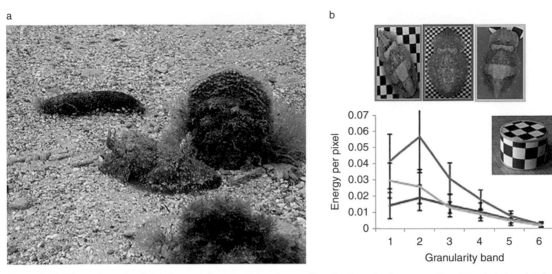

Figure 3.16 Three-dimensional objects can influence choice of camouflage tactic and pattern more than 2D substrates. (**a**) *Sepia officinalis* (centre of image) masquerading as a rock as it takes its visual cues from the adjacent 3D *Pinna* shell; photo taken in 3 m water near Izmir, Turkey (credit R. Hanlon). (**b**) A cylinder with large checks elicits a Disruptive pattern (green box and graph line) on a uniform substrate; controls are the same cuttlefish showing Disruptive on large chequers and Mottle on small chequers with no object present (modified from Buresch *et al.*, 2011: *Vision Research* **51**, 2362–2368, The use of background matching vs. masquerade for camouflage in cuttlefish *Sepia officinalis*, Buresch, K. C., Mäthger, L. M., Allen, J. J. *et al.*, © (2011) with permission from Elsevier).

Disruptive even when the spatial frequencies were very large. These experiments are consistent with the idea that the body pattern deployed attempts to match the spatial frequency of the substrate.

Deliberately introducing patterns on the wall surrounding a test animal revealed that a vertical visual field, as well as the substrate, could also influence body patterning (Barbosa, Litman & Hanlon, 2008). Recent experiments demonstrated that visual background features in the vertical dimension exert a powerful influence on cuttlefish camouflage (Ulmer *et al.*, 2013). Moreover, in another set of experiments, Buresch *et al.* (2011) showed that the pattern on a vertical cylinder (a 3D object on a 2D background) introduced into the experimental tank will override the influence of the pattern on the substrate (a 2D background) in determining the body pattern chosen, if its contrast is higher (Fig. 3.16).

Dynamic arm posture and *skin texture* in camouflage patterns are also guided visually in *Sepia*. Based upon observations in the sea, Barbosa *et al.* (2012) presented *Sepia officinalis* with a series of black and white stripes on the wall of their tank, oriented horizontally, at 45°, or vertically, and found that they altered their arm postures accordingly (Fig. 3.17). Three-dimensional skin texture is dynamically changed when the papillae are physically erected in the skin of cuttlefish and octopuses. Allen *et al.* (2009) placed cuttlefish on one of three substrates: natural, natural under glass (which removes tactile information), and a laminated photograph (which additionally removes depth-of-field information). There were no fewer than nine distinct sets of skin papillae expressed in *Sepia officinalis*, and the authors established that two of these sets (the Small dorsal papillae and the Lateral mantle papillae; Fig. 3.4b)

Caption for Figure 3.15 (*cont.*) disruptive camouflage in cuttlefish, Chiao, C. C., Chubb, C. & Hanlon, R. T., © (2007), with permission from Elsevier). (**b**) Cuttlefish change from uniform/stipple or mottle patterns to disruptive as the same substrate increases through nine size scales (reprinted from Chiao *et al.*, 2010: *Journal of Experimental Biology* **213**, 187–199. Mottle camouflage patterns in cuttlefish: quantitative characterization and visual background stimuli that evoke them, Chiao, C. C., Chubb, C., Buresch, K. C. *et al.*, © (2010) with permission from Elsevier).

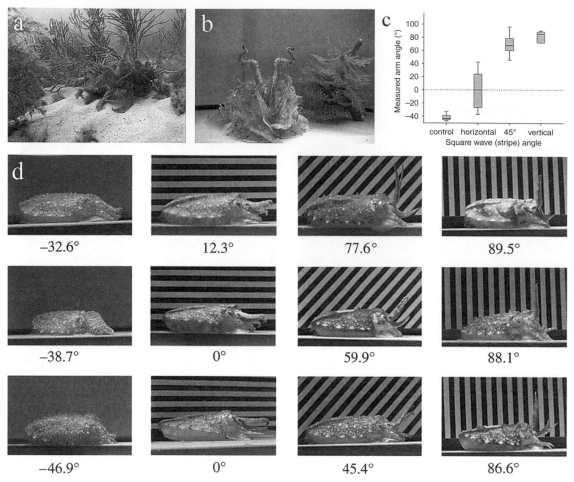

Figure 3.17 Cuttlefish adjust their arm postures according to the vertical orientation of background stimuli. (**a**) Cuttlefish raise their arms often when sitting or hovering adjacent to vertical structures with tree-like forms. (**b**) In the laboratory, cuttlefish settle next to vertical structures in the tank and raise their arms accordingly. (**c**) Data supporting the photo sequences in (**d**) on arm posture orientation relative to background stripes. Numbers in (**d**) denote arm angles in those individuals (from Barbosa *et al.*, 2012: *Proceedings of the Royal Society B: Biological Sciences* **279**, 84–90, Cuttlefish use visual cues to determine arm postures for camouflage, Barbosa, A., Allen, J. J., Mäthger, L. M. & Hanlon, R. T., © (2012), with kind permission of the Royal Society).

were erected on all the substrates, suggesting strongly that visual cues alone were responsible for their expression.

Osorio and his colleagues at Sussex have also provided additional insights into body patterning and visual perception in cuttlefish. Using chequerboard stimuli similar to those of Chiao and his colleagues, Kelman *et al.* (2007) obtained further evidence for the importance of edge detectors in eliciting Disruptive in *Sepia*. In these experiments, the authors compared the appearance of cuttlefish on standard chequerboards and chequerboards with randomised spatial phase: for humans, this removes visual edges, and it apparently does so for cuttlefish as well, which then produce Mottles rather than Disruptive body patterns. Zylinski, Osorio and Shohet (2009c) similarly emphasise the role of edge perception for camouflage, and their principal component analyses reveal, perhaps not surprisingly, how complex visual perception is in these advanced molluscs: certainly comparable to that of their vertebrate predators. The authors suggest that the optic lobe has to deal with first-order information (area, contrast, edges, visual depth) and subsequently second-order information ('visual texture') as in humans. Zylinski and her collaborators stress, too, the complex nature of the varied backgrounds that cuttlefish encounter in the waters off Europe and North Africa (Zylinski *et al.* 2009c). That cuttlefish use multiple cues to select a body pattern for camouflage seems incontrovertible, but it may be that the optical conditions found in such a turbid visual environment make edge detection especially important (Zylinski *et al.*, 2009a).

Kelman, Osorio and Baddeley (2008) also carried out an experiment that placed juvenile cuttlefish on chequerboards with light and dark squares layered one above the other or in the same depth plane. This showed that cuttlefish use information about visual depth to fine-tune the expression of the disruptive pattern, and suggested that they process visual information in a way similar to humans. Zylinski, Darmaillacq and Shashar (2012) presented cuttlefish with fragments of circles in different configurations, and reported that cuttlefish appeared capable of 'contour completion' by perceptually filling in the gaps in an incomplete circle and recognising it as a whole object (Fig. 3.14b). Shohet *et al.* (2006) allowed *Sepia pharaonis* to settle on a series of striped substrates: they found that although the animals did not produce stripes on the body, they oriented themselves with the body-axis perpendicular to the stripes, making it clear that they perceive stripes although they did not align their body in parallel with the stripes. Zylinski, Osorio and Shohet (2009a) also considered camouflage during movement: they found that if cuttlefish are on a high-contrast background before and after moving, their Disruptive patterns become reduced in intensity during movement. On the other hand, if the animals are on a low-contrast background, their Mottle patterns are retained. This suggests that the best tactic for protection against predators during movement is to use low-contrast patterns. Field studies of motion camouflage would be welcome in the future.

An attempt to summarise all this is presented in Fig. 3.18, which provides an overview of how this visual sensorimotor for dynamic camouflage may work in *Sepia officinalis* based on current experimentation (brief reviews have also been provided by Zylinski & Osorio,

2011, and Hanlon *et al.*, 2011). Cuttlefish acquire visual information by actively sensing surrounding environments with vertebrate-like eyes (visual input). Significant information processing then takes place in the optic lobe to extract multiple visual features of substrates (visual processing). Essential visual cues known to affect cuttlefish camouflage include spatial scale, background intensity, background contrast, object edges, object contrast polarity, and object depth. Cuttlefish then integrate all of these visual features and determine the appropriate body pattern types for camouflage, presumably in the optic lobe (pattern selection). To generate the selected body pattern, various skin components responsible for the composition of these body pattern types are selectively activated, perhaps under neural control in the lateral basal lobes and the chromatophore lobes (pattern generation). Finally, the coordinated neural signals from the brain excite radial muscles controlling the expansion of chromatophore organs on the skin and generate appropriate body coloration and patterning for camouflage (motor output).

What general principles, if any, will emerge from these studies remains to be seen. We have discussed these findings at length because the cephalopod chromatophore system offers a unique way of studying visual perception in animals. Clearly, the cephalopod visual system is advanced and the CNS highly developed, so that studying such a system in a group remote from vertebrates could be very rewarding.

3.3 The Significance of Neural Control: Polyphenism and Speed

From the viewpoint of behaviour, there are two important consequences arising from the fact that cephalopod chromatophores are neurally driven. The first is that these animals can expand some groups of chromatophores while leaving others retracted, i.e. they can generate patterns (in the general sense of that word) in the skin. The second is that the patterns can appear, disappear or be modified almost instantly.

The ability to make patterns in the skin is important because the animal can generate a variety of body patterns for concealment. This is a powerful technique for cryptic coloration in all animals (Cott, 1940) and as we shall see in Chapter 5, cephalopods employ it widely. Figure 3.8

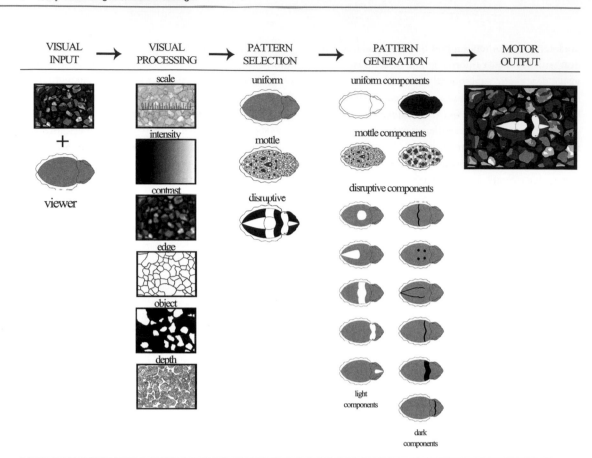

| VISUAL INPUT | → | VISUAL PROCESSING | → | PATTERN SELECTION | → | PATTERN GENERATION | → | MOTOR OUTPUT |

EYE → OPTIC LOBE → OPTIC LOBE → LATERAL BASAL/ CHROMATOPHORE LOBES → CHROMATOPHORE ORGANS

Figure 3.18 A schematic representation of known neural processing stages in cuttlefish camouflage behaviour. See text §3.2.2.

illustrates how different a cuttlefish can appear phenotypically for adaptable camouflage. Rapid adaptive coloration also enables a cephalopod to generate visual signals, either to a conspecific or to another animal. Such signals are used extensively in defence (Chapter 5) and during reproduction (Chapter 6), and it seems likely that the visual channel of communication may be the most significant in the lives of cephalopods (Chapter 7).

The second consequence of having neurally driven chromatophores is speed: it takes less than a second for a retracted chromatophore to expand fully (Hill & Solandt, 1935). Whole patterns vary in speed of appearance. The blue-ringed octopus flashes its blue rings in 0.3–0.5 seconds (Mäthger *et al.*, 2012). *Octopus vulgaris* can change from a fully cryptic Mottle pattern to a highly conspicuous Deimatic Display in about

2 seconds (Fig. 5.2 and Hanlon, 2007); this represents fast control of ca. 30 million chromatophores and thousands of skin papillae. On a behavioural scale, these components or whole body patterns appear or disappear instantaneously. When faced with a predator, cephalopods sometimes exploit this ability by showing four or five body patterns within a few seconds, presumably to startle or confuse it as part of their secondary defence strategy (Chapter 5).

The ability to generate body patterns and to change them instantly is a powerful combination that we term *rapid neural polyphenism* (Hanlon, Forsythe & Joneschild, 1999; Barbato *et al.*, 2007). It allows an individual cephalopod to change its appearance (or phenotype) from moment to moment so that a potential predator will be confronted by quite different-looking forms over a short time span. Like genetic polymorphism, which occurs in a population over a long timescale, this neurally controlled polyphenism, occurring in the individual over a very short time span, may be important as an anti-predatory tactic (§5.1.5).

The polyphenism is based on the many chromatic components, which can be combined (or re-combined) to produce different body patterns. Table 3.2 gives an impression of how many chromatic components there are in various cephalopods. Because cephalopods can grade the intensity of expression of components, or express them unilaterally, in practice an individual may seem even more transiently polyphenic. However, it is important to note that the number of combinations classified in Table 3.2 as body patterns lies between 2 and 16 according to species (Table 3.2; Fig. 3.19).

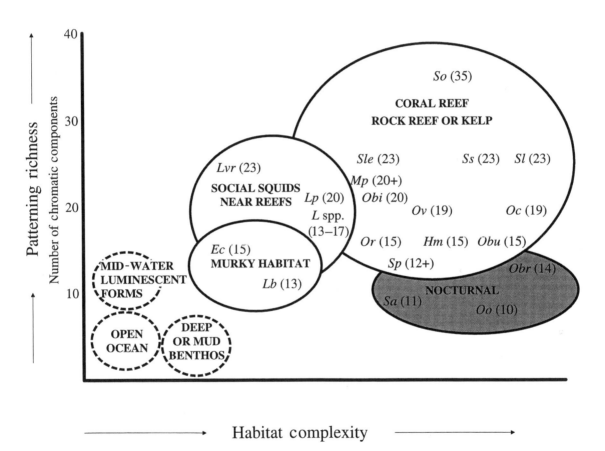

Figure 3.19 Ecological correlates of body patterning. Each species' abbreviation is followed by the number of chromatic components in parentheses (see Table 3.2). Dotted circles indicate no data available. See text. Abbreviations are alphabetical:

Ec Eledone chirrhosa	*Mp Metasepia pfefferi*	*Ov Octopus vulgaris*
Hm Hapalochlaena maculosa	*Obi Octopus bimaculoides*	*Sa Sepiola affinis*
Lb Lolliguncula brevis	*Obr Octopus briareus*	*Sl Sepia latimanus*
Lp Loligo plei	*Obu Octopus burryi*	*Sle Sepioteuthis lessoniana*
L. spp. Loligo opalescens,	*Oc Octopus cyanea*	*So Sepia officinalis*
L. vulgaris, L. pealeii	*Oo Octopus ornatus*	*Sp Sepia papuensis*
Lvr Loligo vulgaris reynaudii	*Or Octopus rubescens*	*Ss Sepioteuthis sepioidea*

These are conservative estimates, because some of the chronic patterns listed here include Mottles and Disruptives that can be graded in intensity of expression so that some authors would split them into separate patterns. Even if we were to treble our estimate of the number of patterns, however, the maximum number would be fewer than 50. The important point is that the number of body patterns is not infinite (see Chapter 7).

3.4 Body Patterns and Behaviour

Perhaps the most important principle of body patterning set forth by Packard and his collaborators is that 'body patterns are themselves components of particular behavioural sequences'. In this book, we present numerous examples supporting the view that cephalopods communicate mainly with visual signals (Chapters 6 and 7) and that the chromatic components of body patterns carry much of the visual information passed to other animals, both intra- and interspecifically. Thus, it follows that body patterns, defined as the combination of postural, textural, locomotor and chromatic components, can themselves be considered behaviours. We now consider two aspects of such behaviours.

3.4.1 How Body Patterns Are Used Behaviourally

Function, causation and development of body patterns are themes explored in later chapters, but it is necessary first to emphasise the central role of body patterning in cephalopod behaviour. Again and again, we find the behaviour of many cephalopods is associated with rapid changes in appearance, a situation rare among animals, and associated with the presence of neurally controlled chromatophores. Essentially, however, the body patterns of cephalopods are used either for crypsis or for communication.

Camouflage, or crypsis, is an important feature in the life of many cephalopods, and it is dealt with in detail in Chapter 5. Here, it is only necessary to recall that for effective crypsis an animal needs to achieve 'general resemblance' to its background, to countershade itself, and to conceal its shadow near the substrate (Cott, 1940). Many animals also adopt 'disruptive coloration' that hinders recognition by breaking up the outlines of the body so the overall form is lost, or they adopt 'deceptive

resemblance' (also known as 'masquerade') to plants or animals to deceive predators. Cephalopods utilise body patterns to achieve multiple methods of crypsis both as primary and secondary defences.

Obviously, some body patterns are potentially useful for communication, and in many cephalopods, there are patterns that have become ritualised into signals and displays (see definitions in Chapter 7). Not surprisingly, intraspecific signalling is especially well developed in some shallow-water, shoaling squid, which show elaborate body patterns during agonistic and courtship behaviour (Chapter 6). One species that has been studied in detail, notably by Moynihan and Rodaniche (1977, 1982), is the Caribbean reef squid *Sepioteuthis sepioidea*, which is remarkable for the range and complexity of its signals (e.g. Fig. 3.20b). Cuttlefish, too, exhibit striking visual signals during agonistic and courtship behaviour, but in most octopuses intraspecific signalling is rather poorly developed.

Interspecific visual signalling is shown by many cephalopods, to prey and to predators, for example the Passing Cloud Display (Chapter 4) and Deimatic Display (Chapter 5).

3.4.2 Ecological and Social Correlates of Body Patterning

Not all cephalopods have body patterning repertoires of the same magnitude. As we would expect, since body patterns are behaviours, cephalopods have evolved patterning repertoires that are related to habitat complexity, activity cycles, predators, social system, reproductive strategies and other ecological factors.

In our first edition, we attempted to develop this idea (Fig. 3.19). Taking just two factors – habitat complexity and activity cycle – we posited a working hypothesis that patterning richness may relate to 'habitat complexity'. The estimates of patterning richness are based on the numbers of chromatic components of body patterning known from the literature (Table 3.2) and are also based on our personal knowledge of the species.

Habitat complexity is difficult to define precisely, but recent progress has been made in this regard (Gratwicke & Speight, 2005; Mellin *et al.*, 2012). The physical diversity of the habitat (the multiple shapes, forms, vertical relief, colours and textures in the

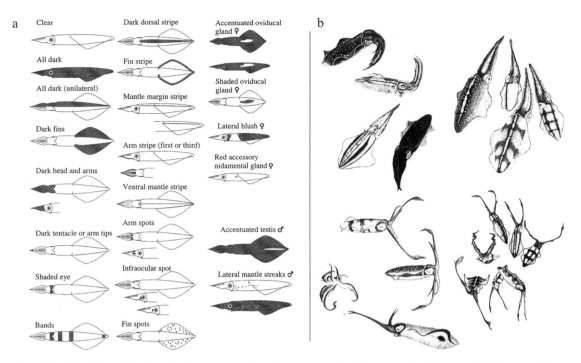

Figure 3.20 Neurally controlled polyphenism. (**a**) The chromatic components in *Loligo reynaudii* that are used to build up body patterns (Hanlon *et al*., 1994: *Biological Bulletin* **187**, 363–372, An ethogram of body patterning behaviour in the squid *Loligo vulgaris reynaudii* on spawning grounds in South Africa, Hanlon, R. T., Smale, M. J. & Sauer, W. H. H., © (1994), with permission from the University of Chicago Press). (**b**) Some body patterns of *Sepioteuthis sepioidea* (from Moynihan & Rodaniche, 1982: *Advances in Ethology*, **25**, 1–150, The Behavior and Natural History of the Caribbean Reef Squid Sepioteuthis sepioidea, Moynihan, M. & Rodaniche, A. F., © (1982), with permission from Wiley).

background) will be very different in a murky habitat such as in the North Sea, for example, and on a coral reef in the Caribbean Sea. Other factors include how clear and well-lit the waters are, whether the cephalopod is active at night or in the day, and how closely it is associated with the bottom for all or part of its life cycle.

The first obvious point made in Fig. 3.19 is that species inhabiting tropical coral reefs, temperate rock reefs or kelp habitats all have rich patterning repertoires; all of these species, including the squid *Sepioteuthis sepioidea* and *S. lessoniana*, live on or near the substrate where they use crypsis and many types of secondary defence to avoid predation (Chapter 5).

The second point is that nocturnal species of octopuses in the same complex habitats have more limited repertoires of patterning. Furthermore, species living in murky habitats have limited patterning diversity that corresponds roughly to that of nocturnal

species inhabiting coral reefs. Presumably, vision is limited for predators under these conditions.

The third point is that many loliginid squid have relatively large repertoires of patterning even though they do not live directly on or near complex habitats such as coral reefs. This brings up a disadvantage of using these numbers: authors may not always have differentiated between components used for crypsis and those for communication. Some cephalopods, especially certain sepioids and the social teuthoids, will have evolved a substantial number of chromatic components for intraspecific communication, and to some extent these will be independent of habitat.

The final point from Fig. 3.19 is that we can only make educated guesses about species living on the deeper continental shelf, the deep fore reef, on bland substrates like mud, or in oceanic and deep-water habitats. From some published and personal observations on live and preserved specimens of such species as the sepioid *Sepia elegans* and the octopus *Euaxoctopus pillsburyi*, we surmise that they have limited patterning repertoires. For neritic squid such as *Illex*, *Todarodes* and *Ommastrephes* that live in open water, countershading is probably the major form of crypsis, but they may have some surprisingly complex courtship and agonistic displays. We also predict that cephalopods with light organs could signal with bioluminescent body patterns in the oceanic mid-waters (see §9.7.1).

Incidentally, despite the suggestion that small cephalopods tend to have fewer body patterns than larger ones in the same habitat (Holmes, 1940; Moynihan & Rodaniche, 1982), evidence suggests otherwise. Small adult *Metasepia pfefferi* have complex repertoires (Roper & Hochberg, 1988), as do the small sepioids *Euprymna scolopes* and *Idiosepius pygmaeus* (Moynihan, 1983a, b).

Two recent investigations call into question some of the above assumptions, however. In the first, Hanlon *et al.* (2007) used ROVs to video *Sepia apama* in red light (a wavelength visible to us but invisible to cuttlefish) in their shallow (2–4 m) spawning grounds in South Australia and followed their behaviour during the night. They discovered that the cuttlefish ceased sexual signalling (Chapter 7) at dusk and then settled on the bottom, effectively camouflaged in a variety of body patterns: Uniform, Mottle or Disruptive. This suggests not only that their nocturnal predators (dolphins, stingrays, various teleosts) possess excellent night vision, but also that the cuttlefish themselves are able to detect fine details of the substrate to assess the most suitable body pattern for camouflage. Laboratory experiments (Allen *et al.*, 2010b) with *S. officinalis* confirmed that cuttlefish respond with appropriate changes in their camouflage patterning when the background changes, even at light levels equivalent to that of starlight.

In the second, conducted over 15 years off Monterey (California), Bush, Robison and Caldwell (2009) used ROVs to photograph *Octopoteuthis deletron*, a solitary mesopelagic squid, at depths of between 344 m and 1841 m. Surprisingly, they recorded no fewer than 59 components of body patterns from these squid: 17 chromatic, 17 postural, 22 locomotor and 3 bioluminescent components. These numbers are comparable to those of many near-shore loliginids living in well-lit waters, and although lights from the ROVs themselves may have influenced patterning, it is clear that we need to be cautious when generalising about body patterning in cephalopods in relation to ecology and life style.

3.5 Summary and Future Research Directions

A distinguishing feature of many cephalopods, especially those living in shallow well-lit waters, is that an individual animal can change its appearance with a speed and diversity unparalleled in the animal kingdom, creating what we term *rapid neural polyphenism*.

What a cephalopod shows to the world is termed its *body pattern*. This is constructed out of a combination of chromatic, textural, postural and locomotor *components*: these are constructed out of *units*, which in turn are constructed out of *elements*: chromatophores, reflecting cells and muscles. The idea that body patterns are built up hierarchically continues to be invaluable for description and behavioural analysis. The number of body patterns varies from species to species: some are able to show 20–30 body patterns, and there may be as many as 36 chromatic components contributing to body patterns, but careful analysis shows that the number of body patterns is not infinite.

The most important of these elements are the chromatophores, which are unique in the animal kingdom in being neuromuscular organs innervated directly from the brain. Consequently, they can be expanded or retracted instantly and, more importantly, they can be differentially expanded to create patterning in the skin. Dynamic iridophores have also been discovered, and there is now a good deal of new information on the spectral interactions among pigments and reflectors in the skin. There is a great deal more to learn about this functional morphology of the dermis and associated behaviours that depend on colour and pattern.

Although we know which lobes in the brain are involved in the control of body patterning, there is still no evidence about precisely how the chromatic components are organised centrally. Neuroethologists might consider the possibility of applying new non-invasive techniques to address this fascinating problem: the skin of a cephalopod can be thought of as a retina-in-reverse, a kind of 'motor retina', yet the exact details of how precise, finely tuned patterns are generated there remain unknown (e.g. Packard, 1995; Messenger, 2001; Laan *et al.*, 2014).

Because the chromatophores are controlled principally by the eyes, another way of gaining insights into how cephalopods select particular body patterns has been to manipulate the visual background systematically and observe the patterning responses for camouflage. Extensive experimentation with cuttlefish has begun to elucidate which features in the visual input are critical for the selection of a particular camouflage pattern. Moreover, the finding of visual opsins in the skin suggests the exciting possibility that the skin may sense light and have some unforeseen influence on body patterning. For students of visual perception in general, the body patterning of cephalopods may offer a unique opportunity for studying the general principles of visual processing by looking at animals phylogenetically remote from humans.

The body patterns of cephalopods presumably evolved for camouflage, although signalling undoubtedly served as a selective pressure as well. Having lost the shell as they evolved into more mobile creatures (Chapter 1), coleoids became more vulnerable to sharp-eyed predators, so that efficient concealment mechanisms would confer a huge selective advantage. As we discuss in Chapter 5, there are many ways of achieving camouflage, and neurally controlled chromatophores and skin papillae are pre-eminently suited to generate different types of patterns that enable fast camouflage in an enormous variety of visual microhabitats. Comparative studies of the sensorimotor system of rapid neural polyphenism with respect to camouflage are needed to understand this unique capability in an evolutionary context as well as that of neural ecology.

The chromatophores are also important for communication, both intra-specific and interspecific, as we show in Chapters 5, 6 and 7. Among the advantages arising from having chromatophores under direct control from the brain are rapid signalling, finely graded intensity of signalling and even bilateral signalling.

An attempt has been made to relate the richness of body patterning repertoires in cephalopods to their habitats. Many species active during the day and living in complex environments such as coral reefs have developed rich patterning repertoires, both for camouflage and communication. Conversely, some that live in murky habitats or are nocturnally active may have poorer repertoires. However, recent reports suggest that the hypothesis proposed in Fig. 3.19 is too simplistic and draw attention to the need for more field studies that document body patterns and their components; incidentally, such descriptions will also be useful for taxonomists. Details of oceanic species would be extremely welcome, although obviously not easy to obtain, but the shallow-water cephalopods of Indo-Pacific and Australian waters, which are relatively more accessible, will surely have much to tell future researchers (see, for example, Roper & Hochberg, 1988; Norman, 1992a).

CHAPTER FOUR

Feeding and Foraging

Coleoid cephalopods are voracious carnivores, feeding by day or by night on a wide variety of live prey that they detect mainly with their eyes. Some feed on large fish or other cephalopods, while others feed on tiny creatures suspended in the water, and they hunt with diverse techniques that can be modified according to prey type. Whatever their diet, cephalopods grow extremely quickly.

4.1 Morphological Adaptations for Feeding

These can be considered under three headings: the arms and tentacles, the mouth and buccal mass, and the gut.

The combination of muscular arms and powerful suckers means that cephalopods can exert strong pulling forces (Dilly, Nixon & Packard, 1964). This allows quite large prey to be seized and restrained; it also permits food to be stockpiled. In the sea, both *Octopus vulgaris* and *Sepioteuthis sepioidea* have been observed storing prey while eating (Cowdry, 1911; Moynihan & Rodaniche, 1982), and in the laboratory, too, octopuses will hold as many crabs as they are offered. The suckered arms are also important in that they enlarge the functional 'gape' of a cephalopod (Packard, 1972), allowing an animal with a small mouth to seize and overpower prey much larger than itself (Fig. 4.1).

In squid, the suckers contain a proteinaceous ring with small teeth, or, in some deep-sea forms, a large hook; both of these are adaptations for gripping active, slippery prey (Mizerez *et al.*, 2009). In octopuses, the suckers are important in other ways (Chapter 2). For example, they constitute a refined chemotactile system for finding food, and those at the end of the slender tapering arms permit crevice feeding in an animal

lacking a terminal mouth. Such feeding is done 'blindly', insofar as the arm is inserted into a cavity not seen by the eyes. In the cirrate octopods, there are cirri on the arms that may detect the small prey animals on which these animals feed, functioning as chemo- or mechano-receptors (Aldred, Nixon & Young, 1983; Villanueva & Guerra, 1991).

The web between the arms is important in prey capture in octopods. In octopuses, it is spread and used to envelop prey as the animal pounces (Fig. 4.2), and several prey or a single large prey individual can be trapped effectively in this way. In the finned octopods and *Vampyroteuthis*, the web is very well developed although it is not clear how it functions in prey capture. In some pelagic octopods such as *Argonauta*, there is a web extending from each of the first arms that is important in feeding (§4.6). *Vampyroteuthis* has two retractile filaments, and Hoving and Robison (2012) speculate that these aid in the capture of detrital matter of various sizes.

Decapodiform cephalopods have two long tentacles that can be extended well beyond the arms to seize prey and bring it back to the arms, which hold it during ingestion. Tentacle ejection is rapid: in *Sepia officinalis*, at 25 °C, the tentacles reach the prey in less than 15 ms, which is faster than the strike of a praying mantis (Messenger, 1968). Such speed is brought about by fast contractions of specially developed cross-striated fibres in the transverse muscles, which differ from the obliquely striated muscles found elsewhere in cephalopods and other molluscs (Kier, 1991; Kier & Van Leeuwen, 1997; Kier & Curtin, 2002).

The tentacles (Fig. 4.3) generally bear large suckers on the club, and sometimes one or more hooks, but in some genera the suckers are minute, for example *Neoteuthis*. In *Mastigoteuthis*, there are numerous tiny

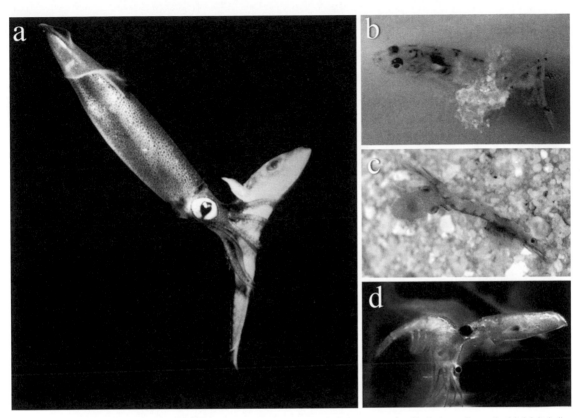

Figure 4.1 The elongate, suckered arms allow cephalopods to seize relatively large prey despite their small mouth. (**a**) Adult *Loligo opalescens* seizing a large fish in the sea (from Hanlon, 1990: *Squid as an Experimental Animal*, 35-62, Maintenance, rearing and culture of teuthoid and sepioid squids, Hanlon, R. T., © (1990), with permission from Plenum Press). (**b**) Hatchling *Octopus maya* eating a much larger fish in the laboratory (original by J.W. Forsythe). (**c**) Paralarva of *Euprymna scolopes* (15 days old) subduing a very large mysid shrimp, which it ate successfully (from Hanlon *et al.*, 1997: *Biological Bulletin* **192**, 364–374, Hanlon, R. T., Claes, M. F., Ashcraft, S. E. & Dunlap, P. V., © (1997), with permission from the University of Chicago Press. (**d**) Pygmy squid *Idiosepius paradoxus* seizing a *Palaemon* shrimp, injecting a toxin and paralysing it within 5 min, and eating it (from Kasugai, Shigeno & Ikeda, 2004: *Journal of Molluscan Studies* **70**, 231–236, Feeding and external digestion in the Japanese pygmy squid *Idiosepius paradoxus* (Cephalopoda : Idiosepiidae), Kasugai, T., Shigeno, S. & Ikeda, Y., © (2004), by permission of Oxford University Press).

suckers along the whole length of the whip-like tentacles, and Nixon and Dilly (1977) speculate that the tentacles may function as a kind of 'fly-paper' to catch small planktonic organisms. The exceptionally long tentacles of such bathypelagic squid could be considered as the functional equivalent of the wide mouths of the fish that inhabit the sparsely populated mid-water zone (Packard, 1972). The deep-sea squid *Grimalditeuthis bonplandi* has thin, fragile tentacles with no suckers, hooks or photophores, but they undulate and are speculated to lure crustacean and cephalopod prey (Hoving *et al.*, 2013). It is possible that the photophores along the tentacles in some species act as lures (Voss, 1956; see §7.3.1).

The various components of the buccal mass have been described in Chapter 2 (see also Mangold, 1989). Inside the buccal mass, the cephalopod mouth is small, and the oesophagus, which passes through the brain, is usually narrow. This means that prey is not swallowed whole but is broken up by the combined actions of the beak and the radula and by the secretions of the salivary glands (Chapter 2). Despite this, the passage of

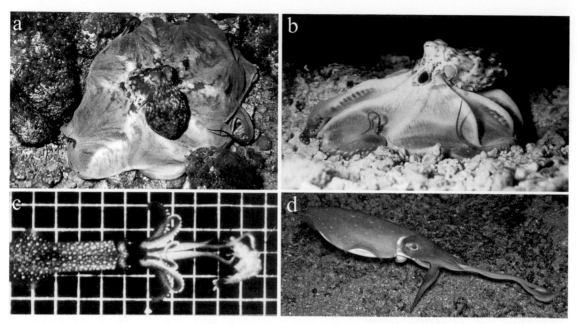

Figure 4.2 Different feeding methods. Octopuses often spread the web before pouncing on a spot on the sea floor, and then feeling underneath for food. (**a**) *Octopus briareus*, subsequently found to have captured and eaten a large parrotfish; (**b**) *Octopus cyanea* covering a small coral, before capturing small molluscs and crustaceans (original by R.T. Hanlon; see also Fig. 4.10); (**c**) a squid, *Loligo pealeii*, and (**d**) a cuttlefish, *Sepia latinanus* attacking shrimps with the extended tentacles. (**a**, **b**, originals by R.T. Hanlon; **c**, from Kier & VanLeeuwen, 1997: republished with permission of the Company of Biologists, from *Journal of Experimental Biology* **200**, 41–53, A kinematic analysis of tentacle extension in the squid *Loligo pealei*, Kier, W. M. & VanLeeuwen, J. L., © (1997); permission conveyed through Copyright Clearance Center, Inc.; **d**, original by Ned De Loach).

food through the oesophagus must be potentially hazardous, and in *Octopus vulgaris*, crustacean exoskeletal fragments, as well as harpoon-like setae from polychaete worms, have been found embedded in the brain (Young, 1971; Nixon & Budelmann, 1984). In the deep-sea octopus, *Graneledone*, the beak is sufficiently large and powerful to be able to crush the shells of gastropods associated with hydrothermal vents (Voight, 2000).

Cuttlefish, some squid and some octopuses secrete toxins from the posterior salivary glands to help immobilise their prey while rendering it suitable for ingestion. The extent to which this occurs throughout the cephalopods is unknown; however, it is possible that squid such as *Loligo forbesi* and *L. vulgaris* do not produce toxins (Koueta & Boucaud-Camou, 1989). Clarke (1966) reports that the bite of *Onychoteuthis banksi* resembles a wasp sting, while that of *Ommastrephes caroli* has no such effect;

Williamson (1965) notes that *Illex illecebrosus* bites are not toxic.

In some octopuses, we are beginning to understand what happens when the animal seizes a crab. Boyle (1990) has shown that when a crab is grasped by an *Eledone cirrhosa*, usually head on, it may show signs of paralysis after a minute and be paralysed irreversibly within three minutes. This paralysis is brought about by the injection of a salivary toxin (after puncture by the toothed salivary papilla), and there is evidence that the commonest site of injection is the corneal surface of the eye (Grisley, Boyle & Key, 1996). This is a region where the exoskeleton is at its thinnest and penetration gives access to a blood sinus close to the eye and CNS. *Eledone* sometimes chooses to drill the crab's carapace, in which case it targets the posterior region of the carapace in the midline, above the heart and thoracic ganglia (Boyle & Knobloch, 1981). But such drilling usually takes 10–20 minutes, so that paralysis of the crab is not nearly so rapid.

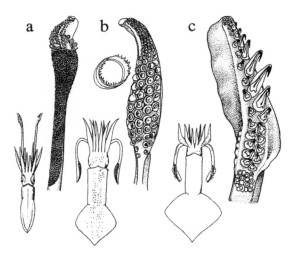

Figure 4.3 The tentacles of some squid. (**a**) *Neoteuthis* with many very small suckers (from Roper, Young & Voss, 1969). (**b**) *Loligo surinamensis* with large suckers, each with a supporting proteinaceous ring (insert). (**c**) *Onychoteuthis borealijaponica* with hooks on the club. (**b**, **c** from Roper, Sweeney & Nauen, 1984. Source: Food and Agriculture Organization of the United Nations, Original Scientific Illustrations Archive. Reproduced with permission.)

It is not until about 40 minutes after seizure that ingestion of crab meat begins: drilling also introduces digestive enzymes into the carapace to destroy musculoskeletal attachments (Nixon, 1984; Grisley & Boyle, 1987, 1990; Grisley, 1993). It is remarkable how clean the disarticulated pieces of a crab skeleton appear after an octopus has discarded them at the end of its meal, and the evidence for external digestion of the musculoskeletal attachments is strong in *Eledone* and *Octopus*. It has yet to be shown whether other octopuses use eye puncture to paralyse their prey.

There is no external digestion in the cuttlefish, *Sepia*. The prey is dealt with summarily: prawns are bitten and paralysed within 6 seconds of capture and crabs are paralysed in about 10 seconds (Messenger, 1968). Many pieces of exoskeleton are ingested (Guerra & Nixon, 1987). Another sepioid, however, does practise external digestion. Kasugai, Shigeno and Ikeda (2004) have good evidence about feeding in the tiny Japanese pygmy squid, *Idiosepius paradoxus* (ML 7–12 mm). This species feeds on shrimps by biting them at the junction of the carapace and the first abdominal segment, where the

cuticle is weakest, and injecting a toxin so that they are paralysed within a minute. It then extrudes the buccal mass into the exoskeleton without biting and ingests the flesh, presumably having injected digestive enzymes, so completely that the discarded remains of the shrimp resemble a complete moult. *Idiosepius* will also attack fish, and although it cannot paralyse the fish, it is able to digest muscle. It is not known if there is external digestion in squid. The gut of cephalopods is well differentiated but short, as in most carnivores, and may contain provision for bulk storage in the form of a crop (in *Vampyroteuthis,* in octopods and in *Nautilus*) and a stomach with a spiral caecum: such arrangements permit opportunistic feeding. There are differences in the speed of food processing in different cephalopods that can be related to their habits and life style. Thus the rate of digestion of food varies from 20 to 16 hours in *Sepia* and *Octopus*, both bottom dwellers, to as little as 3 to 6 hours in *Loligo* and *Ommastrephes*, which actively swim and move throughout the day and night (Bidder, 1966; Boucaud-Camou & Boucher-Rodoni, 1983).

Finally, it should be noted that the funnel can also aid feeding. Hunting cuttlefish direct a jet of water in gravel to make a half-buried prawn more visible (Verwey, quoted in Tinbergen, 1951).

4.2 Detecting and Recognising Prey

Cephalopods can sense prey by sight, by scent, by 'distant touch' via the lateral line analogue, possibly by 'hearing' or presumably by any combination of these (Chapter 2).

Obviously, in any kind of predation the prey must come within range of the sense organs. In an octopus, this might involve full extension of the arms and slow searching movements: quite large areas can be searched in this way, and at night or in conditions of low visibility this must be an important way of locating food. In epipelagic squid, searching is probably done entirely by visual scanning, carried out as the animal swims along a reef or over a sand bank. The panoramic field of vision of most squid must ensure that a large volume of water can be searched, but no estimates have ever been carried out in cephalopods of the kind made in fish (Bone & Marshall, 1982). In this connection, we should recall that it has been calculated that, in fish larvae, eyes on stalks have a considerably larger visual field volume than do sessile eyes (Weihs & Moser, 1981). This could explain why stalked eyes are found in many juvenile cephalopods but not in adults (Fig. 4.4a).

Visual detection of prey may involve movement, contrast, size, shape, orientation and perhaps oddity. Movement is especially important in eliciting visual attacks in sepioids, teuthoids and octopods, although in octopuses attacking crabs, movement is not always absolutely necessary, and octopuses also feed on stationary bivalves and gastropods (Wodinsky, 1969; Ambrose & Nelson, 1983). Contrast with the background has been shown to be important for shape recognition in *Octopus vulgaris* (Bradley & Messenger, 1977). Octopuses and cuttlefish can discriminate size, shape and orientation (Chapter 2), and in the laboratory they exhibit strong preferences for elongate shapes moving along their long axis or for pointed shapes moving in the direction of the point (Sutherland, 1959). Perhaps these responses are related to properties of feature detectors in the visual system. Hatchlings of *Sepia officinalis* respond to horizontally moving lines, which to a human observer look quite like their mysid prey (Wells, 1958, 1962b; Table 8.1). The importance of visual cues in feeding in this species has been shown by experiments in which prey have been presented to the animal behind glass (Wells, 1962b; Messenger, 1973b). Finally, it is known that octopuses visually discriminate between hermit crabs with and without actinian anemones on their shells (Boycott, 1954; Ross, 1971; Ross & Boletzky, 1979; Chapter 8). Cephalopods hunting visually in the deep sea may respond simply to the presence of light, and *Pterygioteuthis* is known to prey heavily on the copepod *Pleuromamma*, which is bioluminescent (Passarella & Hopkins, 1991). There is also experimental evidence that cephalopods can use bioluminescence from dinoflagellates to detect non-luminous prey. Fleisher and Case (1995) presented *Sepia officinalis* and *Euprymna scolopes* with mysids, grass shrimps or fish as prey; they found that both species attacked prey in the dark and that predation increased significantly as the concentration of luminescent algae increased.

Chemical recognition of prey is not well understood in cephalopods. Perhaps it is because they are such visual animals that comparatively little attention has been paid to it. Nevertheless, the experiments of Boyle (1986a), Chase and Wells (1986) and Lee (1992) suggest

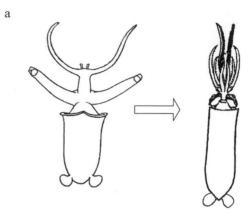

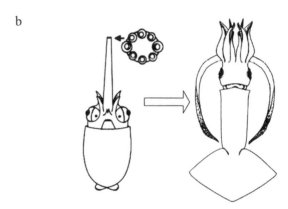

Figure 4.4 (**a**) Juvenile (left) and adult *Bathothauma lyromma*. The stalked eyes of the juvenile (ca. 10 mm ML) may increase the size of its visual field (from Young, 1970: *Journal of Zoology* **162**, 437–447, Stalked eyes of *Bathothauma* (Mollusca, Cephalopoda), Young, J. Z., © (1970), with permission from Wiley); and Okutani, 1974, Tokai Regional Fisheries Research Laboratory. (**b**) 'Rhynchoteuthion' paralarva (left; 2.3 mm ML) and adult *Sthenoteuthis oualaniensis*; the juvenile proboscis is thought to be for feeding because of the suckers at the tip (from Sweeney *et al.*, 1992).

that chemoreception may be used sometimes to guide octopuses to their prey. Moreover, Joll (1977) has evidence in the field that *Octopus tetricus* moves towards bait or lobsters invisible inside closed traps.

Presumably, neither the chemical nor the mechanical channel can be very selective in differentiating between prey species. This is the great advantage of vision: there are several examples in the literature to show that squid

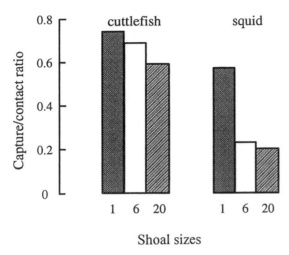

Figure 4.5 Larger shoal sizes lower the success rate of attacking cuttlefish and squid (from Neill & Cullen, 1974: *Journal of Zoology (London)* **172**, 549–569, Experiments on whether schooling by their prey affects the hunting behaviour of cephalopods and fish predators, Neill, S. R. S. J. & Cullen, J. M., © (1974), with permission from Wiley).

and octopuses differentiate between prey animals visually and even adopt different attack tactics accordingly. Thus, cuttlefish seize prawns and crabs in a different way (Messenger, 1968; Duval, Chichery & Chichery 1984), and *Illex* approach trout and mummichogs differently (Foyle & O'Dor, 1987).

The actual *selection* of prey, by which is meant the choice of a particular individual from among a group of potential prey, has not been studied extensively in cephalopods. The cues involved in such choice are likely to be subtle, for example the identification of an odd or conspicuous individual (Curio, 1976). Neill and Cullen (1974) observed that when cuttlefish or squid were allowed to attack groups of fish, they invariably took the straggler. Cuttlefish and squid, even in the confines of an aquarium tank, are less effective predators of large shoals of fish than of small shoals; they spend more time hunting and their catch rate is lower (Neill & Cullen, 1974; Fig. 4.5).

4.3 Capturing Prey

4.3.1 Visual Attack Sequences

Many cephalopods search for and find prey using their eyes, with motion as a key stimulus for the initiation

and maintenance of the attack. Messenger (1968) analysed the visual attack by cuttlefish on prawns as a three-stage sequence involving *attention, positioning* and *seizure*, and this scheme will be followed here, since it has been adopted widely (see various references in Boyle, 1987).

Cuttlefish are ambush predators, lying camouflaged and half buried in the substrate. When a prawn approaches, the animal shows *attention*: there are changes in body patterning, raising of the first (and sometimes the second) pair of arms, and movements of the eyes, head and body, so that the prey is brought onto a forward extension of the body axis with the eyes directed towards it. During *positioning*, the cuttlefish swims towards, or occasionally away from, its prey until it achieves its attacking distance, which is equivalent to about one mantle length (ML). Finally, during *seizure*, the tentacles are ejected to strike the prawn, which is seized and brought back to the arms and mouth. Experiments have shown that the first two stages of the attack are under closed-loop visual control ('guided pursuit') but that the high-speed strike is under open-loop control (i.e. ballistic: Messenger, 1968).

This kind of attack is reserved for prawns, shrimps and small fish. If the prey ambushed is a crab or a *Squilla*, the first two phases of the attack proceed in the same way but the third phase is a 'jump' with partially open arms to seize the prey; the tentacles are not used (Boycott, 1958; Messenger, 1968; Duval, Chichery & Chichery, 1984). Cuttlefish therefore exhibit what Curio (1976) terms 'prey specific hunting'. Moreover, their tentacular strikes on prawns and fish are made from any direction (Neill & Cullen, 1974), while their 'jumps' on crabs are made from behind (Boycott, 1958). Some squid can also exhibit different attacking tactics. *Sepioteuthis* usually makes direct (head first) tentacle attacks on fish, but it can swim backward towards the prey and then suddenly turn round, flaring the arms out prior to seizure by the tentacles (Moynihan & Rodaniche, 1982). *Illex* shows comparable behaviour and also exhibits prey specific hunting (see §4.4.3; Foyle & O'Dor, 1987). Incidentally, many squid sometimes use the arms rather than the tentacles to capture small and slow prey. The mesopelagic squid, *Taningia danae*, which lacks tentacles, has recently been photographed attacking bait in the North Pacific at depths of between 240 and 940 m (Kubodera, Koyama & Mori, 2007). Their remarkable images show that this squid uses its large fins to swim forward rapidly toward the bait, with arms widely outspread, twisting round it and seizing the bait with the ventral arms.

Visual attacks by octopuses, including young animals, have been interpreted by some authors as conforming to the three-stage 'cuttlefish model' (Boucher-Rodoni, Boucaud-Camou & Mangold, 1987) but this is not entirely satisfactory as there often seems to be a single continuous sequence, or a two-phase sequence (approach/jump). The visual attack of *Octopus vulgaris* has been most fully described, and its time course analysed in detail, by Maldonado (1964). In captivity, octopuses presented with a crab bob their heads up and down as if they are estimating the distance to it by monocular parallax (they generally attack with one eye leading); they then launch a swimming attack towards it, using jet propulsion with the arms trailing behind. As they near the crab, they decelerate and drop over it with the interbrachial web expanded. Maldonado showed that there are two time delays before the final acceleration and that this last phase is ballistic (or open-loop) so that it will be successfully completed by an octopus even if the lights are switched off after the attack has been launched.

In *Octopus briareus*, Hanlon and Wolterding (1989) recognised three distinct types of attack. The commonest was the *parachute attack* (or pounce), similar to that just described (Fig. 4.2a). Another kind of attack was the *side arm attack*, made on nearby prey, which involved rapidly extending the rolled arms closest to the prey, which was seized and hauled under the web. The *pincer feeding approach* was the rarest form of attack, being used when one of the other types of attack had failed: it involved the octopus creeping forward on its suckers, with the first two pairs of arms extended.

Another species of octopus whose attack has been studied is the California species, *Octopus rubescens*. Warren, Scheier and Riley (1974) recognised no fewer than seven phases in the attack by this species on crabs: resting, detection, attack, landing, capture, withdrawal, and after capture. They described several features noted in other octopuses, such as bobbing head movements, jetting or crawling to prey, and the outstretched web; they also addressed the issue of body pattern changes during the attack. It is well known that cephalopods often change body pattern while capturing prey

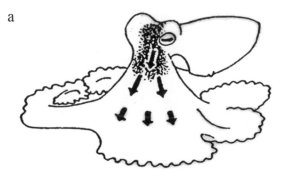

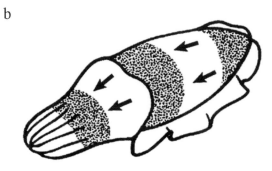

Figure 4.6 The Passing Cloud display associated with predation. (**a**) *Octopus vulgaris* showing it over the head and arms (from Packard & Sanders, 1971: *Animal Behaviour* **19**, 780–790, Body patterns of *Octopus vulgaris* and maturation of the response to disturbance, Packard, A. & Sanders, G. D., © (1971), with permission from Elsevier). (**b**) *Sepia officinalis* showing this display over the head and mantle.

(e.g. Packard, 1963), but the reasons for this are not always clear. Warren and her collaborators sought to establish whether the changes shown by *O. rubescens* during the attack were systematic and whether they could be associated with a conditioned stimulus. The body patterns shown to crabs and to the stimuli were indistinguishable, however, and the authors concluded that some colour and pattern changes are always shown with locomotor activity and are not learned.

This is an appropriate place in which to make some general comments on colour and pattern changes associated with prey capture in cephalopods. Several species have been observed changing body patterns during the attack sequence. Packard and Sanders (1969) described how *Octopus vulgaris* sometimes shows the Passing Cloud display (Fig. 4.6a) when the crab it is stalking stops. They therefore suggested that in this species the Passing Cloud carries the message 'Move, you other animal!' The Passing Cloud is sometimes shown by *Sepia officinalis* (Hanlon & Messenger, 1988) during the attack (Fig. 4.6b) with the possible message of 'Stop and watch me'. During the attack sequence, cuttlefish often show other body patterns, for example Uniform Darkening (Holmes, 1940), perhaps to divert attention from the emerging tentacles. In a more detailed recent study, Adamo *et al.* (2006) have shown that cuttlefish may also show Stipple, Light/Dark mottle, Uniform blanching body patterns at different stages of an attack, whether on crab, shrimp or fish. Moreover, the patterns shown can be suppressed by exposing hunting cuttlefish to a threatening stimulus (a model bird shown overhead).

In *Sepioteuthis sepioidea* in the natural environment, Moynihan and Rodaniche (1982) have also described pattern changes during hunting. They noted that a particular body pattern may persist throughout the attack or be changed for others in a variety of different sequences. It seems clear that the various stages of the attack in cephalopods are not rigidly associated with one body pattern or with certain components.

4.3.2 Tactile Feeding

In octopuses, tactile feeding is of major importance, and many species live almost exclusively on crabs, bivalves and gastropods that are seized with the arms after exploration with the suckers. Such creatures can hardly be said to be 'attacked', and the tactile feeding tactics adopted by octopuses cannot be related to the visual attack scheme. In some species (e.g. *Octopus cyanea* and *O. briareus*) the octopus appears to use vision to find likely feeding places and then makes speculative pounces with the interbrachial web spread widely (§4.4.5; Yarnall, 1969; Hochberg & Couch, 1971; Forsythe & Hanlon, 1997). Other octopuses extend their long arms blindly into crevices. Mather (1991a) described how *O. vulgaris* in Bermuda 'pokes' its arms into areas likely to contain prey; *Octopus macropus* also inserts its very long arms into holes in the sand to extract crustaceans (Hochberg & Couch, 1971; Hanlon *et al.*, 1979). *Octopus briareus* feeds on lobsters in crevices, yet in some cases the stridulation of the lobsters can improve its escape from the octopus; it would be interesting to determine what sensory system

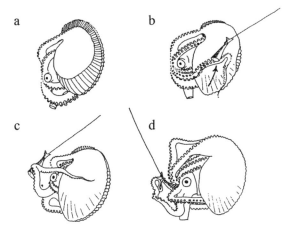

Figure 4.7 The shell of *Argonauta* is covered by the web extending from the first arm (**a**). When food touches the web (**b**), the fourth arm responds and draws it to the mouth (**c, d**) (from Young, 1960a: *Proceedings of the Zoological Society of London* **133**, 471–479, Observations on *Argonauta* and especially its method of feeding, Young, J. Z., © (1960), with permission from Wiley-Blackwell).

of the octopus is being affected by such stridulations (Bouwma & Herrnkind, 2009). *Octopus cyanea* has been observed on a Tahiti atoll extending its arms far into deep crevices in coral to locate food (Forsythe & Hanlon, 1997) and this kind of 'speculative groping' is probably very common in most octopuses. Such a foraging technique obviously has its attendant dangers, and on one occasion at this site an octopus was seen to have its arm tip bitten by a moray eel (*ibid.*).

Tactile feeding is also important in deep-water octopods such as the Bathypolypodinae, the finned octopods (Villanueva & Guerra, 1991), and also in *Vampyroteuthis* (Boucher-Rodoni, Boucaud-Camou & Mangold, 1987). Voight (2008) suggested that *Benthoctopus* and *Graneledone* may forage on infauna by sweeping the mid section of their arms through the sediment. The pelagic octopods such as *Argonauta* and small *Tremoctopus* are also touch feeders. In both genera, there is a well-developed web (Fig. 4.7). Young (1960a) describes how in *Argonauta* this extends from the first arm and can be spread over the thin shell; when food touches the web, the fourth arm sweeps out to seize it and pass it to the mouth. Young never saw the animal attack prawns or fish in the aquarium. These animals are slow-moving creatures with relatively weak suckers, and they presumably feed on the plankton among which they drift; in the sea they are known to feed on pteropods (Okutani, 1960).

Tremoctopus violaceus apparently feeds in a similar way, with tactile stimuli to the mantle eliciting arm sweeping movements (Thomas, 1977), but more remarkable is the fact that small individuals of this species, which is often found associated with epipelagic cnidarians, carry fragments of *Physalia* tentacles in two rows along the first and second pair of arms, corresponding to sucker rows (*ibid.*). Jones (1963), who was badly stung when handling *Tremoctopus*, suggested that the nematocysts could be used mainly as offensive weapons, to capture plankton or small fish that collided with the outspread arms. This remains to be established; it is curious that *Physalia* fragments are found only in males and in young females.

4.4 Modes of Hunting

In this section we find it convenient to follow the classification of Curio (1976), making the proviso that although different cephalopods may have different hunting behaviours, there is often flexibility and overlap among the techniques adopted by a particular species.

4.4.1 Ambushing

Ambushing, a 'lie and wait' predation, is common among cephalopods. The fact that *Octopus vulgaris* can be an ambush predator, hiding behind stones to swoop out and attack a passing crab or a novel object, was cleverly exploited by Boycott and Young (e.g. 1950) in

their entire series of experiments on visual learning (Chapter 8). Hanlon and Hixon (1980) described such behaviour in *Octopus burryi*, and *Octopus joubini* is known to hide in empty bivalve shells until prey come near, before reaching out and grasping them (Hanlon, 1983a). *Sepia officinalis* ambush prey when partially buried. Boletzky and Boletzky (1970) described how *Sepiola affinis* rise out of the substrate and strike passing prey, and Shears (1988) described how *Euprymna scolopes* even retain a sand coat on the mantle as they ambush shrimp. The squid *Sepioteuthis sepioidea* sometimes wait until shoals of small fish approach them before initiating a tentacular strike (Fig. 4.11; and Moynihan & Rodaniche, 1982). Verrill (1880–81) quoted a fascinating account, written in 1872 by S.I. Smith and O. Harger, of ambush behaviour by the squid *Loligo pealeii*: 'Sometimes, after making several unsuccessful attempts [at attacking young mackerel], one of the squid would suddenly drop to the bottom, and, resting upon the sand, would change its colour to that of the sand so perfectly as to be almost invisible. In this position it would wait until the fish came back, and when they were swimming close to or over the ambuscade, the squid, by a sudden dart, would be pretty sure to secure a fish.'

4.4.2 Luring

There is no undisputed example of a cephalopod using a lure but both *Sepia* and *Sepioteuthis* probably do so (Messenger, 1968: Moynihan & Rodaniche, 1982). *Sepia officinalis* commonly darken and raise the first, and sometimes second, pair of arms prior to attacking a shrimp, waving them slowly from side to side (Messenger, 1968). *Sepia latimanus* shows another variation of this (Fig. 4.8a–d). It is possible, however,

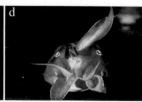

Figure 4.8 *Sepia latimanus* in the laboratory showing the display it makes to small shrimps, in which the first pair of arms is raised (a) and swayed left and right (b–d). Simultaneously, the second arms darken and twirl, and the third arms may be flared (d). The moving arms may draw the shrimp's attention away from the tentacles, which are still hidden but about to be ejected (original by R.T. Hanlon).

that this coloration functions as what Cott (1940) terms a 'directive mark': that is, it directs the prey's attention away from the more dangerous tentacles about to be ejected from below the arms. *Sepioteuthis* also exhibits a similar behaviour with the arms but may also blanch the tentacle tips and hold them in a curved posture, either separately or in combination with the arms. Wickstead (1956) makes a similar claim for *Sepia*. Photophores on the tentacles may act as lures in some deep-sea squid (Voss, 1956).

4.4.3 Stalking

By this is meant a gentle, slow approach to fast prey, prior to the sudden final assault; this contrasts with 'ambush' where it is the prey that moves towards the predator. A good example of stalking is the attack of the squid *Illex illecebrosus* on trout (Fig. 4.9). This involves a head-first attack after a long tracking phase, in which the squid gradually closes in on the trout from behind, where its visibility is poorest. Interestingly, this particular behaviour is reserved for large, fast prey: smaller, slower fish are caught without stalking, by tail-first attacks. These differences in hunting tactics can be explained in terms of the mechanical limitations of the squid's jet propulsion system: head-first acceleration is poor in *Illex* compared with the trout, which could not be caught unless the squid got close to it (Foyle & O'Dor, 1987). *Illex* also attack jigs head first (Williamson, 1965). The squid *Sepioteuthis sepioidea* (and many decapodiformes) also stalk fish and shrimp (Fig. 4.11a), mostly at night (Forsythe & Hanlon, unpublished data); there are no reported observations of octopuses stalking and capturing prey in the wild.

4.4.4 Pursuit

Generally, prey that has been detected flees, and the predator follows. Two kinds of pursuit can be recognised: guided pursuit and ballistic attack (McFarland, 1981). In the former, the pursuit path is continually modified by visual feedback so that it changes according to the position of the prey, as for example when *Illex* pursues a trout (Foyle & O'Dor, 1987). In the latter, the predator establishes the

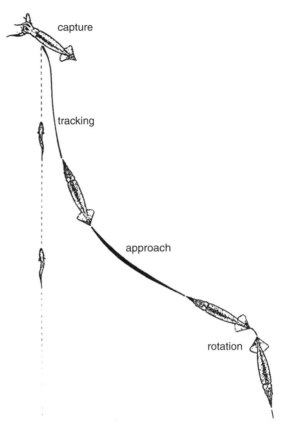

Figure 4.9 The squid *Illex illecebrosus* stalking a trout. Thickness of lines indicates relative speed, showing the slow tracking phase prior to final attack (from Foyle & O'Dor, 1987: *Marine Behaviour and Physiology* **13**, 155–168, Predatory strategies of squid (*Illex illecebrosus*) attacking small and large fish, Foyle, T. P. & O'Dor, R. K., © (1987), with permission from Taylor & Francis Ltd, www.informaworld.com).

position of the prey in space and launches itself at it on the assumption that it will not move. Maldonado (1964) showed that this is how octopuses catch crabs, and the final (seizure) phase of the cuttlefish attack on prawns is also ballistic (Messenger, 1968).

Pursuit may also entail interception of the prey's flight path: this has never been demonstrated reliably in a cephalopod, although cuttlefish may anticipate the flight path of a fish by attacking from the front (Messenger, 1968; Neill & Cullen, 1974). Cuttlefish will pursue prey that has moved out of sight, but such

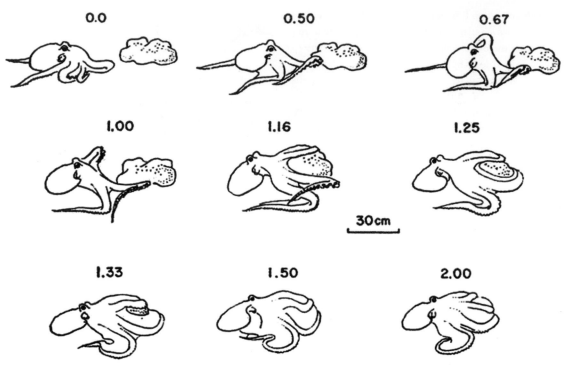

0.0   0.50   0.67

1.00   1.16   1.25

30 cm

1.33   1.50   2.00

Figure 4.10 'Speculative hunting': *Octopus cyanea* pounces and searches for prey as it crawls forward; times (in seconds) are from movie frames (reprinted from Yarnall, 1969: *Animal Behaviour* **17**, 747–754, Aspects of the behaviour of *Octopus cyanea* Gray, Yarnall, J. L., © (1969), with permission from Elsevier). (See text, and Fig. 4.2b.)

behaviour has not been reported in any other cephalopods (Sanders & Young, 1940; see Chapter 8).

4.4.5 Speculative Hunting

Although octopuses have keen vision and undoubtedly make visual attacks on prey in the laboratory, many field observations indicate that shallow-water octopuses regularly indulge in what Yarnall (1969) termed 'speculative hunting', that is, they first pounce with outspread web, and then feel around under the web for food (Fig. 4.10). Similar speculative pouncing with outspread web has been seen in *Octopus briareus* (Hochberg & Couch, 1971; Hanlon & Wolterding, 1989; Fig. 4.2a), *Octopus cyanea* (Yarnall, 1969; Forsythe & Hanlon, 1997; Fig. 4.2b), *Octopus vulgaris* (Mather, 1991a), *Octopus bimaculatus* (R.F. Ambrose, pers. comm., 1994) and *Enteroctopus dofleini* (Hartwick, 1983; Cosgrove, 2003). The squid *Sepioteuthis sepioidea* may also engage in a form of speculative hunting (see Fig. 4.11c and §4.5).

4.4.6 Hunting in Disguise

Moynihan and Rodaniche (1982) described how *Sepioteuthis sepioidea* mimics sargassum weed as it hunts. This involves the individual making the appropriate locomotor, postural, textural and chromatic adjustments, and also associating itself with clumps of weed, from which it can dart out at prey and to which it will subsequently return (Fig. 4.11b; see also §4.4.1).

The Flamboyant body patterns shown by *Hapalochlaena* in association with algae (Roper & Hochberg, 1988; their Fig. 14) could also be an example of hunting in disguise.

Hanlon and Forsythe (unpublished) have filmed *Sepioteuthis sepioidea* swimming backwards while displaying two false eyespots, so that it looked very like a herbivorous parrotfish common in the habitat, especially as the arms and tentacles were held together and waved from side-to-side like the caudal fin of a fish. Swimming in this way, the squid swam amidst small reef

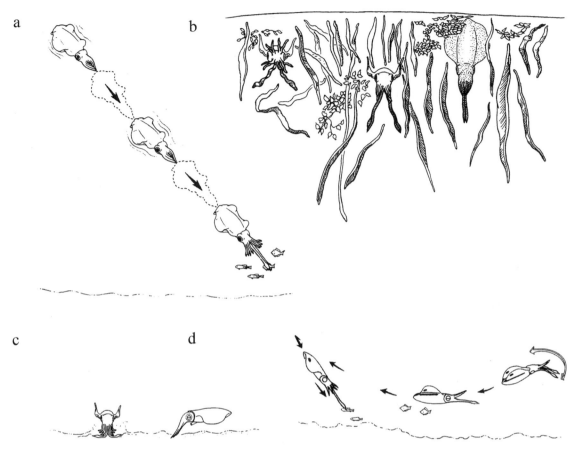

Figure 4.11 Different modes of hunting by the squid *Sepioteuthis sepioidea*. (**a**) Active stalking, with forward attack; (**b**) hunting in disguise in floating seaweed and seagrasses; (**c**) a possible example of speculative hunting by disturbing sand with the arms to flush out buried shrimps; (**d**) mimicking a parrotfish to approach fish more closely (from Hanlon video at Little Cayman Island, 1993 and 2012).

fish and occasionally made short forward thrusts to grasp fish (Fig. 4.11d).

4.4.7 Cooperative Hunting?

By this is meant the combined action of two or more individual predators to secure prey that might otherwise escape. There is no unequivocal evidence for such behaviour in any cephalopod. In the unusual conditions of the laboratory, Hurley (1976) and Yang *et al.* (1983) noted that very young *Loligo opalescens* fought over food, and that if the prey were large, several squid (up to four) would attack it even if there were excess prey in the tank. On the other hand, no such behaviour has been observed in the sea. Group feeding has been reported in

squid, for example in *Symplectoteuthis* (Wormuth, 1976), but it appears that individuals are always attacking on their own account and any benefit to a conspecific may be incidental. In *Illex illecebrosus*, too, Nicol and O'Dor (1985) observed head-first attacks on swarms of the krill *Meganyctiphanes norvegica* in the Bay of Fundy, but noted that individuals in a small shoal captured the euphausiids repeatedly but independently of each other, attacking the body of the swarm rather than peripheral members and showing no signs of cooperative hunting. *Illex* behave similarly in the laboratory (Foyle & O'Dor, 1987). Benoit-Bird & Gilly (2012), using sophisticated acoustic analyses, have described the curious ascending spiral pathways followed during foraging by the jumbo squid, *Dosidicus gigas*, in the Gulf of California

(Chapter 9). However, they found no evidence that this grouping by the squid was conferring a direct foraging benefit.

It is especially interesting that there is no sign of cooperative hunting by the highly developed Caribbean reef squid, *Sepioteuthis sepioidea*. As Moynihan and Rodaniche (1982, p. 23) make clear, 'the behavior of one individual may alert its companions to the presence of possible prey in the neighborhood. The members of a group do not, however, seem to cooperate to drive, chase, or trap their prey by any purposeful communal effort.'

Voight (2005) has a unique record of a feeding frenzy being carried out by *Vulcanoctopus hydrothermalis* at a depth of 2600 m in the Pacific. Twelve individuals, photographed from a submersible, were seen feeding on dense swarms of a small amphipod (*Halice*), using their arms and web to engulf the prey, which they presumably detected by touch. During the course of the feeding some octopuses interacted with others, but it was clear that this 'simultaneous foraging' was definitely not cooperative.

4.5 Foraging Strategies

So far, we have described some of the ways in which cephalopods hunt and capture prey. However, their world is not littered with conspicuous animals waiting to be consumed, any more than is the world of other predatory animals. Prey has to be sought for, and other members of the community may be searching for the same resource. What kinds of strategies do cephalopods adopt in their searches? We do not know. Despite the wide interest in foraging in vertebrates and the development of elegant models of optimality (Krebs & Davies, 1993), there have, for obvious reasons, been few field studies of these problems in cephalopods: only octopuses have been studied so far.

In the site studied by Ambrose (1983), Santa Catalina Island, California, *Octopus bimaculatus* was found occupying shelters in shallow water, some of which were marked by conspicuous 'middens', or refuse heaps around their dens. Others were not, however, because either water currents or hermit crabs had carried away the discarded shells, which were mostly from gastropods. These comprised 75% of the diet. Yet in

laboratory experiments it emerged that they were the least preferred prey in the series:

crabs $\gg$ bivalves = sedentary grazers $>$ gastropods

where the term sedentary grazers includes limpets and abalones (Ambrose, 1984). The strong preference for crabs agrees with the findings of other workers for other species of octopuses (e.g. Taki, 1941), but the preponderance of gastropods in the diet despite their unattractiveness stemmed from the fact that they were extremely common in the Catalina site while crabs were rare. Clearly, diet will depend not only on preference but also on availability, although the relationship may be complex. Closer examination of the data reveals that even among four species of gastropods, the numbers consumed do not reflect the availability. Ambrose (1984) showed that *Norrisia norrisi* was the most preferred species and *Tegula eiseni* the least, despite the fact that the former was very rare and the latter extremely common. In short, the diet of *O. bimaculatus* at this site was shown to be a compromise between preference and availability.

It has been suggested (Curio, 1976; Hughes, 1980) that 'searching predators' foraging over a wide range of habitats should be opportunistic feeders, attacking anything physically possible. Ambrose (1984) found no fewer than 59 species in the diets of over 100 individuals of *O. bimaculatus,* so that it is highly opportunistic; later Ambrose (1986) classified it as a 'generalist predator', but it can clearly be highly selective. One can only speculate on the reasons for this selectivity: is it that crabs and shrimps can be eaten within minutes whereas it may take several hours to drill and eat a snail? Or have crabs a higher dietary value? Perhaps they simply taste nicer, to octopuses as to humans!

Whatever the answer, the foraging strategies of this particular cephalopod are not straightforward, and Ambrose (1984) has made the interesting suggestion that because the number of prey species in the diet was correlated positively to the length of time spent in the same den, *O. bimaculatus* could have been learning the distribution of prey around the den and foraging accordingly. In support of this, he noted that most individual octopuses at Santa Catalina stayed in the same den for at least a month, which would have given them ample opportunity to learn where the rarer prey was likely to be found (see Chapter 8).

Octopus vulgaris could also be described as an opportunistic feeder, taking 17 bivalve species, five gastropods and several species of crabs, in the study of Ambrose and Nelson (1983) at Banyuls in the Mediterranean. Again, molluscs made up the bulk of the diet (about 80%), but bivalves were not only the most frequently encountered prey species in the field, they were also the commonest species in the diets, together with the abalone *Haliotis*. However, in Bermuda, *O. vulgaris* preferred crabs and bivalves (*Lima*) in the laboratory, but its taste in the field was catholic (Mather, 1991a). In one sample, 12 octopuses were found to have consumed 28 prey species over a 4-week period: 11 bivalves, 8 gastropods, 8 crustaceans and 1 chiton. Mather also found that some prey were consumed in the den and some during foraging. Octopuses were more likely to dine out if they were far from the den, but it was not clear what determined such a decision: it was not prey size. Mather (*ibid*.) noted that in this daytime study the octopuses apparently hunted entirely by chemotactile exploration over what she termed 'home ranges' of between $120\,\text{m}^2$ and $200\,\text{m}^2$. However, she suggested that they might have learned about their surrounds visually, for after foraging they generally returned to their dens along different routes (Mather, 1991b; see also Fig. 8.11). In their careful study of *O. vulgaris* middens in the Caribbean Island of Bonaire, Anderson, Wood and Mather (2008) concluded that this species is a 'specializing generalist'. By this they mean that the population as a whole takes a wide range of prey species, but individual animals differ strongly in their preferences and may make quite narrow choices. Scheel and Anderson (2012) also noted great variation in the diet of *Enteroctopus dofleini* (formerly *Octopus dofleini*) along the Pacific seaboard of North America, from California to the Aleutian Islands. In a detailed investigation, again based on midden contents, they found that most individuals of this species were generalist feeders and that dietary species richness was very high. Specialists were common within a population only where the dominant prey (and the midden) was large.

For *O. vulgaris* in Bermuda, the study of Mather and O'Dor (1991) has shown that foraging behaviour involves a clear compromise between gaining food and being eaten. It does not seem to be influenced by the behaviour of conspecifics. Because of the wide choice of prey in the Bermuda habitat, octopuses easily found food, and their high food conversion rate allowed them to grow adequately with even a small food intake. The trade-off of a slower growth rate appeared to be acceptable to octopuses in this habitat. Hence, octopuses spent only about 12% of the daylight period foraging each day.

Foraging by *Octopus cyanea* has also been studied in a semi-natural coral setting in Hawaii (Yarnall, 1969) and on a coral atoll near Tahiti (Forsythe & Hanlon, 1997; Fig. 4.12). *Octopus cyanea* also tended to forage and return to its den over different routes. Octopuses usually foraged twice each day, once early in the morning and once in late afternoon; single forages on the coral atoll usually covered 40–60 m in linear distance and took from 1 to 3 hours, but one covered 120 m and lasted 6 hours. Octopuses spent 65% of the time away from the den actively foraging, 21% moving and 14% sitting. Their overall strategy was to forage slowly on all five of the surrounding habitat types, but not in the same area from day to day (Fig. 4.12a). Forsythe and Hanlon (*op. cit.*) were able to relate hunting tactics to the type of habitat structure: octopuses 'pounced' on small corals and rocks (i.e. smaller in diameter than their arm spread; see Fig. 4.2b), and 'groped' in and around larger corals and rocks (Fig. 4.12b). The octopuses appeared to use vision to determine the forage path, to scan for predators and to find their way back to their den, and they always made long and accurate backward swims (up to 20 m at once) at the end of the forage, landing precisely at their den.

Curiously, in this study *O. cyanea* were not cryptic during the whole time that they were exposed while foraging. Hanlon, Forsythe and Joneschild (1999), using continuous video recording, established that octopuses were only fully cryptic about 40% of the time, and in partial crypsis for another 25% of the time. During the remaining 35% they were highly conspicuous. Presumably, these times were when they were most vulnerable to visual detection by predators, and they showed a variety of changing body patterns, two of which appeared to be mimicry of herbivorous parrotfish. It was noteworthy that the octopuses changed body pattern nearly 1000 times during 7 hours of foraging, which is more than two changes per minute. It would surely be difficult for any predator to form a search image of this, and neurally controlled polymorphism may play an important role in successful foraging.

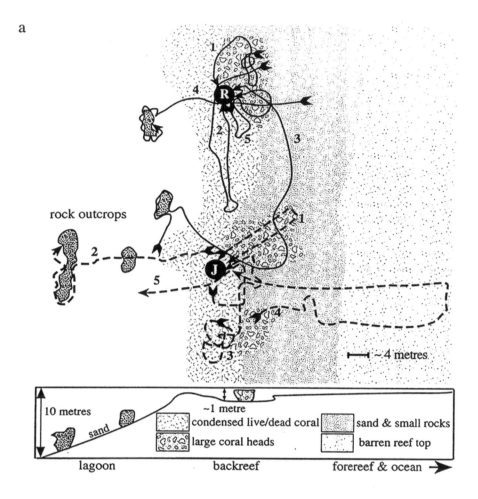

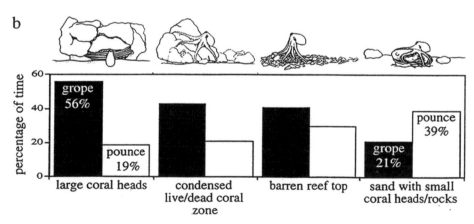

Figure 4.12 Foraging behaviour of *Octopus cyanea* on a Pacific coral atoll (see text). (**a**) Forage paths of two octopuses, R (solid lines) and J (dashed lines), observed over a 6-day period. (**b**) Types of behaviour shown by octopus J in different habitats (see immediately above). Data reconstructed from video (from Forsythe & Hanlon, 1997: *Journal of Experimental Marine Biology and Ecology* **209**, 15–31, Foraging and associated behavior by Octopus cyanea Gray, 1849 on a coral atoll, French Polynesia, Forsythe, J. W. & Hanlon, R. T., © (1997), with permission from Elsevier).

Another octopus, the small *Octopus digueti* of the Gulf of California, exhibited a simple form of prey selectivity but one based on size (Kobayashi, 1986). When given a choice, it attacked and consumed large rather than medium or small hermit crabs. However, when the experimenter placed medium-sized crabs in small, medium or large snail shells and offered them to the octopuses, they preferentially selected those that were in the small shells, presumably because the tight fit made them more accessible.

Finally, the remarkable net-climbing behaviour of *Enteroctopus dofleini* (formerly *Octopus dofleini*) provides further evidence of the flexibility of cephalopod behaviour, in foraging as in other activities. A radio acoustic tagging study (Chapter 1) of the Giant Pacific Octopus in Northern Japan, combined with SCUBA diving, has shown that this species can move from a den towards a commercially deployed gill net, climb it, and feed on the ensnared fish, before descending and returning to the den (Rigby & Sakurai, 2005).

Much less is known about foraging strategies in squid. The only direct observations are for the Caribbean reef squid *Sepioteuthis sepioidea* (Moynihan & Rodaniche, 1982; Hanlon & Forsythe, unpublished data). This species mostly 'rests' during the day, even though it has ample feeding opportunities then. Near dusk, the shoaling squid move to shallow water and slowly split up into progressively smaller groups until they are alone throughout the night. They forage and feed until dawn, when they aggregate into shoals. *Sepioteuthis* has flexible tactics according to circumstance. During the day, they will occasionally take a small shoaling atherinid fish by ambushing it as it approaches too closely; small squid also hunt in disguise by imitating floating sargassum during the day and at twilight and sunrise (Fig. 4.11b: Moynihan & Rodaniche, 1982). At dusk, individual squid stalk prey by moving about ten body lengths above the substrate and swiftly attacking small reef fish (Fig. 4.11a). If this fails, they drop close to the substrate to mimic a parrotfish (Fig. 4.11d), which presumably allows them to approach and capture small fish (Hanlon & Forsythe, unpublished data). Finally, these squid have been observed to descend to the open sand substrate and use the arms to disturb the sand, flushing out small snapping shrimps (Fig. 4.11c); this may be a form of speculative hunting, and it is possible that the squid use the lateral line

analogue to find the general location of the shrimps. It was noteworthy that these squid, like *Octopus cyanea* in Tahiti, foraged with a 'stop and go' (saltatory) search pattern (O'Brien, Browman & Evans, 1990).

Our knowledge of the foraging strategies of the more inaccessible cephalopods and their changes during ontogeny is woefully inadequate. The careful analyses of gut contents of different size-classes of oceanic squid, *Sthenoteuthis*, although valuable, can only give us hints about how these animals forage (Shchetinnikov, 1992). Field *et al.* (2013) sampled 900 *Dosidicus gigas* over 20 degrees of latitude in the eastern Pacific and found that when foraging in open ocean offshore areas they prey upon mesopelagic fish and cephalopods, yet in more coastal waters (and on the continental shelf) they shift to a much broader mix of coastal pelagic fish (herring, anchovy, salmonids, groundfish). This is a highly mobile and generalist predator that exhibits exceptional flexibility in predation.

4.6 Prey Types and the Ontogeny of Feeding

We know about the diets of cephalopods from three sources: from direct observations in the sea or in the laboratory; from examination of their gut contents; and, for some octopods, from studying their 'middens': the refuse heaps around their dens. The two latter methods must exclude many soft-bodied prey animals, but this seems unlikely to alter the main conclusions. However, modern methodology may soon be able to overcome this problem: serological techniques that use antibodies to identify prey proteins, or DNA barcoding, can be used to identify the species of origin of ingested material in any part of the cephalopod gut (Grisley & Boyle, 1985, 1988; Braid & Bolstad, 2014).

Cephalopods feed mainly on molluscs, crustaceans and teleost fish (Box 4.1). The major molluscan prey are bivalves, other cephalopods, even of the same species, and to a lesser extent gastropods. All manner of crustaceans are taken: pelagic or benthic according to habit, juvenile or adult according to age and size. Fish of many kinds are eaten: clupeids, flatfish, scombrids, gadids, eels, myctophids and so forth. Curiously, there seem to be no records in the literature of cephalopods feeding on elasmobranchs or any other non-teleost fish,

Box 4.1

What cephalopods eat

A. According to phylum

PROTOZOA	Foraminiferans	*Major food sources are underlined*
CNIDARIA	Notably siphonophores	
MOLLUSCA	Chitons	
	Gastropods, including heteropods and pteropods	
	Bivalves, a great variety	
	Cephalopods, other species and cannibalism	
ANNELIDA	Errant polychaetes	
CRUSTACEA	Ostracods	
	Copepods	
	Cirripedes	
	Malacostracans, including stomatopods, mysids, cumaceans, tanaidaceans, isopods, amphipods, euphausiids, decapods of all kinds (including shrimps, prawns, lobsters, crabs)	
CHAETOGNATHA	Arrow worms	
ECHINODERMATA	Sea-urchins, ophiuroids	
CHORDATA	Teleosts, an enormous variety (no records for any other kind of fish)	

Remains have also been found of animals from the phyla PARAZOA, NEMATODA, NEMERTEA, BRACHIOPODA, BRYOZOA and SIPUNCULOIDEA

B. Diets[1] of some representative cephalopods

CUTTLEFISHES	Main foods: shrimps, crabs and fish
Sepia	Crustaceans: including mysids, shrimps, prawns, amphipods, isopods, copepods, ostracods, stomatopods
	Teleosts: numerous, including gobies, mullet, whiting, flatfish
	Molluscs: including pteropods, bivalves, other cephalopods and cannibalism
	Polychaetes and nemertines
SQUIDS	Main foods: fish, crustaceans and squid
Loligo	Teleosts: *Trachurus, Boops, Anchoa, Anguilla* and many others
	Crustaceans: including euphausiids, shrimps, prawns, copepods, decapod larvae
	Cephalopods: including cannibalism
Todarodes	Teleosts: *Scomberesox*, myctophids, herring, cod, *Maurolicus* and many others
	Cephalopods: enoploteuthids, ommastrephids and cannibalism
	Crustaceans: including *Parathemisto*, decapods, phronomids heteropod molluscs and polychaetes

Ommastrephes	Teleosts: myctophids, *Tetragonurus*, *Hygophia* and many others
	Cephalopods: enoploteuthids, ommastrephids, onychoteuthids, histioteuthids and cannibalism
	Crustaceans: including copepods, cirripedes, amphipods, euphausiids, decapods
	Gastropods: including larvae, pteropods and heteropods chaetognaths and planktonic annelids (e.g. *Tomopteris*)
OCTOPODS[2]	Main foods: crabs, bivalves and gastropods
Octopus vulgaris[3]	Crabs: *Carcinus*, *Pachygrapsus*, *Sesarma* and many others
	Other crustaceans: *Squilla*, squat lobsters, brachyurans, anomurans, natantians, isopods, stomatopods, ostracods, amphipods
	Bivalves: *Pitaria*, *Venus*, *Tapes*, *Prothotaca*, *Paphia* and many others
	Gastropods: *Haliotis*, *Rapana*, *Chicoreus*, *Polynices* and many others
	Cephalopods: including cannibalism
	Teleosts: many
	Others: foraminiferans, polychaetes (e.g. *Hermione*), ophiuroids, fish rarely

[1] Food listed in order of importance.

[2] Octopods comprise pelagic and bathypelagic forms whose diets differ considerably from *O. vulgaris*.

[3] There are differences in diet according to geographical locality. For example, in the Catalan Sea, *O. vulgaris* feeds on 80% crustaceans but in the Algarve, Portugal, 80% bivalves (K. Mangold, pers. comm., 1992). From data in Boletzky & Hanlon, 1983; papers in Boyle, 1983b & 1987; Nixon, 1987; Nixon & Maconnachie, 1988; Villanueva & Guerra, 1991.

although DNA barcoding of stomach contents from the large mastigoteuthid squid *Idioteuthis cordiformis* suggests that they consume dogfish sharks (Braid & Bolstad, 2014).

Representatives of several other phyla, notably polychaetes, chaetognaths, siphonophores and echinoderms, have been reported regularly. For example, in a study of the nocturnal feeding of the small sepiolid, *Euprymna scolopes*, Shears (1988) found that over 90% of prey were polychaete worms. In a study of *Bathypolypus arcticus*, nearly 60% of prey in the stomachs were ophiuroids (O'Dor & Macalaster, 1983).

Because cephalopods are so varied in size and life style, they take a correspondingly large variety of prey, and it would be inappropriate here to give any kind of comprehensive list. Instead, we summarise in Box 4.1 some of the prey that have been identified in the guts of some common cephalopods. Among recent accounts with detailed lists of the prey of some octopus species, we can note the studies of Bouth *et al.* (2011), Anderson *et al.* (1999), Anderson, Wood and Mather (2008), and Scheel and Anderson (2012), all based on remains found in middens or in beer-bottle dens! Parry (2006) presents detailed data on the diet of two genera of ommastrephid squid off Hawaii on the basis of otoliths and beaks found in their stomachs; and Markaida (2006) lists data from nearly 300 *Dosidicus* stomachs, taken in different regions of the Gulf of California.

Many cephalopods are generalist rather than specialist feeders so that they can feed opportunistically and take whatever is available. Thus, although the Humboldt squid *Dosidicus gigas* feeds mainly on lanternfish and other squid (including its own species), it has been described as a 'schooling nektonic predator, which eats any prey that moves, provided only that it is abundant and of convenient size ... they even devour garbage thrown from a ship, cucumber rinds, banana peels, and the like' (Nesis, 1970, 1983). There is also a first-hand account of attacks on a human SCUBA diver (Hall, 1990). The deep-sea cirrate octopod,

Opisthoteuthis, is also a catholic feeder but takes predominantly small prey, crustaceans, polychaetes, even foraminiferans (Villanueva & Guerra, 1991). In captivity, cephalopods will feed on a variety of prey, including species that they would never encounter in their natural habitat. The fact that many squid are relatively indiscriminate attackers is, of course, exploited by fishermen the world over: 'jigging' for squid involves moving an elongate unbaited lure, of any colour, up and down in the water.

What a particular cephalopod eats will depend on prey availability and prey size, as well as on its inherited programmes of feeding and individual past experience. The availability of prey will change in space and in time: potential prey may enter or leave the vicinity, or the cephalopod may encounter new prey as it travels. Prey numbers may fluctuate on a daily or seasonal basis, and hence many cephalopods will exhibit feeding cycles reflecting this availability: *Illex illecebrosus*, for example, tends to feed around sunrise (Amaratunga, 1980); *Dosidicus gigas* feeds after sunset and before sunrise (Nesis, 1970). Nearshore squid also show a marked diel feeding pattern: *L. opalescens* from 1000 to 1400 hours (Karpov & Cailliet, 1978); *L. pealeii* from 1600 to 2000 hours (Vovk, 1974); *L. vulgaris reynaudii* late at night or in the early morning (Lipinski, 1987). According to Nigmatullin and Ostapenko (1976), *Octopus vulgaris* in northwest Africa eat fish at night and crabs during the day; and Amaratunga *et al.* (1980) noted that *Illex illecebrosus* were more cannibalistic during the night but shifted to fish in the day. Seasonal variation in food availability is also well documented (e.g. Macy, 1982).

Boyle and Rodhouse (2005) draw attention to the importance of predation by cephalopods on commercial stocks of fish and crustaceans, and also discuss the role of cephalopods on the major marine ecosystems.

4.6.1 Paralarval Feeding

Some cephalopods hatch as miniature replicas of the adult and feed in a similar way. Thus, the response of a hatchling cuttlefish to a mysid is not very different from an adult's to a prawn, and the ensuing sequence of events is essentially similar (see §4.3.1). The same seems to be true of large-egged octopuses such as *O. briareus*, *O. joubini* and *O. maya* (see papers in Boyle, 1983b). The prey taken by hatchlings of these species differs from

that taken by adults for reasons of size, but nonetheless the natural diet is apparently entirely crustacean (Boletzky & Hanlon, 1983).

Many juvenile cephalopods, however, differ in form as well as in size from their parents (Sweeney *et al.*, 1992), and consequently their feeding behaviour must be substantially different. The paralarvae of *Enteroctopus dofleini*, for example, have been observed hanging upside down from the surface of the water, presumably supported by surface tension, feeding on *Artemia* or frozen krill, which they seize with the arm tips; in fact, Marliave (1981) speculates that they may feed on neuston (tiny organisms attached to the surface film) in the sea for the first month of life. There may also be neustonic feeding in young *O. vulgaris* (Boletzky, 1987b).

The young of *Macrotritopus defilippi* (the so-called 'macrotritopus larvae') are unusual in having elongated third arms that can be extended to enable the animal to drift in the water column: the spread arms may also serve to catch large plankton passively (Nesis & Nikitina, 1981; Hanlon, Forsythe & Boletzky, 1985).

Among squid, the paralarvae of *Loligo vulgaris* are known to feed on zooplankton: they can be laboratory-reared on the zoeae larvae of decapod crustaceans such as the hermit crab, *Pagurus prideaux* (Villanueva, 1994).

Other differences in feeding habits between young and adult can be inferred from morphology alone. For example, juveniles of the pygmy cuttlefish, *Idiosepius pygmaeus*, have no tentacles, although in this species, as in many others, the actual diet is still not known (Natsukari, 1970). O'Dor, Helm and Balch (1985) speculate that the curious little 'rhynchoteuthions', or 'snouted' hatchlings of ommastrephid squid (Fig. 4.4b) are suspension feeders, using cilia and mucus like so many other molluscs to trap small prey.

4.6.2 Diet Changes During Ontogeny

The size of prey obviously changes as the individual grows. Larger cephalopods are not only faster and more mobile but may have acquired new effectors suitable for grasping and managing stronger prey (Figs. 4.2 and 4.3). For example, *Gonatus fabricii* does not have tentacular hooks until it reaches about 25 mm ML and can then feed upon fish larger than itself, instead of on smaller organisms (Nesis, 1965; Kristensen, 1983).

Changes in diet during development have been followed in several species of squid: together they provide good evidence that prey size is critical. Vovk and Khvichiya (1980) recognised four stages of feeding in *Loligo pealeii:* from 11 to 80 mm ML they fed on a variety of food, notably copepods; from 80 to 120 mm they fed on crustaceans and young fish; from 120 to 160 mm, on fish and squid but also some euphausiids; and, over 160 mm, on fish and squid. In *L. opalescens* reared in the laboratory, Yang *et al.* (1983) followed the changes in size of prey taken from hatching to maturity. During the first 70 days, squid selected copepods less than 4 mm in length, as well as *Artemia* and chaetognaths; from days 60 to 130 they ate mysid and shrimp larvae up to 10 mm long; from day 100 they ate various shrimps (up to 25 mm) and especially fish (up to 70 mm).

In *Sepioteuthis sepioidea,* LaRoe (1971) noted that 3-week-old juveniles stopped attacking mysids, now relatively smaller. Incidentally, even before this stage, the young showed clear signs of selectivity. Mysids and young fish were preferred over copepods, amphipods, various other crustacean larvae, and polychaetes. Many other examples of diet change are known from laboratory studies (e.g. Boletzky & Hanlon, 1983) and from stomach content analyses (Shchetinnikov, 1992). Both types of study have limitations, however, and we are in need of field observations to understand what juvenile cephalopods feed on.

The development of feeding behaviour in *Sepia* has been studied in detail (e.g. Darmaillacq *et al.,* 2004a, b); this work has implications for learning studies and is considered fully in Chapter 8.

One of the interesting general findings about diet is that the same species may have different eating habits in different parts of the world. Nixon (1987) points out that the ability of *Octopus vulgaris* to drill shells was discovered in Japan because of predation on cultured pearl oysters: yet this species does not drill oysters in the English Channel, the Bahamas or the Bay of Naples. Such 'cultural' differences demonstrate that feeding programmes in cephalopods are flexible and suggest that the ability to learn, so characteristic of these animals, may be important for feeding. For example, octopuses in the laboratory soon stop attacking sea anemones (Chapter 8) and quickly come to avoid the venomous gastropod *Conus* (Nixon &

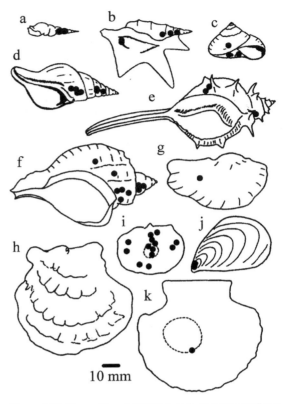

Figure 4.13 The position of drill holes made on various molluscs offered to *Octopus vulgaris* in Naples. The holes are consistently drilled over the site of muscle attachment: above the columella in gastropods (a–f), and over the myostracum in bivalves (g–k). All the species shown were successfully drilled and eaten, except for the oyster (h), which was drilled but never penetrated (modified from Nixon & Maconnachie, 1988: *Journal of Zoology (London)* **216**, 687–716, Drilling by *Octopus vulgaris* (Mollusca: Cephalopoda) in the Mediterranean, Nixon, M. & Maconnachie, E., © (1988), with permission from Wiley).

Maconnachie, 1988); and adult cuttlefish attack large crabs from behind to avoid being nipped. Even more remarkable, adult octopuses inject saliva through holes that they drill in the shells of crustaceans, bivalves and gastropods (see §4.1), but these holes are made over, or close to, the site of muscle attachment, so that the saliva can act most effectively to separate the muscle from the shell and render the victim accessible. The experiments of Nixon and Maconnachie (1988) and Cortez, Castro and Guerra (1998) leave no doubt that hole drilling is not a random process (Fig. 4.13). This is also true of *Octopus rubescens* feeding on the clam *Venerupis*, although it is fascinating that senescent animals no longer accurately drill over the precise area of the shell that lies over the adductor muscle (Anderson, Sinn & Mather, 2008; Anderson, Mather & Sinn, 2008).

It would be especially interesting to see how such drilling behaviour develops in the individual. Very young *Eledone* drill holes in crabs as soon as they encounter them (Boyle & Knobloch, 1981); adult *Octopus vulgaris* have preferences for drill sites on molluscs according to Arnold and Arnold (1969); and in the same species, Wodinsky (1969) reports that a few individuals use force to extract molluscs before resorting to drilling, although the majority commence by drilling.

The way that octopuses handle another common type of prey – bivalve molluscs – has also attracted attention recently. Fiorito and Gherardi (1999) were the first to show that *Octopus vulgaris* offered clams, scallops or ear-shells (*Pinna*) begin by attempting to pull the valves apart but then switch to drilling and injecting venom to weaken the adductor muscles. Anderson and Mather (2007) drew attention to the flexible behaviour shown by *Enteroctopus dofleini* (formerly *Octopus dofleini*) when feeding on three different clams. This octopus pulls apart the shell valves of thin-shelled species with weak adductor muscles, such as *Mytilus trossulus* and *Venerupis philippinarum*; however, it chips or bores into the thick shells of *Protothaca staminea*, presumably to admit toxin. When these three clam species are offered together to the octopuses, they consume more of the first two; if they are presented after the shells have been opened, they consume significantly more *Protothaca*. Moreover, when octopuses were offered intact *Venerupis* with their valves wired together, they switched to chipping and drilling. Ebisawa, Tsuchiya and Segawa (2011) have measured the energy consumption of *O. ocellatus* during feeding on clams and shown that low-energy pulling plus drilling is the most efficient method of feeding for young octopuses.

Incidentally, female octopuses may cease feeding while brooding eggs (the literature is contradictory on this point: Wodinsky, 1978); in one study of *O. vulgaris*, brooding females that had previously drilled gastropod shells began to pull snails from their shells rather than drill holes. It is curious that these females would expend unnecessary energy like this, for they ate little or nothing of the snails they extracted (Wodinsky, *ibid.*).

Finally, one of the most remarkable features of cephalopods and their diets, and one that influences much of their ecology and behaviour, is that their growth rate is extraordinarily high, whatever the diet (Fig. 4.14; see Forsythe & Van Heukelem, 1987 or any of the papers in Boyle, 1983b; Boyle & Rodhouse, 2005). This is due to the spectacular conversion rate: on a wet weight basis, between 30 and 60% of ingested food is used for growth (Mangold & Boletzky, 1973; Pascual, 1978; Mangold, 1983b; Van Heukelem, 1983a).

4.7 Cannibalism

Cannibalism seems to be widespread in cephalopods (Quetglas *et al.*, 1999) but, as with so many other features of their behaviour, data are scarce. Neither cannibalism observed in captivity nor attacks by squid on conspecifics injured by commercial jigging machines provide evidence that cannibalism occurs normally in that species.

Following Wooton (1990), we recognise two types of cannibalism: intracohort and intercohort. Intracohort cannibalism, which occurs between conspecifics of approximately the same age, is common in fish larvae and is also known among juvenile cephalopods. It has been reported, in the field, in *Octopus dofleini* (Hartwick, 1983); in the field and in the laboratory in *Octopus briareus* (Hanlon, 1983b; Aronson, 1989); and in the laboratory in *Euprymna scolopes* (Singley, 1983). Cannibalism extends into adult life in *O. dofleini* and *O. briareus* and also occurs in adult *O. vulgaris*. Intercohort cannibalism by *O. vulgaris* has been documented in the field by Hernandez-Urcera *et al.* (2014). In the Naples Zoological Station, and the Laboratoire Arago in Banyuls, France, intercohort cannibalism is common in this species: larger, older animals regularly eat smaller, younger individuals. This may be because they have not been adequately fed, but the fact that the males of this species have some extra-large suckers on the arms that may signal their sex (Packard, 1961; Chapter 6) suggests that small males may be vulnerable to cannibalism by larger females in the natural habitat.

There is one well-documented report of sexual cannibalism by an octopus. While studying *O. cyanea* on a coral reef in the Pacific, Hanlon and Forsythe (2008) observed a small male repeatedly mating a large female out of her den and in the open.

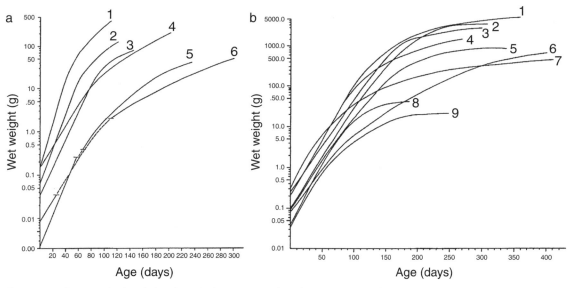

Figure 4.14 Fast growth of cephalopods in the laboratory. (**a**) Cuttlefish and squid: 1 *Sepia subaculeata*; 2 *Sepia esculenta*; 3 *Sepioteuthis sepioidea*; 4 *Sepia officinalis*; 5 *Loligo opalescens*; 6 *Loligo forbesi*. (**b**) Octopuses: 1 *Octopus vulgaris* (South Africa); 2 *Octopus maya*; 3 *Octopus tetricus*; 4 *Octopus vulgaris* (Mediterranean); 5 *Octopus briareus*; 6 *Octopus bimaculoides*; 7 *Eledone moschata*; 8 *Octopus digueti*; 9 *Octopus joubini* (reprinted from Forsythe & Van Heukelem, 1987: Growth, In *Cephalopod Life Cycles*, Vol. II, *Comparative Reviews*, ed. P. R. Boyle, © (1987), with permission from Elsevier.

After the 13th mating she attacked, then suffocated him and spent two days cannibalising him in her den. The authors speculate that cannibalism may explain many other reports of octopuses mating in the open, where presumably the male will often manage to escape.

No examples of egg cannibalism have been reported in coleoid cephalopods, although this is known to occur in *Nautilus* (Arnold & Carlson, 1986).

Cannibalism is frequent in the natural environment in *Sepia officinalis*, and intracohort cannibalism is also known among several species of shoaling squid, including *Illex illecebrosus*, *Dosidicus gigas*, *Loligo opalescens*, *L. pealeii* (see the papers by O'Dor, Nesis, Hixon, and Summers, respectively, all in Boyle, 1983b) and *Symplectoteuthis* (Wormuth, 1976). In *Dosidicus*, cannibalism may account for as much as 30% by weight of the diet in months when sardines and shrimp are scarce (Ehrhardt, 1991). Sanchez (2003) also emphasises the importance of cannibalism in this species, especially by larger animals. In *Illex*, cannibalism may make an important contribution to mortality. It increases as the food supply diminishes,

with the largest squid being the most cannibalistic (Amaratunga, 1980) and small males falling victim to larger females (O'Dor, 1983). In captivity, this species becomes cannibalistic after only 3 days' starvation (*ibid.*). By contrast, *Sepioteuthis sepioidea* is the only known squid that commonly has all size classes (and ages) represented in its shoals, and no cannibalism has ever been observed during extensive studies (Arnold, 1965; Moynihan & Rodaniche, 1982; Hanlon & Forsythe, unpublished data).

It has also been suggested that, in some squid, cannibalism might be important for the survival of at least part of a shoal during its long-distance migration through poor feeding grounds (O'Dor & Wells, 1987; Clarke, Rodhouse & Gore, 1994; see §9.5).

4.8 Summary and Future Research Directions

Coleoid cephalopods are carnivores, many of them generalised predators with preferences for certain prey but feeding opportunistically according to circumstances. The main prey are crustaceans, teleost fish and molluscs (gastropods, bivalves and other cephalopods). Octopuses tend to feed mainly on crabs and bivalves; cuttlefish on shrimp-like crustaceans and fish; and squid on fish, other cephalopods and crustaceans. What a particular individual consumes seems to be a compromise between preference and availability. Yet abundant species are often actively ignored (Vincent, Scheel & Hough, 1998), and as more data emerge it is becoming clear that feeding, like other behaviours, is both complex and flexible in cephalopods.

Cephalopods are well equipped to detect and pursue prey under differing conditions, and they have an arsenal of weapons for capturing and subduing relatively large prey. Not the least of these are subtle behaviours, for example the stalking of strongly swimming fish, or the drilling of mollusc shells in the most appropriate place. Trial and error learning may be involved here, and more evidence about how such behaviours develop would be most welcome. It would be interesting to explore the extent to which different populations of a species differ in their feeding behaviour.

In the great majority of cephalopods, the details of how prey is immobilised and digested are not known. In squid, there are almost no data on the posterior salivary gland, in particular whether it produces toxins, and if so what kind.

Foraging strategies are poorly understood in octopuses and have not been explored systematically in decapodiformes. In many animals, searching for prey involves quite complex behaviours, for example the formation of 'search images' (L. Tinbergen, 1960; Dawkins, 1971; Pietrewicz & Kamil, 1981; Ishii & Shimada, 2010), but unfortunately nothing is known about this in cephalopods. Once it has located prey, a predator often tends to stay in the vicinity, for a short while at least, and possibly returns to the same area later (Curio, 1976), suggesting that learning is involved. Again, no data are available for cephalopods. Nor is anything known about whether there is social facilitation during foraging by shoaling squid.

More field data will undoubtedly reveal new types of interaction with prey, and it is especially to be hoped that more information will soon be forthcoming about how mesopelagic and bathypelagic cephalopods feed. More studies on feeding by shoaling squid would also be welcome, and it will be interesting to learn how widespread a phenomenon cannibalism is in the field. The role of learning (Chapter 8) in many aspects of feeding and foraging will surely repay study.

CHAPTER FIVE

Defence

All animals must defend themselves from predation, and they have evolved a great variety of morphological and behavioural devices to do so. Many authors have provided schemes for classifying defences against predators (e.g. Cott, 1940; Robinson, 1969; Edmunds, 1974; Hailman, 1977; Driver & Humphries, 1988; Endler, 1991; Tollrian & Harvell, 1999; Ruxton, Sherratt & Speed, 2004; Caro, 2005; Stevens, 2013), and we have borrowed from these classifications as appropriate to discuss cephalopod defences.

Cephalopods, like other animals, have two general defensive strategies: *primary defence* and *secondary defence* (Fig. 5.1). Primary defence functions to decrease the chances of the predator encountering prey, detecting it as an object distinct from the background, or recognising it as edible (Endler, 1991). The common primary defences of most animals, including cephalopods, are the various forms of camouflage and cryptic behaviour. Secondary defences, which are brought into play when the prey has been detected or recognised, interfere with the approach or attack sequences of the predator, or force the predator to release the prey during capture or consumption. Secondary defences are many: in cephalopods, they include either flight or some form of 'deimatic' behaviour (threat, startle, frighten or bluff), followed by highly unpredictable ('protean') behaviour. It is not always possible to categorise behaviours as being strictly primary or secondary because predators hunt in different ways (cf. Curio, 1976; Ruxton, Sherratt & Speed, 2004); thus our classification will not be appropriate for every circumstance.

Two noteworthy features of cephalopod defences are that they have keen senses and quick movements, which give them time to assess a predator's behaviour, and that they have various defence options from which to choose. In this respect they are like some of the higher vertebrates.

5.1 Primary Defence: Camouflage

Since coleoids are soft-bodied and vulnerable, they rely to a great extent on camouflage and cryptic behaviour as their primary defence. Camouflage seems simple to understand as a concept, but it is difficult to define and quantify because the mechanisms of visual concealment are varied and depend on the multiple visual capabilities of diverse predators. There are several historical classifications of camouflage, with differing emphasis (e.g. Robinson, 1969; Edmunds, 1974; Hailman, 1977; Endler, 1978, 1984), and we have generally followed that outlined by Cott (1940) in his classic work on adaptive coloration. There has been a great deal of research on animal camouflage in the past decade, and the interested reader should consult this literature beginning with two volumes dedicated to the subject (Stevens & Merilaita, 2009; 2011a). Here we blend our previous classifications of camouflage in cephalopods with these new data and perspectives to enable general comparisons of the mechanisms and functions of camouflage across animal taxa.

The general approach in current literature is to consider *camouflage* as encompassing all strategies involved in concealment, and the two principal mechanisms are to retard or prevent **detection** or **recognition** of the prey organism. In the most recent review, Stevens and Merilaita (2011b) proposed a revision of camouflage classification; four categories are acknowledged: *crypsis, masquerade, motion camouflage* and *motion dazzle*. Basically, crypsis and motion camouflage retard detection, masquerade retards

PREDATOR: CEPHALOPOD'S DEFENCE:

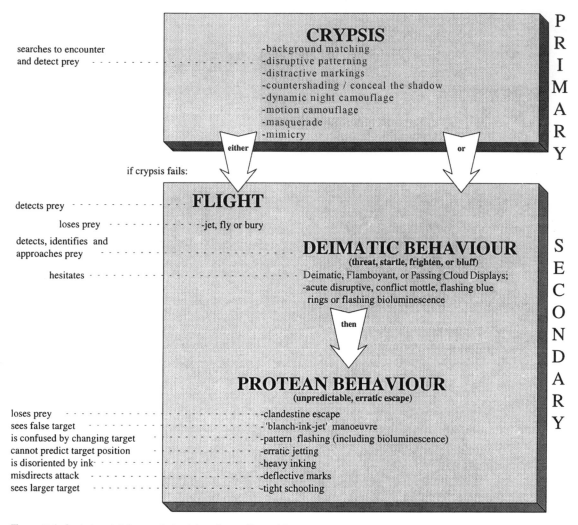

Figure 5.1 Cephalopod defences during interactions with predators.

recognition, and motion dazzle impedes estimates of speed and trajectory. 'Crypsis' as a category is considered by Stevens and Merilaita as encompassing eight subcategories (see their Table 1 in both 2009 and 2011b). There are not always clear distinctions among some of these strategies as to their mechanism of visual deception, as the examples that follow will make clear, as will the diverse literature from other taxa. Thus we have tailored our chapter overview to what is currently known about cephalopod behaviour.

This organisation and the set of definitions below are based on two factors. One is the 'function' that broadly describes what the camouflage adaptation may do, and the other is the 'mechanism' of visual perception that deceives the predator. However, little is known about how predators visually perceive prey, and this hinders studies of camouflage mechanisms. For example, according to Edmunds (1974), 'animals which are camouflaged to resemble part of the environment are said to be cryptic', but for Endler (1991), 'a colour or

Figure 5.2 Rapid neural polyphenism: dynamic transitions between primary and secondary defence. (**a**) *Sepia* spp. in a conspicuous Disruptive pattern, then changing to a Uniform pattern while partially burying to match the background (video frame grabs, Lembeh Straits, Indonesia; Gus Campbell). (**b**) *Octopus vulgaris* changing from primary defence (crypsis) to secondary defence (Deimatic Display) upon approach of the cameraman (video frame grabs, Grand Cayman Island, BWI; R. Hanlon; reprinted from *Current Biology* **17**, R400–404, © (2007), with permission from Elsevier).

pattern is cryptic if it resembles a random sample of the visual background as perceived by the predator at the time and place at which the prey is most vulnerable to predation'. This latter definition, although not fully accepted, rightly draws attention to the importance of the predator's visual capabilities, and Endler emphasises that our intuitive feel for crypsis in patterns can be wrong if our vision and viewing conditions are different from those of the predator, as they generally are. Many recent studies acknowledge the complexities of studying and understanding mechanisms and functions of camouflage (e.g. Stevens *et al.*, 2006; Troscianko *et al.*, 2009; Stevens & Merilaita, 2011b). Substantial progress has been made during field and laboratory experiments and observations with cephalopods, and these are highlighted in this chapter.

To begin, we draw attention to the unique capability that enables dynamic transitions between primary and secondary defence in cephalopods: rapid neural polyphenism (Fig. 5.2).

5.1.1 Crypsis

Crypsis makes detection difficult or impossible and includes features of physical appearance (pattern, brightness, contrast, colour, 3D skin texture) as well as behavioural traits. The following six kinds of crypsis are described briefly and examples from cephalopods are reviewed.

Background matching

In this form of crypsis, often referred to as *general background resemblance*, the cephalopod takes on the appearance of the substrate or background by ensuring that the brightness of the skin, its contrast, colour and its pattern are similar. Equally important for crypsis are the textural and postural components that ensure blending with the environment. Cephalopods, with their flexible arms and dynamic skin papillae (the latter are unique in the animal kingdom), are particularly adept at this. Figure 5.3 illlustrates the effectiveness and

Figure 5.3 Background matching. Note the general brightness, texture, pattern, shape and colour resemblance to the backgrounds. (**a**) *Sepia officinalis* in a uniform pattern. (**b**) *Sepia officinalis* in mottle pattern amidst mossy rocks (Çeşmealtı, Urla, Turkey).

refinement of background matching in cephalopods and shows how well they conceal the appendages, the eyes and the contours of the body.

It is important to realise that there are different degrees of 'matching' to the background and, in fact, there are very few studies in any animal group demonstrating quantitatively that specific background features are truly matched, or to what degree they are matched. For cephalopods, 'background matching' takes two general forms: a 'high fidelity match' and a 'generalist match' (Hanlon et al., 2009). The former is rare and the latter is common. This makes sense given that on a Caribbean coral reef, for example, there are literally a thousand animal, plant and rock backgrounds that would have to be matched. The currently popular term 'match' is unfortunate because we and others who have used spectrometers and other optical measurement devices in the natural environment have not found strict matches to brightness, contrast or colour in the surrounding background (Akkaynak et al., 2013; Hanlon et al., 2013b). The point is: to achieve general background resemblance to avoid detection, the cephalopod needs only to deceive the visual perception of the predator (Hanlon et al., 2009); it does not have to 'look just like the background' or match various items with high fidelity, but only match to a degree that fools the predator. Thus, the 'generalist match' is common in cephalopods as it is in most animals. As a general rule in other taxa, an animal's fixed appearance can match one background type (a 'specialist pattern') or several (a 'compromise pattern') (Stevens & Merilaita, 2011b).

Brightness and *contrast* of the cephalopod must resemble the adjacent background, and laboratory experiments on cuttlefish show that they can control the skin to produce similar brightness and contrast (see Chapter 3). Indeed, the description of the chromatophores as 'neutral density filters' that help regulate brightness and contrast emphasises this point

(Packard & Hochberg, 1977). Octopuses can respond to small brightness differences in the background (Messenger, Wilson & Hedge, 1973). Many of the images in this chapter bear this out.

Colour resemblance is also crucial because many of the common predators on cephalopods have keen colour perception. The images in Fig. 5.3 show that cephalopods can resemble the *colour* of the background, and this is often quite precise. Since cephalopods appear to be colour blind (Chapter 2), it is not obvious how they might do this. As Box 5.1 explains, one passive contribution to colour matching is the reflecting cells (iridophores, reflector cells and especially the leucophores) that enable the skin to take on the colour of the background. When these cells are revealed by retraction of the overlying neutral-density screen of chromatophores, they reflect many of the ambient wavelengths (Messenger, 1974, 1979a; Froesch & Messenger, 1978; Mäthger et al., 2013).

Patterns provide a major contribution to general background resemblance (Fig. 5.3). The 'Uniform' patterns of cephalopods are characterised by little or no contrast; that is, they can be uniformly light or dark or of any uniform colour. Cuttlefish generally match slightly variegated substrates such as fine or moderately coarse sand (Figs. 5.2a & 5.3a). 'Mottle' patterns are characterised by small to moderately sized light and dark splotches of moderate contrast; these are deployed on substrates with larger substrate particles of variable size and shape. Most of the shapes are roundish, and the sizes of the Mottles correspond roughly to the particles in the substrate (Figs. 5.2b and 5.3 b–h) (Chiao et al., 2010). The dark components of mottles are groups of expanded chromatophores (i.e. physiological units) all over the body; in ethograms they are listed collectively as 'Mottle'. The light or white components of these Mottle patterns are as varied and as important as the dark components. The majority of light chromatic

---

Caption for Figure 5.3 (*cont.*) (**c**) Background matching using translucency and countershading on a Caribbean coral reef; 12 well-camouflaged *Sepioteuthis sepioidea* are in the image. (**d**) *Sepioteuthis sepioidea* in a mottle pattern near the substrate adjacent to a soft coral. (**e**) *Octopus vulgaris* in a mottle pattern at Saba, Netherlands Antilles (directly in middle of image). (**f**) *Octopus burryi* in a mottle pattern while attached to a swaying calcarious alga in Saba. (**g**) *Octopus vulgaris* in a mottle pattern in Puerto Rico. (**h**) *Octopus cyanea* in a mottle pattern on a coral head at Manihi Atoll, Polynesia (octopus on right side). All images are field shots by R. Hanlon except (**a**), which is a lab shot by Kendra Buresch.

Box 5.1 How might cephalopods match the background colour?

a
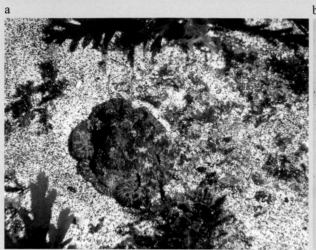

b
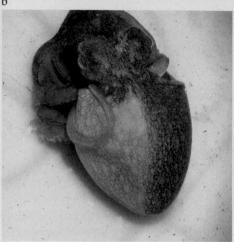

If cephalopods are colour blind (Chapter 2), how could they change colour to match the background? Although the mechanism is unknown, part of the answer may be that they do not change colour solely with the chromatophores but rather through an interactive process that occurs when they match the *brightness* (or luminance) of their skin. In the process of adjusting the brightness, they reveal more or fewer of the underlying reflecting cells, and these assist the animal in resembling the background in colour.

Octopuses cannot use the chromatophores to match all the colours in their world because their pigments are restricted; they are black, brown, red, orange or yellow, never green, cyan, blue or violet. Such chromatophores could never match the blues of a limestone grotto or the greens of algae in a shallow pool. Yet, in life, octopuses can take on the colour of their surroundings. Note in (a) how *Octopus bimaculatus* matches the nearby green algae in its natural habitat, and in (b) how part of the body of an *Eledone cirrhosa* matches the blue light filtering through a plastic bowl (image a by R.T. Hanlon; image b by J.B. Messenger). The reason for this effect in (b) is that the chromatophores on the right-hand side of the mantle area are retracted as a result of being denervated.

The point of this experiment is to demonstrate that skin *without* active chromatophores can in some circumstances achieve a better colour match than skin with them.

The reason that the denervated side appears bluish in these artificial surroundings is that the operation has maximally exposed the reflecting cells in the skin beneath the chromatophore layer, notably the leucophores, which, being broadband reflectors, faithfully reflect the predominant environmental wavelengths. In the sea, of course, an octopus would rarely find itself exposed to such a uniform environment, but when there is bright light, the ensuing retraction of chromatophores reveals reflecting cells, especially at the tips of the papillae, so that the predominant wavelengths in the environment will be reflected from the skin. This is what is occurring with the octopus in (a).

Detecting the brightness of the surrounds will pose no problem for the colour-blind eye of an octopus, and there is ample evidence that octopuses can match the background in brightness. The CNS presumably estimates the albedo and regulates the expansion of the chromatophores via the optic and chromatophore lobes (Fig. 3.9).

components of sepioids and octopods are white spots or white splotches, which are distributed all over the dorsal body. Many of these are enhanced by underlying leucophores, and usually all of the spots and splotches are expressed in a Mottle pattern.

The third broad category of cryptic patterns is 'Disruptive'. These comprise large light and dark components of different shapes, orientations and scale, and some of them are typically of high contrast (Fig. 5.5). In some cases these can retard detection, and thus they are listed here under background matching. In other cases, they are thought to impede recognition by a predator, and thus this is explained in the next section.

Skin papillae add dynamic 3D physical and optical texture to the skin of octopuses and cuttlefish. This is a capability unique in the animal kingdom. The papillae are controlled by visual input in cuttlefish, and there are no fewer than nine sets of independently controlled papillae in the skin of *Sepia officinalis* (Allen *et al.*, 2009). The papillae can be completely flat (and thus invisible) or can expand to any degree from flat to fully extended; this capability is due to their construction as muscular hydrostats (Allen *et al.*, 2013, 2014). Cuttlefish and octopus appear to resemble the rugosity of the adjacent backgrounds (e.g. spiky soft corals, algae, hydrozoans) but the degree to which they accomplish this has not been measured quantitatively in the field nor experimentally manipulated in laboratory experiments. Expansion of papillae when an octopus is on a background with spiky material is important in disguising its outline; that is, if the octopus appeared with rounded edges or body outline (as when the papillae are not expressed) then it would stand out compared with the 3D edges of its background (Fig. 5.2b).

The *very flexible body shapes* of octopuses also aid crypsis. For example, by placing their arms in different configurations, they can produce a disruptive effect that helps to hinder both detection and recognition. See for example Figs. 5.3g, 5.5h. The arm shapes and postures of cuttlefish and squid produce similar benefits (Fig. 5.13; Barbosa *et al.*, 2012).

Some cephalopods use extraneous material to improve crypsis. Octopuses sometimes hold pebbles, sponge-covered shells or even large objects with the suckers (Finn, Tregenza & Norman, 2009a), and the

Figure 5.4 Camouflaged eggs of the Japanese cuttlefish *Sepia esculenta*, with sand grains adhering to provide crypsis (R. Hanlon).

small sepiolid *Euprymna scolopes* has a specialised epidermis to which sand adheres, and this is used for crypsis during the day, especially when hunting prey (Shears, 1988). This sand coat can be dropped quickly as a unit (Singley, 1982), presumably for other behavioural manoeuvres such as escape. Some species of *Sepia* and *Euprymna* lay their eggs individually, and the 'sticky' exterior accumulates local sediment or debris to help conceal them (Figs. 5.4; 6.12b).

Disruptive patterning

Disruptive patterning and disruptive coloration are different. With cephalopods thus far, we have no experimental proof that disruptive coloration is a visual mechanism that impedes predator vision (see some details in Hanlon *et al.*, 2009). Stevens and Merilaita (2011b) define disruptive coloration as 'a set of marking that creates the appearance of false edges and boundaries and hinders the detection or recognition of an object's, or part of an object's, true outline and shape'. Disruptive coloration disrupts the natural outline of the animal; formerly, distractive markings were considered part of disruptive coloration, but they have since been recognised as operating by a different mechanism (see next section). Cott (1940) has described in detail the many different ways by which animals achieve disruptive coloration: cephalopods appear to possess all of the basic markings and behaviours to achieve disruptive coloration. We summarise these in Box 5.2, while acknowledging that experimental proof remains to be demonstrated and should be a future goal.

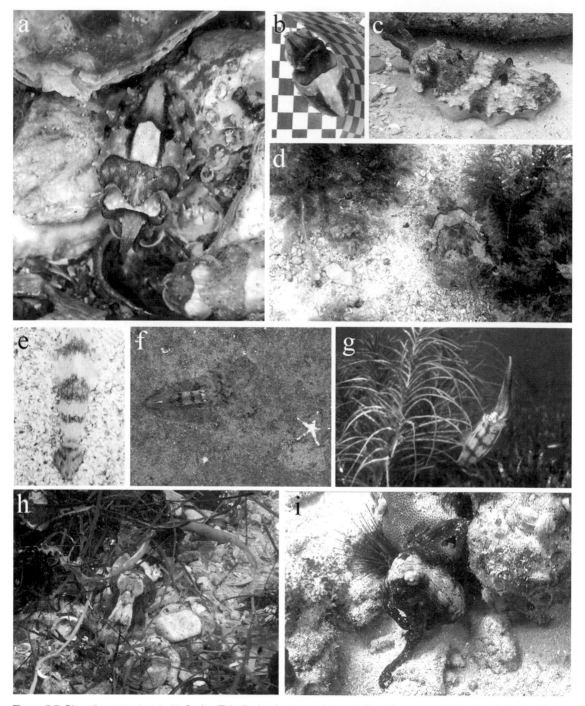

Figure 5.5 Disruptive patterning. (a, b) *Sepia officinalis* showing two variations of Disruptive, one amidst oyster shells, the other on high contrast chequerboards. Note the second cuttlefish in (a) at bottom right. (c, d) The giant Australian cuttlefish *Sepia apama* in the field showing variations of Disruptive. The Transverse mantle bar is effective in breaking up the longitudinal outline of the animal, especially as viewed from above. (e) The squid *Loligo pealeii* in a strong Disruptive pattern that presumably disguises its overall shape. (f) *L. pealeii* deep on the continental slope east of Massachusetts USA (NOAA photo). (g) Reef squid *Sepioteuthis sepioidea*

Thus in cephalopods we can only discuss disruptive patterning (Fig. 5.5), which we will define loosely here as pattern designs that are characterised by many of the features in Box 5.2, particularly transverse lines and bars that extend to the outline of the animals (see additional details in Stevens & Merilaita, 2009b), and various shapes and sizes of large-scale, high-contrast lines, bars, bands, stripes, spots, etc. that occur on the body as a result of chromatophore, iridophore and leucophore expression. Note, for example, in *Sepia*

Figure 5.6 Distractive markings in *Sepia officinalis*. (**a**) A cuttlefish showing a weak White head bar and a bright White square that could perhaps be considered as a random sample of the white rocks in the background (see text). (**b**) Underwater image (Vigo, Spain) of a feeding cuttlefish with very bright White mantle bar (compare pure white standard disc in foreground). (**c**) Adult and (**d, e**) juveniles showing dark and light chromatic components expressed in various degrees of contrast. Photo credits: **a** by Lydia Mäthger; **b–d** by R. Hanlon; (**e**) by Alex Barbosa.

Caption for Figure 5.5 (*cont.*) showing a Disruptive pattern on its ventral mantle as it hovers next to a soft coral. (h) *Octopus rubescens* at 3 m depth at Pacific Grove, California with a large white area between the eyes and first pair of arms, disrupting the body form. (i) *Octopus vulgaris* showing a Disruptive pattern similar to that of *O. rubescens* while darkening the rest of the mantle and arms on a coral reef in Puerto Rico. All photos by R. Hanlon except **b** by Kendra Buresch and **f** by National Oceanic and Atmospheric Administration (NOAA).

Box 5.2 How cephalopods utilise disruptive patterning

Cuttlefish have perhaps the greatest repertoire of disruptive patterns. They spend most of their life on, or partially buried in, various substrates, and because of the rigid cuttlebone they cannot distort the mantle very much; thus, the chromatic components in the patterns must create the optical illusion of disruption. *Sepia officinalis* exhibit a variety of disruptive patterns involving a large number of chromatic components. This species has 11 chromatic components (6 dark, 5 light) that can be expressed in different combinations to create disruptiveness of appearance (see Figs. 3.6, 3.7, 3.8 and 5.5). An extensive range of disruptive components shown by other sepioids has also been recorded by Corner and Moore (1980), Roper and Hochberg (1988), Mauris (1989), Lee, Yan and Chiao (2010), and others.

Elongate cephalopods, such as squid and cuttlefish, tend to use *transverse* components (bars, bands or lines) to apparently disrupt the outline. Most *Loligo* spp. can produce three to five Bands around their bodies, one of which usually passes across the head and eyes. Stevenson (1934) and Macy (1982) noted that, when fish predators approached under natural conditions, *Loligo pealeii* dropped to the substrate and showed the Band pattern. Moreover, images taken offshore on the continental shelf also show this species sitting on the substrate with Band patterns, thus achieving good camouflage under natural conditions (Fig. 5.5f). This behaviour was tested with live predatory bluefish in laboratory experiments and squid that dropped to the substrate and expressed the Band pattern: the fish swam repeatedly over them without detection (Fig. 5.28; Staudinger, Hanlon & Juanes, 2011). As seen in Fig. 5.5e, this same disruptive effect is highly effective to a human observer even on a uniform substrate in laboratory tanks. *Sepioteuthis sepioidea* is remarkable among teuthoids because it spends a great deal of time near the substrate and amidst various corals; thus, it has the greatest repertoire of disruptive components for a teuthoid. One of its novel patterns includes Mantle bars across the ventral mantle when it orients upwards and exposes the ventral body surface to the predator, often conforming with some variegated structure in the visual background (Fig. 5.5g).

In benthic octopuses, the body is highly malleable and can be contorted, flattened, or even partially buried to help break up the outline of the animal. These postures are complemented by several chromatic components that visually break up body parts such as the arms, which cannot always be concealed by other means. Octopuses tend to employ a mixture of general background resemblance and disruptive patterning to achieve crypsis: components such as Mottle, Arm bars, Mantle bars, Head bars and Frontal white spots are all important disruptive devices (Fig. 5.5h,i).

Three groups of unusual octopuses also show a form of disruptive patterning. The so-called 'harlequin' octopuses (*Octopus chierchiae, O. spilotus, O. zonatus* and *O. mototi*) have bold, semi-permanent black-and-white bands or spots all over the body and may be cryptic on appropriate backgrounds (Moynihan & Rodaniche, 1982; Moynihan, 1985a). Those bold basic patterns generally persist much of the time, although they can be obliterated so that the animal becomes uniformly dark or pale. Several of the sand-dwelling, long-armed mimic octopuses commonly show high-contrast bands, streaks and splotches as a primary defence (Norman, Finn & Tregenza, 2001; Hanlon, Conroy & Forsythe, 2008; Huffard *et al.*, 2008). Thirdly, the 'blue-ringed' octopuses of the genus *Hapalochlaena* have many dark bands and patches that are interposed regularly between patches of yellow or white all over the body, and it seems possible that this could be disruptive on the appropriate background. It is noteworthy that the iridescent blues of these octopuses are not usually expressed when the dark/light pattern is being used chronically for crypsis (Mäthger *et al.*, 2012). When they are expressed, however, they become an acute pattern that may be aposematic.

officinalis (Fig. 5.5a, b), how the cuttlefish's body is broken up optically into a highly heterogeneous pattern design with different shapes and contrasts, all in different orientations and sizes. Some of these markings do not extend to the outline or margin of the body, and such markings have been termed 'surface disruption' because they create false edges that are sometimes more visually salient than the true body form (Stevens *et al.*, 2009). The giant Australian cuttlefish occasionally produces a wide transverse bar across the midsection of its mantle, which gives the illusion that the front and back of the animal are different objects (Fig. 5.5c, d), and this sort of disruption of body orientation is also used by tube-shaped loliginid squid, which deploy dark bars of different broadness transversely across the entire body from arms to head to mantle (Figs. 5.5e, f, g).

Octopuses deserve special consideration with regard to disruptive patterning because they have a flexible body that can contort its shape dramatically and thus (hypothetically) disrupt recognition by predators. Such body shape control may thus reduce the need for optical disruptive patterns. Disruptive body patterns in octopuses are relatively rare compared with cuttlefish and squid; two examples are illustrated in Fig. 5.5h, i). In both examples, the octopuses have produced large areas of whiteness along the centre of the mantle–head–upper-arm axis, while other parts of the body are darker and patterned. In many cases, the arms of camouflaged octopuses are spread out with small transverse bars expressed along their lengths, thus hindering their linear unit shape. Those arms are also often twisted and contorted to further enhance visual concealment.

Distractive markings

Distractive markings, once thought to be part of disruptive coloration, direct the attention or gaze of the receiver from traits that would give away the animal (such as its outline) (Thayer, 1909; Cott, 1940; Dimitrova et al., 2009; Stevens & Merilaita, 2011b). A key defining feature of distractive markings is that they comprise colours not found in the background or have contrasts in excess of features of the background; that is, they appear conspicuous to a human observer. Some researchers report that these markings are not a form of background matching (Stevens et al., 2008). Results of experiments with bird predators and computer-based images with human observers are equivocal: one study finds that they aid camouflage effectiveness (Dimitrova et al., 2009) and another concludes that they reduce it (Stevens et al., 2013). These experimental contradictions have led Merilaita, Schaefer and Dimitrova (2013) to suggest a more detailed definition: 'distractive markings are markings that through (and despite) their relative salience compared to the rest of the coloration or morphology of an animal make it more difficult for a viewer to perceive the characteristics useful for detection or recognition of the animal, hence increasing

its net camouflage'. These arguments are not trivial: they will help to sort out the visual mechanisms of camouflage – detection versus recognition – and should be an aim in cephalopod research as well as throughout all taxa.

Cuttlefish in particular have distractive markings such as the White head bar, White triangle and especially White square, which has been measured in the field to be sometimes higher contrast than other nearby objects (Hanlon et al., 2009; their Fig. 4). Several other examples are illustrated in Fig. 5.6. We draw attention to Fig. 5.6b because in this circumstance the cuttlefish was feeding repeatedly on small schooling fish that would eventually descend to the substrate right in front of the stationary cuttlefish, which then rapidly preyed upon them with fast strikes of the tentacles. This is the same circumstance in which bright white conspicuousness on part of a disruptive pattern occurred during feeding on small fish as observed by Hanlon et al. (2009, 2011). Apparently the unusually bright white can act as a distraction (or an attraction?) that enhances feeding.

Countershading, Concealment of the Shadow, and Transparency

The principles of countershading, which are well established (Thayer, 1909; Cott, 1940; Denton, 1970; Rowland, 2009), essentially require elimination of any silhouette or shadow created by downwelling light. One of the chief visual effects of countershading is that rounded surfaces appear flat. Thus, a slightly darker dorsal surface grading into a lighter undersurface will render a squid in the water column less visible when viewed laterally, even from high or low oblique angles (Fig. 5.3c). During countershading in cephalopods, the dorsal chromatophores are expanded, the ventral chromatophores are retracted, while those along the side of the mantle and arms are beautifully graded between the two extremes. There are fewer chromatophores ventrally and numerous iridophores to enhance reflection. Only a few species are known to have specific iridophore arrangement for this: *Sepia officinalis* (Hanlon & Messenger, 1988), *Lolliguncula brevis* (Hanlon et al., 1990), *Loligo vulgaris* and *Alloteuthis subulata* (Mäthger & Denton, 2001; Mäthger, 2003). Yet many cephalopods are pale ventrally (e.g. *Octopus vulgaris* and *Eledone* spp.).

Countershading is so important for cephalopods that they have a 'countershading reflex' (CSR) to help to maintain it even when the animal moves out of normal orientation (Fig. 5.7). Experiments have shown that *Sepia officinalis*, *Loligo vulgaris* and *Octopus vulgaris* all have a roll CSR, whereby the upper half of the ventral mantle turns dark when the animal is rotated more than 90° in the roll plane (Holmes, 1940; Ferguson & Messenger, 1991). *Sepia officinalis* also has a pitch CSR, such that the upper part of the ventral mantle darkens as the animal rotates in the pitch plane, either head down or head up (Ferguson, Messenger & Budelmann, 1994). The reflex is driven mostly by the macula system of the statocyst (§2.1.2) using information about the direction of gravity, but light has some influence on the roll CSR (*op. cit.*). Mäthger (2003) demonstrated a 'flexible countershading' chromatophore pattern in the squid *Alloteuthis subulata* that darkened the side of the body facing the brightest light. This was not considered to be a reflex because, unlike the CSR, expression of the pattern appeared to be under the control of the squid. This squid countershading could also function as background matching to conceal it from above when viewed against the darker deeper water or substrate, and from below against the downwelling light (cf. Fig. 5.3c).

Counter-illumination is a physiologically active mechanism by which some pelagic cephalopods (particularly mid-water squid that migrate vertically) adjust the amount of light they produce with their ventral photophores to match the downwelling light, so that predators below cannot detect them (see §9.7.1). Some species also have compressed or tubular eyes to minimise the silhouette of the eye as seen from below (Young, 1975a; Land, 1992). Ventral counter-illumination in a shallow-water nocturnal species, the Hawaiian bobtail squid *Euprymna scolopes*, has been studied by Jones and Nishiguchi (2004), who showed that these small sepioids modified the intensity of light produced in their relatively large light organ by the symbiotic bacterium *Vibrio fischeri* as downwelling light intensity was changed under experimental laboratory conditions (Fig. 5.8a–c).

Benthic cephalopods in shallow, well-lit environments eliminate their shadows against the substrate by partially burying themselves in the substrate (various sepiolids, cuttlefish, octopuses) or by flattening their mantles or arms against or into the substrate. The cuttlefish *Sepia officinalis*, when sitting

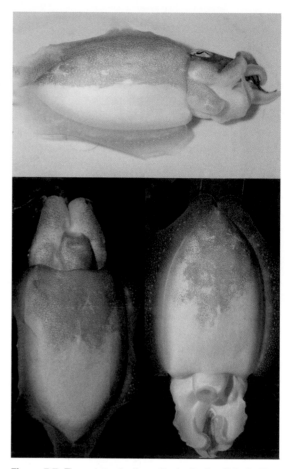

Figure 5.7 The countershading reflex in *Sepia officinalis*. The uppermost side always bears expanded chromatophores (after Ferguson, Messenger & Budelmann, 1994; republished with permission of the Company of Biologists from *Journal of Experimental Biology* **191**, 247–256, Gravity and light influence the countershading reflexes of the cuttlefish *Sepia officinalis*, Ferguson, G. P., Messenger, J. B. & Budelmann, B. U., © (1994); permission conveyed through Copyright Clearance Center, Inc.).

on the substrate, eliminates shadows by using the fin to form a 'flange' joining the mantle with the substrate; they do the same with the natural flange on the ventro-lateral part of the fourth arm (Hanlon & Messenger, 1988). Several sepioids also ripple the fin to flip sand or gravel over it, in the manner of flatfish.

Transparency is a common tactic of aquatic crypsis that reduces the likelihood that an animal will be detected by a predator (Johnsen, 2000; Stevens &

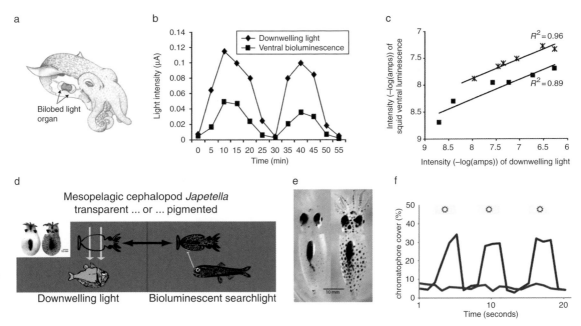

Figure 5.8 Counter-illumination via the ventral light organ in the Hawaiian bobtail squid. (**a**) Schematic diagram of the bilobed light organ in the ventral mantle. (**b**) Changes in the intensity of light produced by the squid during a 1-hour trial in which downwelling light was slowly increased and decreased. (**c**) Two examples of the correlation between overhead light intensity and ventral bioluminescence. **a–c** modified from Jones and Nishiguchi (2004): *Marine Biology* **144**, 1151–1155 Counterillumination in the Hawaiian bobtail squid, *Euprymna scolopes* Berry (Mollusca : Cephalopoda), Jones, B. W. & Nishiguchi, M. K., © (2004), with permission of Springer. (**d**) Dynamic switching between transparency and pigmentation for crypsis; transparency hypothetically minimises detection by hatchetfish in mid-waters, while pigmentation interferes with detection by fish with bioluminescent flashlights. (**e**, **f**) The squid *Onychoteuthis banksii* expands pigmented chromatophores in response to blue light (blue line) but not to red light (red line). Yellow boxes with symbol show onset and offset of light. Modified from Zylinski and Johnsen (2011): *Current Biology* **21**, 1937–1941, Mesopelagic cephalopods switch between transparency and pigmentation to optimize camouflage in the deep, Zylinski, S. & Johnsen, S., © (2011), with permission from Elsevier.

Merilaita, 2011a). Total transparency in cephalopods is not possible because of internal organs and eyes, yet many small or mid-water cephalopods exhibit a high degree of transparency or translucency in their tissues (Fig. 5.3c). Changing light conditions greatly affect the efficacy of transparency crypsis (Johnsen, 2003) and this led Zylinski and Johnsen (2011) to investigate whether some cephalopods actively switch between transparency and pigmentation to achieve crypsis. As illustrated in Fig. 5.8d–f, mesopelagic cephalopods such as *Japetella heathi* and *Onychoteuthis banksii* appear capable of a dynamic strategy of going transparent when downwelling light is present (to avoid detection from predators looking upwards to detect silhouettes) and switching to red and black chromatophore pigmentation when no downwelling light is present (to avoid detection by predators using bioluminescent searchlights).

Dynamic Night Camouflage

Night camouflage is largely an unstudied subject in biology. Yet a great deal of predation occurs at night, and many animals have keen night vision (Warrant, 2004; Land & Nilsson, 2012), including cephalopods (§2.1.1). Thus the question arises: do cephalopods, with their adaptive camouflage, continue to use camouflage at night, and can they change their body patterns under very low light levels comparable to those in nature (e.g. starlight, ca. 0.003 lux)? Two studies on cuttlefish confirm both possibilities.

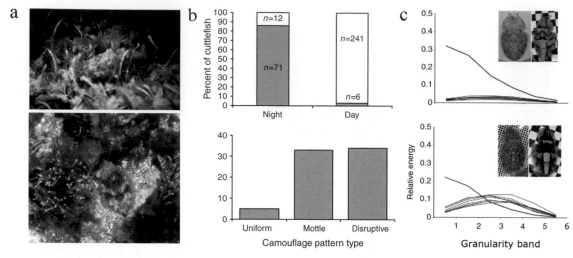

Figure 5.9 Dynamic night camouflage. (**a**) *Sepia apama* in a mottle pattern amidst seagrass (top) and a disruptive pattern amidst algae and rocks (bottom). (**b**) Percentage of camouflaged (shaded area) versus non-camouflaged cuttlefish during day and night, and counts of which patterns were shown at night (from Hanlon *et al.*, 2007: *American Naturalist* **169**, 543–551, Adjustable night camouflage by cuttlefish, Hanlon, R. T. *et al.*, © 2007 the University of Chicago). (**c**) Active pattern change in the lab under starlight levels (0.003 lux): from Uniform to Disruptive (top) and from Mottle to Disruptive (bottom). The red line is the average of ten *Sepia officinalis* after the substrate was switched (modified from Allen *et al.*, 2010b: adapted with permission of Company of Biologists Ltd from *Journal of Experimental Biology* **213**, 3953–3960, Night vision by cuttlefish enables changeable camouflage, Allen, J. J., Mäthger, L. M., Buresch, K. C. *et al.*, © (2010); permission conveyed through Copyright Clearance Center, Inc.).

The giant Australian cuttlefish, *Sepia apama*, aggregates yearly in massive spawning aggregations, and during the day most of the animals are displaying sexual signals (§6.2.1). To investigate whether this sexual signalling continued during the night, Hanlon *et al.* (2007) used a video-equipped ROV (with red filters on lights; Williams *et al.*, 2009) to conduct 16 visual surveys (8 at day, 8 at night) in 2–4 m of water on the spawning grounds. At night, 71 cuttlefish were found throughout the area in either Uniform, Mottle or Disruptive camouflage patterns, and each pattern was tailored to the specific microhabitat in which the cuttlefish had settled for the night (Fig. 5.9a, b). The implication is that cuttlefish night vision is very good, and that nocturnal visual predators actively apply the selective pressure for round-the-clock camouflage in this temperate rock reef habitat. Otherwise the cuttlefish would be found in no camouflaged patterns or a single camouflaged pattern regardless of background features.

To address the second question experimentally, Allen *et al.* (2010b) tested whether cuttlefish (*Sepia officinalis*) that have been exposed to a particular artificial substrate could change their camouflage body pattern when the substrate was changed during darkness (down to starlight, 0.003 lux). Indeed they did (Fig. 5.9c), thus demonstrating not only habitat-tailored camouflage at night, but dynamic night camouflage patterning as well.

Motion Camouflage

Motion camouflage is moving in a fashion that decreases the probability of movement detection (Stevens & Merilaita, 2011a). Camouflaged cephalopods have to move as they forage, and several octopus species have combined slow stealth movement with camouflaged body patterns. *Octopus cyanea* on coral reefs near Tahiti and Palau often performed quite a remarkable behaviour called the 'moving rock' wherein the octopus took on the body pattern and shape of rocks and small coral heads that were typical in the habitat (known as masquerade; §5.1.2), then in a stealthy manner used the tips of its arms to move slowly across open substrates (Fig. 5.10; Hanlon, Forsythe & Joneschild, 1999). Unless one was watching very carefully, the octopus's slow

Figure 5.10 'Moving rock' behaviour used by *Octopus cyanea* for motion camouflage on a Polynesian coral reef. The octopus moved from a coral head (a) across an open area (b–d) at a speed similar to that of the artificial motion caused by sunlight dapple from waves. Time lapse indicated in seconds; arrows indicate the octopus. Video frame grabs courtesy of Dana Berquim.

movement was indistinguishable from the stationary rocks in the background because (1) the octopus strongly resembled other rocks and coral heads, and (2) attention was drawn towards the illusion of movement caused by light from rippling waves and by the movement of swaying algae and small fish, rather than to the slowly moving octopus. Motion, per se, was thus camouflaged.

The cognitive aspects of this behaviour are particularly noteworthy because the speed of the octopus was generally similar to the speed of rippling light in that environment, suggesting that the octopus consciously regulated its stealth speed in accordance with ambient motion such as dappled sunlight from waves. This concept requires experimentation and data, yet the moving rock trick appears to be conserved among benthic octopuses. To wit, it has been observed and photographed in the field for *O. vulgaris* and *O.*

burryi in the Caribbean, and *O. tetricus* in Australia (Fig. 5.11). Furthermore, *O. vulgaris* in the Caribbean sometimes modifies its body posture and skin papillae to conduct a 'moving algae' walking behaviour when the surrounding objects on a sand plain are spiky forms of algae and soft corals (Fig. 5.12). This latter behaviour has also been observed – in modified form – in *Macrotritopus defilippi* in Florida, *O. rubescens* in California (while swimming), and *Abdopus aculeatus* in Australia (Fig. 5.11). The biomechanics of using the fourth pair of arms for walking while elevating the other arm pairs for crypsis have been described by Huffard, Boneka and Full (2005), and Huffard (2006). A complementary tactic used by *O. cyanea* to overcome the motion handicap while foraging is to change their appearance constantly, up to 177 times per hour, presumably to impede search image formation by predators (Hanlon, Forsythe & Joneschild, 1999).

Figure 5.11 Motion camouflage by octopuses in different habitats. (**a**) *Octopus vulgaris* in the 'moving rock' behaviour on a coral reef in Puerto Rico. (**b**) *O. burryi* showing the same kind of behaviour on a sand plain at Saba Island, Lesser Antilles. (**c**) *O. tetricus* moving across a rock reef in New South Wales, Australia. (**d**) *Macrotritopus defillippi* performing the 'moving algae' behaviour on a sand plain in south Florida. (**e**) *O. rubescens* doing a swimming 'moving algae' manoeuvre near a kelp bed at Pacific Grove, California. (**f**) *Abdopus aculeatus* walking in the 'moving algae' behaviour at Lizard Island, Australia. Photo **d** by Chelsea Bennice; **f** from Huffard (2006), republished with permission of the Company of Biologists, from *Journal of Experimental Biology* **209**, 3697–3707, Locomotion by *Abdopus aculeatus* (Cephalopoda: Octopodidae): walking the line between primary and secondary defenses, Huffard, C. L., © (2006), permission conveyed through Copyright Clearance Center, Inc.; remaining images by R. Hanlon.

Cuttlefish motion camouflage has not been studied in the field. In two laboratory studies, *Sepia officinalis* on chequerboard backgrounds showed low-contrast body patterns while moving, which rules out high-contrast motion dazzle (Zylinski, Osorio & Shohet, 2009), and in *Sepia pharaonis*, Shohet *et al.* (2006) speculated that when they settle orthogonally to stripes (which might be proxies for sand ripples in nature) this may minimise involuntary motion displacement from wave surge.

5.1.2 Masquerade

Masquerade is distinguished from mimicry as follows: masquerade is resembling objects that are not animals, while mimicry is resembling another animal (sometimes toxic). Masquerade prevents recognition of a prey organism by resembling an uninteresting or inedible object such as a leaf, stick, rock or alga (Endler, 1981; Stevens & Merilaita, 2009b). Previously it has been referred to as deceptive resemblance (e.g. Cott, 1940;

Hanlon & Messenger, 1996) and concealing imitation (Hailman, 1977). Masquerade is considered to be different from crypsis because it acts against recognition rather than detection and is therefore a different form of concealment (*op. cit.*; Skelhorn *et al.*, 2010). Yet there are cases in which prey animals seem also to resemble the background so that the distinction is difficult to demonstrate experimentally. Masquerade is also different from Batesian mimicry, in which the species that they mimic may experience demographic or evolutionary change as a result of the presence of the mimic (Skelhorn, Rowland & Ruxton, 2010). These same authors presented an additional definition of masquerade: 'the appearance of a masquerading species may cause its predators or prey to misclassify it as a specific object found in the environment, causing the observer to change its behaviour in a way that enhances the survival of the masquerader'.

Masquerade is widespread among shallow-water octopuses, cuttlefish and squid (Fig. 5.13). To a human

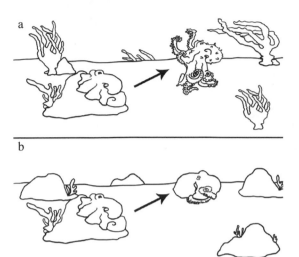

Figure 5.12 Two motion-camouflage tactics by *Octopus vulgaris*. (**a**) The 'moving algae' trick that is deployed when masquerading as nearby objects that are spiky, (**b**) The same octopus has changed into a 'moving rock' with a rounded body shape like nearby rocks and coral heads. Drawings based upon video by R.T. Hanlon.

observer, cephalopods can sometimes look like stones, algae, seagrasses and other objects such as soft and hard corals (which are animals, but are plant-like in appearance). Various specific examples and explanations of cephalopod masquerade are provided in Hanlon *et al.* (2011). In Fig. 5.13, note the key roles played by posture, colour, 3D skin texture and pattern, and how they mask the recognition of the cephalopod. For example, the tiny sepiolids *Idiosepius pygmaeus* attach themselves to the underside of a seagrass blade with a dorsal adhesive organ (Natsukari, 1970; Norman, 2000) to mask their body form (Fig. 5.13c) and they sometimes drift in a head-down posture and resemble floating algae, seagrasses and flotsam (Moynihan, 1983b). Octopuses can resemble encrusting sponges, bryozoans and other common benthic organisms when camouflaged (Fig. 5.13e, f), and it is difficult to distinguish whether this is general background matching or distinctive masquerade of those particular organisms (this is a visual perception challenge; see Chapter 3). Josef *et al.* (2012) also noted how octopuses often used distant objects as the basis of body pattern choices in a manner similar to that of the 'moving rock' of *Octopus cyanea*, which itself is a form of masquerade (see previous section and

Fig. 5.11 on motion camouflage). Squid make use of arm and body postures to produce masquerade amidst vertical objects (Fig. 5.13g, h).

In laboratory experiments on cuttlefish, Buresch *et al.* (2011) showed that *S. officinalis* prefer masquerade of 3D objects over substrate background matching when visual cues on nearby 3D objects are high-contrast and the surrounding substrate is low-contrast (see Fig. 3.16). However, field observations and images suggest that other factors (e.g. number, size, texture and shape of nearby objects) are involved in this visual decision-making process.

5.1.3 Mimicry

Mimicry is commonly understood to imply the resemblance of one animal (the mimic) to another animal (the model) such that a third animal (e.g. a predator) is deceived by their physical similarity into confusing the two (cf. Wickler, 1968; Edmunds, 1974; Curio, 1976). There is a good deal of conjecture about cephalopod defensive mimicry but little regimented study of it in nature.

There are three cases of defensive mimicry authenticated with field data. These three cases involve octopuses that live in open sand plains and mimic the swimming appearance and behaviour of flatfish that are common in such habitats (Norman, Finn & Tregenza, 2001; Hanlon, Conroy & Forsythe, 2008; Hanlon, Watson & Barbosa, 2010). The three species are: the so-called 'mimic octopus' *Thaumoctopus mimicus* of tropical Indonesia; an undescribed species, possibly that reported by Norman (2000) as 'white V octopus, *Octopus* sp. 18' of tropical Indonesia; and the Caribbean longarm octopus, *Macrotritopus defilippi*. In each case, the octopuses use mimicry of flatfish only when they are moving (Fig. 5.14b, d); thus, we consider this as primary defence that may aid misidentification by predators when octopuses are actively swimming some distance and are readily detectable. Before and after flounder mimicry, these octopuses are camouflaged and either motionless or crawling slowly (Fig. 5.14a).

There have been suggestions that *T. mimicus* mimics other organisms. Norman, Finn and Tregenza (2001) stated confidently that, in addition to a toxic sole, this octopus was mimicking lionfish and banded sea snakes. No measurable data (morphometrics, locomotion,

Figure 5.13 Masquerade (= deceptive resemblance). (**a**) *Sepia apama* in Australia resembling algae. (**b**) *Sepia latimanus* in Bali resembling algae. (**c**) *Ideosepius pygmaeus* attached to algae and matching its coloration and shadowing. (**d**) *Sepia apama* slowly crossing a sand plain and masquerading as distant dark objects. (**e**) *Octopus vulgaris* resembling adjacent algae in coloration

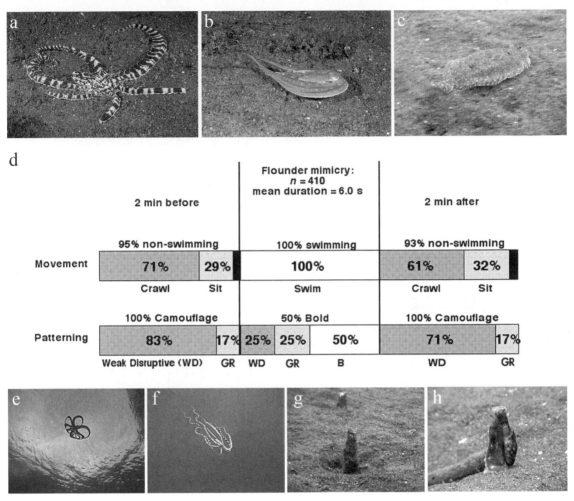

Figure 5.14 Mimic octopus. (**a**) The 'mimic octopus' *Thaumoctopus mimicus* in a common posture and conspicuous Disruptive body pattern. (**b**) *T. mimicus* mimicking a swimming flounder. (**c**) The locally abundant flounder *Bothus mancus* swimming in its typical uniform coloration. (**d**) Percentage of time spent by *T. mimicus* in different movements and coloration (body patterning) before and after flounder-like swimming. GR, general resemblance; B, bold high-contrast Disruptive patterning. (**e**) *T. mimicus* actively swimming to the surface from 10 m, then (**f**) gliding slowly in a coordinated swimming manoeuvre with arms highly contorted. (**g**) Tubes of retracted polychaete 'tube worms', and (**h**) an example of a very common posture of stationary mimic octopuses, which often confuses divers. (Modified from Hanlon, Conroy & Forsythe, 2008: *Biological Journal of the Linnean Society* **93**, 23–38, Mimicry and foraging behaviour of two tropical sand-flat octopus species off North Sulawesi, Indonesia. Hanlon, R. T., Conroy, L. A. & Forsythe, J. W., © (2008), with permission from Wiley. **a** and **b** © Fred Bavendam.)

Caption for Figure 5.13 (*cont.*) and skin texture in the Cayman Islands. (**f**) *Octopus rubescens* resembling red and lavender-coloured encrusting algae in California. (**g**) *Sepioteuthis sepioidea* using upward pointing posture and striped patterning to imitate soft coral in the Cayman Islands. (**h**) *S. sepioidea* splaying its arms and masquerading amidst soft corals. (**i**) *Sepia apama* sitting beneath branching algae and raising its arms to imitate it. (Photo credits: **b**, **c** © Fred Bavendam; **f** © Douglas Mason; others R. Hanlon.)

behaviour) were provided, unfortunately. In addition, they suggested sea anemone and jellyfish mimicry. Hanlon, Conroy and Forsythe (2008) suggested that, when stationary, *T. mimicus* deployed body patterns and postures that resembled the few sessile organisms that lived in this open sand habitat: small sponges, tubes of polychaete worms, and colonial tunicates (single photograph comparisons only). Certainly, all of this is a testament to the extreme phenotypic plasticity of soft-bodied octopuses.

There are 'good mimics' and 'poor mimics' among animals, as determined by a very precise or a less precise resemblance (Edmunds, 2000), and the three octopus flatfish mimics span this range (Hanlon, Watson & Barbosa, 2010). *M. defilippi* is a good – if not excellent – mimic of the flounder *Bothus lunatus* in all aspects of coloration as well as most aspects of swimming. The undescribed Indo-Pacific octopus dubbed 'blandopus' (or White V Octopus sp. 18) is also a good – or high fidelity – mimic of *Bothus mancus*, while *Thauoctopus mimicus* was judged to be a more general (perhaps even poor) mimic of *B. mancus* in Indonesia because its coloration pattern did not often match that of the flounder (Hanlon, Conroy & Forsythe, 2008). Huffard *et al.* (2010) have commented on the evolutionary and behavioural significance of this mismatch in coloration patterning.

There are shortcomings in all of these analyses; at present, none of them are compelling examples of mimicry *sensu stricto* (Wickler, 1968; Edmunds, 2000). First, there is no air-tight case for defensive mimicry because no common predators are known for any of these mimics. Second, there is no solid evidence for Batesian mimicry, which requires that the model for the mimic be unpalatable (i.e. poisonous). Third, some of the claims for mimicry are far too subjective (especially those suggesting sting rays, feather stars, nudibranchs, giant crabs, snake eels and seahorses).

There are other examples of cephalopods exhibiting highly unusual body patterns while swimming. On Pacific coral reefs, *Octopus cyanea* exhibits several bold patterns in rapid succession when swimming over large coral heads (Fig. 5.15a–d). In other instances, the mantle and head are dark brown and the trailing arms white, so that it looks rather like the local parrotfish (Fig. 5.15e, f; Hanlon, Forsythe & Joneschild, 1999). *Sepioteuthis sepioidea* and *Octopus vulgaris* (Fig. 5.15g, h) on

Caribbean coral reefs often show a bold pattern of longitudinal stripes (Cowdry, 1911; Moynihan & Rodaniche, 1982). These confer a striking resemblance to the striped parrotfish (*Scarus taeniopterus*) found on the same reefs. It is interesting that Longley (1918) reported that many reef fish with control of patterning also commonly wear longitudinal stripes when in motion; he speculated that this is 'an arrangement which makes for concealment in that it tends to mask forward movement'. In today's parlance, that equates roughly to a form of motion camouflage.

There are several other suggestions in the literature about potential mimicry involving cephalopods, but all are very preliminary in nature (Krajewski *et al.*, 2009; Rocha, Ross & Kopp, 2012; Warnke, Kaiser & Hasselmann, 2012; Hoving *et al.*, 2013). Nevertheless, collectively the pioneering studies described early in this section lay the groundwork for future research on defensive mimicry.

5.1.4 How Many Camouflage Patterns Are There?

This broad question has not been posed for animal phyla in general. Nor has it been addressed until recently in cephalopods. Since cephalopods are the fastest-changing animals with the greatest diversity of appearances, it raises the question of how they can change so quickly when there is so much visual information surrounding them (e.g. a coral reef, kelp forest, etc.). Concentrated study on the cuttlefish *Sepia officinalis*, including the accumulation of thousands of camouflaged images both in the field and the laboratory, suggests that very few pattern templates are required to achieve camouflage (Hanlon, 2007; Hanlon *et al.*, 2009). It appears as though this species has only three basic pattern templates (Uniform, Mottle, Disruptive; see Chapter 3) in its camouflage repertoire. We stress that there is variation on each pattern design, but argue that the variations are relatively minor, and that the remarkable speed of change is enabled by the cuttlefish having only to seek one or a few background cues to turn on each of the three pattern types. This sort of parsimonious computational solution would explain how they can achieve this extraordinarily fast and complex feat without having a large brain. The idea needs additional field work. Alternative approaches that are based on

Figure 5.15 Mimicry, deceptive resemblance or rarity? (**a–d**) *Octopus cyanea*: a sequence at Bali in which the octopus showed markedly different body patterns every few seconds as it swam over live corals (Photos © Fred Bavendam). (**e**) *Octopus cyanea* at

computer vision and other computational approaches to pattern design are beginning to be brought to bear on this issue (Anderson *et al.*, 2003). There is controversy about this concept of few pattern designs for camouflage, expressed most strongly by Stevens and Merilaita (2009a), yet oddly enough no one has a competing or alternative hypothesis concerning how many camouflage patterns exist either in cephalopods or other animals. This seems to be an area of evolutionary biology and behavioural ecology that has been ignored.

5.1.5 Rarity Through Rapid Neural Polyphenism

Many animals select the appropriate background to achieve crypsis with their single (or slowly changing) colour or body pattern, yet some cephalopods can settle on any substrate and select one of several patterns to achieve crypsis. We term this 'rapid neural polyphenism' (Hanlon *et al.*, 1999) because the physical appearance of the individual cephalopod (its phenotype) can look so different and can be changed within a second or two (Fig. 5.2).

Polymorphism is extremely widespread in animals and plants; it is the presence of two or more distinct forms (morphs) in the same population of a single species. The term is usually applied to genetically distinct, permanent or semi-permanent morphs (e.g. Ford, 1975). Various selective forces have led to the development of different mechanisms of polymorphism (Maynard Smith, 1970), and Curio (1976) has pointed out that an important influence on polymorphism may have been the development of search images by predators. The advantages of polymorphism or polyphenism are: (1) crypsis to reduce detection, (2) confusion to reduce either detection or identification, and (3) increased

'apparent rarity' that would reduce encounters with predators, because many predators prey differentially upon more common phenotypes (or morphs), and this 'apostatic selection' can be offset by polymorphism or polyphenism (Curio, 1976; Endler, 1991). Both mechanisms presumably make it difficult for predators to learn which morphs they prefer.

The advantages of rapid neural polyphenism in each individual cephalopod are presumably the same as for genetically regulated or hormonally controlled polymorphism, with the added advantage of speed and greater diversity. Note, for example, in Fig. 3.8 how two hatchling cuttlefish use very different body patterns to hide on the same substrate. One cuttlefish is in a Uniform dark body pattern, thus achieving masquerade of adjacent rocks of similar size and appearance. The other cuttlefish shows a Disruptive pattern that hinders recognition of its body outline. These represent two of the three basic pattern templates that *Sepia officinalis* uses for crypsis, the other being Mottle (§5.1.4 and Chapter 3). This may, at first glance, seem like limited polyphenism for crypsis, but we have stressed here and elsewhere (Hanlon *et al.*, 2009) that there is substantial variation in each pattern template (note many images in this chapter). Moreover, with body contortions of octopuses and the many arm and body postures of squid and cuttlefish, collectively there are many polyphenisms for crypsis.

Squid on coral reefs can also show different cryptic body patterns while hovering over the same species of soft coral. Some octopuses show considerable polyphenism during short swims, as illustrated in Fig. 5.15a–f.

A new form of primary defence was discovered in field work on *Octopus cyanea* (Hanlon, Forsythe & Joneschild, 1999). *Octopus cyanea* changed body pattern

---

Caption for Figure 5.15 (*cont.*) Manihi Atoll, Polynesia, seemingly mimicking a local parrotfish by swimming forwards, undulating the trailing arms like the caudal fin of a fish, and expressing the ocellus like a large eye on the body shape of a fish. (**f**) *Octopus cyanea* at Manihi swimming backwards and closely resembling a bicolour parrotfish *Sparus rubroviolaceous*. (Both images from video; Hanlon, Forsythe & Joneschild, 1999: *Biological Journal of the Linnean Society* **66**, 1–22, Crypsis, conspicuousness, mimicry and polyphenism as antipredator defences of foraging octopuses on Indo-Pacific coral reefs, with a method of quantifying crypsis from video tapes, Hanlon, R. T., Forsythe, J. W. & Joneschild, D. E., © (1999), by permission of Oxford University Press. (**g**) Longitudinal stripes in a fast backward-swimming squid *Sepioteuthis sepioidea* and (**h**) a fast backward-swimming *Octopus vulgaris*, both in the Caribbean (R.T. Hanlon).

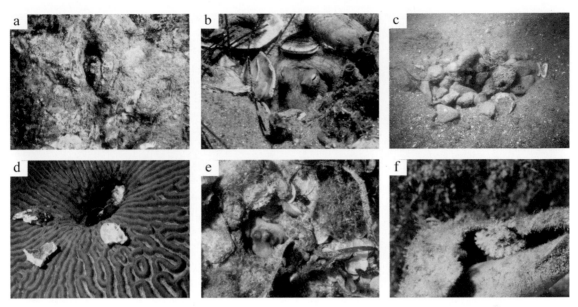

Figure 5.16 Dens of octopuses. Note the discarded bivalve shells that mark dens in some species. (**a**) *Octopus bimaculatus* in a vertical rock crevice in a kelp habitat. (**b**) *Octopus vulgaris* in a natural rock crevice in western Turkey. (**c**) *Octopus bimaculoides* burrow in soft mud, constructed with stones. (**d**) *Octopus vulgaris* in a brain coral in Bonaire, Netherlands Antilles. (**e**) *Octopus maya* in a natural depression, off Yucatan, Mexico. (**f**) *Octopus burryi* juvenile occupying a conch shell in the Caribbean. Photos by R. Hanlon, except **c** by J.W. Forsythe.

(or phenotype) an average of 177 times per hour while foraging (§4.5). The noteworthy tactic is that it involved both cryptic and conspicuous body patterns, the latter of which were shown during movement. This must confer a considerable advantage for a foraging octopus that may travel several hundred metres each day.

Genetic polymorphism and clinal variation in chromatic expression due to environmental factors have not been investigated in cephalopods. Packard and Hochberg (1977) raised the possibility that the colours shown by adult octopuses could be influenced by the nature of the algae they were exposed to as they grew up, but we know nothing about this. *Octopus vulgaris,* which has extremely wide distribution, would be an obvious species in which to investigate this possibility.

5.1.6 Cryptic Behaviour and Vigilance

Cryptic body patterns are useful in primary defence, but cryptic behaviour is equally important, and there are many other ways of minimising the risk of detection. *Immobility* is crucial if crypsis is to be effective, and

most predators are exquisitely sensitive to motion in their visual fields (Curio, 1976; Troscianko *et al.*, 2009). Benthic cephalopods all practise immobility, which can be assisted by suckers in octopuses and by adhesive mechanisms (suction or chemical) on the ventral surfaces of *Sepia* or the dorsal surfaces of *Euprymna* and *Idiosepius* (von Byern & Klepal, 2006; von Byern *et al.*, 2011). Buoyancy mechanisms in cuttlefish and some squid also aid immobility.

Nocturnal and crepuscular activity diminishes the usefulness of vision for some predators. Vertical migration, which keeps cephalopods in darker mid-waters during the day and in surface waters at night, could be considered under this heading, although this is not its primary function (Chapter 9).

Anachoresis or *living in crevices or holes* (Edmunds, 1974) is used by many benthic octopuses, which usually spend much of their daily cycle out of sight in dens (sometimes called 'lairs' or 'homes') of various sorts (Fig. 5.16; §9.4). Most of these dens are natural crevices, or take advantage of such crevices or depressions so the octopus can modify them to varying degrees (e.g. de Beer & Potts, 2013). *Octopus vulgaris* is reported to use

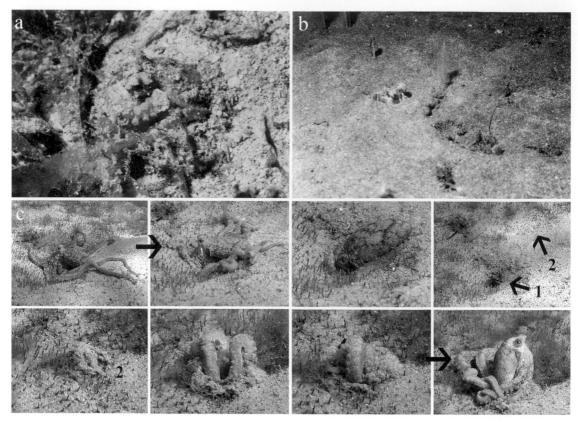

Figure 5.17 Burying behaviour in octopuses. (**a**) *Octopus burryi* emerging from the sand after burying itself completely for several seconds when being pursued by a diver (Hanlon & Hixon, 1980: *Bulletin of Marine Science* **30**, 749–755, Body patterning and field observations of *Octopus burryi* Voss 1950, Hanlon, R. T. & Hixon, R. F., © (1980), with permission from Rosenstiel School of Marine and Atmospheric Science/Allen Press, Inc.). (**b**) *Octopus rubescens*, which partially buries itself for long periods during the night on sand plains (J. Forsythe). (**c**) Video sequence of the Atlantic mimic octopus, *Macrotritopus defilippi*, rapidly burying itself in a very small hole (top series of images; arrow 1) then moving underground (arrow 2) and emerging (bottom four images) minutes later where no hole existed previously (from R. Hanlon video).

four types of dens on soft sediment: 'well', 'rock/stone', 'shell' and 'human origin'; thus the dexterity and strength of octopuses enable diverse adaptations for anachoresis (Katsanevakis & Verriopoulos, 2004). In the pelagic octopods *Ocythoe* and *Argonauta*, small juveniles have been found hiding in pelagic salps (Jatta, 1896; Hardwick, 1970; Banas, Smith & Biggs, 1982; Okutani & Osuga, 1986). *Argonauta nouryi* have been observed at sea surface to form long chains of up to 18 individuals (Rosa & Seibel, 2010b); the protective value of this behaviour is unknown yet intriguing.

Burying is defined here as the act of covering oneself with the substrate (or diving into it), resulting in temporary concealment. Sepioids bury themselves

partially or completely (Fig. 5.2a) to conceal themselves from predator and prey (e.g. Boletzky & Boletzky, 1970; Boletzky, 1977; Singley, 1983; Mather, 1986b; Hanlon & Messenger, 1988). Some benthic octopuses bury themselves quickly in the substrate for immediate predator avoidance (Fig. 5.17; Hanlon & Hixon, 1980). In a most fortuitous filming, the Atlantic mimic octopus, *Macrotritopus defilippi*, buried itself in the sand where only a very small hole existed, and then re-emerged minutes later about 0.6 m away, where no previous hole existed (Fig. 5.17c). How the animal breathed during that 5-minute period is a mystery. The excavation capability of this species is extraordinary. The mimic octopus, *Thaumoctopus mimicus*, was also seen performing a similar excavation and burying (Hanlon, Conroy & Forsythe, 2008).

Burrowing is defined here as the construction of a semi-permanent den or refuge where none existed previously; protracted excavation and residence longer than a few minutes are implied in this definition. Burrowing is performed only by octopuses and occurs in a variety of habitats. For example, *Octopus dofleini* burrows in sand, or sand and mud mixtures (Hartwick, Thorarinsson & Tulloch, 1978), *O. macropus* and *Macrotritopus defilippi* make narrow deep holes in sand (Hochberg & Couch, 1971; Hanlon, 1988; Hanlon, Watson & Barbosa, 2010), *O. cyanea* burrows dens in coral rubble (Yarnall, 1969; Forsythe & Hanlon, 1997), *O. bimaculoides* burrows in soft mud by constructing elaborate stone-lined holes (Fig. 5.16c), and *O. vulgaris* builds dens by collecting stones and empty shells to block the entrance to a depression (Woods, 1965). The time taken to construct a burrow varies greatly. Yarnall (1969) reported 5 minutes and Forsythe and Hanlon (1997) 20 minutes for *O. cyanea* digging in coral rubble. Ambrose (1982) reported 'a matter of minutes' for small *O. bimaculatus* (ca. 75 g), but R.T. Hanlon (pers. comm., 1988) observed 25 minutes.

Shoaling could also, perhaps, be said to aid crypsis in the sense that shoaling condenses the population to a smaller mass in a more restricted area (Chapter 9); that is, it would contribute to 'apparent rarity' to reduce encounters.

For cryptic behaviour to be effective, animals must exert *vigilance*, in which the senses are kept attuned for potential danger. This allows evasive action before a predator gets dangerously close. One way of improving vigilance is to associate with other individuals in a group so that there are more senses to detect danger. Fish often follow octopuses while they are foraging (§9.3). Obviously, squid in a shoal have more individuals to maintain vigilance, and Moynihan and Rodaniche (1982) reported several associations of *Sepioteuthis sepioidea* with other squid species and even with reef fish. Moynihan and Rodaniche (*op. cit.*) suggested that *Sepioteuthis sepioidea* on Caribbean (Panama) coral reefs have vigilant 'sentinels' (§9.2.2) amidst their shoals (see also Fig. 5.24a, b), and field observations at Little Cayman Island have corroborated this (Hanlon & Budelmann, 1987; Hanlon & Forsythe, unpublished data). Mather (2010) working at Bonaire did not report

sentinels. Variations of school structure were observed in *S. lessoniana* in Okinawa by Sugimoto *et al.* (2013) and despite obvious vigilance by squid, there was no report of sentinels. Adamo and Weichelt (1999) did not observe obvious sentinels in *Sepioteuthis lessoniana* at Lizard Island, Australia. Thus, there may be species- and location-specificity in sentinel behaviours. Squid often retreat before the predator arrives, perhaps because of alerting actions from the sentinels; thus, shoaling may allow some members to 'rest' while others stay vigilant. It is not known how sentinels transmit alarm signals to other members of the shoal nor whether they are more susceptible to predation (Chapter 9).

Are bottom-dwelling foraging octopuses cryptic all the time when out of their dens? This question was addressed in a field study in which *O. cyanea* were filmed continuously over several days in different habitats and locations in the western Pacific Ocean (Hanlon *et al.*, 1999). Surprisingly, they were only 'highly cryptic' about 30% of the time, moderately cryptic another 20% and conspicuous the rest of the time (Fig. 5.18). Some conspicuousness is accounted for by mimicry when moving, but sometimes octopuses were conspicuous even when moving slowly, particularly in Palau where (1) the octopuses were larger, (2) there was a high degree of mating, and (3) there were fewer signs of predation evident compared with the Polynesian field site. Such unexpected findings are certainly species-specific; we are confident from our extensive diving experience that a similar study on *Octopus vulgaris*, for example, would show much higher percentages of time spent cryptic while foraging (§5.3.6; see also hints of this in Woods, 1965; Mather, 1988; Mather & O'Dor, 1991; Josef *et al.*, 2012; de Beer & Potts, 2013). However, the *O. cyanea* study draws attention to the trade-offs that animals have to make regarding signalling (e.g. sexual selection) and camouflage (natural selection).

5.2 Secondary Defences

Secondary defences are brought into play once an animal has been detected: that is, when crypsis, masquerade or mimicry have failed. In cephalopods, as in most animals, there are numerous secondary defences that are both age- and context-dependent. Basically, however, cephalopods either flee immediately, or show

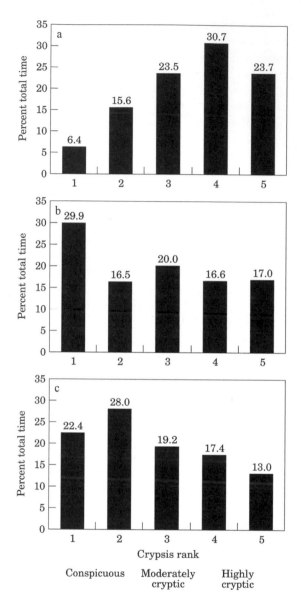

Figure 5.18 Are foraging *Octopus cyanea* camouflaged all the time? No. Percentages of time that they were conspicuous, moderately cryptic or highly cryptic according to ranks assigned by observers. Data based upon (**a**) 268 min of video from Manihi Atoll, Polynesia, (**b**) 23 min from Palau live coral habitats and (**c**) 79 min from Palau dead coral habitats. Modified from Hanlon, Forsythe and Joneschild (1999); *Biological Journal of the Linnean Society* **66**, 1–22, Crypsis, conspicuousness, mimicry and polyphenism as antipredator defences of foraging octopuses on Indo-Pacific coral reefs, with a method of quantifying crypsis from video tapes, Hanlon, R. T., Forsythe, J. W. & Joneschild, D. E., © (1999), with permission from Wiley.

deimatic behaviour followed by some form of protean behaviour (Fig. 5.1). We emphasise that the sequence of events depicted in Fig. 5.1 is not absolutely fixed and is substantiated for only a few species. For example, there may be cases in which crypsis is followed by protean behaviour, or in which flight is followed by protean behaviour, as when, for example, a pursuing predator once again encounters or detects a fleeing cephalopod.

Field data on secondary defences were presented initially by Hanlon and Messenger (1988) studying *Sepia officinalis* at Banyuls, France. These remain the only field data published to date. However, several laboratory studies have been conducted on *S. officinalis* (Adamo *et al.*, 2006; Langridge, Broom & Osorio, 2007; Langridge, 2009; Cartron *et al.*, 2013; Staudinger *et al.*, 2013a) and *Loligo pealeii* (Staudinger, Hanlon & Juanes, 2011), and we include unpublished field data on *Octopus vulgaris* to inspire future studies.

5.2.1 'Wait and See' Assessment

Crypsis and vigilance enable many cephalopods to 'wait and see' what an approaching predator is going to do before selecting an appropriate response. Their rapid and varied escape mechanisms, aided by the chromatophore system that permits an individual to show rapid neural polyphenism, enable cephalopods to respond quickly and in many ways after they ascertain that secondary defences are required to avoid predation. In other words, they have the capability for many behavioural choices, and thus decision-making is a key component of this critical process. They do not react in the same way to all approaches. Some specific examples of how this process functions in nature are given in §5.3.

5.2.2 Flight

Rapid locomotion away from an approaching predator is perhaps the most common secondary defence among animals (Edmunds, 1974). The free-swimming squid and cuttlefish commonly *jet* away from potential predators as their immediate secondary defence, sometimes ejecting ink and making specific body pattern changes (Major, 1986; Hanlon & Messenger, 1988). A striking real-life example is provided in Fig. 5.19a. The coral reef squid

Figure 5.19 (**a**) Escape from predation from *Serranus cabrilla* underwater at Banyuls, France by a very small cuttlefish *Sepia officinalis* that has swiftly performed the Blanch–Ink–Jet manoeuvre. See Fig. 5.25 for the next behaviours. (**b**) *Octopus cyanea* resembling a dead coral head and remaining undetected as a predatory jack swims overhead at Manihi Atoll, Polynesia. Inset shows the conspicuous pattern the octopus was in a few seconds before the fish arrived. Photos by R. Hanlon.

a

Figure 5.20 Flying squid. (**a**) Drawing of an ommastrephid squid exiting the water in streamlined body form while jetting forcefully, then continuing to jet water while airborne and extending the fins and arms to create temporary wings to generate lift for gliding. (**b**) Live squid in full aerial flight. (**c**) View from the bow of a ship showing a school of about 100 squid 'flying' just above the water. Jets of water can be seen as light streaks behind some of the squid (modified from Muramatsu *et al.*, 2013: *Marine Biology* **160**, 1171–1175, Oceanic squid do fly, Muramatsu, K., Yamamoto, J., Abe, T., Sekiguchi, K., Hoshi, N. & Sakurai, Y., © (2013), by permission of Springer).

Sepioteuthis sepioidea uses rapid jetting as the only defence when its most dangerous predators approach (§5.3.3). Jetting after deimatic behaviour is common among cephalopods, and the way it is used as protean behaviour is considered in §5.2.4.

'Flying squid' (Fig. 5.20) are well known and particularly common among oceanic oegopsids that live near the surface. One of the best documented accounts is that of Muramatsu *et al.* (2013) who filmed a school of ca. 100 ommastrephids simultaneously exit the water for about 3 seconds, then continue to jet water when airborne and to glide for more than 30 m (individual speeds of 9–11 m s⁻1). The gliding was accomplished by spreading the fins and the arms, the latter of which have a thin membrane that increases the 'wing surface' and provides more lift than the fins (Azuma, 2006). Macia *et al.* (2004) provided new data on *Sepioteuthis sepioidea* flying and reviewed many reports of flying squid (e.g. Arata, 1954; Cole & Gilbert, 1970; Packard, 1972; Azuma, 1981). Rather than simple gliding (as previously thought) after incidental exit from the water, this behaviour involved jet propulsion, generation of lift force (arms and fins),

and control of different body postures in different phases of flight. This form of protean escape is likely to be effective against their aquatic visual predators. Although impressive, squid do not glide nearly as far as flying fish.

Octopuses, although lacking giant fibres, can also jet very quickly (Fig. 5.15h), far faster than a human can swim. Octopuses and cuttlefish also combine flight behaviour with sudden *burying*. Hanlon and Hixon (1980) described a sequence in which *Octopus burryi* on an open sand plain used both tactics: the octopus first swam quickly away from the divers for about 4 m, then abruptly dived into the coarse sand substrate, burying itself quickly and completely before emerging slowly several minutes later (Fig. 5.17a).

Young cuttlefish often show a bold body pattern – Passing Cloud – while swimming in the open. *Sepioteuthis sepioidea* and *Octopus cyanea* both often show bold uniformly dark brown patterns, especially when moving over white sand (Fig. 5.15g, h). Both these patterns make the animals conspicuous, and it is not clear why exposed, moving animals should behave like this.

5.2.3 Deimatic Behaviour

Young (1950) coined the term 'dymantic' to describe animal behaviour that startles or bluffs. This category of behaviour has gained acceptance in behavioural literature, although the alternative term 'deimatic' from the same Greek root δειματόω – I frighten (Maldonado, 1970) – is more widely used (Edmunds, 1974; McFarland, 1981; Barrows, 1992), and some authors use the term 'startle display'. *Deimatic behaviour* is threat, startle, frightening or bluff behaviour, and in most cases it serves to make a predator hesitate during the close approach phase of attack (e.g. Cott, 1940; Young, 1950; Edmunds, 1974; Endler, 1991; Ruxton, Sherratt & Speed, 2004; Olofsson *et al.*, 2012). Deimatic behaviour and patterning can be intraspecific or interspecific, and such displays can be shown unilaterally depending on the behavioural context (Hanlon & Messenger, 1988;

Langridge, 2006). Interspecific behaviour may intimidate a naive predator or provide a set of ambiguous signals to an experienced one (Edmunds, 1974); it may also have the additional function, suggested by Moynihan and Rodaniche (1982), of simply informing the predator that it has been detected by its intended prey (as in stotting by antelope). Deimatic behaviour may be a genuine warning that the cephalopod can attack and harm the predator, or it may be dishonest signalling – a bluff. However, since no clear cases of retaliation to attack have been described in cephalopods, bluff seems more likely.

Box 5.3 outlines various forms of deimatic expression in cephalopods and Fig. 5.21 illustrates a few of them. We would emphasise that the broad category of deimatic behaviour should not be confused with the specific behaviour variously called 'Dymantic Display' or 'Dymantic Pattern' in most of the

Box 5.3 Different expressions of deimatic behaviour in cephalopods

Deimatic behaviour is threat, startle, frightening or bluff behaviour and is exhibited in widely different forms by ccphalopods.

Deimatic (or Dymantic) Displays (see Fig. 5.21 and Box 7.2). Deimatic Displays are characterised by the sudden appearance of bold light/dark chromatic components, usually with spreading of arms, web or fins to create the illusion of largeness. In *Sepia officinalis*, the display comprises six signals: (1) Flattened body posture; (2) paling of the skin; (3) Paired mantle spots that look like eyes; (4) Dark fin line; (5) Dark eye ring; and (6) Dilated pupil. In *S. latimanus*, the display is similar but without false eyespots. Sometimes young cuttlefish show a cross between the Deimatic Display and the Acute Conflict Mottle. Cuttlefish will orient themselves appropriately to show the mantle spots to fish, and usually inflate their mantle, so the postural component is important.

Among teuthoids, only the loliginids *Sepioteuthis sepioidea* and *S. lessoniana* show a 'typical' Deimatic Display, with false eyespots on the mantle and a pale or lightly mottled coloration. *Lolliguncula brevis* (Dubas *et al.*, 1986) and *L. panamensis* (Moynihan & Rodaniche, 1982) tend to show displays intermediate between the Deimatic

and the Flamboyant (described below); that is, they have arms flared in Upward V-curl, light bodies with some sort of dark border (on the fins or the flared arms), but they do not have conspicuous false eyespots.

In *Octopus vulgaris*, the Deimatic Display comprises six components: (1) paling of the skin; (2) arms curved in wide arc and web spread maximally; (3) Dark eye ring, (4) Dilated pupil; (5) Dark edged suckers to create a dark margin to the octopus's outline; and (6) jetting water. Deimatic Displays in other octopuses differ slightly. For example, *Octopus maya* has diffuse 'spots' but no distinct ocelli and *Octopus bimaculoides*, *O. bimaculatus* and *O. filosus* have prominent ocelli at the bases of their arms that are used with lightly graded mottles to produce a Deimatic Display. Quite different Deimatic Displays are shown by *Octopus macropus* and *Octopus ornatus*. The arms or web are spread to enlarge the appearance of the body, and there is maximal contrast of light/dark on the body: both species turn bright reddish-brown with bright white ovals all over the body. The deimatic response in the oceanic octopod *Argonauta argo* (the paper nautilus) is different again and occurs when the brightly silver iridescent web (part of the first arm pair) that covers the

Continued

shell is quickly withdrawn by the octopus (Young, 1960a).

Each species shows gradations of the Deimatic Display that can be deployed in response to different stimuli. For example, the Deimatic Display of *Octopus vulgaris* can include the second arms being lashed out at the intruder. In octopuses, one important adjunct to the chromatic signals of Deimatic Display is the forcible blowing of water at an intruder, a response that has startled many divers (e.g. Hartwick, Thorarinsson & Tulloch, 1978).

Flamboyant Displays (Fig. 5.22) are highly stereotyped and well developed in cuttlefish, squid and octopuses. The term 'flamboyant', introduced by Packard and Sanders (1971), refers to the wavy, undulating lines of the window tracery in some late Gothic architecture. See text for details. Very young animals also show this display (Fig. 5.22e); in *Octopus vulgaris*, for example, Flamboyant comes to be replaced by the Deimatic as the animals grow larger (see also Fig. 8.3; Packard & Sanders, 1971; Hanlon & Messenger, 1988). The *Passing Cloud Display* is characterised by waves of transiently expanded chromatophores flowing over the body in a coordinated manner. In octopuses, they originate from the head and radiate forward and backwards over the body; in cuttlefish, they originate from the posterior mantle and move forward to the arm tips (Fig. 4.6).

Some patterns, especially Mottles, Disruptives or All Darks, may function deimatically during 'Pattern Flashing'. Acute Disruptive and All Dark are used by *Sepia officinalis* and *Sepioteuthis sepioidea* after inking and jetting away from a predator. Cuttlefish occasionally show a dark mottle during some Flamboyant Displays that may also indicate conflicting behavioural interactions (Hanlon & Messenger, 1988). Octopuses often show a brief but strong Mottle Pattern just prior to the more typical Deimatic Display (Fig. 5.21g). The flashing blue iridescent rings of *Hapalochlaena* spp. are shown during Mottle and Disruptive Patterns (Fig. 5.21h) and may function deimatically, although they may be flashed as aposematic coloration.

cephalopod literature, but which we term the Deimatic Display; each is the name of only one type of deimatic behaviour. As noted in Fig. 5.1, some form of deimatic behaviour is usually one of the first reactions made to predators that have approached closely. The *Deimatic Display* (Fig. 5.21) is typically shown after the animal has been discovered by a predator; it is usually held for 2–10 seconds before the next defensive manoeuvre is begun (i.e. some kind of protean behaviour; see following section). The *Acute Conflict Mottle Pattern* (Fig. 5.21a, b, g, h) or *Acute Disruptive Pattern* (Fig. 5.21e, f) is also shown by some species in similar circumstances. All of these are bold patterns, of high contrast, that serve to make the cephalopod appear larger (see also Box 7.2); the deimatic effect is enhanced because of the swiftness with which the pattern is switched on (i.e. <1 second). The Deimatic Display is shown upon very close approach of a predator (or a human, or a camera) to the cephalopod.

Octopuses add another behaviour to Deimatic: they often blow water forcibly at the approaching animal, adding a tactile signal to the visual display. They also perform this from within their den.

Flamboyant (Fig. 5.22) is another category of deimatic behaviour: we define it as any display that includes widely flared or splayed and contorted arms or mantle, usually in conjunction with mottled or disruptive patterning. In all genera, the Flamboyant Displays seem to function in one of two ways: either as masquerade to floating weeds (primary defence: see §5.1.4), or as a threat posture in which the cephalopod is 'showing the weapons' (Eibl-Eibesfeldt, 1975), i.e. the arms and beak. When *Sepioteuthis sepioidea* flares the arms and shows a disruptive pattern with Bars and Streaks, it is thought that it may be signalling another message, 'You have been detected and I am preparing to flee successfully', at least according to Moynihan and Rodaniche (1982). Since this pattern (and its many variations; Mather, Griebel & Byrne, 2010) is shown regularly to approaching fish (*op. cit.*) that subsequently do not attack, the message is likely to be successful. Vecchione and Roper (1991) described 'J-postures' and 'cockatoo postures' in some deep-sea squid, and these appear to be other forms of Flamboyant Displays, as do some of the postures and bioluminescence reported for *Octopoteuthis deletron* (Bush, Robison & Caldwell, 2009).

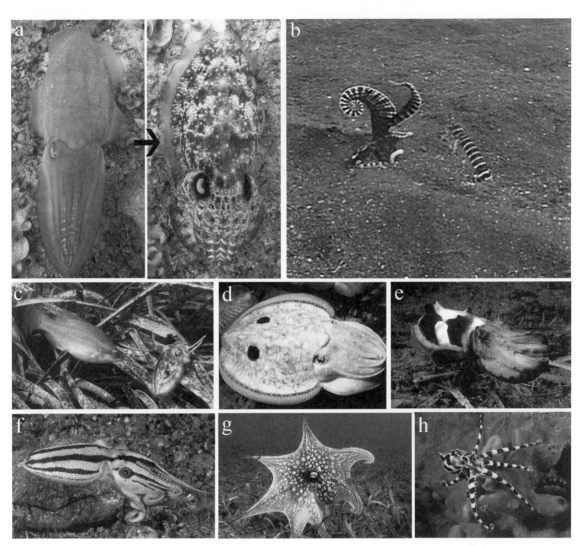

Figure 5.21 Deimatic Displays in cephalopods. (**a**) Photos taken moments apart in Bali of the broadclub cuttlefish *Sepia latimanus* as the photographer approaches. (**b**) The mimic octopus, *Thaumoctopus mimicus*, displaying to the large stomatopod (right) that had just attacked it (Indonesia, from Hanlon, Conroy & Forsythe, 2008: *Biological Journal of the Linnean Society* **93**, 23–38, Mimicry and foraging behaviour of two tropical sand-flat octopus species off North Sulawesi, Indonesia, Hanlon, R. T., Conroy, L. A. & Forsythe, J. W., © (2008), by permission of Wiley). (**c**) Juvenile *Sepia officinalis* (40 mm ML) orienting its mantle towards an approaching herbivorous fish. (**d**) Typical Deimatic Display of adult *S. officinalis*. (**e**) Flamboyant cuttlefish *Metasepia pfefferi* in New Guinea, showing high-contrast white markings with bright yellow and red pigmentation. (**f**) *Amphioctopus siamensis* in Anilao, Philippines, alarmed by a photographer. (**g**) *Octopus vulgaris* in Puerto Rico 'standing' on its four left arms to orient vertically to a threat stimulus. (**h**) Blue-ringed octopus *Hapalochlaena maculosa* in South Australia showing deimatic response to a photographer. Photos **e**, **h** © Fred Bavendam; image **f** by Patrice Marker; others by R. Hanlon.

Passing Cloud of cephalopods is a bizarre display perhaps unique among animals: dark waves pass rapidly across the body, generated by the action of the chromatophores (Fig. 4.6). It was first noted (but unnamed) in the laboratory by Holmes (1940) in *Sepia officinalis* adults; later, in *Octopus vulgaris*, it was called 'passing cloud' (Packard & Sanders, 1969) and 'moving flush' (Wells, 1978). Direct

Figure 5.22 Flamboyant Displays used as deimatic behaviour. (**a**) Juvenile *Sepia officinalis* (right) showing Flamboyant upon approach of the predatory *Serranis cabrilla* (left) at Banyuls, France. (**b**) The fish's view of the cuttlefish Flamboyant, face-on. (**c–f**) Flamboyant Displays by *Sepioteuthis sepioidea* (on a Cayman reef at night), *Lolliguncula brevis* (in the lab), young *Octopus vulgaris* (Azores), and unidentified octopus in western Turkey. Photo **c** by Eliot Ferguson, **e** by J.W. Forsythe; others by R. Hanlon.

observations in the sea (Hanlon & Messenger, 1988) have shown that Passing Cloud can be shown by hatchling and juvenile *Sepia officinalis* to small fish that approach too closely or during protean escapes (Fig. 5.25). Passing Cloud is also shown to prey (§4.3.1) and, as we have seen, it is sometimes used by young cuttlefish as they swim across the substrate (§5.2.2).

Flashing Bioluminescence has been reported in mid-water squid and may function as a form of deimatic behaviour (see Chapter 9); and it is possible that *Flashing Blue Rings* in *Hapalochlaena* spp. octopuses may also function this way.

Aposematism is defined as conspicuousness associated with some unpleasant attribute such as poison or distastefulness or general unprofitability (Edmunds, 1974; Endler, 1991; Ruxton, Sherratt & Speed, 2004; Stevens, 2013). A classic terrestrial example is the skunk. Predators either learn this association and avoid attacking prey with this attribute, or have an inborn aversion to the pattern. Aposematism has never been described in cephalopods, but it seems possible that the yellow and black markings – the latter with their flashing blue rings – of the venomous blue-ringed octopuses *(Hapalochlaena* spp.) or the bright white, yellow and red of the Flamboyant cuttlefish *Metasepia pfefferi* (Fig. 5.21e, h) could be aposematic coloration, and this is worthy of experimentation.

5.2.4 Protean Behaviour

Chance and Russell (1959) drew attention to 'displays which are sufficiently unsystematic to be quite unpredictable in detail' and they referred to them as 'protean' after the mythical Proteus who frustrated would-be captors by continually changing his shape, so that they had nothing systematic to which to react. The authors used body patterning changes in the cuttlefish *Sepia officinalis* (as documented by Holmes, 1940), as one of their prime examples of this type of behaviour. Subsequently, numerous examples drawn from many phyla have described erratic and unpredictable escape behaviour (Humphries & Driver, 1970; Driver & Humphries, 1988; Edut & Eilam, 2004; Jones, Jackson & Ruxton, 2011; Briffa, 2013). Random zig-zag escape routes are an obvious example.

'Protean behaviour is that behaviour which is sufficiently unsystematic in appearance to prevent a reactor predicting in detail the position or actions of the actor' (Driver & Humphries, 1988). That is, it makes a predator hesitate or upsets its target prediction. Protean behaviour is thought to have evolved by a process rather like ritualisation in reverse, a form of 'behavioural anarchy' that produces random irregularity and confusing complexity. This impairs learning by the predator, and

Curio (1976) has noted that even the most experienced predator can be defeated by protean defences.

Cephalopods show many types of protean behaviour (§5.3). One of the most 'typical' responses is the *Blanch–Ink–Jet manoeuvre* that is performed when attack is imminent: the cephalopod blanches white or dark (usually the opposite of what it was showing previously) and ejects ink as it jets away, leaving an ink blob in its place (Fig. 5.19a; see also Packard, 1988a). Erratic jetting, in combination with a variety of body patterns, is used by many cephalopods to escape predators; jet propulsion provides the additional advantage of allowing the animal to move forwards or backwards because the funnel is flexible. When this erratic jetting is accompanied by an emission of ink, the results are very confusing for humans and perhaps for other animals. For example, squid that had been singled out from their shoal by dolphins survived attack only if they used the tactic of changing direction erratically as they also changed body patterns and inked (Major, 1986).

Another type of unpredictable behaviour is what we term 'clandestine escape'; it is shown in Fig. 5.25b by young cuttlefish under natural conditions underwater (Hanlon & Messenger, 1988). This is typical protean behaviour, and is a good example of the use of rapid neural polyphenism: this type of escape requires the cuttlefish to show different body patterns, each at a critical time, and requires substantive cognitive processing.

Pattern Flashing in cephalopods is an attribute of transient individual polyphenism: different body patterns are exhibited in rapid succession, providing the predator with deceptive images flashed in an unpredictable order. The distinction between Pattern Flashing and Flashing Blue Rings (above) is that the former involves different patterns whereas the latter involves flashing the same pattern. Although there is undoubtedly a protean element to this flashing of patterns, it is possible that in some circumstances it could have an effect of startling the predator. Again, there is not always a simple way to classify behaviours solely into one category such as protean or deimatic.

5.2.5 Other Defences

Inking

Inking is a characteristic behaviour shown by nearly all coleoid cephalopods. As a visible defence, it is dispersed in multiple forms. One form is the production of a 'pseudomorph', a blob of ink held together by mucus that hangs in the water while the emitter jets away (Hall, 1956). The pseudomorph is usually about the size of the cephalopod and probably serves to hold the visual attention of the predator while the cephalopod escapes, for example during the Blanch–Ink–Jet manoeuvre. In this sense, the ink may function as deceptive resemblance to the cephalopod that was just there, so that predators attack the ink (Fig. 5.30a). At other times, the mucousy secretion is called a 'rope' because it is longer and thinner than a pseudomorph (Fig. 5.19a). Small *Octopus bocki* produced ink pseudomorphs upon approach of green turtle hatchlings and escaped consumption (Caldwell, 2005).

Cephalopods can also produce a large cloud of ink like a 'smoke screen' behind which they can hide; small cuttlefish do this during protean escape from fish predators in the wild (Fig. 5.25d; Hanlon & Messenger, 1988). Moynihan and Rodaniche (1982) observed small *Sepioteuthis sepioidea* high in the water column on coral reefs ejecting small blobs of ink, then turning all dark themselves and hovering amidst the blobs in some sort of imitation or general resemblance to the ink. Deimatic and protean inking behaviours have been seen in laboratory experiments in response to computer-generated visual stimuli (including fish images; Cartron *et al.*, 2013), low-frequency sound at high intensities (Samson *et al.*, 2014) and live predatory fish (Staudinger *et al.*, 2013a).

Chemical defence is another function of inking (reviewed by Derby, 2014; also Derby 2007). Moray eels attacking octopuses (MacGinitie & MacGinitie, 1968; Forsythe & Hanlon, 1997) and teleost fish attacking young reef squid or cuttlefish (Fig. 5.19a; Moynihan & Rodaniche, 1982; Hanlon & Messenger, 1988) all react to a large cloud of ink by hesitating and shaking their heads for several moments. Other suggestions come from observations of octopuses squirting ink at snails or crabs approaching their eggs (Eibl-Eibesfeldt & Scheer, 1962) and cuttlefish such as *S. officinalis* coating their egg capsules with ink. Recent experimental tests support the idea that cephalopods use ink as anti-predatory chemical defence. Ink from two species of squid – *Sepioteuthis sepioidea* and *Loligo pealeii* – was shown to be unpalatable to predatory fish and, as such, might provide a defence during attacks (Wood *et al.*, 2010; Derby *et al.*, 2013).

Intraspecific effects of inking have been reported as well. Inking can function as an alarm cue among squid such as *Sepioteuthis sepioidea* (Wood, Pennoyer & Derby, 2008). In the squid *Loligo opalescens*, dopamine at biologically relevant concentrations can evoke jetting (Gilly & Lucero, 1992; Lucero, Farrington & Gilly, 1994). As illustrated in Fig. 6.7, ink can be used aggressively during escalated male–male agonistic contests in cuttlefish. In *Octopus cyanea*, a male that was attacked and cannibalised by a large female mate inked profusely attempting to escape (Hanlon & Forsythe, 2008). In the cuttlefish *Sepia officinalis*, Boal and Golden (1999), studying chemical perception, showed experimentally that ink caused the largest increase in ventilation rate (compared with predator odour, conspecific odour or food odour) and suggested that ink could function similarly to alarm substances as in fish.

Inking behaviours are not confined to shallow well-lit environments. Bush and Robison (2007) reported six types of ink shapes videotaped from ROVs in a variety of mesopelagic squid species: pseudomorphs, pseudomorph series, ink ropes, clouds/smokescreens, diffuse puffs and mantle fills. Bush, Robison and Caldwell (2009) reported several of these shapes for the deep-sea squid *Octopoteuthis deletron*. Dilly and Herring (1978) reported the only known case in which a mesopelagic squid – the sepiolid *Heteroteuthis dispar* – releases luminous fluid with its ink and mucus, thus creating luminous clouds that might function as visual defence in a manner similar to the way cephalopods use black ink in well-lit environments. The curious inking behaviour of the oceanic squid *Teuthowenia* (formerly *Taonius*) *megalops* may fall into the category of protean behaviour. Upon initial disturbance in a shipboard aquarium, small individuals will ink and jet away, but upon further disturbance they begin a 'balling up' sequence, in which the fin and eventually most of the mantle and head are inverted into the mantle cavity; the two tentacles are left extended but can be withdrawn too, and the animal will then ink inside the balled-up mantle (Dilly, 1972). R.E. Young (1972b) noted that the oceanic squid *Cranchia scabra* reacts in much the same way.

Clearly there is much more to learn regarding the widespread and unusual defensive capabilities of inking in cephalopods.

Figure 5.23 Deflective marks (**a**) giving the impression of eyes and head at the posterior mantle in a young juvenile *Sepia officinalis* (lab image from Hanlon & Messenger, 1988, by permission of the Royal Society), and (**b**) paired darker spots in middle of mantle of an otherwise cryptic adult (original image, Çeşmealtı, Urla, Turkey, R. Hanlon).

Deflective Marks

Deflective marks are found in many animal groups and function to direct an attack to a less vulnerable part of the body or to deceive the predator about the direction of escape. The Paired mantle spots of *Sepia* and *Sepioteuthis* are often shown before the full Deimatic pattern and could function as 'false eyespots' (§5.2.3, §5.3 and Box 5.3), although validation of this function is difficult to prove (Vallin *et al.*, 2011; Stevens & Ruxton, 2014). Yet when differently expressed, these spots may give the false visual impression that the head, and thus the direction of locomotion, is in the other direction. There is no direct evidence for this, although *Sepia officinalis* also has a pattern suggesting a 'head-in-reverse' (Fig. 5.23). The paired Dark ocelli on the proximal part of the arms of certain octopuses (e.g. Fig. 5.21f) may help to divert attacks away from the head and mantle. We know little about the nature of the signal that the ocelli produce, nor their use in nature (cf. Vallin *et al.*, 2011). Why, for example, do some ocellated octopuses have iridescent ocelli and some not? In Cott's (1940) terminology, these deflective spots all 'misrepresent the posture' of the prey, and Hailman's (1977) term 'symmetry deception' implies the same visual effect. Wickler (1968) and Curio (1976) pointed out that, as a rule, a 'false head' or 'false eye' does not resemble the particular head or eye of its possessor. Future students of deflective marks in cephalopods should bear in mind that the marks and postures used may suggest the wrong direction of escape, but they may also serve to resemble a different animal, which could impede the predator anywhere along the sequence of encounter, detection, identification, approach (attack) and subjugation (Endler, 1991). If these marks affected one of the first three of this sequence, then they would be operating as primary defence.

Defensive Postures and Autotomy

Octopuses attacked by moray eels pull all eight arms tightly up and over the head and mantle, exposing only the suckers (MacGinitie & MacGinitie, 1968); octopuses in an aquarium react like this to humans (Packard, 1963) and sometimes to fighting conspecifics (Hanlon & Wolterding, 1989). Such a posture may deter a smaller predator, but would not appear to be very effective with a larger one. Some octopuses may be able to detach an arm during an engagement with a predator. Such *autotomy* of a non-essential limb is a common form of defence in many phyla (Edmunds, 1974). Nesis (1982) described female *Tremoctopus* that autotomise the first arm pair, and Hanlon and Wolterding (1989) mention that *O. briareus* sometimes lose an arm during an intraspecific fight. However, in neither case has histology or experimentation shown that this is actually autotomy. In *Ameloctopus litoralis*, all eight arms have been found on occasion to have been autotomised (Norman, 1992a), and these arms have been found histologically to possess specialised 'break' regions. Alupay, Hadjisolomou and Crook (2014), studying nociception, induced nocifensive behaviours including arm autotomy in *Abdopus aculeatus* by administering a calibrated arm crush. Autotomy occurred at a consistent site in four octopuses, around suckers 5–7 on the injured arm, and induced wound-grooming behaviour for up to 20 minutes and general guarding behaviour of the injured arm up to 24 hours. Bush (2012) reported that

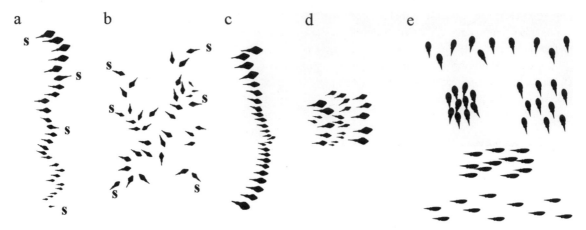

Figure 5.24 Shoaling and schooling in squid. (**a–d**) Shoal structure, sentinels (S) and reactions to teleost predators by the Caribbean reef squid *Sepioteuthis sepioidea*. (**a**) and (**c**) are typical 'picket lines' during the day, with larger individuals at either end or only one end; presumed sentinels are always on the ends and often near the middle. (**b**) Another typical arrangement, with sentinels facing in all directions. (**d**) Extreme reaction to a predator, with tight school that has jetted away from approaching fish (original based on Moynihan & Rodaniche, 1982: *Advances in Ethology* **25**, 1–150, The behavior and natural history of the Caribbean reef squid *Sepioteuthis sepioidea*. With a consideration of social, signal and defensive patterns for difficult and dangerous environments, Moynihan, M. & Rodaniche, A. F., © (1982), with permission from Wiley; Hanlon & Forsythe, unpublished). (**e**) Five formations of the Pacific reef squid *Sepioteuthis lessoniana* (Okinawa, from Sugimoto *et al.*, 2013: © Marine Biological Association of the United Kingdom, 2013, with kind permission of Cambridge University Press).

the mesopelagic squid *Octopoteuthis deletron* can autotomise an arm anywhere along its length, although only a few well-defined fracture planes were found in histological sections.

Retaliation?

Retaliation, or aggressive defence, although common in some animals, has not been described in cephalopods. Stories of battles between sperm whales and giant squid (cf. Roper & Boss, 1982) imply some form of retaliation, but until someone manages to fix a video camera on the head of a sperm whale and record an encounter with squid, we can say nothing! During intraspecific encounters (for example between males in agonistic contests; see §6.2 and 9.4), biting has been observed, and it seems likely that cephalopods of different species may bite each other during aggressive encounters. There are reports of lethal bites on humans by blue-ringed octopuses, but not on other species of octopus (Sheumack *et al.*, 1978; §2.2.7). Inking might be considered a form of retaliation, since it appears to have some mild deleterious chemical effect on a predator (Wood *et al.*, 2010; Derby *et al.*, 2013). Finally, we must

list the stinging weapons carried on the arms of *Tremoctopus*: pieces of the Portuguese Man-of-War (§2.2.6).

Schooling

It has long been recognised that tight schooling is a defensive tactic in many fish and crustaceans (Shaw, 1978; Driver & Humphries, 1988; Pitcher & Parrish, 1993). When a predator approaches a shoal of squid in nature (see a variety of shoal structures in Fig. 5.24), the squid close ranks to form a tight polarised school; see §9.2.2 for the distinction between a school and a shoal. The 'dilution effect' of having many potential prey available for the predator may lessen the probability of attack on any one squid (but see discussion by Pitcher & Parrish, 1993). Our limited observations, while diving in daytime or watching squid around night lights, suggest that tight schools, which often change shape and move erratically, are not pursued so vigorously as loose ones. As an aquatic comparison, Fig. 9.1b shows some of the many ways in which fish move when grouped together (Pitcher & Parrish, 1993). Descriptions of schooling behaviours as 'flash expansion', 'ball' and 'hourglass'

(Pitcher, 1993) apply to squid as well as teleost fish. For example, Sugimoto *et al.* (2013) described 'ball' shapes as well as 'belt' and 'sheet' shapes in *Sepioteuthis lessoniana* schools of 8–100 individuals at three coral locations in Okinawa, but these formations were not correlated with interactions with predators.

5.3 Some Defence Sequences

5.3.1 Young Cuttlefish, *Sepia officinalis*, in the Field

Hanlon and Messenger (1988) subjected 32 cuttlefish to natural predation in an underwater ecohabitat at Banyuls, France, and 28 attacks by fish predators were recorded. Cuttlefish were of three age groups (1, 7 and 17 weeks old) and had never seen a fish during their laboratory rearing from eggs. They were taken underwater one at a time in a plastic bag and released onto a sand plain (5 m deep, 10 m diameter) surrounded by seagrass beds and large rocks. The hunting territories of two predatory fish, *Serranus cabrilla* (14 and 20 cm standard length), overlapped this sand plain so that cuttlefish encountered one of the foraging predatory fish within 5–60 min. The natural predator–prey interactions were then observed and are summarised in Fig. 5.25.

A striking finding was that the cuttlefish recognised predator versus non-predator among the fish: different deimatic responses were shown to closely approaching non-predators than to *Serranus* (Fig. 5.25a vs c). Primary defence of crypsis was effective: on more than 40 occasions, hunting fish failed to detect a motionless camouflaged cuttlefish on the substrate as they swam a few centimetres over them. On three occasions when young cuttlefish detected the predatory *Serranus* at a distance (about 2 m), the cuttlefish made a slow stealthy 'clandestine escape' by moving unpredictably with the wave surge. As shown in Fig. 5.25b, individuals used rapid neural polyphenism to alternately resemble the substrate, then the loose floating algae, as they slowly and erratically moved away from the predator in protean fashion. When moving across the substrate to forage for mysids, young cuttlefish frequently showed the Passing Cloud pattern and would stop every metre or so and show a cryptic body pattern before moving again. *Serranus* detected cuttlefish only as a result of movement

(e.g. swimming or partially burying into the substrate). When *Serranus* then moved directly towards the cuttlefish in an attack posture, cuttlefish responded at distances of up to 3 m, even when other fish were nearer. In these cases Deimatic Displays (as towards non-predators) were not shown by the smallest cuttlefish (age groups 2 and 10 weeks) but Flamboyant Display was shown (Figs. 5.22a,b; 5.25c) followed by one of two protean tactics: jet upwards in the water column to the surface (11 of 28 attacks) or jet ca. 1 m and show Flamboyant masquerade amidst seagrass and algae (17 of 28 attacks). *Serranus* never pursued the *Sepia* more than 1 m above the substrate, nor was any second attack ever performed. Overall, seven successful attacks were achieved by *Serranus* (six of them on the 1-week-old cuttlefish), and most of these occurred when the cuttlefish appeared to have been distracted by the diver (R.T.H.). Overall, excluding attacks in which the diver distracted the young *Sepia*, the cuttlefish successfully avoided capture 29 out of 36 times (81% survival).

Comparable field studies on adults have not been undertaken, but when disturbed in the laboratory (Fig. 5.25d), adults first show an incomplete Deimatic Display to disturbance (i.e. Paired mantle spots or four Mantle stripes), followed by a complete Deimatic Display, followed by various sequences of protean behaviour such as Blanch–Ink–Jet, erratic jetting and Pattern Flashing (All Dark, Clear, Passing Cloud Display), or eventually erratic jetting and profuse inking (Holmes, 1940; Boycott, 1958). These tactics may be characteristic of adults in nature, but there are no data available. It is worth emphasising that the reactions of young cuttlefish to disturbance in the laboratory (i.e. to large objects) were similar to those shown to *non-predators* in the field, so caution must be invoked when extrapolating natural behaviour from laboratory observations.

5.3.2 *Sepia officinalis* in Laboratory Experiments

Four laboratory approaches have been taken to test anti-predator responses of *S. officinalis* to threat stimuli of: (1) models of bird predators, (2) computer images of different shapes, (3) live fish but separated by plastic barrier (i.e. no predation) and (4) live fish predator with actual predation and repeated interactions. Only the last

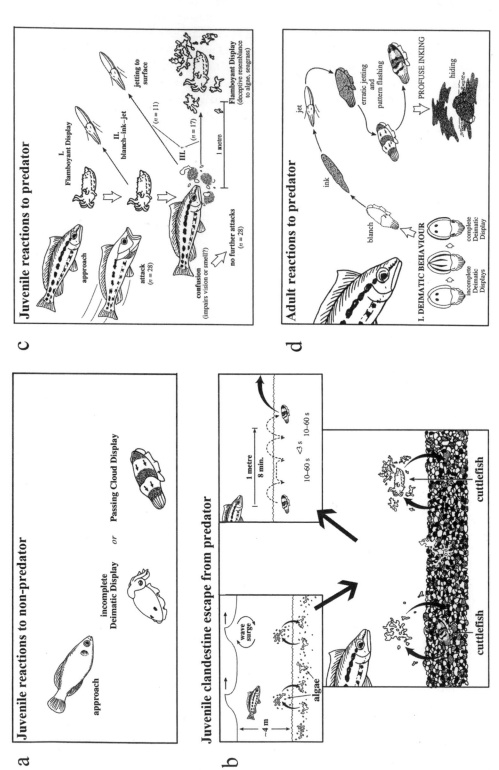

Figure 5.25 Cuttlefish *S. officinalis* defences in the field. (**a**) Young cuttlefish display one of two forms of deimatic (startle) behaviour to non-predators. (**b**) To predators, young cuttlefish sometimes showed complex 'clandestine escape' by remaining cryptic and alternately releasing from the substrate and lifting away with the wave surge while resembling surging seaweed. (**c**) Upon close approach by larger predatory fish, a Flamboyant Display was shown by young cuttlefish; upon attack, the Blanch–Ink–Jet manoeuvre was performed by all 28 cuttlefish. Thereafter, 11 jetted to the surface and 17 deployed Flamboyant amidst floating algae. The fish never followed. (**d**) Secondary defences by adults. Deimatic Displays (variations) were shown first, followed by erratic unpredictable blanching, inking, jetting, pattern flashing, etc. Modified from Hanlon and Messenger (1988, 1996), by permission of the Royal Society.

134

approach provided protean behaviours as well as initial deimatic behaviours.

For the first approaches, Adamo *et al.* (2006), using bird models, found that all ten cuttlefish ceased foraging and showed a Deimatic Display as soon as a bird model was moved just above the water surface. King and Adamo (2006), using a similar set-up, demonstrated that this display included 'behavioural freezing' and hyperinflation of the mantle, which increased the apparent size of the cuttlefish yet also readied it for rapid jetting as the next phase of defence. In the second approach, Cartron *et al.* (2013) used video displays and computer images to present five different geometrical shapes to ten cuttlefish (Fig. 5.26a). Three of the shapes (circle, rectangle and lozenge) elicited a transient Deimatic Display of low intensity followed by a swift return to the camouflaged state and extinction of the response after a few presentations. The star and fish shapes elicited a strong Deimatic Display that even escalated to fleeing and inking. The return to camouflage after presentations of the fish image was slower than that of the geometrical shapes. The summary of this experiment is that: (1) these naive cuttlefish can discriminate a potential predator's shape (i.e. fish) from other non-biological geometrical shapes, (2) the response is robust, and (3) it can be processed through the polarisation vision channels to work even in turbid waters. These two approaches only addressed initial defence – deimatic behaviours – because of the stationary nature of the visual stimuli that they used.

For the third approach, Langridge, Broom and Osorio (2007) subjected young cuttlefish to the presence of live sea bass, dogfish and crabs, but with a transparent partition between predator and prey. They reported that cuttlefish used the Deimatic Display in response to teleost fish (sea bass) but not to the others, which have poorer vision and use chemoreception and electroreception for hunting (Fig. 5.26b). Their result is different from that found in Hanlon and Messenger's field study (Fig 5.25a) and could be explained by the similar size of cuttlefish and sea bass in the Langridge paper (see their Fig. 1). Cuttlefish and squid both tend to show Deimatic Displays to fish that are relatively small – i.e. about the size of the cuttlefish or smaller. Langridge (2009) then subjected young cuttlefish to small and large sea bass, and found different responses for each

(Fig. 5.26c); only small fish predators elicited the Deimatic Display while large fish predators elicited Flamboyant Displays, as found in field studies by Hanlon and Messenger (1988; compare with Fig. 5.25a).

In the fourth approach, Staudinger *et al.* (2013a) carried out actual predator–prey experiments with three teleost predators: bluefish (*Pomatomus saltatrix*), summer flounder (*Paralichthys dentatus*) and black sea bass (*Centropristis striata*). This allowed observations of both primary and secondary defences. Over the course of 25 predator–prey trials, 44 behavioural sequences were evaluated, and 3 primary and 15 secondary defences were observed (Fig. 5.27a). Defences to bluefish and sea bass were very different. When sea bass (sit and wait predator) initiated an attack by making a direct approach, cuttlefish alternated between Stay (60%, 12/20) and Flee (40%, 8/20) tactics as initial responses (Fig. 5.27b). By comparison, cuttlefish predominantly used Stay tactics (92.9%, 13/14) as initial responses to approaches and attacks by bluefish, which are active swimming predators. The authors were unable to evaluate secondary defences to flounder – a lie and wait predator – because flounders did not pursue cuttlefish. Cuttlefish appeared unaware of a flounder's presence, and thus may be especially vulnerable to ambush predators, particularly those that are well camouflaged like flounders. Overall, Deimatic Displays ($n = 51$) were used more than any other defence tactic with sea bass and bluefish (Fig. 5.27c) and were highly effective in disrupting the attack sequence. Given the substantial differences in studying young *S. officinalis* in the ocean versus in small laboratory tanks, it is surprising that some of the basic tenets of primary and secondary defence seem quite similar. Nevertheless, as noted in all of these papers, studies in nature are crucial to proper ethological discoveries of defence.

5.3.3 Squid, *Sepioteuthis sepioidea*, in the Field

The Caribbean reef squid lives in a coral reef environment with numerous predators (Moynihan & Rodaniche, 1982), and field observations have shown that individuals may have, on average, about seven interactions with predators per hour throughout the daylight period (Hanlon & Forsythe, unpublished field

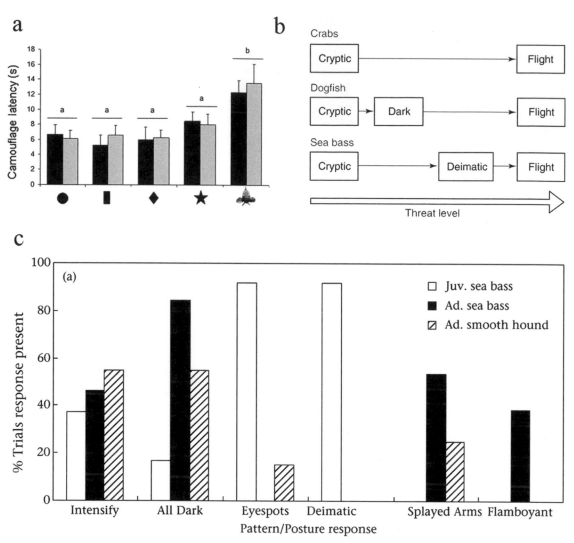

Figure 5.26 Cuttlefish *S. officinalis* defences in lab trials with no predation. (**a**) Young cuttlefish show a Deimatic Display significantly more to a fish symbol (right) on a video monitor than to other symbols. Black boxes are intensity, grey boxes are polarisation contrast (modified from Cartron *et al.*, 2013: *Invertebrate Neuroscience* **13**, 19–26, Effects of stimuli shape and polarization in evoking deimatic patterns in the European cuttlefish, *Sepia officinalis*, under varying turbidity conditions. Cartron, L., Shashar, N., Dickel, L. & Darmaillacq, A. S., © (2013), with permission of Springer. (**b**) Young cuttlefish show Deimatic Display to a live sea bass predator but not to live crabs or dogfish, showing specificity of responses to predators vs non-predators (reprinted from Langridge, Broom & Osorio, 2007: *Current Biology* **17**, R1044–R1045, Selective signalling by cuttlefish to predators, Langridge, K. V., Broom, M. & Osorio, D., © (2007), with permission from Elsevier). (**c**) Young cuttlefish show different deimatic responses to live juvenile vs adult sea bass and to adult smooth hound fish; Deimatic Display is not shown to large predators whereas Flamboyant Display is (reprinted from Langridge, 2009: *Animal Behaviour* **77**, 847–856, Cuttlefish use startle displays, but not against large predators, Langridge, K. V., © (2009). with permission from Elsevier).

data; Mather, 2010), which is very high compared with most animals (Curio, 1976). Since these squid do not rest on the bottom, their primary defence cannot be hiding on or in the substrate like octopuses and cuttlefish. During the day, this species shoals over seagrass beds adjacent to the reefs, where there are fewer predators.

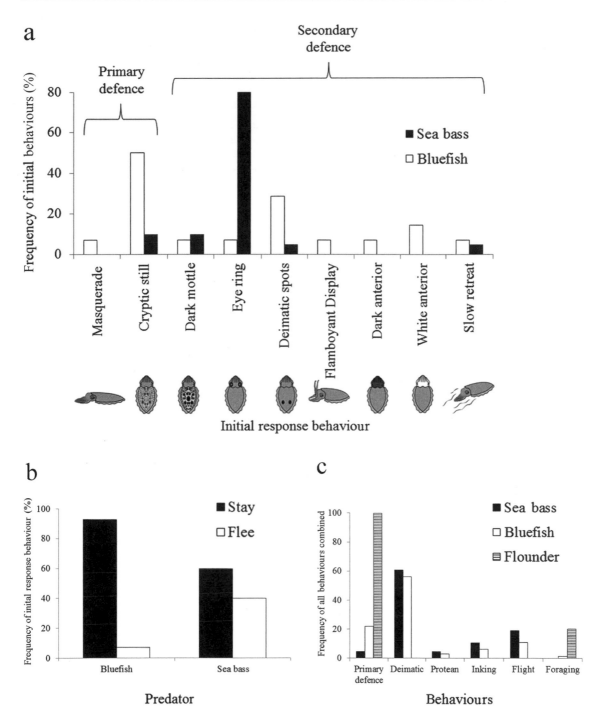

Figure 5.27 Cuttlefish *S. officinalis* defences in lab trials with predation. (**a**) Relative frequency of initial response behaviours shown by cuttlefish to sea bass (filled bars, $n = 20$ behavioural sequences from 6 trials) and bluefish (open bars, $n = 14$ sequences from 9 trials) predators during predator–prey trials. Both primary and secondary responses were very different for each predator. (**b**) Percentage of Stay or Flee tactics shown by cuttlefish as initial responses to approaches by swimming bluefish and sea bass. (**c**) Relative frequencies of all primary and secondary defences, as well as cuttlefish foraging behaviours, shown

They maintain countershading and translucency when in the water column, and this form of crypsis can be very effective (Fig. 5.3c), yet they have many body patterns to help conceal themselves amidst seagrasses, corals and sponges (Fig. 5.3d; Fig. 5.13g,h). For much of the time the squid are in linear shoals showing a dark body pattern (termed Basic by Moynihan & Rodaniche, 1982) that is not cryptic. They have apparent 'sentinels' in each shoal, and usually there are squid facing in all directions at any one time (Fig. 5.24). Thus, primary defences are (1) shoaling with some squid facing in all directions but not in apparently cryptic coloration, or (2) when single or in pairs, deploying cryptic body patterns for background matching, disruptive patterning and masquerade.

When fast, large predators such as the Cero mackerel (*Scomberomorus regalis*) and the yellow jack (*Caranx bartholomaei*) approach, the squid's secondary defence is consistently to jet away quickly and for long distances as a tight school, usually without inking if the predator is detected far away, as it often is (Moynihan & Rodaniche, 1982; Hanlon & Budelmann, 1987; Hanlon & Forsythe, unpublished field data). Sometimes they have been observed to jet out of the water in response to predators (Macia *et al.*, 2004).

Smaller bar jacks, *Caranx ruber*, elicit a 'wait and see' secondary defence tactic from the reef squid. Loosely aggregated squid close ranks, face the intruder, and assume a straight or crescent-shaped line (Fig. 5.24c). Squid often deploy deimatic spots on the mantle (Figs. 3.20b, 5.22c, Box 7.2) towards these small predators (which can only inflict bites rather than consume the whole squid) and to herbivorous fish such as parrotfish (Mather, 2010). *Sepioteuthis sepioidea* appear to assess the size and behaviour of approaching fish and react accordingly. A wide range of Flamboyant Patterns is shown to bar jacks and also to barracudas;

these include many different kinds of arm postures and coloration (Figs. 3.20b and 5.22c), usually some Bars or Streaks on the body combined with various arm Vs or Curls. Distinctive forms of protean behaviour are sometimes shown to closely approaching fish. For example, Moynihan and Rodaniche (1982) observed what we term Pattern Flashing to a fish: the squid flashed its ventral mantle with three dark stripes (while in Upward pointing posture), then abruptly changed posture to horizontal or Downward pointing with Basic coloration while flashing the Paired mantle spots (which are part of the Deimatic Display). It continued to do this several times. The fish apparently never attacked.

To less intense threats, squid showed an equally large variety of body patterns. Moynihan & Rodaniche (1982) reported that they saw 279 permutations of seven components (not specified) for all types of behaviour over their lengthy field study; even allowing for misinterpretation, it is clear that this species can show many patterns. We estimate from our own extensive field observations and video recordings that *Sepioteuthis* can show perhaps several dozen body patterns to predators during defence interactions. These tactics are effective: no successful predation – and only a few unsuccessful attacks – has ever been seen during many hundreds of hours of observation by divers and hundreds of approaches by predators (Hanlon & Forsythe, unpublished data; Mather, 2010). Thus, the many body patterns serve two purposes in this context: they interrupt the approach/attack sequence of fish predators, and, as suggested by Moynihan and Rodaniche (1982), they possibly aid conspecifics in recognising alarm signals from sentinels. The body patterns may also tell the predator that it has been detected. By any measure, the behaviour of *Sepioteuthis sepioidea* is sophisticated and highly evolved.

Caption for Figure 5.27 (*cont.*) during trials with sea bass (filled bars, $n = 22$ behavioural sequences from 6 trials), bluefish (open bars, $n = 17$ sequences from 9 trials) and flounder (striped bar). Note that secondary defence behaviours were not observed during trials with flounder. See text. Reprinted from Staudinger *et al.* (2013a): *Biological Bulletin* **225**, 161–174, Defensive responses of cuttlefish to different teleost predators, Staudinger, M. D. *et al.*, © (2013) Marine Biological Laboratory, with permission from the University of Chicago.

5.3.4 Squid, *Loligo pealeii*, in Laboratory Experiments

These experiments with live fish predators and actual predation provide insight into the way that squid tailor their defence tactics to specific behaviours of predators with very different predation tactics. The longfin squid is preyed upon by more than 50 fish and 6 marine mammal species (Bowman *et al.*, 2000). Staudinger, Hanlon and Juanes (2011) chose two primary predators that represent contrasting foraging tactics: bluefish *Pomatomus saltatrix*, and summer flounder, *Paralichthys dentatus*, which represent cruising and ambush foraging modes, respectively (Staudinger & Juanes, 2010a). With bluefish, 86 predator–prey interactions were evaluated in the course of 35 trials. With flounder, there were 92 interactions during 29 trials. The results were very different. For the initial response (i.e. Flee or Stay) to bluefish, the squid predominantly used Stay tactics (68%, 59/86; Fig. 5.28b), the most common behaviour being quick deployment of a disruptive body pattern and remaining motionless on the substrate; the bluefish repeatedly failed to detect the cryptic squid even though they were swimming just a few centimetres above them (Fig. 5.28a). With flounders, which were camouflaged on the substrate, squid did not detect the flounder in 34 of 92 interactions and showed no reaction to being attacked. When they were aware of an impending attack by the flounder, Flee was the common reaction 43% of the time (40/92). Flee tactics included flight (with or without inking), scattering, and moving to the surface. After initial tactics of Stay or Flee, squid showed varying protean sequences of behaviours to avoid or deter attacks, and these were substantially different for bluefish versus flounder (Fig. 5.28c). When all defence behaviours were considered collectively, flight was identified as the strongest predictor of squid survival during interactions with each predator. For example, with bluefish, there was a high probability of survival (65%, 20/31) when squid fled at least ten body lengths away (owing to abandoned attacks by bluefish), and conversely a high probability of mortality (64%, 35/55) when squid did not flee from bluefish. The most important deimatic/protean behaviour used by squid was inking. Inking caused bluefish to be startled (deimatic) and abandon attacks (probability of survival = 61%, 11/18), and caused flounder to misdirect (protean) attacks towards ink plumes rather than towards squid (probability of survival = 56%, 14/25). Collectively, these experiments – with actual full-length predation or escape – have allowed development of more detailed sequences of secondary defence tactics between this particular squid species and two of its major predators. These combinations of deimatic and protean behaviours are illustrated in Fig. 5.28d.

5.3.5 Influence of Injury on Squid Defence Tactics

Multiple laboratory experiments with live fish predators have enlightened our understanding of nociception and anti-predator behaviour in *Loligo pealeii*. First, it was shown that minor injury to one arm produces long-term sensitisation of behavioural and neuronal responses (Crook *et al.*, 2011; Crook, Hanlon & Walters, 2013). To determine the adaptive value of this sensitisation, Crook *et al.* (2014) subjected injured and uninjured squid to predatory black sea bass. The sequences of behaviours by fish and squid are illustrated in Fig. 5.29a. Surprisingly, these fish selectively targeted injured squid. Once targeted, injured squid began defensive behavioural sequences earlier than uninjured squid (Fig. 5.29b). This effect was blocked by brief anaesthetic treatment that prevented development of nociceptive sensitisation. Importantly, the early anaesthetic treatment also reduced the subsequent escape and survival of injured, but not uninjured, squid (Fig. 5.29c). The 'benefit' or adaptive value of nociceptive sensitisation can be measured by the difference in survival between these two groups (e.g. 45% for injured squid vs 19% injured anaesthetic in these experiments) (Fig. 5.29c). Thus, while minor injury increases the risk of predatory attack, it also triggers a sensitised state that promotes enhanced responsiveness to threats, increasing the survival of injured animals during subsequent predatory encounters.

5.3.6 *Octopus vulgaris* in the Field

Secondary defences in octopuses have not been investigated experimentally although laboratory observations on animals in the 100–3000 g range suggest

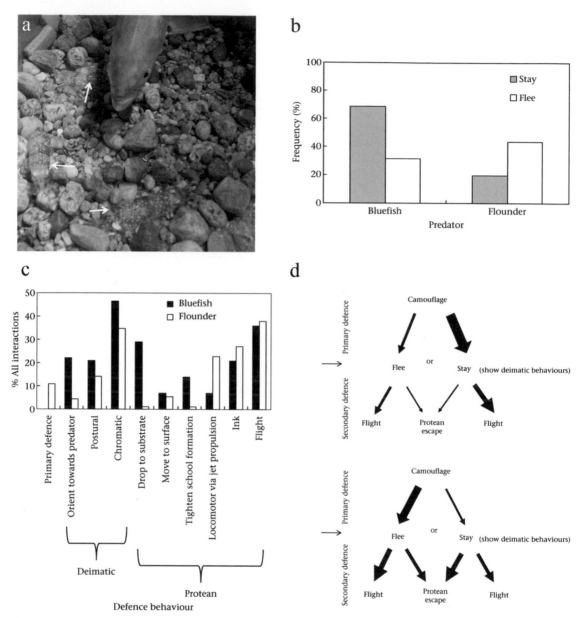

Figure 5.28 Secondary defences of squid *Loligo pealeii* in lab trials with predation. (**a**) Squid showing a disruptive banded body pattern as initial reaction to predatory bluefish, which did not detect the three camouflaged squid (arrows). (**b**) Frequency of Stay and Flee tactics as initial responses to predatory fish. Low percentages for flounder reflect the high mortality of squid inflicted by that predator, which itself was camouflaged on the substrate. (**c**) Frequency of occurrence (as a percentage of all interactions) of all squid behavioural defences shown during predatory–prey interactions with bluefish and flounder. Includes initial responses (deimatic) and subsequent reactions (protean). (**d**) Refined model of anti-predator defence sequences shown by squid during interactions with pelagic cruising predator (top) and benthic ambush predator (bottom). Thicker arrows indicate stronger emphasis. (Modified from Staudinger, Hanlon & Juanes, 2011: *Animal Behaviour* **81**, 585–594, Primary and secondary defenses of squid to cruising and ambush fish predators: variable tactics and their survival value, Staudinger, M. D., Hanlon, R. T. & Juanes, F., © (2011), with permission from Elsevier.)

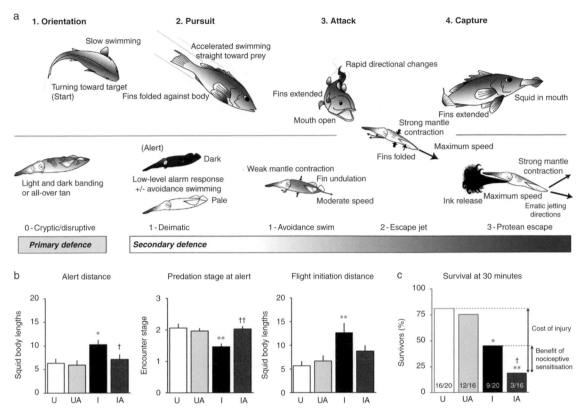

Figure 5.29 Influence of injury on squid defence tactics: nociceptive sensitisation enhances survival in injured squid. (**a**) Top: four stages of predator behaviour by black sea bass. Bottom: primary and secondary defences of squid shown during live predator/prey laboratory experiments. (**b**) Squid in the injured group (I) had longer alert distances (i.e. earlier initiation of secondary defences in **a**), were alerted at earlier encounter stages, and initiated flight (escape jetting with or without ink) at greater distances compared to the other groups. (**c**) At the conclusion of a 30-min trial with free interaction of squid and fish, squid in the I and IA groups had lower overall survival than in the U group, and IA group squid were most likely to be killed. See text. Reprinted from Crook *et al.* (2014): *Current Biology* **24**, 1121–1125, Nociceptive sensitization reduces predation risk, Crook, R. J., Dickson, K., Hanlon, R. T. & Walters, E. T., © (2014), with permission from Elsevier. U, uninjured; UA, uninjured with anaesthetic treatment; I, injured; IA, injured with anaesthetic treatment.

multistage responses to artificial disturbance (Packard, 1963; Packard & Sanders, 1969, 1971; Packard, 1988a). Octopuses initially show a Flamboyant Display or the Defensive Posture in which suckers are shown and water is forcibly blown at the intruder (this particular Defensive Posture is shown to approaching fish; Packard, 1963). Further disturbance leads to the Deimatic Display and finally to the Blanch–Ink–Jet manoeuvre.

To acquire initial data on octopus secondary defences in the wild, Hanlon and colleagues first allowed an *Octopus vulgaris* to habituate to the diver's presence, then tested its responses *in situ* by 'attacking' it repeatedly in rapid succession and recording its behaviours with a video camera. Fifteen octopuses were tested at Grand Cayman Island in the Caribbean. Overall, 15 behaviours were observed in these trials: 10 were Flee behaviours and 5 were Stay behaviours (Fig. 5.30b). There was great variation in the number, duration and sequences of secondary defence behaviours used by the 15 octopuses. For the reactions to first attack, the initial deimatic behaviours were: eight octopuses used a Flee tactic (e.g. jet escape) and seven used a Stay tactic (e.g. conspicuous Deimatic

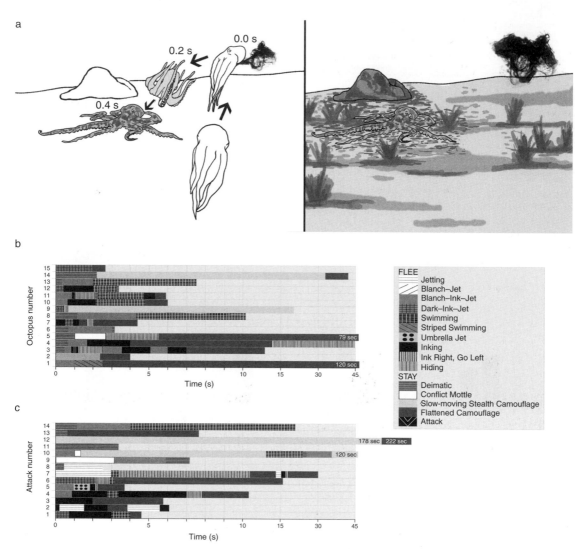

Figure 5.30 Sequences of secondary defence behaviours of *Octopus vulgaris* in the field. (**a**) 'Fake right, go left' protean escape manoeuvre illustrating extremely fast neural polyphenism in which it changes its body pattern twice within 400 ms after changing swim direction and simultaneously ejecting ink in the opposite direction. Drawing (right) shows the visual deception, with the ejected ink as a high-contrast black while the octopus is cryptic amidst the substrata. (**b**) Responses of 15 octopuses to first attack; each of 15 behaviours is coded on the right. (**c**) All responses of Octopus 3 (from **b**) during 14 serial attacks. See text. Based on video by R. Hanlon at Grand Cayman Island, BWI.

Display). For all subsequent attacks, the results showed similar proportions: 62 Flee and 77 Stay. Thus, it would be difficult for a predator to predict Stay or Flee upon attack.

Subsequent to the first attack, various protean escape behaviours were exhibited in rapid succession; the complexity and diversity of these are shown in

Fig. 5.30b. Note, for example, that Octopus 3 used six behaviours to implement ten changes within 13 seconds! To further illustrate the concept of unpredictable erratic escape (i.e. protean defence), Fig. 5.30c shows all responses by Octopus 3 during 14 serial attacks. Put another way, this octopus varied its sequence of behaviours in every attack.

Here we see rapid neural polyphenism being exercised at its best. The cognitive aspects of these complex behaviours deserve future study because, while rapidly fleeing, the octopuses were tailoring their next behaviours to the benthic surroundings. An example is the 'fake right, go left' behaviour seen often in these trials (Fig. 5.30a) in which the swiftly swimming octopus suddenly stops, inks to the right, blanches and descends left amidst a patch of mottled algae (all in 0.4 seconds), and matches the mottled algal background. Obviously there is still much yet to learn about secondary defences in octopuses and how their decision-making is so precise.

5.4 Predators of Cephalopods

Cephalopods are preyed upon by a wide variety of vertebrates, particularly the highly diverse and numerous teleost fish, which are renowned for their keen visual senses. Hence the importance of the body patterns that are described in this chapter. Yet teleosts have other senses that are important in detecting prey, such as olfaction and electro-sensing, against which visual crypsis is ineffective. Elasmobranch fish (sharks, rays, eels) are key predators on some cephalopod species, and they rely on electro-sensing and olfaction more than vision. Marine mammals are also primary predators of cephalopods, some using vision (e.g. dolphins and seals) and many using sonar (e.g. toothed whales). Many diving birds consume cephalopods in the upper 20 m, and those species have keen vision; these include albatrosses, auks, petrels and terns. Penguins can dive to more than 200 m, and they feed heavily on squid throughout those depths. Some general reviews on predators of cephalopods have been published and the interested reader may consult those (e.g. Clarke, 1996 on cetaceans; Croxhall & Prince, 1996 on seabirds; Klages, 1996 on seals; Smale, 1996 on fish; Boyle & Rodhouse, 2005).

Predation by birds and mammals is summarised in Box 5.4. However, we know few of the behavioural aspects of predator–prey interactions. For example,

---

Box 5.4 Bird and mammal predation on cephalopods

Many species of birds are important predators of cephalopods. For example, king penguins (*Aptenodytes forsteri*), which can dive to more than 260 m, have been reported to capture as many as 90 squid, each weighing 159–200 g, per dive (Kooyman *et al.*, 1982). It is interesting that they hunt by day and by night (e.g. Klages, 1989). Other species of birds, such as the sooty albatross (*Phoebetria fusca*) hunt mostly at night when squid are near the surface (Imber, 1973; Berruti & Harcus, 1978); 37 species of cephalopods have been recorded from the guts of these birds, most of them being bioluminescent forms. Great-winged petrels (*Pterodroma macroptera*) prey heavily on *Spirula spirula*, which may constitute as much as 25% of their diet (Imber, 1973); this is another bioluminescent cephalopod (Bruun, 1943) emitting a pale yellowish-green light from the light organ at the tip of the mantle, which is uppermost as the animal swims (Denton, Gilpin-Brown & Howarth, 1967). Other species of albatross (e.g. Rodhouse *et al.*, 1990) and shearwaters (Morejohn, Harvey & Krasnow, 1978) are among other birds known to consume large quantities of squid. Some albatrosses feed by scavenging dead squid, but others probably feed on live squid, which must come near the surface because these birds can only dive to a few metres' depth. Croxall and Prince (1996) review many of these interactions.

Cape clawless otters (*Aonyx capensis*) forage not only by vision, but by using their forepaws, which have a highly developed tactile sense, to feel underneath and between loose stones (Maxwell, 1965). In one study, *Octopus granulatus* was found to constitute about 15% of its diet by weight (Verwoerd, 1987). The Eurasian otter *Lutra lutra* forages in marine habitats and takes octopuses as a minor constituent of its diet (Beja, 1991). The North American sea otter (*Enhydra lutris*) eats *Octopus dofleini* in the 1–2 kg size range (Kenyon, 1975). McCleneghan and Ames (1976) report that sea otters regularly search for and tear apart aluminium cans to prey upon octopuses hiding inside!

Northern and southern elephant seals (*Mirounga* spp.) feed on at least 12 species of sepioids, squid, and octopuses. These animals make repeated dives to depths

Continued

as great as 1250 m; energetics calculations indicate that they must capture many squid per dive (e.g. Condit & Le Boeuf, 1984; Clarke, 1985; Antonelis *et al.*, 1987; Le Boeuf *et al.*, 1989; Sakamoto *et al.*, 1989; Boyd & Arnbom, 1991). The California sea lion (*Zalophus californianus*) consumes large quantities of the squid *Loligo opalescens* (Morejohn, Harvey & Krasnow, 1978), and in Alaska and British Columbia sea lions and harbour seals prey upon *Octopus dofleini* (Kenyon, 1975; Hartwick, 1983). During their breeding season in Hawaii, monk seals forage mainly at night on coral reefs (usually 19–40 m deep) and capture benthic octopuses, which form approximately 25% of their diet (DeLong *et al.*, 1984).

Whales must certainly be the prime consumers of cephalopods in terms of biomass. For example, sperm whales have been estimated to consume over 150 million tonnes of squid annually (Denton, 1970; Clarke, 1977, 1983, 1996). By comparison, the total catch by human fisheries is only about 60 million tonnes annually, with the total annual catch for cephalopods being about 1.5 million tonnes.

Dolphins (*Delphinus delphis*) have been seen attacking *Illex illecebrosus* being fished from a boat under night lights (Major, 1986). The dolphins aligned themselves ten abreast and swam swiftly into a very large shoal under the hull of the boat, breaking up the shoal until some squid were isolated and could be chased into the illuminated area by the dolphins. Three captures were seen, and in each the squid were captured despite inking, changing body pattern and jetting away in a straight-line trajectory. Other squid evaded capture by changing direction erratically as they changed body pattern and inked (protean behaviour, see §5.2.4). The dolphins seemed to be hunting visually, and their tactic of breaking up the shoal into individuals is a common one used by vertebrates for many shoaling prey (e.g. Keenleyside, 1979; Würsig, 1986). Bottlenose dolphin (*Tursiops aduncus*) sometimes feed on spawning giant Australian cuttlefish by a complex handling routine in which they herd them out of the seaweeds (from camouflage) to sandy areas, then pin the individual cuttlefish to the substrate with their snout and kill them with a downward thrust. Thereafter they thrash the dead cuttlefish to get rid of ink and smash it on the substrate to remove the skin and release the cuttlebone, to finally ingest a 'clean meal' of cuttlefish meat (Finn, Tregenza & Norman, 2009b).

Humans could also be said to be predators of cephalopods in the sense that they harvest them in many parts of the world. Commercial fisheries for cephalopods are growing steadily (Worms, 1983; Boyle & Rodhouse, 2005). Jigs and jigging machines combined with bright lights seem to be very effective predatory devices. We do not know, however, why squid are attracted to lights at night, and it does not seem that jigs and their motion mimic the movements of natural prey. Since most cephalopods can swim or jet quickly they avoid small, traditional, bottom trawls, but modern trawls are sufficiently large that they capture many cephalopods, although not large ones. Sonic detection and other aspects of fishing technology have improved as well. Oceanic gill nets, strung over hundreds of miles, are extremely efficient at entangling many squid species that apparently cannot detect the clear monofilament netting.

although sperm whales consume huge quantities of squid, we do not know exactly how they catch them or how some squid avoid predation. Norris and Mohl (1983) hypothesised that toothed whales might stun prey with intense bursts of sound from their sonar apparatus. However, it has now been shown that squid such as *Loligo pealeii* are not debilitated by sound bursts, nor are they able to sense the intense ultrasonic clicks of approaching toothed whales that are using sonar to locate squid (Madsen *et al.*, 2007; Wilson *et al.*, 2007). However, squid and cuttlefish have very acute low-frequency hearing that would allow them to detect and respond to the head waves of approaching predators such as other cephalopods, fish, seals and toothed whales (Mooney *et al.*, 2010; Samson *et al.*, 2014). Furthermore, in the case of the giant squid *Architeuthis*, their extremely large eyes are thought to be capable of detecting the large diffuse bioluminescence caused by approaching toothed whales; it is calculated that they can detect this at a distance of 120 m, thus hypothetically giving them the opportunity to deploy secondary defence tactics (Nilsson *et al.*, 2012, 2013). Clearly, many of these predators are using vision, even at night or in very deep water.

Among teleosts, swordfish (*Xiphias gladius*) prey on a great variety of cephalopods; apparently they use their long bills to slash shoaling squid with swift horizontal head thrashes while they move up or down through the school (Scott & Tibbo, 1968; Stillwell & Kohler, 1985; Bower & Ichii, 2005). Toll and Hess (1981) and Bello (1991) found decapitated heads and slashed mantles in the stomachs of swordfish, and estimate that squid provide their main diet in the epipelagic and upper mesopelagic zones. Various billfish (blue marlin, white marlin, sailfish) also eat large quantities of oceanic squid (including strong swimmers such as *Thysanoteuthis* and *Ommastrephes*), the octopod *Argonauta* and, as they venture inshore, *Loligo*. Staudinger *et al.* (2013b) identified 13 species of fish that feed collectively on 22 species of squid and 4 species of octopuses in the northwest Atlantic Ocean.

On reefs and seagrass beds in tropical and temperate zones, a very wide variety of teleosts consumes cephalopods; a complete list would probably include every carnivore, but major predators include the flatfish, scorpion fish, serranids, scombrids, sphyraenids and lutjanids (cf. Randall, 1967; Taylor & Chen, 1969; Klumpp & Nichols, 1983). The common Caribbean reef squid, *Sepioteuthis sepioidea*, has been observed being attacked by several teleosts: yellow jacks (*Caranx bartholomaei*), bar jacks (*Caranx ruber*), mutton snapper (*Lutjanis analis*), cero mackerel (*Scomberomoras regalis*), great barracuda (*Sphyraena barracuda*) and houndfish (*Tylosurus crocodilus*) (Moynihan & Rodaniche, 1982; Hanlon & Forsythe, unpublished data; Mather, 2010). The fish may approach from high or low in the water column from anywhere in the 360° field of view, sometimes slicing through the middle of the linearly arranged squid shoal to break it up into small groups. Barracuda, in particular, often remain motionless near the squid for long periods of time so that the squid must remain vigilant towards them while simultaneously observing other passing predators.

In the Mediterranean, hatchling and juvenile cuttlefish *Sepia officinalis* are preyed upon by the teleost *Serranus cabrilla*, which patrol 'hunting territories' looking for movement or other signs of potential foods, or lie amidst *Posidonia* grass blades and ambush passing prey (Hanlon & Messenger, 1988).

Sharks exploit cephalopods in many habitats; for example, the blue shark (*Prionace glauca*) feeds heavily on oceanic squid in the epipelagic and upper mesopelagic zones but also consumes neritic *Sepia*, *Rossia*, *Alloteuthis* and the octopus *Eledone* when it migrates inshore (Clarke & Stevens, 1974). The deep-water shark *Centroscymnus* takes 15% of its diet as squid and occasionally consumes octopuses (Yano & Tanaka, 1984). In the neritic waters off southern Australia, four shark species have been estimated to obtain 25–60% of their diet from benthic octopuses; their diet also includes cuttlefish and oceanic squid (Coleman, 1984). Alonzo *et al.* (2002) reported that the squid *Illex argentinus* is a primary prey of spiny dogfish in Patagonian waters. Pyjama sharks lie hidden in egg masses of the squid *Loligo reynaudii* in South Africa and prey on females as they descend to lay egg fingers (Smale, Sauer & Hanlon, 1995). Sharks and rays perhaps locate benthic cephalopods by electroreception or by olfaction.

Octopuses are thought to be especially vulnerable to eels, which may use olfactory cues to find them (Lane, 1957; Randall, 1967; MacGinitie & MacGinitie, 1968). Ambrose (1988) reported that brooding females of *Octopus bimaculatus* were subject to a high mortality rate and that this was sometimes due to predation by moray eels (*Gymnothorax mordax*).

In oceanic and Arctic regions, the cetaceans, pinnipeds and birds, as well as teleost fish, are major predators (especially of squid). In the mid-water and deep benthic habitats, some sharks and teleosts, such as the lancetfish *Alepisaurus ferox*, consume a variety of cephalopods. In neritic habitats of temperate regions, the pinnipeds, birds, teleosts and sharks are the main predators. Large oceanic squid eat other smaller squid of many genera (cf. species accounts in Boyle, 1983b; Nixon, 1987). Size seems to be the important factor regulating this behaviour. Perhaps for this reason it is unusual to see squid shoals with differently sized individuals, although *Sepioteuthis sepioidea* is an exception. Octopuses prey on *Nautilus* and other octopuses.

5.5 Summary and Future Research Directions

Defence is of critical importance to the soft-bodied cephalopods, and they have evolved elaborate behaviours to accomplish this, both in the context of primary defence and, once discovered, by a complex array of secondary defences. The cognitive aspects of decision-making during these endeavours are particularly noteworthy, especially during secondary defence when a predator is upon them and they are rapidly assessing the behavioural context and the habitat structure around them to conduct the next critical evasive move. We have learned that some cephalopods recognise predator types and tailor their secondary defence responses to different predators. The predators of cephalopods are mainly vertebrates (teleost fish, marine mammals, birds) with highly refined and diverse senses, and most cephalopods are prey to many predators; thus their defences have to be diverse. Rapid neural polyphenism provides much of that defensive diversity. Cephalopods are predominantly visual animals living in a visual world, and most of their known defences are directed at animals that hunt by sight, even at night.

One of the greatest advances in the past decade has been the renaissance in the study of animal camouflage as a primary defence strategy in natural selection. Widespread studies in many animal taxa have helped us to refine our approaches to describing and understanding cephalopod camouflage. Conversely, cephalopods, with their unique capability of rapid neural polyphenism, have provided a valuable comparative group with which to study the principles of visual camouflage.

Figure 5.1 summarises primary and secondary defences and some typical interactions between cephalopods and their predators. Substantial progress has been made in supporting this basic flow of events since we first presented it in 1996. For primary defences, a great deal has been learned about the multiple tactics of crypsis, particularly the details of background matching and the possible roles of disruptive patterning. An important future goal would be to prove or disprove disruptive coloration experimentally; that is, we need one or several predator models (preferably in the field) to test whether recognition or detection of the cephalopod is occurring. This is an ongoing challenge for all animal studies. New information has been acquired on dynamic night camouflage and motion camouflage, and this can lead to future studies on visual perception of both the cephalopod and its predators.

Octopuses seem capable of adjusting their speed of stealth movement to the motion around them (due to wave flicker, water surge, etc.), and the cognitive skills required to coordinate such behaviours would seem advanced and worthy of study. Among other priorities, it will be important to quantify crypsis in the eyes of predators, and recent developments with new hyperspectral imaging cameras will enable analyses to be made of spectral as well as spatial information on animal versus background.

Substantial strides have been made on masquerade and mimicry in cephalopods. Yet how and when does a cephalopod choose to use background matching versus masquerade in a given habitat? The former tactic requires analysing the background substrate immediately around them, and the latter tactic requires analysis of visual information in the distance. And when does mimicry offer a better solution for primary defence? The three species of mimic octopus all live on open sand plains where fast movement will be detected from long distances, so these octopuses have imitated the swimming behaviours and coloration of flatfish. Yet we are ignorant of the mimic models (and whether they are toxic) and the common predators that are driving the evolution of these mimicry behaviours. Extensive field studies are needed for this and many of the issues raised in this chapter.

For secondary defences, various laboratory experiments have enlightened our understanding of when and how specific tactics are used during deimatic and protean behaviours. For example, it is now clear that deimatic behaviours (startle, threat) are a useful initial 'Stay' tactic once a cephalopod has been detected or recognised and attack is imminent. We now have the first metrics on the survival value of the Deimatic Display and we have also learned, unexpectedly, that minor injuries to a squid can sensitise it and render it more attentive to predation. One common 'Flee' tactic is the 'Blanch–Ink–Jet' ('disappear, reappear and divert attention') manoeuvre that occurs either during initial flight, or in variable ways after deimatic behaviour. Inking is another unique characteristic of cephalopods, and its role in chemical defence has been partially uncovered with studies on squid. The overall strategy seems to be to maintain diversity and unpredictability (protean behaviour). It is also important to recall the advantage that accrues from having good vision, a large brain and fast escape responses: the cephalopod can often 'wait and see' what a predator is likely to do before choosing one of its many defence options.

Various refinements to the overall schematic flow of secondary defences have been determined through full-scale live predator–prey experiments on squid, and these more realistic trials help to quantify the costs and benefits of different defence tactics. A great deal more of this enquiry is needed in the future, but it needs to be corroborated with field studies. This sort of research is difficult but highly informative.

CHAPTER SIX

Reproductive Behaviour

Reproductive behaviour includes agonistic behaviour (i.e. the complex of behaviours that includes fighting, threat, appeasement and avoidance, usually among males), courtship behaviour (i.e. behaviours that coordinate the activities of sexual partners in time; orient partners in space; and increase sexual motivation), mating behaviour (i.e. copulation, which in cephalopods does not usually include fertilisation), and any form of parental care if it occurs.

In cephalopods, there are considerable differences between the reproductive behaviour of decapods and octopods. Shoaling squid have ample opportunity for social interactions with conspecifics throughout most of their lives, and some species have developed elaborate agonistic and courtship behaviour. Cuttlefish, which seldom or never shoal, have intraspecific interactions for part of their lives, and some have intricate agonistic and courtship behaviour. Octopuses, on the other hand, are mostly solitary and thus have relatively fewer interactions with conspecifics. Most octopuses studied thus far exhibit little courtship behaviour and male agonistic behaviour, yet mating is prolonged compared with that in cuttlefish and squid, and female octopuses invest large amounts of time and energy in brooding eggs. Nearly all cephalopods, apart from the pelagic octopod *Ocythoe*, are oviparous, the sexes are separate, and there are no cases of sex reversal as in other molluscs. Probably all cephalopods of both sexes have multiple mates.

6.1 Anatomy, Sexual Dimorphism and Reproductive Physiology

In cephalopods, the sexes are separate, and there are no hermaphrodites or sex reversals as in other molluscs.

The reproductive organs of a squid are shown in Fig. 6.1 and those of an octopus in Fig. 6.2. Males produce large quantities of sperm stored in elaborate spermatophores. For example, squid spermatophores each contain 7×10^6 sperm in *Loligo pealeii* (Austin, Lutwak-Mann & Mann, 1964), and *Octopus dofleini* spermatophores may be 1 m long and contain billions of sperm (3.7×10^{10} (Mann, Martin & Thiersch, 1970). Other species have variable sizes of spermatophores (Mann, Martin & Thiersch, 1970; Mann, 1984; Ikeda, Sakurai & Shimazaki, 1993; Voight, 2001; Naud & Havenhand, 2006; Hoving *et al.*, 2010; Marian, 2012) that are stored in Needham's sac and are extruded during mating. Extrusion occurs through a protrusible terminal organ (sometimes called a penis) normally located near the excretory duct. Most males possess a modification of one arm – the hectocotylus – that helps to transport spermatophores from the penis to the female. The structure and position of the hectocotylus are species-specific and are useful taxonomic features. Many oceanic squid have no hectocotylus, and it is absent from *Vampyroteuthis* and the finned octopus *Cirroteuthis* (Murata, Ishii & Osako, 1982; Arnold, 1984). Some sepioids and teuthoids have two arms hectocotylised. In cuttlefish and some squid, the hectocotylus is evident only by the reduction or absence of suckers on the arm (usually one of the fourth arms); in other squid, the entire arm may be modified. In octopuses, the whole arm is usually modified by a groove along its ventral surface, terminating near the arm tip, which is modified and often spoon-like in shape. During copulation, the hectocotylised arm acquires the spermatophore (either by reaching in with the arm tip, as in squid and cuttlefish; or by extension of the penis to the base of the hectocotylised arm as in octopods) and transfers the spermatophore to the female

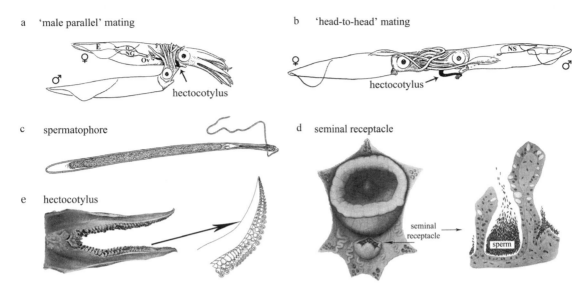

Figure 6.1 Mating and reproductive behaviour anatomy in squid. (**a**) In the 'male-parallel' position, the male places spermatophores in the mantle cavity usually near the distal end of the oviduct. (**b**) In the 'head-to-head' position, sperm from the spermatophores of 8–16 mm length (**c**) are placed for storage in a seminal receptacle (**d**) usually near the mouth (based on Drew, 1911, with permission from Wiley) by the hectocotylus (**e**) of the male. Ov, oviduct. NG, nidamental gland. O, ovary. E, eggs. NS, Needham's sac. T, testis.

in a variety of ways (see following sections). Males of most species mature early in life (e.g. in 3–6 months) and are capable of passing ripe spermatophores to females for roughly two-thirds of their life span (for details, see the species accounts in Boyle, 1983b and Rosa, O'Dor & Pierce, 2013).

Seminal receptacles of females store viable sperm, and in most decapods these receptacles are distant from the oviducts. Cephalopods have a single ovary and either one oviduct (in decapods) or two (in octopods). In *Sepia*, *Idiosepius* and *Loligo*, as well as some other decapods, the seminal receptacle is situated just below the mouth (Figs. 6.1d, 6.9d, 6.11g, 6.17d, 6.29) or in a ring around the mouth (e.g. Drew, 1911; Ikeda, Sakurai & Shimazaki, 1993; Naud *et al.*, 2005; Sato *et al.*, 2010; Iwata *et al.*, 2011). In some sepioids (e.g. *Rossia, Euprymna, Spirula*), females have a modification of the oviducal openings, called a pharetra, that serves as the seminal receptacle (e.g. Naef, 1923; Arnold, 1984). The squid *Lolliguncula brevis* has a spermatophoric 'pad' located on the inner mantle wall (near the left gill) where males attach spermatophores during copulation, and other squid have adaptations for embedding spermatophores near the neck, head or on the mantle (Norman & Lu, 1997;

Marian, 2012). In the oceanic squid of the Octopoteuthidae, Onychoteuthidae and the Cranchiidae, the spermatophores are deposited in cuts made in the exterior mantle skin (possibly by the male's tentacular hooks) (Hoving *et al.*, 2010; Hoving, Bush & Robison, 2012; Marian, 2012). Most octopuses have compartmented oviducal glands (one in each of two oviducts) in which sperm can be stored for long periods (Fig. 6.2b), although in *Eledone* the spermatophores ascend straight into the ovary (Froesch & Marthy, 1975). In general, females mature later (ca. 4–8 months) than males, but they too can mate and store sperm for about half to two-thirds of their life (Boyle, 1983b).

The extent of external sexual dimorphism, which is often associated with sexual selection, varies among cephalopods. The main characters are: size, body patterns, enlarged suckers, gonad shape or colour and, in a few genera, the photophores.

Among cuttlefish, males of the small sepioids *Euprymna*, *Sepiola* and *Sepietta* have enlarged suckers on some of the arms and, in some species, the gonads are easily distinguished through the integument (Boyle, 1983b). In *Sepia officinalis*, dimorphism is evident only in mature males, which have the fourth arms

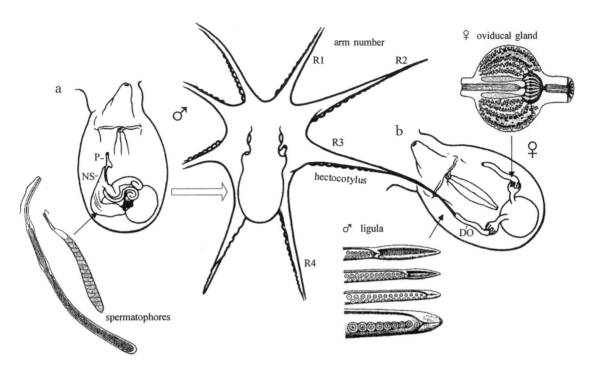

Figure 6.2 Octopus reproductive system. (**a**) Male internal anatomy. Two examples of spermatophores are shown. *Eledone moschata* (8mm) and *Eledone cirrhosa* (15mm). Spermatophores are extruded by the penis (P) from Needham's sac (NS) and passed along the hectocotylised third arm during mating. (**b**) Female internal anatomy. The hectocotylus is presumably inserted during mating into the distal oviduct (DO), and sperm are stored in the oviducal gland until eggs are fertilised as they pass from the ovary. Examples of the ligula (distal end of the hectocotylus) of hectocotyli of *Octopus salutii*, *O. macropus*, *O. vulgaris* and *Eledone moschata* are shown (after Naef, 1923). Diagram of the oviducal gland is from Froesch and Marthy (1975): *Proceedings of the Royal Society of London B* **188**, 95–101, The structure and function of the oviducal gland in octopods (Cephalopoda), Froesch, D. & Marthy, H.-J., © (1975), by permission of the Royal Society.

emboldened by the chromatic components White and Black zebra bands and White arm spots (cf. Hanlon & Messenger, 1988); this is essentially a visual signal (see below). Males of *Sepia latimanus* have Zebra bands on the mantle and arms that are lacking in females (Corner & Moore, 1980), and in *Sepia apama*, males have much larger fourth arms than females, which becomes important during sexual mimicry (Hanlon *et al.*, 2005; §6.2.1).

In a few squid, there are marked size differences between the sexes. For example, *Loligo plei* males are nearly twice as long as females (Hanlon, 1982), whereas females of oceanic species such as *Dosidicus gigas* are considerably larger than males (Nesis, 1983). Many loliginid squid are translucent so that the testis is often highly conspicuous in males (Fig. 6.24b); in females, the

red accessory nidamental gland is often visible, as in *Lolliguncula*, *Alloteuthis* and some species of *Loligo*. Body patterns are sometimes different in males and females, each having distinctive patterns used during agonistic and courtship encounters (see below). Photophores are arranged differently in males and females of some genera such as *Lycoteuthis*, *Chtenopteryx* and *Leachia* (Fig. 6.3).

In most octopuses, sexual dimorphism is not obvious. However, there are four genera exhibiting extreme dimorphism. Males of *Bathypolypus lentus* have a large ligula (Fig. 6.3a). Males of the secondarily pelagic *Argonauta*, *Tremoctopus* and *Ocythoe* are exceptionally small (ca. 5% the length of the female) and carry a hectocotylus that is huge by comparison (Fig. 6.3b); the arm becomes charged with sperm and breaks off to remain in the mantle cavity of the female (Wells &

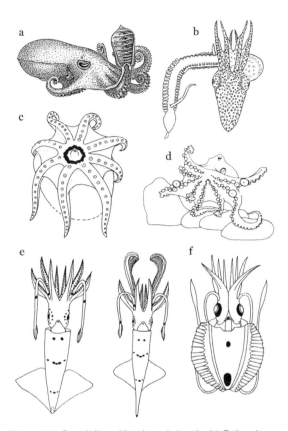

Figure 6.3 Sexual dimorphism in cephalopods. (**a**) *Bathypolypus lentus* male with enlarged ligula (from Verrill, 1880–81). (**b**) Male *Tremoctopus violaceus*, only 13.8 mm ML, with very large hectocotylus (from Thomas, 1977; *Bulletin of Marine Science* **27**, 353–392, Systematics, distribution, and biology of cephalopods of the genus *Tremoctopus* (Octopoda: Tremoctopodidae), Thomas, R. F., © (1977), with permission from Rosenstiel School of Marine and Atmospheric Science/Allen Press, Inc.). (**c**) Female *Japetella diaphana* with circumoral photophores (Robison & Young, 1981: *Pacific Science* **35**, 39–44, Bioluminescence in pelagic octopods, Robison, B. H. & Young, R. E., © (1981), with permission from University of Hawaii Press). (**d**) Male *Octopus vulgaris* with enlarged suckers on the second and third arms (reprinted from Packard, 1961, by permission from Macmillan Publishers Ltd: *Nature* **190**, 736–737, Sucker display of Octopus, Packard, A., © 1961). (**e**) Female (left) and male *Lycoteuthis diadema* showing differences in photophore arrangement (black spots) and in the structure of the second pair of arms (from Herring, 1988). (**f**) Male *Chtenopteryx sicula* with posterior photophore (from Herring, 1988). Panels (**e**) and (**f**) reprinted from *The Mollusca*, Vol. 11. *Form and Function* (ed. E.R. Trueman & M.R. Clarke), 449–489, © (1988), with permission from Elsevier.

Wells, 1977; Mangold, 1989; Norman *et al.*, 2002). Unfortunately, nothing is known about mating in these genera. Males of some octopuses have a few enlarged suckers on the arms used in a 'sucker display' meant possibly for sexual identification (§6.6.3). Many oceanic cephalopods have dimorphic photophore patterns and modifications in the arms (hooks or tubercles) but their behavioural functions are unknown. For example, females of the deep-sea octopods *Eledonella pygmaea* and *Japetella diaphana* develop circumoral photophores that luminesce green (Fig. 6.3c); these are thought to be important in attracting mates (Robison & Young, 1981; Chapter 9).

The physiological control of the reproductive system in cephalopods is not well understood, and even less is known about the regulation of sexual behaviour in these animals. Several sex hormones have been identified in *Octopus*: the ovary produces 17β-oestradiol and progesterone, as does the testis, which also produces testosterone (D'Aniello *et al.*, 1996; Di Cosmo, Di Cristo & Paolucci, 2001). However, no behavioural functions for these hormones have been identified. Neural mechanisms initiating reproductive behaviour in cephalopods have not been identified, but a hormonal influence on maturation of the gonads has long been known. In *Octopus vulgaris*, Wells and Wells (1959) performed classic experiments indicating that the pathway of control is:

light → eye → optic lobe → subpedunculate lobe → optic gland → gonad

The first four links were thought to be neural and inhibitory, the last hormonal and excitatory. When the optic gland is freed from inhibition, it releases a hormone, whose nature is unknown, that causes the gonad to grow. Wells and Wells (1972) showed experimentally that it was not possible to alter the sexual behaviour of *Octopus vulgaris* or *O. cyanea* by altering the supply of hormone from the optic gland or by removing the gonads. They did point out, however, that their experiments do not preclude the possibility that active optic glands or ripened gonads have some more generalised effect on behaviour. Even earlier, Callan (1940) had demonstrated that the development of the male hectocotylus was not under hormonal control from the testis: when the tip of the third arm was amputated, the hectocotylus regenerated whether or not the animal had the testis removed. Wells (1978) and Mangold (1987) provide detailed reviews of these earlier data.

Recent research indicates that the olfactory lobe plays a key role in reproduction. The subpeduncule lobe produces a peptide (FMRF-amide) that influences the optic gland (Di Cosmo & Di Cristo, 1998). The presence of gonadotropin-releasing hormone in the olfactory lobe suggests that it might exert an excitatory action on optic gland activity; other neuropeptides have now been localised in the olfactory lobe, as well as steroidogenic enzymes and an oestrogen receptor orthologue (Di Cristo, 2013). The presence of neuropeptides in the system regulating reproductive development suggests some parallels with insects and vertebrates, but this subject requires future study. The olfactory lobe of *Octopus* receives distant chemical stimuli and would appear to be an integrative centre containing a variety of neuropeptides involved in the onset of sexual maturation via the optic lobe. These new data are reviewed by Di Cristo (2013) who presents a modification of the pathway listed above. This research draws even more attention to the optic gland hormone, whose structure and function remain unknown. Thus, fundamental questions remain unstudied: for example, what is the hormone responsible for gametogenesis and egg laying in *Octopus*, and what is its influence on male sexual maturation and behaviour?

6.2 Agonistic, Courtship and Mating Behaviour in Cuttlefish and Other Sepioids

Cuttlefish fall between the shoaling teuthoid squid and the solitary octopods in that some species form loose aggregations in nature. Information on reproductive behaviour is available for about 14 species of sepioids (Table 6.1). The giant Australian cuttlefish, *Sepia apama*, which inhabits temperate rock reefs and forms the only known large spawning aggregation, is the most thoroughly studied (e.g. Hall & Hanlon, 2002 and others below). The Pacific coral reef species *S. latimanus* and the European cuttlefish *Sepia officinalis* are two others that have been studied in some detail, the former in a prescient field study by Corner and Moore (1980), and the latter primarily in the laboratory (Bott, 1938; L. Tinbergen, 1939; Adamo & Hanlon, 1996 and others below).

Table 6.1 outlines mating behaviour in various sepioids. In addition to these, partial accounts can be found for *Sepia pharaonis* (Gutsal, 1989; Minton *et al.*, 2001), *S. elegans* (Bouligand, 1961), *S. bandensis* (Van Heukelem cited in Arnold, 1984), *S. officinalis* (Grimpe, 1926; Bott, 1938; Tinbergen, 1939), *Sepiella japonka* (Natsukari & Tashiro, 1991), *Sepiola* and *Sepietta* spp. (Racovitza, 1894) and several others in Norman (2000).

6.2.1 *Sepia apama*

The giant Australian cuttlefish aggregates annually into a unique and spectacular mass spawning event (Fig. 6.4a) in South Australia each austral winter (May to August) in which up to 170 000 cuttlefish amass along an 8-km rock reef shoreline in 1–10 m depth (Hall & Hanlon, 2002). Densities peak at 105 cuttlefish per 100 m^2. The cuttlefish habituate swiftly to divers, and more than 150 research SCUBA dives have been made to study this spawning population from 1999 to 2004, rendering this the best-studied mating system among all cephalopods (Norman, Finn & Tregenza, 1999; Hall & Hanlon, 2002; Naud *et al.*, 2004, 2005; Hanlon *et al.*, 2005; Naud & Havenhand, 2006). With such an ideal shallow study site, substantial progress has been made in understanding sexual selection processes in cuttlefish, and cephalopods in general.

Competition for females is intense because of the skewed sex ratio, which averages 4M:1F but ranges from 11:1 at the start of the season down to 1:1 at the end (Fig. 6.4b). Males are sexually dimorphic by size and possess an unusually large pair of fourth arms used in intraspecific signalling (Fig. 6.5a). The largest males (370 mm ML) are often twice the weight of the largest females (up to 250 mm ML). Moreover, the males are generally of two size classes (the smaller class similar to average female size), which influences the mating tactics of males.

Pair formation is temporary and occurs diurnally. Each night, the sexual activities cease, and all cuttlefish segregate and camouflage themselves (see §5.1.1; Hanlon *et al.*, 2007) until dawn, at which time the cuttlefish commence reproductive behaviours once again.

Male tactics to achieve copulations are diverse and highly flexible according to the behavioural context, which can change rapidly. Large males (consorts) pair with females and guard them both pre- and postcopulation (Fig. 6.4c, d). These large consort males engage frequently in elaborate agonistic bouts with other

Table 6.1 Mating by sepioids

Species	Duration (minutes)	Mating position	Spermatophore placement	Temporary mate guarding	References
1. *Sepia apama	2–3	Head-to-head	bmv	Yes	Norman, Finn & Tregenza, 1999; Hall & Hanlon, 2002; Naud *et al.*, 2004, 2005; Hanlon *et al.*, 2005
2. *Sepia officinalis*	2–20	Head-to-head	bmv	Yes	Richard, 1971; Boletzky, 1983b; Hanlon *et al.*, 1999
3. *Sepia latimanus	0.5–1.5	Head-to-head	bmv	Yes	Corner & Moore, 1980
4. *Sepia esculenta*	2	Head-to-head	bmv	Yes	Natsukari & Tashiro, 1991
5. *Sepia lycidas*	2	Head-to-head	bmv	Yes	Wada *et al.*, 2010
6. *Rossia macrosoma*	–	♂ to ♀ neck	phar?	–	Racovitza, 1894; Mangold-Wirz, 1963
7. *Rossia pacifica*	–	♂ parallel	phar?	–	Brocco, 1971; Summers, 1985
8. *Euprymna scolopes* (♂ large suckers)	25–80	♂ to ♀ neck	phar	–	Moynihan, 1983b; Singley, 1983; Hanlon *et al.*, 1997
9. *E. tazmanica*	120–180	♂ parallel	?	?	Franklin, Squires & Stuart-Fox, 2012
10. *Sepiola robusta*	–	♂ to ♀ neck	phar	Yes	Boletzky, 1983a
11. *Sepiola rondeletii*	8	♂ to ♀ neck	phar	–	Racovitza, 1894
12. *Sepietta oweniana* (♂ large suckers)	'In short period of time'	Head-to-head	phar	–	Bergstrom & Summers, 1983
13. *Idiosepius paradoxus*	3–6 seconds	Head-to-head	?	No	Kasugai, 2000; Sato *et al.*, 2010
14. *Sepiadarium austrinium*	3–45	Head-to-head	outer buccal membrane	?	Wegener *et al.*, 2013a

* Observations in the sea; phar: pharetra, internal seminal receptacle near opening of oviduct; bmv: buccal membrane ventrally under mouth; – no information available; ? not certain

large consorts and large lone males (Fig. 6.5b–g). These bouts usually begin with the Frontal Display, where the signaller orients face-first towards the opponent, arms are typically whitened and flaring, and the mantle is down and not visible to the rival. Bouts can then progress to the Lateral Display, where the signaller is oriented laterally to the opponent, arms are typically darkened and flaring, and high-contrast dark bands of chromatophores move across the mantle (i.e. Passing Cloud; Fig. 6.5d).

Alternatively, males can progress to the Shovel Display, where the orientation is the same as the Frontal Display but the mantle is raised and visible to the rival, and the arms are whitened, extended and rigid in a shovel-like shape. Sometimes these fights escalate to lunging (Fig. 6.5f), flipping over an opponent (Fig. 6.5g) and rarely biting. Details are explained in §6.6.2.

Lone large males actively search for lone females or challenge consorts with agonistic displays. Small lone

Figure 6.4 Spawning dynamics of the giant Australian cuttlefish, *Sepia apama*. (**a**) Common assemblage of cuttlefish in which a large consort male is guarding a female in between egg laying; 11 other males surround the mating pair. (**b**) Operational sex ratio throughout the spawning season; data from Hall and Hanlon (2002). (**c**) Consort male (centre) defending a female (right) from a challenging large lone male that is displaying the Passing Cloud in an aggressive manoeuvre. (**d**) Mate guarding by large consort male who covers the female from approaching males. (**e**) Mating; male is on left. (**f**) Female holding a large egg amidst her arms (note the bulge). (**g**) Egg laying by the female, who glues the egg to the underside of a rock in which other eggs have been deposited. (**h**) Egg cluster under a large rock; inset is a close-up of eggs with fully formed cuttlefish about to hatch. Images in **a** and **h** © Fred Bavendam; all others by R. Hanlon.

males guard females if there are no large males nearby, or they search for lone females with whom they mate without guarding. Most commonly on the spawning ground, small males (150–250 mm ML) seek extra-pair copulations (EPC) with paired females via three tactics. 'Open stealth' involves hovering and watching a pair until the paired male is distracted by other small males or by a large lone male challenger, and the small male

then attempts an overt 'sneak mating' with the paired female. 'Hidden stealth' occurs when a male remains concealed under a rock and attempts a copulation with a female who searches under the rock for a potential egg-laying site. Copulations by hidden stealth only occur when the female does not have an egg in her arms. Small males also search for unguarded entrances to find females that are under rocks; these females are paired

Figure 6.5 Agonistic bouts in the giant Australian cuttlefish, *Sepia apama*. (**a**) Non-fighting male illustrating the sexually dimorphic large fourth arm. (**b, c**) Display of White flared arms and initial Frontal Display to rival male. (**d**) Lateral Display in which one male shows the dramatic Passing Cloud of moving dark bands across the mantle, while both animals show the arching of the body, flared arms and high-contrast patterning. (**e**) Moderate escalation in which the lower male flares his arms and body maximally. (**f**) Strong escalation in which the male on the left lunges towards the rival male and attempts to grasp and bite it. (**g**) Extreme aggression in which the top male flares backwards over the rival male. All images by R. Hanlon.

but sometimes out of sight of the consort. Third, 'female mimicry' occurs when a small male adopts the coloration and posture of an egg-laying female (Fig. 6.6a, b) to gain unchallenged access to a female being guarded by a large consort (Norman, Finn & Tregenza, 1999; Hall & Hanlon 2002). Visual deception is effective and leads to successful copulations by the sneaker (see details below). It is possible to see most or all of these tactics amidst a concentrated gathering of ca. 15–20 cuttlefish (Figs. 6.4a, 6.6c) within the course of a 1-hr SCUBA dive at Whyalla, South Australia.

Females are choosy, although they reject and accept mates of all sizes and status. Two female behaviours are most common on the spawning ground: moving around (with or without a paired male) repeatedly looking under rocks, or hovering near or under a particular rock, usually with a paired male. Females reject unwanted mating attempts by jetting backwards or forcibly breaking free from a male's persistent grasp on the head or arms and jetting or inking. In one field study (Hall & Hanlon, 2002) involving 122 mating attempts, 70% were rejected, 27% were successful and the remaining 3% resulted in forced copulations (defined here as any mating that followed a prolonged period of the female trying to break free from the male, during

which the female firmly held her arms together and would not open them to accept a mating). Among the 70% rejections, in one-third of those the female had an egg in her arms, and in another third she signalled non-receptivity with a distinct unilateral white stripe along the base of the fin towards the male (Fig. 6.28; see other details in §6.6.3). In another field study involving 49 mating attempts, 43% were rejected, but again females of any size mated with males regardless of their size or status (Naud *et al.*, 2004). Boal (1997) argued that, based on controlled laboratory experiments, female choice in *S. officinalis* was probably based on olfactory cues, and this remains a possibility for other sepioids, and calls for future research on sexual pheromones and the study of other male characteristics not visible in male phenotype, such as major histocompatibility complex variability (e.g. Landry *et al.*, 2001).

Courtship behaviour – as a general definition – encompasses any behaviour exhibited by either sex that increases the receptivity of the partner to mating or fertilisation. More specifically, it coordinates the activities of sexual partners in time; orients partners in space; and increases sexual motivation. Courtship in *S. apama* has hardly been observed from either sex. The mean time interval between a male encounter with a female and

Figure 6.6 Sexual mimicry as a sneaker male tactic to obtain copulations (*Sepia apama*). (**a**) A small male in normal coloration (i.e. with the longer male dimorphic fourth arms obvious) then facultatively morphs into a female look-alike (bottom image) by retracting his fourth arms and bulging his arms as if holding a large egg (compare Fig. 6.4f). (**b**) The male can then approach the female (F) without being challenged by the large consort (C). Note two other female mimics in this image. (**c**) The female mimic (M) obtains an extra-pair copulation with the female as the consort male (centre) fights another male (top left). All images by R. Hanlon.

attempted mating was only 19 seconds ($n = 35$; Hall & Hanlon, 2002), which allowed little time for courtship. Rarely, a female appeared to signal a male by whitening her arms, extending them anteriorly, and waving them like banners towards the male; this evoked a mating attempt by the male, but these observations need directed study. Once paired with a female, a male sometimes ($n = 19$) brushed his fourth arms over the female's mantle (Fig. 6.4d), but this usually occurred postcopulation and is probably mate guarding rather than courtship.

Mating occurs in the head-to-head position (Fig. 6.4e) and is generally initiated by the male who spreads his arms and grasps the side of the female's head. The female then accepts the mating attempt by opening her arms and overlapping them with the male's. Mating lasts 2.4 min on average and can be broken into three stages: a 'flushing period' of ca. 70%; rapid 1–2 second transfer of spermatophores; and 30% time spent breaking open the spermatophores. Usually only one passing of spermatophores occurs, but sometimes there are two transfers (6 of 39 matings; Hall & Hanlon, 2002).

Females lay one large egg at a time (5–39 eggs per day), and they often move frequently and deposit eggs under different flat rocks, often in the immediate vicinity of eggs from other females, resulting in egg clusters deposited by multiple females under the same rock (Fig. 6.4h). No parental care of eggs or offspring has been observed in either sex. There is a stereotyped sequence of egg-laying behaviours involved in the 7–8 min between serial egg depositions. First, the female jets water to clean her arms, then she holds the large egg amidst her arms for fertilisation to occur (see details

'self'. This display is highly stereotyped and ritualised, and contains at least nine separate signals: two postural, one locomotor, one textural and five chromatic (Box 7.1). Two of those appear to be key signals: if the other cuttlefish does not extend a fourth arm (postural signal), or that arm does not have the Zebra bands and White spots on it (these are two chromatic signals that only males possess; Hanlon & Messenger, 1988), then it is presumed to be a female. L. Tinbergen (1939) was able to show with models that the extended fourth arm and the Zebra bands of the mantle were important signals (or 'releasing stimuli') for this behaviour. Messenger (1970) showed that unilaterally blinded males failed to respond to males displaying on their blind side, resulting in an attempted copulation and subsequent fight.

Agonistic bouts between males begin with the Intense Zebra Display and can escalate to strong physical fighting (Fig. 6.7) and biting in much the same way as seen in *S. apama* and *S. latimanus*. If neither male withdraws after some minutes of performing an Intense Zebra Display, males begin to push against each other, usually using their fourth arms and mantle (Fig. 6.8a). Larger males win most fights; i.e. they forced smaller males to retreat in 11 of 14 laboratory trials (Adamo & Hanlon, 1996). Fighting is frequent: during a 4-hr period, 63 bouts (one every 3.8 min on average) were video-recorded in a very large indoor tank (Adamo et al., 2000). In general, if neither male withdraws after one or more bouts of pushing and displaying, one male leaps upon the other and bites him on the dorsal mantle while the other male tries to bite as well. Our field observations in Turkey (Fig. 6.7), involving a mating pair and one challenging male, showed dramatic and successive agonistic bouts between the consort male and his challenger. During these escalated fights, the males often jetted about and inked profusely for several minutes. Bite marks are sometimes found on fighting males, but these are not mortal wounds.

One component of the Intense Zebra Display – the 'Dark face' (Fig. 6.8a, c) – was found to be a predictor of which male–male encounters would escalate to physical contact (Adamo & Hanlon, 1996). Some males show Intense Zebra Display without the Dark face, and these males withdraw from the bout. As noted in Fig. 6.8b, males that previously lost a bout to another male, but then copulated with a female, came back to the male they previously lost to, showed a Dark face, and did not withdraw from bouts. Thus, the motivational state of the male can be influenced positively after mating and lead to an aggressive, escalated fight.

Male dominance hierarchies are established in laboratory tanks based upon these agonistic bouts (e.g. L. Tinbergen, 1939; Adamo & Hanlon, 1996; Boal, 1996, 1997). The dominant males tend to occupy the centre of small round tanks (2–3 m diameter) while subordinate males stay near the walls. Curiously, males that are deemed to be dominant in such hierarchies do not necessarily achieve the most copulations (see Table II in Boal, 1996), and this is based, in part, on the nature of female choice (see below).

Mating in *Sepia officinalis* seems to be initiated by males (on rare occasions females will spread their arms and approach males), and there has been no evidence of courtship in extensive laboratory experiments (Boal, 1997; Hanlon, Ament & Gabr, 1999; Adamo et al., 2000). The male swims parallel to the female and shows the Intense Zebra Display; females show various patterns (e.g. a drab Light Mottle) or sometimes a Zebra pattern, but never the extended fourth arm. In a few cases, Boal (2007) noted that a female spread her arms towards a male when her choice of mate had been made, and the transparent barrier separating them was lifted; this could be a sign of receptivity to mating. Eventually the male will grasp her and mate in the 'head-to-head' position (Fig. 6.9a). Mating commonly lasts about 10 minutes in the laboratory (range 2–20). Based upon 20 staged matings, Hanlon, Ament and Gabr (1999) described a typical sequence in which apparent 'flushing' occurred for ca. 6 min, followed by swift transfer of a large bundle of spermatophores (ca. 150–300), followed by a 3–4 min period of spermatangia placement in which the male used part of the hectocotylised arm to break open spermatophores and manipulate them in the vicinity of the paired seminal receptacles (Fig. 6.9b–d). Subsequent histology revealed that the seminal receptacles, which comprise a series of sperm storage bulbs connected by a central duct, were mostly full of sperm (Fig. 6.9e). Single mating pairs can mate several times in succession, sometimes intermixed with egg laying. Thus, on some occasions mating and egg laying are linked, but in others they are not. Temporary mate guarding by the male has been observed in the field (Fig. 6.7a, d, e) and is common in the laboratory (Fig. 6.8c). There is no evidence for

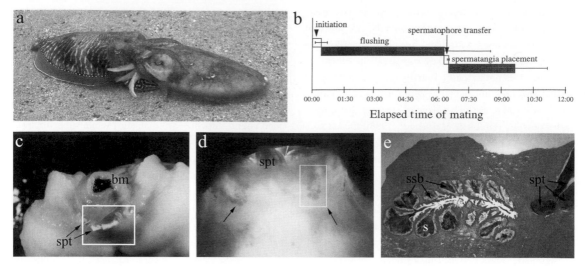

Figure 6.9 Mating and sperm storage in *Sepia officinalis*. (**a**) Mating in the head-to-head position; the male is on the left. (**b**) Sequence of behaviours during matings (ca. 12 min.). (**c**) The buccal mass of the female (bm), which contains the beak and mouth, and attached spermatangia (spt). Box indicates approximate area of image in (**d**). (**d**) Back-lighted outlines of paired seminal receptacles (arrows), one of which is outlined and illustrated in (**e**). (**e**) Light micrograph of seminal receptacle, and of spermatangia (spt) embedded in connective tissue. Note the numerous sperm storage bulbs (ssb) and central duct with some sperm (s) present. From Hanlon, Ament and Gabr (1999): *Marine Biology* **134**, 719–728 Behavioral aspects of sperm competition in cuttlefish, *Sepia officinalis* (Sepioidea: Cephalopoda). Hanlon, R. T., Ament, S. A. & Gabr, H., © (1999), with permission of Springer.

anything other than temporary pairing, and both sexes have multiple mates.

Female choice of mates was shown during extensive controlled laboratory trials ($n = 92$) not to be based on dominant males or the winner of male–male bouts, but rather on males that had mated most recently, thus implicating olfactory rather than visual cues as key sensory stimuli (Boal, 1997). Overall, females mated often and with multiple males, and currently there are neither experimental laboratory data nor extensive field observations to know how often females reject copulation attempts by males.

Chemoattractants in the eggs of *S. officinalis* were first reported by Zatylny *et al.* (2000), and Boal *et al.* (2010) used Y-maze choice experiments to provide the first evidence that sexually mature cuttlefish are attracted to odours from conspecific eggs. These results suggest that *Sepia* eggs could be a source of reproductive pheromones. Enault *et al.* (2012) reported a complex of molecules thought perhaps to act as water-borne sex pheromones, but this suggestion requires future behavioural testing. A detailed study by Cummins *et al.* (2011) on the squid *Loligo pealeii* showed that a contact

pheromone in the egg tunic produced immediate and extreme male–male aggression even in the absence of females. Thus, there may be undiscovered chemical cues that help to resolve some of the unexplained aspects of reproductive behaviour in this and other cephalopods.

6.2.3 *Sepia latimanus*

Near Guam, in Micronesia, this species was observed during 75 SCUBA dives over a 3-month period by Corner and Moore (1980). The habitat was between 15 and 36 m deep on a reef slope characterised mainly by rubble, some live coral mounds and tall colonies of dead coral, *Lobophyllia* sp. Generally, these divers found three to four lone males positioned next to prominent coral heads, each with a commanding view towards deeper water. Unaccompanied females came from deeper water straight to *Lobophyllia* coral heads, which were strongly preferred for egg laying. Rarely, the female would go to the coral and immediately lay eggs that had been fertilised previously; more often, one male that had out-competed the other males would join her temporarily.

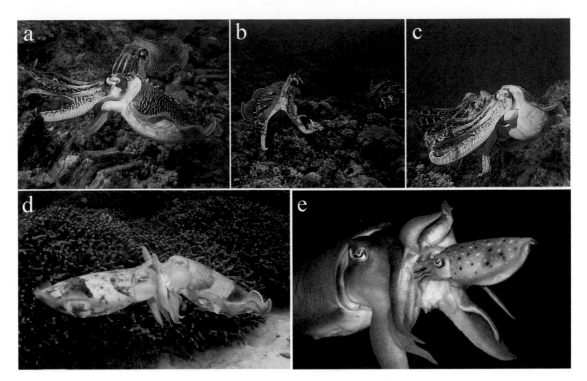

Figure 6.10 Reproductive behaviour in the Indo-Pacific cuttlefish *Sepia latimanus*. (**a**) Males in the early stage of Intense Zebra Display during an agonistic encounter. (**b**) Male in full Intense Zebra Display. (**c**) An escalated agonistic contest with physical contact of the rival male. (**d**) Mating, which lasts 0.5–1.5 minutes in the head-to-head position. (**e**) A small sneaker male placing spermatophores in the buccal area of a large female. Photos **a**–**c** from Corner and Moore (1980): *Micronesica* **16**, 235–260, with permission from University of Guam Press; **d, e** courtesy of A. Crawley.

Competition for females was keen and involved highly stereotyped agonistic displays. When a female approached, one to three males would advance rapidly, often with the larger male arriving first and swimming alongside her. Males then engaged in agonistic contests characterised by an Intense Zebra Display, with the arms either stretched forward or arched towards each other (Fig. 6.10a). Males dashed towards one another showing this display, and typically the smaller male was the one to terminate the contest. Males engaged in these contests whether they had a temporary female mate or not; any time males got close enough they would display. In one encounter, two male–female pairs were approached by a lone male, and the three males drifted off several metres as they engaged in displaying; upon return, the same two males came back to the same two females, but they switched partners. Apparently, it was the female 'resource' that was being fought for, not one specific individual female. Contests between males appeared to

escalate when females changed from a cryptic body pattern to a 'precopulatory pattern', which was overall light grey in both sexes. Males then fought more frequently and intensely, and would escalate the contest by flaring their arms upward and outward in an exaggerated posture (called the 'Umbrella pattern' by Corner & Moore, 1980), exposing the mouth and suckers to the rival male (Fig. 6.10b). Ultimate escalation involved moderate physical contact, with the larger male positioning himself immediately parallel to the rival male, displaying highly Intense Zebra, and arching his arms over the head of the other in an apparent attempt to bite him (Fig. 6.10c; 'exaggerated Umbrella', *ibid.*). Invariably the rival male retreated several metres after this, returning to a normal dark or cryptic Mottle body pattern and presenting the rear portion of its body to the attacker. This sort of behaviour and posture are reminiscent of submissive behaviour in many animals; that is, the posture, orientation and colour are the opposite to those shown during threat behaviour (e.g. Tinbergen, 1959; Eibl-Eibesfeldt, 1975).

Courtship was elaborate and seemed to be initiated by the male. Females generally went straight to the *Lobophyllia* coral head; at the arrival of the female, males then moved swiftly through the phase of agonistic contests among themselves, and the male that won the contest began to accompany the female as she showed her grey 'precopulatory pattern'. The male first attempted to rub his curled arm tips across the female's forehead (between the eyes and arm bases). If no response was evoked from the female (or she responded negatively with an Acute Disruptive pattern), he would move between the female and the coral and then puff a water jet at her (or at her eggs if she was laying, as she sometimes was) and jet backward 1 m or so. This continued until the female accepted the rubbing on the head, at which time the female then opened her arms and the couple would mate for 0.5–1.5 minutes in the 'head-to-head' position (Fig. 6.10d). If another male intervened during courtship or mating, the male switched on the Intense Zebra Display. On one occasion during courtship, a male was observed displaying Intense Zebra to a male on one side while contralaterally displaying the uniform grey precopulatory pattern to a female. Females accepted mating after egg laying, as well as before, so that it is probable that sperm from different copulations were used for different series of eggs (see §6.6.4).

Sneaker mating by a very small male was observed by A. Crawley (Fig. 6.10e) in a mating group of five males and one female. This small male successfully mated her while the large consort male was fighting another male; later, the sneaker was rejected by the female in a subsequent attempt. This example of female choice is similar to that seen in *S. officinalis* and *S. apama*.

Other references to reproductive behaviour in the Family Sepiidae can be found in Norman (2000) and Wada *et al.* (2005b, 2006, 2010). These papers report behaviours similar to those described in the previous three sections, but their details should be consulted for future research.

6.2.4 Small Sepioids: Pygmy, Bobtail and Bottletail

The so-called 'pygmy cuttlefish' or 'pygmy squid' *Idiosepius paradoxus* and *I. thailandicus* have adhesive

glands on their dorsal mantle and they adhere to seaweed or seagrass leaves such as *Zostera*. Mating requires an unusual position because the female stays attached to seagrass while the male is roughly in a 'head-to-head' position (Fig. 6.11a) and deposits spermatophores around the buccal mass of the female (Figs. 6.11f–h). Kasugai (2000) reported no signs of courtship in *I. paradoxus*: the male darted towards a female, grasped her at the base of the arms and deposited spermatophores mainly at the base of her arms, not directly into the seminal receptacle. These copulations lasted only 5 seconds. The female then picked up the spermatangia with her flexible buccal mass (the 'mouth') and manipulated them somehow. When females were actively laying eggs, males would mate her and deposit spermatophores in the buccal area even when she was actively moving eggs from her oviduct in the mantle to her arms. Interestingly, often more than one male engaged in this mating behaviour at the same time, sometimes mating her serially and other times simultaneously (Fig. 6.11b, c). No agonistic behaviour among males was observed, nor any postcopulatory mate guarding.

Sato *et al.* (2010) observed 76 copulations in laboratory trials of *I. paradoxus*, and the mean mating duration was 3.6 seconds. No courtship or male–male agonistic behaviours were observed, and females mated many times in succession (up to 20 per day) and with different males, with no signs of rejecting males. Males passed one to nine spermatophores to the base of the arms but did so in an extraordinary manner (Fig. 6.11d): the left hectocotylus, with its two flaps (Fig. 6.11e), grasped the spermatophores and was guided by the right hectocotylus, which is grooved, to various locations at the base of arms (Sato, Kasugai and Munehara, 2013). Females immediately elongated their highly protrusable buccal mass towards them (Fig. 6.11f) and either blew them away (= female rejection) or ate them (presumably for nutrition) (Sato *et al.*, 2013). Females did not actively place sperm into the seminal receptacles, because no sperm were found there 10 min after these behaviours. However, sperm were found in the seminal receptacles 30 and 120 min after copulations (Fig. 6.11g), suggesting that free-swimming sperm somehow find the seminal receptacles. The seminal receptacles were sectioned, and sperm were found to be distributed throughout multiple sacs

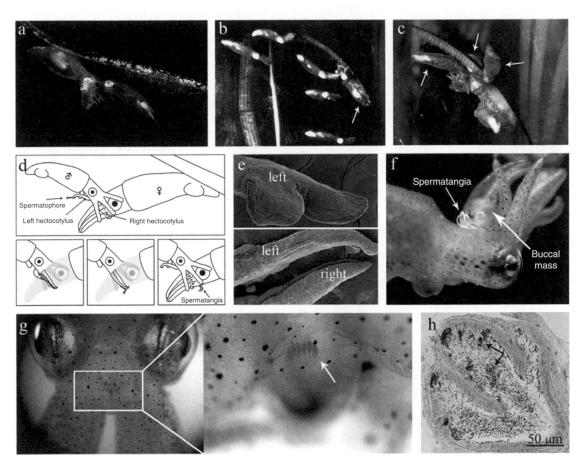

Figure 6.11 Mating and sperm storage in pygmy cuttlefish *Ideosepius* spp. (**a**) Mating in the head-to-head position while the female adheres to the seagrass leaf (from Sato *et al.*, 2010: *Journal of Zoology* **282**, 151–156, with permission of Wiley). (**b**) Five males who mated the female (arrow) serially (from Kasugai, 2000: *Venus* **59**, 37–44). (**c**) Three males (arrows) mating the female simultaneously (from Kasugai, 2000: *Venus* **59**, 37–44). (**d**) Details of spermatophore transfer; the left hectocotylus, with its two flaps (shown in scanning electron micrograph image in **e**), grasped the spermatophores while the right hectocotylus guided the hectocotylus to the base of the arms (from Sato *et al.*, 2013: *Journal of Molluscan Studies* **79**, 183–186, © 2013, by permission of Oxford University Press). (**f**) Female elongated her protrusible buccal mass to blow away or eat the spermatophores (from Sato, Kasugai and Munehara, 2013b: *Marine Biology* **160**, 553–561, with permission of Wiley). (**g**) The seminal receptacle near the beak holds sperm (**h**; arrows). See text (from Sato *et al.*, 2010: *Journal of Zoology* **282**, 151–156, with permission of Wiley).

(Fig. 6.11h). Sato *et al.* (2010) found that the volume of sperm in the seminal receptacles increased linearly from the first through eight copulations, indicating that sperm were being stored from every mating.

Similar behaviours were reported for *Idiosepius thailandicus* (Nabhitabhata, 1998). Mating occurred all day in the shade with a peak activity from ca. 1500 to 1700. Pair formation was not observed, males and females had several partners, and females were not observed to reject males. Copulation lasted only 0.5 to

1 second, occurred in the head-to-head position, and the female sometimes opened her arms in umbrella fashion. Mating could occur in the water column or with the female attached to seagrass and the male hovering. Curiously, Nabhitabhata and Suwanamala (2008) observed interbreeding between *I. thailandicus* and *I. biserialis* in laboratory trials; however, it is not clear that these are separate species.

Hawaiian bobtail squid *Euprymna scolopes* mated for 30–50 min in the 'male-parallel' position during

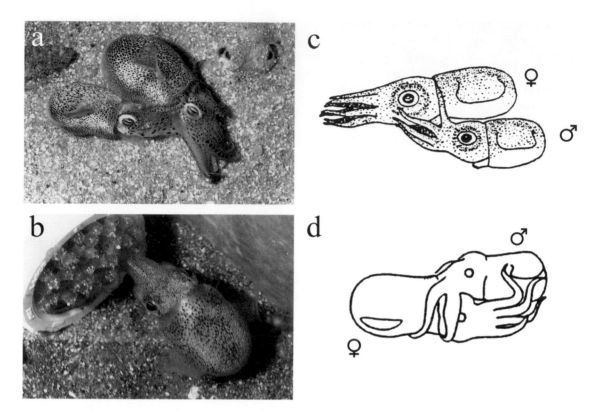

Figure 6.12 Mating in sepiolids. (**a**, **b**) The Hawaiian bobtail squid *Euprymna scolopes* mates at night in the male-parallel position for 30–50 minutes, and the next morning the females lays ca. 30 eggs and coats them with sand for protection (images from Hanlon *et al.*, 1997, by permission of University of Chicago Press). (**c**) *Rossia pacifica*. The first left arm is the hectocotylus and is inserted into the female's mantle (from Brocco, 1971). (**d**) Typical 'male to female neck' mating position in the genera *Sepiola* and *Sepietta* (from Mangold-Wirz, 1963: *Vie Milieu*, Sup. No. 13, 1–285, Biologie des cephalopodes benthiques et nectoniques de la Mer Catalane, Mangold-Wirz, K., © (1963), by permission of Elsevier).

laboratory trials (Fig. 6.12a; Hanlon *et al.*, 1997). Sixteen matings were observed with the aid of night-vision goggles, and all occurred in the first few hours of darkness. Courtship was not seen, and the males initiated mating. In one case, two males grabbed a female simultaneously, but the two males fought briefly and mating did not ensue. Females laid eggs in the morning, affixing each egg for 10 seconds and laying ca. 1 egg/minute for clutches of about 30 eggs. The females somehow coated the eggs with sand, which imparted camouflage on the eggs (Fig. 6.12b); this makes sense given that in nature this species is a sand dweller (Shears, 1988).

The dumpling squid *Euprymna tazmanica* also mates in the 'male-parallel' position but for 2–3 hours, and

both the male and female are reported to suffer physical exhaustion for 30 min thereafter, thus presumably implying a cost to mating in terms of reduced energy for predator avoidance or foraging (Franklin, Squires & Stuart-Fox, 2012). In a subsequent study, it was found that polyandrous females produced eggs faster and had larger hatchlings relative to egg mass compared with females that had mated only once, implying some fitness benefits to polyandry in this species (Squires *et al.*, 2012).

The southern bottletail squid *Sepiadarium austrinum* is a small species similar in appearance to *Idiosepius* and *Euprymna*, although it mates 'head to head'. Males tended to mate with females that were proportionally larger than other available females, and with egg-carrying (i.e. mature) females they mated for longer duration and were more likely to transfer spermatophores (Wegener et al., 2013a). Females in the same study consistently ate spermatophores that were transferred by males. Subsequently, Wegener et al. (2013b) demonstrated with ^{14}C radiolabeling that those spermatophore nutrients were directed into both somatic maintenance as well as egg production, suggesting that such behaviour by females (combined with short-term sperm storage) has potential to influence male ejaculates and mating tactics.

The small *Rossia pacifica* (Fig. 6.12c) mates in a position similar to *Euprymna scolopes*, and the first left arm is the hectocotylus that inserts into the female's mantle cavity (Brocco, 1971). The mating position in the genera *Sepiola* and *Sepietta* has the male grasping the female's neck (Fig. 6.12d; from Mangold-Wirz, 1963).

6.3 Agonistic, Courtship and Mating Behaviour in Squid

Many coastal and epipelagic squid shoal for a large proportion of their short life and engage in more extended social behaviour. Currently we have descriptions (some of them only fragmentary) of mating behaviour in 16 species of teuthoids (Table 6.2). Some of the most elaborate reproductive behaviour known so far in cephalopods occurs in the coastal genera *Sepioteuthis* and *Loligo*. These have evolved complex behavioural interactions and diverse displays. There are only two species for which we have both field observations and data combined with laboratory experimentation: *Sepioteuthis australis* and *Loligo pealeii*. In addition to substantial progress in understanding mating systems and sexual selection mechanisms in coastal squid, there have been new observations in several oceanic and deep-sea squid.

6.3.1 *Sepioteuthis sepioidea*

The reproductive behaviour of this Caribbean coral reef squid was reported first by Arnold (1965), and subsequently in extensive field observations by Moynihan and Rodaniche (1982) and by R.T. Hanlon & J.W. Forsythe (unpublished data). Aside from complexity, the outstanding feature of *S. sepioidea* is

that they spend a good deal of time (and presumably energy) in reproduction, which seems to occur throughout the year within the population. During early stages of courtship, 'courting parties' of one female and several males begin to interact in complex ways. Later on, a successful male pairs temporarily with the female and mating occurs. In ways similar to *Sepia apama*, *S. latimanus* and *S. officinalis*, males compete for females, and both sexes are promiscuous.

Typical shoals range from 10 to 30 individuals (Fig. 6.13a), and the first signs of reproductive behaviour are changes in the social and spatial relationships between the squid. One or several subgroups (courting parties) are formed, usually of one female and two to five males (Fig. 6.13b, c). Unlike *Sepia latimanus* on coral reefs in Guam, *Sepioteuthis sepioidea* shoal continually throughout the daylight period, and the same shoal (or many individuals) may remain together for a few days. In courting parties within shoals, it is useful to speak of one 'near male' and several 'far males' because one male (often the largest, but not always) stays closer to the female and, in later stages of courtship, attempts to segregate her from the far males in the courting party. While males are more active in the courtship process, it is not clear which sex actually controls the pace or sequence of events that unfold.

'Mutual rocking' (swimming to and fro together) in the same direction is the first sign of a relationship between one male and one female. Soon the male begins to approach the female closely (to within a few centimetres), and there ensues a phase in which the female flees a short distance and the male pursues in a continuing and intensifying sequence.

Body patterns then become important. Females show a brightly distinctive Pied Display (Fig. 6.13b), whose primary function appears to be to repel males temporarily, at least the near male; there is a negative correlation between the occurrence of Pied coloration and copulations (Moynihan & Rodaniche, 1982). However, the Pied may also attract the attention of other suitors and contribute to the onset of courtship and agonistic behaviour of males in the party. For example, Hanlon and Forsythe (unpublished data) often observed females that 'paraded' in front of a line of squid, occasionally displaying Pied; courtship activity usually began shortly thereafter.

Table 6.2 Mating by squid

Species	Duration (seconds)	Mating position	Spermatophore placement	Temporary mate guarding	References
Suborder Myopsida					
1. *Alloteuthis subulata*	–	Head-to-head	bmv	–	Lipinski, 1985
2. *Lolliguncula brevis*	3–5	♂ parallel	mc (pad)	–	Hanlon, Hixon & Hulet, 1983
3. *Loligo (Heterololigo) bleekeri*	– ~300	Head-to-head ♂ parallel	bmv mc	Yes	Hamabe & Shimizu, 1957; Natsukari & Tashiro, 1991; Iwata, Ito & Sakurai, 2008
4. *Loligo (Photololigo) edulis*	~2–5	♂ parallel	mc	–	Natsukari & Tashiro, 1991
5. **Loligo opalescens*	– ~30–120	Head-to-head ♂ parallel	bmv mc	Yes	McGowan, 1954; Fields, 1965; Hurley, 1977;
6. **Loligo pealeii*	5–20 5–20	Head-to-head ♂ parallel	bmv mc	Yes Yes	Drew, 1911; Stevenson, 1934; Griswold & Prezioso, 1981; Shashar & Hanlon, 2013
7. **Loligo (Doryteuthis) plei*	5–10 5–10	Head-to-head ♂ parallel	bmv mc	Yes	Waller & Wicklund, 1968; Hanlon, 1978
8. *Loligo vulgaris*	~5 ~30	Head-to-head ♂ parallel	bmv mc	Yes	Tardent, 1962
9. **Loligo reynaudii*	– 2–39 2–11	Head-to-head ♂ parallel Sneaker ♂	bmv mc bmv	– No Yes	Sauer, Smale & Lipinski, 1992; Hanlon, Smale & Sauer, 1994; Hanlon, Smale & Sauer, 2002
10. **Sepioteuthis lessoniana*	3–4	♂ parallel ♂ upturned	mc bmv	Yes No	Larcombe & Russell, 1971; Segawa, 1987; Segawa *et al.*, 1993; Boal & Gonzalez, 1998; Wada *et al.*, 2005a
11. **Sepioteuthis sepioidea*	1	♂ parallel [ms]	bmv [fg]	Yes	Arnold, 1965; Moynihan & Rodaniche, 1982
12. **Sepioteuthis australis*	2–4	♂ upturned ♂ parallel Head-to-head	bmv ? ?	Yes – –	Jantzen & Havenhand, 2003a,b
Suborder Oegopsida					
13. *Gonatus fabricii*	–	Head-to-head	bmr	–	Kristensen, 1983
14. *Illex illecebrosus*	–	♂ parallel	mc[†]	–	O'Dor, 1983
15. *Dosidicus gigas*	–	Head-to-head	bmr	–	Nesis, 1983
16. **Todarodes pacificus*	3–10	Head-to-head ♂ parallel	bmr bmr	–	Murata, 1990; Ikeda, Sakurai & Shimazaki, 1993; Sakurai, Bower & Ikeda, 2003

* Observations in the sea; – no information available; bmr, buccal membrane in ring around mouth; bmv, buccal membrane ventrally under mouth; [ms] male strikes female on head or arms; † no seminal receptacle known; mc mantle cavity; ? not certain; [fg] female grasps spermatophores from arm surface, and places them in bmv herself.

Figure 6.13 Reproductive behaviour in the Caribbean coral reef squid *Sepioteuthis sepioidea*. (**a**) A typical school of squid. (**b**) A female (top) in the Pied Display, and two male suitors. (**c**) A mating pair with a challenging male (right) showing first stages of an agonistic bout. (**d–g**) Various stages of an agonistic bout; note the different patterns expressed on the dorsal versus ventral body of each male. (**h**) Lateral Silver Display shown by a male (compare Fig. 6.14). (**i**) Mating in the male-parallel position; male (bottom) is extending his hectocotylised arm into the mantle of the female. (**j**) Female placing spermatophores in the seminal receptacle near the mouth just after mating. (**k**) Female depositing an egg strand in a hole in the fire coral; egg strands can be seen in bottom photo. All photographs by R. Hanlon except (**i**) by Eliot Ferguson.

Males show two stereotyped and striking displays during agonistic contests with other males: the Zebra Spread Display (Fig. 6.13d–g) and the Lateral Silver Display (Fig. 6.13h; 6.14a, b). Most interactions are between the near male and one of the far males. As a far male attempts to approach the female, the near male intervenes and both animals show the Zebra Spread Display, which lasts from 3 to 40 seconds. Positioning is important. One male is above the other, so that the pattern shown on the dorsal versus ventral body surface is different (Fig. 6.13f). The upper individual has a dark head, dark spots on the ventral mantle and the flared arms. The lower individual assumes the tail-up posture and always has a clear head, Zebra bands on the mantle and widely flared arms with mottling. Often the lower individual jets upward and backward, sometimes bumping the other male; however, this display is almost entirely visual. Moynihan and Rodaniche (1982) suggested that the upper individual is showing 'the most aggressive and/or impressive form'. However, this is not the case, as Hanlon and Forsythe (unpublished data) discovered by analysing videotapes of 45 Zebra

167

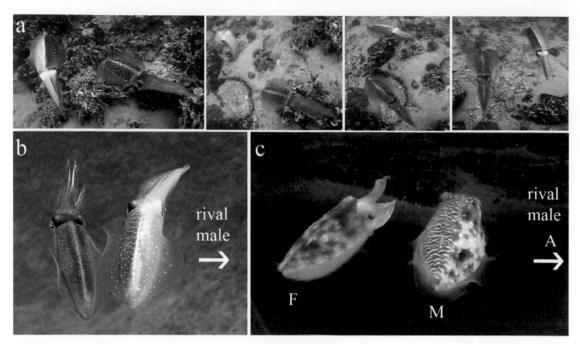

Figure 6.14 Simultaneous dual signalling. (**a**, **b**) *Sepioteuthis sepioidea* on a Caribbean coral reef. Sequence of video frames (**a**) showing how the male (left in frame one) switches his dark 'normal' pattern from one side to the other as the female swims to his other side. The male shows the Lateral Silver Display to approaching male rivals; if he shows it to his temporary female mate, she swims away. (**b**) Mating pair, with the male (M) showing his mate the 'normal' pattern while simultaneously showing the aggressive Lateral Silver Display towards rival males. Video and photograph by R. Hanlon. (**c**) A male mourning cuttlefish *Sepia plangon* (M) showing different body patterns to his female mate (F) and to approaching rival males (from Brown, Garwood & Williamson, 2012: *Biology Letters* **8**, 729–732, It pays to cheat: tactical deception in a cephalopod social signalling system, Brown, C., Garwood, M. P. & Williamson, J. E., © (2012), by permission of the Royal Society).

Spread Displays at Little Cayman Island. The lower individual was always the male that had been with the female and 'won' 43 of 45 contests that escalated to Zebra Spread Display (note: this is an advantage of using SCUBA for observations, rather than snorkelling on the surface as Moynihan and Rodaniche did). There are many gradations of the Zebra Spread Display (Fig. 6.13d–g) manifest by different combinations and intensities of the chromatic and postural signals contained within this display (see Box 7.1), and these form the basis for the degree of escalation of these agonistic contests. These individual signals have not yet been analysed quantitatively as they have in the Lateral Displays of *Loligo plei* (see §6.6.2).

In the later stages of pairing, males show the Lateral Silver Display to approaching males (Fig. 6.14b). The display is shown unilaterally only towards an approaching male, so that the female never sees it. It can

be shown for long periods (many seconds or minutes), and Moynihan and Rodaniche (1982) speculate that it is a warning for other males to 'keep away', whereas the Zebra Spread Display is a warning to 'drive away' males. The Lateral Silver Display is not unlike the female's Pied both in appearance and signal function: they both have an exaggerated silver-white colour and have a mildly repellent effect on the receiver.

Simultaneous dual signalling is occurring during these male displays (Fig. 6.14), which can be quite dynamic. The female is receiving a signal of the standard calm courtship coloration of uniform brown in the male, whereas approaching males receive a highly aggressive all bright white signal. If the male shows the Lateral Silver Display to the female, she is repelled and the male loses his mate. Thus, when the female moves to his other side, he must quickly change sides with his patterning, which is done very smoothly as illustrated in video frame grabs in Fig. 6.14a. Only two other species have been seen to perform this feat: *Sepia latimanus* (Corner & Moore, 1980) and *Sepia plangon* (Fig. 6.14c; Brown, Garwood & Williamson, 2012).

A male with a temporary mate can probably segregate her as long as he is healthy and vigilant and uses the Lateral Silver Display to repel the more opportunistic or aggressive far males. However, the constant presence of competing male rivals throughout all of courtship and most of mating may explain the need for these two dramatic agonistic displays, which are probably dangerous to show because they must be conspicuous to predators as well. Shoals of these *Sepioteuthis* have about seven interactions per hour with potential predators throughout the day (see Chapter 5).

Mating usually begins after one pair becomes fairly isolated within the courting party. Males are the initiators and suddenly accelerate to a position directly next to the female; then they attempt to mate her by placing a spermatophore with a swift 'strike' on the arm or head region of the female. The transfer, or copulation, lasts only a second or so. Video analysis and observations of 42 mating attempts showed that the spermatophore is rarely attached successfully and floats away, or it can stick almost anywhere near the arms or the head region (Hanlon & Forsythe, unpublished data). If it is attached successfully, the chase ends abruptly and the female quickly grasps the spermatophore and transfers it to the seminal receptacle near the mouth (Fig. 6.13j).

If the spermatophores do not stick, the chase continues. Numerous body patterns have been observed during and after copulation, but their functions and relationships are uncertain. A fairly common pattern is the rapid 'pulsating' chromatophores of the male as he approaches the female for copulation. Moynihan and Rodaniche (1982) described several uniform body patterns with pastel colours that seem to appear only near mating; this again raises the question of possible colour vision in some species of cephalopods (see §2.1.1). On the other hand, they noted some examples (from the 27 they witnessed) in which mating occurred without special displays or body patterns, and they suggested that the dullness of these encounters indicated the temporary firmness of the relationship between the male and female.

Mating would appear to require large energetic demands on both partners. Hanlon and Forsythe

(unpublished data) documented dozens of episodes in which the male pursued the female at nearly full jetting speed for one hour continuously, as the pair swam throughout the water column, jetting and zigzagging in every direction, apparently oblivious of predators or divers. In these episodes, which were seen every evening the hour before dark (during five weeks of field observations in different seasons at Little Cayman Island), the female was apparently not receptive while the male behaved as if he were highly motivated to copulate.

A second mating procedure – very different from that described above – has now been seen in three locations in the Caribbean: the northern coast of Venezuela and at Little Cayman Island (R.T. Hanlon, pers. comm. 2012); and in Bonaire (Mather, Griebel & Byrne, 2010). In these cases, one male and one female paired temporarily during the day, then dispersed at night to feed, and paired temporarily the next day. In Cayman, the male positioned himself under the female and hovered gently in concert with her for 4 minutes, and on two occasions he grasped spermatophores with his ventral arms and swept slowly upwards to gently transfer spermatophores into her mantle cavity (duration 2–3 seconds) in the 'male-parallel' position (Fig. 6.13i). In Venezuela, egg laying ensued within 5 minutes, and the male had to consistently guard her from challenging males.

Egg laying is not linked tightly to mating. The few field observations of egg laying show females paired with a male, who vigorously guards her during the multiple times that she carefully ties a string of four to five very large eggs deep within hard corals or other reef structures (Fig. 6.13k).

6.3.2 *Sepioteuthis australis*

Reproductive behaviour in this southern Australian species has been studied in detailed field studies by Jantzen and Havenhand (2003a, b) and complemented by mesocosm studies by van Camp *et al.* (2005). Jantzen and Havenhand observed more than 550 reproductively active squid in 75 hours of underwater observations near Adelaide in South Australia. The main spawning season was September to March, and it occurred in a shallow habitat (2–4.5 m depth) of bare sand and rock interspersed with seagrass and brown macroalgae. Squid shoals comprised 2–45 individuals, and the operational

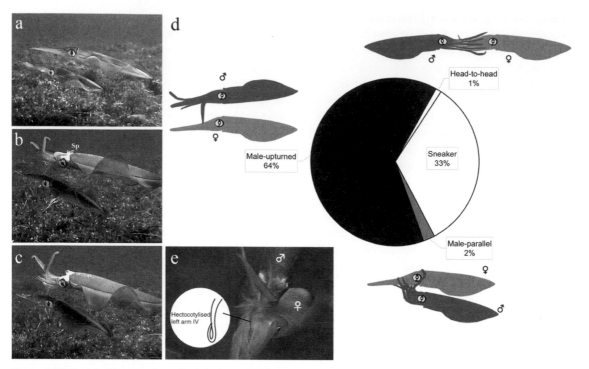

Figure 6.15 *Sepioteuthis australis* mating behaviours. (**a**) Mating pair swimming normally (male is top). (**b**) Male swimming upside down and reaching into his mantle to grasp a spermatophore (Sp). (**c**) Male transporting the spermatophore amidst the arms of the female. (**d**) Percentages of time that different mating positions are used. Sneaker position is not shown but is identical to head-to-head; see text. (**a–d**) from Jantzen and Havenhand, 2003a, *Biological Bulletin* **204**, 290–304, Reproduction behavior in the squid *Sepiotheutis australis* from South Australia: ethogram of reproductive body patterns, Jantzen, T. M. & Havenhand, J. N., © (2003), by permission of the University of Chicago Press. (**e**) A small unpaired male, upside down, inserting the spermatophore into the buccal mass of the female (from Wada *et al.*, 2005a: Fig. 1 and Fig. 4 of *Zoological Science* **22**, 645–651, © (2005), by permission of the Zoological Society of Japan).

sex ratio ranged from 1M:1F (i.e. a single pair) to 3M:1F. However, Moltschaniwskyj and Steer (2004) found sex ratios of up to 5M:1F in Tasmania, so there is variability in this aspect of sexual selection.

Three mating positions were observed during 85 mating attempts. 'Male-upturned mating' (54 of 85 matings; 64%) involved the male swimming above the female and swiftly rotating upside down, extending the right fourth arm to the buccal region of the female, and grabbing spermatophores from the funnel with the hectocotylised left fourth arm and using the right fourth arm as a guide to place three to five spermatophores in the buccal region of the female (Fig. 6.15). The entire sequence lasted 3 seconds. These matings always occurred after the female passed an egg strand from her funnel to the buccal region, and before each egg-laying episode.

'Male-parallel mating' occurred only twice during 85 matings, and this was a typical loliginid mating position in which the male was underneath the female (Fig. 6.15; see also Fig. 6.13i), but placement of spermatophores could not be seen. 'Head-to-head mating' was seen only once. An additional 28 matings (33%) were classified by Jantzen and Havenhand as 'sneaker' matings because those males were unpaired; most of those matings were in the 'male-upturned' position and occurred when a paired female was attempting to lay an egg strand. On two occasions, the sneaker appeared to be mimicking a female by taking on the body pattern coloration of females; this mimicry pattern succeeded insofar as no agonistic behaviour was elicited from the consort male; furthermore, a second unpaired male attempted to mate the female mimic (similar to that seen in *Sepia apama* female mimics; Hanlon *et al.*, 2005).

Mate guarding by consort males was common. The time interval between egg acquisition from the female's funnel and egg laying was 30 seconds, and this is the time during which mating occurred and multiple sperm sources could be accessed by the female for fertilisation. Van Camp et al. (2004), using microsatellite loci as DNA fingerprints, determined that females are using up to four different males to fertilise their eggs in a single capsule; subsequent studies in a mesocosm found that large consort males did not sire proportionally more offspring (Van Camp et al., 2005). Following egg capsule deposition, the female often flared her arms and tentacles and blew water forcibly over them, and small white particles (apparently parts of spermatophores and spermatangia) were expelled. The functional significance of this common behaviour remains unknown.

Male agonistic contests were common, and paired males always tried to place themselves between the female and challenging males (Jantzen & Havenhand, 2003a). Some contests escalated to physical fin beating, which was quite forceful, and the consort males won 100% of 67 contests. Some paired consort males 'charged' approaching males in the forward position and struck the challenger with its tentacles.

Females and males mated with multiple mates but this was not quantified in this study. Females laid eggs in succession – usually every 70 seconds – and were capable of depositing in excess of 50 egg capsules per hour. Thus, mating and mate guarding by consort males was a continual and active set of behaviours. Details of the chromatic and postural components of males and females are carefully and fully explained in Jantzen and Havenhand (2003b).

In Tasmania, S. australis forms large spawning aggregations on the east coast (some egg masses had up to 1241 egg capsules) during spring and summer, but not in autumn or winter. However, on the southeast coast, spawning activity is sporadic, resulting in isolated, low-density egg patches deposited over broad areas (Moltschaniwskyj & Steer, 2004). By adopting different spawning behaviour in different locations and seasons, this species may spread the risk of mortality in both space and time.

Sepioteuthis lessoniana also occurs in the western Pacific but in the northern hemisphere. Two laboratory studies indicate that the reproductive behaviour is similar to S. australis but the frequency of each mating

position can be very different. For example, Wada et al. (2005a) observed that most matings ($n = 174$) were 'male-parallel' whereas Jantzen and Havenhand (2003a) observed this mating position only twice among 85 matings in their field study. 'Male-upturned' mating was observed frequently in both studies, and Wada et al. (2005a) acquired a stunning photograph of the upside-down male inserting his hectocotylus directly in between the female's first pair of arms and into the buccal mass region (Fig. 6.15e). These authors found that the larger paired males (consorts) mated in the 'male-parallel' position and that smaller unpaired males performed the 'male-upturned' position. The latter were considered conditional tactics that are dependent on male size because some males demonstrated both positions (and their respectively different placement of spermatophores) but fewer than half of these attempts were successful, whereas the large-male 'male-parallel' matings were 95% successful. In addition, they found that relatively larger males won agonistic contests (93% of 27 fights). See Boal and Gonzalez (1998) and Wada et al. (2005a) for additional details for Sepia lessoniana.

6.3.3 *Loligo pealeii*

The mating system of L. pealeii is complex and shares many features with other Loligo and Doryteuthis species. It has been documented in a detailed field study by Shashar and Hanlon (2013), who conducted 136 SCUBA dives over 3 years. This account is based upon that field study except where noted with complementary laboratory experiments.

Squid migrate inshore annually from the edge of the continental shelf of the northeast coast of the United States (Jacobson, 2005) to shallow nearshore areas to spawn in spring. When arriving near shore, schools of squid will pause at selected locations where a complex system of courtship, mating and communal egg laying will arise. The spawning aggregations are usually small, involving <100 squid that produce communal egg mops that are usually <1 m diameter (Fig. 6.16a). However, in 2013 a massive spawning event was recorded by recreational divers in Nova Scotia, verifying that on rare occasions this species can aggregate in thousands to produce huge egg masses (Fig. 6.16b).

Figure 6.16a illustrates some of the main features of reproductive behaviour on a typical spawning arena.

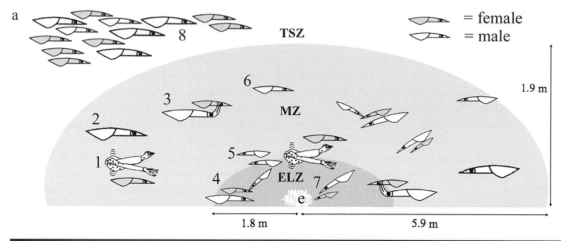

Figure 6.16 Behavioural dynamics on the spawning arena of the longfin squid *Loligo pealeii* near Woods Hole, Massachusetts, USA. (**a**) Typical mating arena and the behaviours associated with each zone: (ELZ) the Egg-Laying Zone; (MZ) the Mating Zone; and (TSZ) the Transient/Schooling Zone. Behaviours illustrated are: (1) fight between a paired consort and lone large male, (2) lone large male, (3) male-parallel mating, (4) head-to-head mating, (5) bold sneakers, (6) surreptitious sneaker, (7) egg laying (e, egg capsules), and (8) schooling squid that are not engaging in reproductive behaviour. Illustration not drawn to scale; modified from Shashar and Hanlon (2013): *Journal of Experimental Marine Biology and Ecology* **447**, 65–74, Spawning behavior dynamics at communal egg beds in the squid *Doryteuthis* (*Loligo*) *pealeii*, Shashar, N. & Hanlon, R. T., © (2013), with permission from Elsevier. (**b**) An extraordinary mass spawning event in summer of 2013 at 4 m depth at Fox Point, St Margaret's Bay, Nova Scotia. Older egg capsules (slightly darker) are in the mound in the foreground as well as the long row in the background, which is ca. 1 m high. Newer, light-coloured egg capsules have been laid around the periphery of the mound in the foreground. (**c**) Diagram of a typical egg capsule (or 'finger') with many ova. (**b** courtesy of Bob Semple.)

Three behavioural zones are common in a mating arena. In the *Egg-Laying Zone* (A), all squid are sexually active and the distances between paired males and females are less than 50 cm. In the *Mating Zone* (B) most squid are sexually active, although other squid pass through the zone, and distances between pairs varies and often is more than 1 m. Most (80–90%) of the male-parallel copulations (described below) occur in this zone. In the *Transient/Schooling Zone* (C), squid are mostly in a school where no pairing or copulation occurs, and it is difficult to distinguish males from females in this zone because they are not showing any of the sex-specific body patterns or behaviours (Hanlon *et al.*, 1999). The operational sex ratio is ca. 3M:2F. Mating-related activities have been recorded at all hours of daylight from dawn until just after sunset.

Four male mating tactics have been observed: *consorts, lone large males* (LLMs), *surreptitious sneaker* and *bold sneakers*, each yielding different mating success. Only consorts pair with a female mate. Generally, the consorts and LLMs are large (ca. 24 cm ML) and sneakers are small males (ca. 10–15 cm ML). Furthermore, mostly large males engage in agonistic behaviour (Fig. 6.17f) and mate guarding (Fig. 6.17e). Consorts mate in the parallel position (Fig. 6.17a) and obtain most of the matings (55% of all successful matings); overall, they succeeded in 89% of their mating attempts. Mating duration was 25.4 ± 9.5 seconds (range 14–42 seconds), and afterwards, in collected females, very white sperm could be seen in the otherwise translucent seminal receptacle (Fig. 6.17d). Curiously, consorts only tried to fight off sneaker mating attempts 5% of the time. Unpaired males used several tactics to obtain access to females. LLMs would challenge a consort male, engage in a fight that could escalate to fin beating, and if successful (10% of the times) become the new consort of the female; their mating success was only 6%. Some LLMs had an alternative tactic: they would forgo the fighting and pairing phase and would act like sneakers, quickly grab a paired female (Fig. 6.17c) and try to mate with her in the head-to-head position (6% success rate). Some of these appeared to be forced copulations, and they were seen only in the Egg-Laying Zone.

Surreptitious sneakers were generally alone and would swim near a pair, occasionally following them throughout all zones of the mating arena, and then jet very rapidly towards the female and try to mate with her in the head-to-head position (24% success rate) (Fig. 6.17b). Sneaker mating duration was 15.5 ± 5.6 seconds (range 4–25 seconds). Alternatively, small males would act as bold sneakers by swimming (often in small groups of three to six males) in the Egg-Laying Zone and attempting a head-to-head mating with a paired female as she approached the egg bed to lay an egg capsule (37% success rate). It was common to see these small groups of sneakers following a pair or swimming over the egg bed. Collectively, sneakers accounted for 45% of all successful matings due to frequent attempts; average success rate was 29%. In the absence of large males, small males acted as consorts. In laboratory trials, small males acted as consorts until a larger male was introduced, and then immediately switched to sneaker behaviour (Hanlon *et al.*, 1997); when the large male was removed, the small male swiftly shifted back to consort behaviour, showing flexibility and a conditional strategy.

Agonistic bouts (Fig. 6.17f) were common and most (77%) occurred within the Mating Zone and were between consort males and challenging LLMs. Fight duration ranged from 2 to 53 seconds (13.3 ± 10.2 seconds). Overall, consort males won 90% of 94 fights. Mate guarding was a busy enterprise: focal sampling indicated that consort males fought a challenging LLM every 6 minutes when spawning was in full swing.

A very unusual discovery revealed a major stimulus to male aggression. A contact pheromone deposited by the female in the outer egg tunic immediately initiates highly aggressive fighting behaviour among males, even in the absence of females. This is described in §7.2 (Cummins *et al.*, 2011). The authors surmise that this critical trigger of agonistic behaviour allows males to target such behaviour to times and places when fertile, receptive females are available for mating and egg laying.

Females are polyandrous. Paired, receptive females mated every 14 min, with either consorts or sneakers, and laid egg capsules every 5 min. Females did, however, demonstrate direct mate choice by rejecting 55% of all mating attempts. Indirect mate choice is likely to operate via postcopulatory sperm choices. Field-sampled egg capsules (ca. 100–300 eggs each) had multiple paternity as determined by DNA fingerprinting

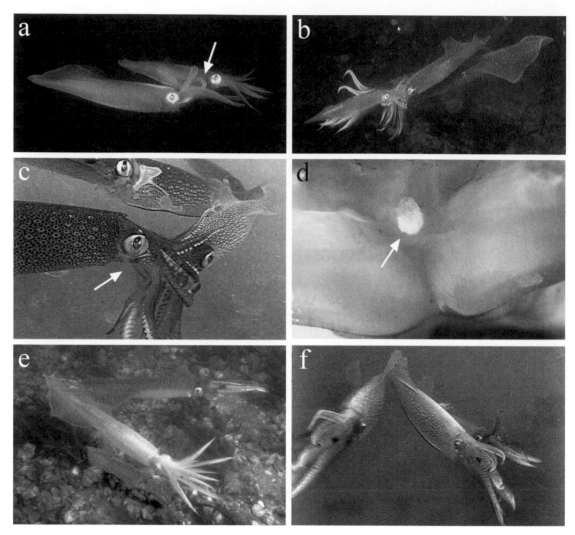

Figure 6.17 Reproductive behaviours of the squid *Loligo pealeii*. (**a**) Mating by a consort male (below) in the male-parallel position; the male is transferring spermatophores with his fourth right arm into the mantle cavity of the female (note his arm insertion just behind the female's eye). (**b**) Mating by a sneaker male (left) in the head-to-head position. The consort male (right) is being cuckolded but not reacting to the sneaker male. (**c**) Head-to-head mating by a large sneaker male (left) inserting his hectocotylised arm with spermatophores into the female's buccal region. (**d**) Close-up photograph of sperm (white mass, arrow) stored in the female's seminal receptacle just below the buccal mass and amidst the bases of the arms. (**e**) Mate guarding by a consort male (middle), who spreads his arms over the female and turns white in response to an approaching lone large male (top). (**f**) Agonistic bout between two males. The consort and his mate are on the right, the challenging male on the left. This Lateral Display includes parallel positioning, transient flame markings on the mantle, arm spots, contorted arms and fin beating. (Photos: **a** by Brian Skerry; **b, d, f** by R. Hanlon; **c** © Nick Caloyianis Productions, Inc.; **e** by William Macy, with kind permission.)

of multiple ova in each capsule (Buresch *et al.*, 2001). Each egg capsule had at least two to four fathers, and further fingerprinting would probably reveal more than this (§6.6.4).

This finding begged the question of which male mating tactics were most successful, and if/how female choice influenced paternity. Controlled laboratory mating experiments (Buresch *et al.*, 2009) indicated that relative paternity of the first egg capsule was typically in favour of the first male to mate in the trial. When the female mated with an additional male before the second egg capsule was laid ($n = 10$ trials), the first male to mate typically continued to achieve high relative paternity in the second capsule when the interval between first mating and second egg laying was relatively brief (i.e. 40 min or less). Great differences in relative paternity were observed in the second capsule when the interval between the first mating and the laying of the second egg capsule was longer than 140 min. In this study and that of Maxwell and Hanlon (2000), females in some trials – or even those maintained in isolation without access to males – laid fertilised eggs over periods of 15 or more days, demonstrating the use of stored sperm. These experiments argue against routine second and later-male sperm precedence in loliginid squid, pointing to other influences on paternity, such as the interval between insemination and egg laying, and perhaps other aspects of cryptic female choice (Eberhard, 1996, 2000).

Multiple spawning events by individual female *L. pealeii* in the lab during many weeks are common (Maxwell & Hanlon, 2000). Multiply ovipositing females exhibited a variety of patterns of oviposition, ranging from relatively small clutches at short intervals to large clutches several weeks apart. The 'spawning strategy' of *L. pealeii* appears to involve multiple ovipositions over weeks or months, with oocytes possibly being developed continually. Thus, *D. pealeii* is not strictly semelparous as often suggested in past literature for squid. Placing the results of this study in a larger context, reproduction by females in this and other loliginids most likely entails copulation with multiple males and the laying of multiple clutches of eggs, possibly in different locations.

In summary, this is a dynamic mating system with many complex behaviours by both sexes. There is a high level of sperm competition and cryptic female choice (Eberhard, 1996), details of which are summarised in §6.6.4. In *L. pealeii*'s southern range in the Gulf of Mexico, there is little seasonality of egg production and no large communal egg masses (Hixon, 1980; Summers, 1983; Vecchione, 1988), emphasising flexibility in reproductive behaviour that is not as evident in *Loligo opalescens* (see below). This short-lived species demonstrates conditional reproductive tactics with high flexibility.

6.3.4 Other *Loligo* and *Doryteuthis* Squid

Several species form inshore spawning aggregations similar to *Loligo pealeii*, and in this section we provide only brief comparisons and highlight notable differences. Table 6.2 lists species for which some mating details are known.

Loligo opalescens in California is notorious for the largest-known nocturnal spawning aggregations (Fig. 6.18a) characterised by enormous fields of eggs (Fig. 6.18b) as well as mass mortality of spawned-out squid (Fields, 1965; Hurley, 1977). However, contrary to that older literature, recent studies with ROVs and echo-sounders show that squid in Monterey often, if not predominantly, spawn during the day in small groups which produce egg beds that are seldom more than 1–2 m in diameter (Fig. 6.18c) (Forsythe, Kangas & Hanlon, 2004). Zeidberg *et al.* (2012), also using ROVs, found most spawning at night in the same region; thus, there are flexible dynamics of spawning in this species. Unlike other loliginids, mate guarding by consort males takes the form of a continual 'nuptial embrace' of the female (Fig. 6.18c) before, during and after mating and egg laying. This tactic is effective: lone male takeovers were rare (14 of 370 attempts; Hanlon, Kangas & Forsythe, 2004 and unpublished). Importantly, these ROV observations confirmed that females actively broke off the nuptial embrace and swam strongly upwards in the water column to re-join the large schools of squid many metres up in the water column. This is contrary to previous reports in which females die soon after spawning. The overall dynamics of reproduction in Monterey are depicted in Fig. 6.18d (constructed from Forsythe, Kangas & Hanlon, 2004; Hanlon, Kangas & Forsythe, 2004; and unpublished). Large schools of thousands of squid are 10–20 m above the spawning grounds, and small groups of lone males (10–100) leave the school and try to intercept lone females ascending from the egg beds. Females that descend from the large school in mid-water are all paired. Note that the Operational Sex Ratio (OSR) is estimated (from fishery catch statistics) at 1M:1F in the large school, but it is ca. 2M:1F at the egg beds, thus setting up a competition gradient in which males are competing for female mates. Overall, the reproductive behaviour of *L. opalescens* differs in several fundamental ways from other loliginids, and comparative studies may be rewarding.

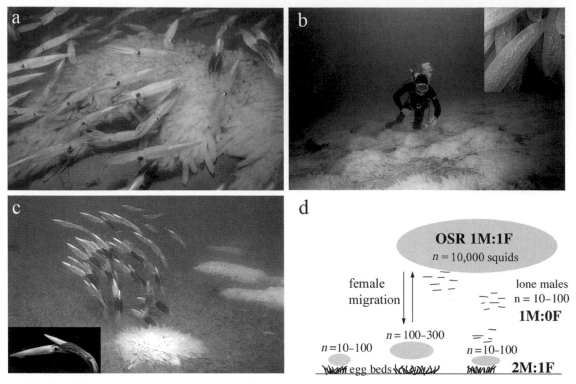

Figure 6.18 'Big Bang' and normal reproduction by *Loligo opalescens* off California. (**a**) A normal-sized egg bed in southern California with two mating pairs in upper right, and various unpaired squid (mainly males) around the white-coloured egg mops. (**b**) An extensive egg bed in southern California (photos **a**, **b** by Howard Hall); insert shows a single egg capsule, each with many ova. (**c**) A small 'normal' school of actively spawning squid (ca. 25 pairs) at 25 m off Monterey, California; inset shows one of the mating pairs, with the male below holding the female with his darkly patterned arms (photos by R. Hanlon). (**d**) A schematic of the dynamics of reproduction in Monterey, California (Hanlon, unpublished). See text for explanation.

Loligo reynaudii in South Africa has been studied via SCUBA diving (Hanlon, Smale & Sauer, 1994, 2002) and telemetry (Sauer *et al.*, 1997). The mating arena zones shared many characteristics seen in Fig. 6.16a. Mating tactics were similar to those of *L. pealeii*, with consorts and sneakers using the same mating positions. Consorts won 16 agonistic bouts, while LLMs won 9; thus, with LLMs winning 1 in 3 contests, there was higher turnover of consorts than found in *L. pealeii*. Sneakers were more choosy, pursuing copulations only with females that had just mated with consorts. Multiple paternity of egg cases was also determined (Shaw & Sauer, 2004). Various other aspects of reproduction that complement behaviour are known, and should be consulted by the interested reader (Sauer, Smale & Lipinski, 1992; Sauer & Smale, 1993; Sauer, 1995; Lipinski *et al.*, 1998; Melo & Sauer, 1999, 2007; Roberts,

Downey & Sauer, 2012). The dynamics of daily formation of mating arenas were determined by using 3D radio-linked acoustic positioning (RAP) to track electronically tagged squid (Sauer *et al.*, 1997). Each dawn, squid navigated several kilometres toward the shore to small, well-defined zones near egg beds. After several hours of circling above the egg beds, a pelagic lek-like aggregation condensed as males and females paired, mated and laid eggs.

Loligo bleekeri in Japan has been studied in considerable detail with respect to sperm competition, which is treated below (§6.6.4 and Fig. 6.31). Like *L. pealeii*, they mate in male-parallel and head-to-head positions, the former being with consorts and the latter with sneakers (Hamabe & Shimizu, 1957; Iwata & Sakurai, 2007). However, this is the only known species in which the male places spermatangia inside the oviduct of the female, whereas other loliginids place them inside the mantle cavity near the oviduct. Females also show a distinctive 'lateral blush' pattern (Fig. 6.28) whose function is unclear, but has to do with receptivity. Moreover, in laboratory experiments, the mating position seems to be affected by temperature, with more male-parallel attempts and successes occurring at the higher temperatures of 8 °C. The effects of commercial fishing on actively spawning squid are considered by Iwata, Ito and Sakurai (2010).

The European squid *Loligo forbesi* and *L. vulgaris* have been studied in various aspects of reproduction, but little of that work involved field behaviour or detailed experimentation. A few items are noteworthy. *Loligo forbesi* has been observed cursorily in laboratory tanks showing male agonistic behaviour and egg laying (Porteiro, Martins & Hanlon, 1990; Pham *et al.*, 2009), and multiple paternity in individual egg capsules (Shaw & Boyle, 1997). Related aspects of reproduction have also been pursued (Holme, 1974; Lum-Kong, Pierce & Yau, 1992; Lum-Kong, 1993; Boyle, Pierce & Hastie, 1995; Collins, Burnell & Rodhouse, 1995; Shaw, 1997). Practically no reproductive behaviour has been reported for *L. vulgaris* (Europe), but several complementary papers may aid future ethological research (Tardent, 1962; Coelho *et al.*, 1994; Rocha & Guerra, 1996; Sifner & Vrgoc, 2004).

Loligo plei in the Caribbean occurs near coral reefs where it occasionally mingles with *Sepioteuthis sepioidea* (Boycott, 1965; Moynihan & Rodaniche, 1982). Waller and Wicklund (1968) observed a large natural spawning shoal in the Bahamas, noting that nearly all squid were paired and that mating in the 'male-parallel' position was followed almost immediately by egg laying. Males initiate pairing in the field and laboratory (Hanlon, Hixon & Hulet, 1983). In a small shoal of 10–20 squid and a 1:1 sex ratio, one male will establish obvious dominance within 1 to 4 days. Males show the Accentuated testis or Lateral flame components of patterning to other squid; failure to respond to these male-only signals presumably indicate 'non-maleness', as in *Sepia*. Females are smaller than males and have a distinctive body pattern with bands and downward curled arms that acts as a signal (or releasing stimulus) for courtship and agonistic behaviour in males; healthy males often copulate with moribund males that curl their arms downward like the female. Details of male agonistic contests are described in §6.6.1.

6.3.5 Oceanic and Deep-Sea Cephalopods

Our knowledge of reproductive behaviour in oceanic (oegopsid) squid lags far behind what is known for nearshore myopsid squid. The enormous diversity of species and habitats will certainly yield differences in reproductive behaviour. Judging from the anatomy of both sexes and the presence of spermatophores in the seminal receptacles of several species (Table 6.2), it is anticipated that the mating systems of at least some oegopsid squid may be somewhat similar to myopsid squid.

Todarodes pacificus in Japan is one of the best-known examples, and Sakurai, Bower and Ikeda (2003) provided a working hypothesis of the reproductive process based on laboratory experiments and field observations (Murata, 1990; Bower & Sakurai, 1996). As depicted in Fig. 6.19a, females descend to the substrate on the continental slope apparently in preparation to produce pelagic egg masses (ca. 80 cm diameter, containing over 200 000 embryos); these are released high in the water column in depths of ca. 70–100 m. Although the place and timing of mating in nature are unknown, in the laboratory a male swims under the female and grasps her in a 'male-parallel' position (Fig. 6.19b) and grasps a large bundle of spermatophores and places them on the buccal membrane (Fig. 6.19c, d) similar to loliginid squid. After a few days, the female ceases feeding and then often rests on the bottom for about 2 days before spawning the gelatinous egg mass. The authors noted that the female's spawning sequence of the gelatinous egg mass is similar to that reported for the oegopsid squid *Illex illecebrosus* in Canada (O'Dor, Balch & Amaratunga, 1982).

Spermatophores have been found embedded in the fourth pair of arms of female giant squid *Architeuthis*, as for a few other squid (Norman & Lu, 1997). The deep-sea squid *Pholidoteuthis adami* has been filmed mating with the male upside down over the female, and implanting spermatangia into her dorsal mantle (Fig. 6.20a; Hoving & Vecchione, 2012). Museum specimens of *P. adami* have large sperm masses on the inside of the mantle, indicating that this ROV observation of mating resulted in successful transfer of spermatangia, which somehow migrate from the outside to the inside of the mantle. The deep-sea squid *Taningia danae* also has been found with spermatangia implanted in the mantle, head and neck,

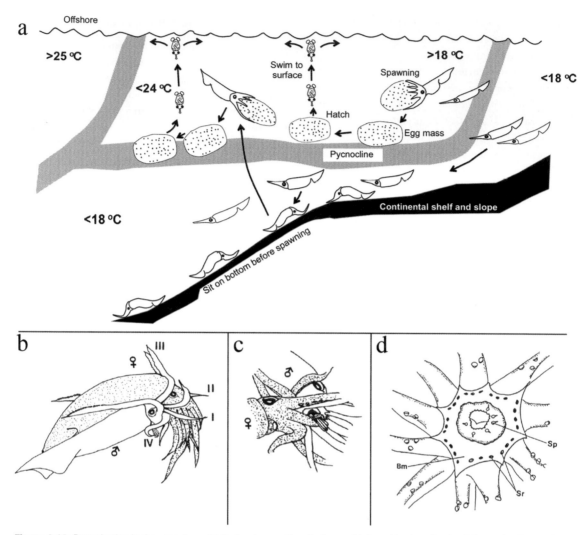

Figure 6.19 Reproduction in the oceanic squid *Todarodes pacificus* in Japan. (**a**) A working hypothesis of the cycle of spawning; see text. (**b**) 'Male-parallel' mating; Roman numerals depict arm number. (**c**) The male (left) places spermatophores within the female's buccal membrane region. (**d**) Buccal region of female showing buccal membrane (Bm) and location of embedded spermatophores from copulation (Sp) and seminal receptacles (Sr). Reprinted from Sakurai, Bower and Ikeda (2003): *Modern Approaches to Assess Maturity and Fecundity of Warm- and Cold-Water Fish and Squids*, Vol. 12, by permission of the Institute of Marine Research; and from Sakurai *et al.* (2013): *Advances in Squid Biology, Ecology and Fisheries*. Part II, 249–271, ed. R. Rosa, G. J. Pierce & R. O'Dor, © (2013), with permission from Nova Science Publishers.

often associated with incisions that are thought to be made by the beak or arm hooks of male mates (Hoving *et al.*, 2010). Males of the deep-sea squid *Octopoteuthis deletron* mate indiscriminately with males and females, and place spermatangia in a variety of dorsal and ventral body locations (Hoving, Bush & Robison, 2012). Male deep-sea incirrate octopods have also been filmed mating another male (Lutz & Voight, 1994), suggesting, as in

O. deletron, that, in sparsely occupied habitats of the deep sea, it may pay cephalopods to mate any conspecific encountered.

Brooding of gelatinous egg masses seems to be a characteristic of deep-sea squid. Female mid-water squid *Gonatus onyx* use hooks on their arms to hold egg masses (ca. 2000–3000 eggs) that extend twice the length of the arms (Fig. 6.20c; Seibel, Hochberg & Carlini, 2000;

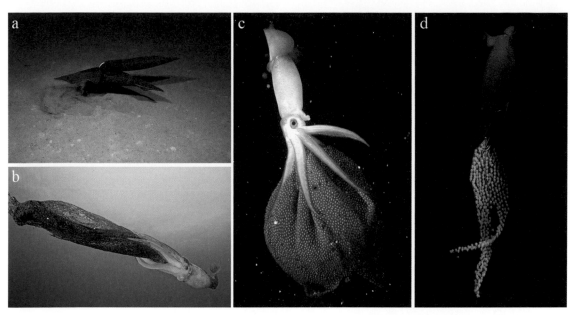

Figure 6.20 Egg brooding and mating in some deep-sea squid. (**a**) Mating by *Pholidoteuthis adami*; the male (top) is upside down and implanting spermatangia on the dorsal mantle of the female (reprinted from Hoving & Vecchione, 2012: *Biological Bulletin* **223**, 263–267, Mating behavior of a deep-sea squid revealed by *in situ* videography and the study of archived specimens, Hoving, H. J. T. & Vecchione, M., © (2012), by permission of the University of Chicago Press). (**b**) *Gonatus madokai* swimming at 20 m off Japan while brooding a large mass of eggs amidst her arms (from Bower *et al.*, 2012: *Biological Bulletin*, **223**, 259–262, Brooding in a gonatid squid off northern Japan, Bower, J. R. *et al.*, © (2012), by permission of the University of Chicago Press). (**c**) The mid-water squid *Gonatus onyx* (145 mm ML) brooding a large mass of eggs amidst her arms at 1539 m (from Seibel, Robison & Haddock, 2005: reprinted by permission from Macmillan Publishers Ltd, *Nature* **438**, 929, © 2005). (**d**) The deep-sea squid *Bathyteuthis berryi* (72 mm ML) brooding a single-layer egg sheet at 1351 m (reproduced from Bush *et al.*, 2012: *Journal of the Marine Biological Association of the United Kingdom* **92**, 1629–1636, Brooding and sperm storage by the deep-sea squid *Bathyteuthis berryi* (Cephalopoda: Decapodiformes), Bush, S. L., *et al.*, © (2012), with permission of the Marine Biological Association).

Seibel, Robison & Haddock, 2005). Low temperatures at these depths of ca. 1500–2500 m suggest that females may be brooding these eggs for 6–9 months, during which time their jet-escape responses are diminished. *Gonatus madokai* off Japan also brood eggs similarly and have been found naturally by SCUBA divers in just 5–20 m depth in spring (Fig. 6.20b). Presumably the females rise to surface waters just before egg hatching to aid paralarval survival in habitats where planktonic prey are more abundant (Bower *et al.*, 2012). The deep-sea squid *Bathyteuthis berryi* has been videotaped by ROV at 1351 m holding a long thin sheet of ca. 360 embedded embryos (Fig. 6.20d), and subsequent histological investigation of the collected female showed a seminal receptacle within the buccal membrane and stored sperm (Bush *et al.*, 2012). Four deep-living pelagic octopuses also brood their eggs at depth by holding egg masses in their arms: *Balitaena pygmaea*; *Haliphron atlanticus*; *Japetella diaphana*; and *Vitreledonella richardi* (Young, 1972b, 1995).

Non-brooded floating gelatinous egg masses are known in several species, such as the Humboldt squid *Dosidicus gigas* (Staaf *et al.*, 2008), *Illex illecebrosus* (O'Dor, 1983), *Ommastrephes sloani pacificus* (Hamabe, 1961), *Thysanoteuthis rhombus* (Guerra *et al.*, 2002a; Biagi & Bello, 2009; Escanez *et al.*, 2012) and *Nototodarus gouldi* (O'Shea & Bolstad, 2004). As more direct observations are performed in the deep sea, and details of anatomy and behaviour of deep-sea cephalopods are revealed (e.g. Nesis, 1995; Roper & Vecchione, 1996; Hoving, 2008), it is expected that brooding and other unique tactics of reproductive behaviour will be discovered.

6.4 Agonistic, Courtship and Mating Behaviour in Octopuses

Octopuses are, for the most part, solitary animals that show little agonistic or courtship behaviour before mating, yet a notable exception has been discovered in the intertidal species *Abdopus aculeatus* (Huffard *et al.*, 2008, 2010). Collectively, mating has been observed in 19 species (Table 6.3), 11 of them in the sea, and mating duration was usually from 30 minutes to 2 hours but varied greatly by species (ranging from 1 minute to 6 hours).

Table 6.3 Mating by octopods

Species	Duration (minutes)	Mating position	Mate guarding	References
1. *Abdopus aculeatus	~5	Distance	Yes	Huffard, Caldwell & Boneka, 2008
2. *Amphioctopus marginatus	?	Distance	?	Huffard & Godfrey-Smith, 2010
3. Eledone cirrhosa	~60	Mounting	No	Orelli, 1962; Boyle, 1983b
4. E. moschata	20–60	Mounting	No	Mangold, 1983c; Mather, 1985
5. *Enteroctopus dofleini	~120–240	Mounting	No	Gabe, 1975; Hartwick, 1983
6. Hapalochlaena maculosa	~60	Mounting?	No	Tranter & Augustine, 1973
7. H. lunulata	160 (25–246)	Mounting	No	Cheng & Caldwell, 2000
8. *Octopus bimaculatus	~10–60	Distance	?	Fox, 1938; Pickford & McConnaughey, 1949
9. *O. bimaculoides	~60 (10–180)	Distance	No	Forsythe & Hanlon, 1988
10. *O. briareus	30–80	Mounting	No	Hanlon, 1983b
11. O. chierchiae	1	Mounting beak-to-beak	No	Rodaniche, 1984; Hofmeister *et al.*, 2011
12. *O. cyanea	~60	Distance or mounting	?	Wells & Wells, 1972; Van Heukelem, 1983a; Tsuchiya & Uzu, 1997; Hanlon & Forsythe, 2008
13. *O. digueti	70	Distance	No	Voight, 1991a
14. *O. horridus	10	Distance	No	Young, 1962a
15. O. joubini	~5 (2–28)	Distance or mounting	No	Mather, 1978; Hanlon, 1983a
16. O. maya	100–240	Distance or mounting	No	Van Heukelem, 1983b
17. *O. tetricus	12–360	Distance or mounting	?	Joll, 1976; Huffard & Godfrey-Smith, 2010
18. *†O. vulgaris	~60–120	Distance or mounting	No	Racovitza, 1894; Orelli, 1962; Woods, 1965; Wells & Wells, 1972; Wodinsky, 1973, 2008
19. O. maorum	20–210	Mounting	?	Anderson, 1999

* Observations in the sea; † spermatophore placement is in oviducal gland, so far as is known; ? not certain.

6.4.1 *Octopus vulgaris* and *Octopus cyanea*

Amongst laboratory studies, the most detailed descriptions are for *Octopus vulgaris* (Atlantic species) and *Octopus cyanea* (Pacific species), both of which are diurnally active and attain large sizes up to 5 kg. Much of this account is based on the study of Wells and Wells (1972), who observed 161 and 46 matings, respectively, in these species. No evidence was found of cohabitation (as in *A. aculeatus* below) and only one example of temporary pairing, and both sexes had multiple mates.

Octopus vulgaris is typical of many octopuses known thus far in that there appears to be little or no prelude to mating: that is, no specific body patterns by either partner, nor any consistently seen ritualised postures or behaviour, although some form of web spreading is occasionally shown by an approaching male. Packard (1961) suggested that small males use a 'sucker display' to identify their sex to larger females. However, his very few observations (and no observations of this behaviour by Wells & Wells, 1972) leave this question open. *Octopus vulgaris* copulates in one of two ways: a male may leap upon a female, mounting her mantle (Fig. 6.21a), or a male may sit near the female and extend the hectocotylised third right arm towards her (Fig. 6.21b). There is considerable variation with both techniques, but despite this the end organ of the hectocotylised arm (with a ligula and calamus) 'seeks' the oviduct via the mantle opening, seldom taking more than a few seconds to find it. Wells and Wells (1972) suggest that there may be a chemical cue for the end organ of the hectocotylus to find its target, but this remains speculative. However, the end organ apparently possesses sensory capability, because Wodinsky (2008) cut off the ligula and reported that the hectocotylised arm moved continuously in and out of the female's mantle cavity, never became quiescent, and did not attach to the oviduct.

Males, which carry about 50 spermatophores, each 2–3 cm long, place a spermatophore at the base of the hectocotylised arm with an 'arched' posture followed by a 'pumping' action that sends the spermatophore down the hectocotylised arm and into the oviduct (see also Wodinsky, 2008). A spermatophore is passed approximately every 15 minutes during a typical hour-long mating period, but there is no information about how many of these sperm are stored by the female. Anecdotal field observations on *O. vulgaris* by Orelli (1962) and Woods (1965) are in agreement with these laboratory observations.

Octopus cyanea males sometimes approach females with spread web, or often by 'standing tall' with the proximal part of the arms held stiff and papillae erected over each eye. There is considerable variation in approach, as with *O. vulgaris*, and there seems to be no obvious sequence of visual signals that could qualify as a courtship display (see our definition in §6.1.2). In both species, observations suggest that physical contact plays a part in sex recognition. For example, in laboratory trials, relatively quiet males of both species become highly active after a female touches them, or vice versa.

Octopus cyanea occasionally flashes a distinctive pattern of dark brown stripes over a white background (Young, 1962a), but this striped pattern is shown during male–male encounters as well as to females that have resisted the advances of the male, and is probably a form of deimatic behaviour (Chapter 5; also Forsythe & Hanlon, 1997). A similar striped pattern has been reported in *Abdopus aculeatus* (next section). During mating (Fig. 6.21d), which occurred in either position and lasted about an hour in Wells and Wells (1972), spermatophores were passed at intervals of two to three minutes. However, great variation can occur in mating duration, as found in the field study by Hanlon and Forsythe (2008), in which durations of 13 consecutive matings 'at a distance' between a large foraging female and a small male were 1, 2, 0.5, 1.5, 14, 5, 7, 1, 23, 12, 5, 5 and 2 seconds. That sequence ended with sexual cannibalism of the male (see below §6.4.3). Moreover, anecdotal field observations report that *O. cyanea* may occupy adjacent lairs on reefs and mate repeatedly over 2–3 day periods (Yarnall, 1969; Van Heukelem, 1983a; Norman, 1991; Fred Bavendam, pers. comm. 2012, Bali; Heather Ylitalo-Ward, pers. comm. 2013, Hawaii). Tsuchiya and Uzu (1997) observed two males simultaneously mating a single female for two hours, each male inserting his hectocotylus on opposite sides of her mantle, presumably to utilise both of her oviducts. Moreover, the second male was reported to have lost a fight to the larger male, then to have sneaked up to the female for this dual mating; this was interpreted as possibly being a 'sneaker male' tactic. Obviously, there is a great deal yet to learn about mating dynamics and sexual selection in this species.

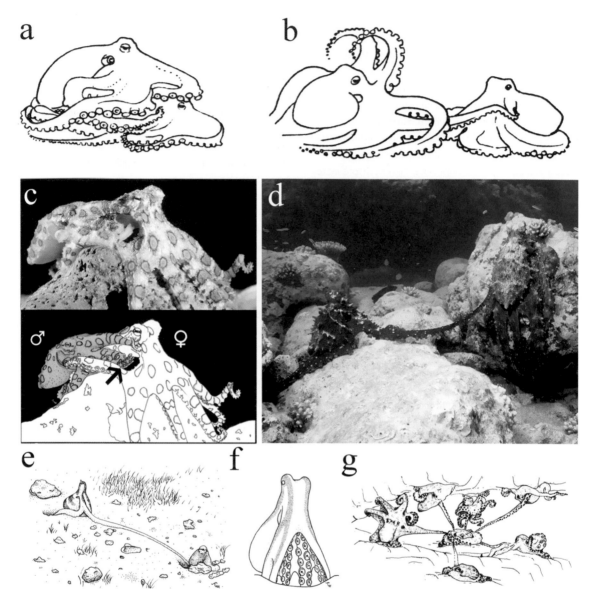

Figure 6.21 Mating in octopuses. (**a**) Male *O. vulgaris* (top) mounting a female. (**b**) Male (left) at a distance, extending his third right arm, the hectocotylus. (**c**) Blue-ringed octopus, *Hapalochlaena maculosa*, male (top left in both the photograph and drawing), mounting a female and inserting his hectocotylus (arrow) into her mantle cavity. (**d**) *Octopus cyanea* mating at a distance on a Pacific coral reef, Palau, Micronesia; male (left) is extending his hectocotylised arm to the female. (**e**) Male *Abdopus aculeatus* (left) extending his hectocotylised arm to a female in her den; (**f**) note the striped pattern on the male (from Huffard, Caldwell & Boneka, 2008: *Marine Biology* **154**, 353–362, Mating behavior of *Abdopus aculeatus* (d'Orbigny 1834) (Cephalopoda: Octopodidae) in the wild, Huffard, C. L., Caldwell, R. L. & Boneka, F., © (2008), with permission of Springer). (**g**) Six male *O. bimaculatus* attempting to mate a single female at Catalina Island, California (unpublished from R.F. Ambrose). Drawings in (**a**, **b**) from Wells & Wells, 1972: *Animal Behaviour* **20**, 293–308, Sexual displays and mating of *Octopus vulgaris* Cuvier and *O. cyanea* Gray and attempts to alter the performance by manipulating the glandular condition of the animals, Wells, M. J. & Wells, J., © (1972), with permission from Elsevier; photograph in (**c**) by Roy Caldwell; drawings in (**c**, **e**, **f**) by Christine Huffard; photograph in (**d**) by R. Hanlon; drawing in (**g**) by R. Ambrose.

6.4.2 *Abdopus aculeatus*

Unlike any other octopus studied thus far, *A. aculeatus* has a complex mating system involving male–male competition, frequent copulations by temporary pairs, mate guarding, several tactics for copulation and visual displays observed only in the context of mating (Huffard, 2007). This octopus is a diurnal intertidal species in the Indo-Pacific, small in size (up to 7 cm ML), and relatively abundant, which enabled detailed observations in the wild (Huffard, Caldwell & Boneka, 2008, 2010). Focal animal observations (789 hours) were conducted via snorkelling at three locations in Sulawesi, Indonesia, and the operational sex ratio averaged 1.8M:1F. More than 160 octopuses were located, and the study concentrated on animals greater than 3 cm ML, the approximate lower size limit of mating.

Large males occupied lairs within arms' reach of large females, and they copulated frequently over successive days, during which time they successfully guarded the female from other males and thus temporarily monopolised mating. Both 'guarding' males and 'guarded' females also foraged away from their dens and obtained 'transient' copulations with other individuals they encountered. Intense agonistic bouts between males occurred, but only in defence of a female. Such bouts involved aggressive behaviours including chase, touch, 'whip' with one straight arm, and grapple. Larger males won 26 of 29 bouts. Male–female aggression was common but did not appear to be a consistent form of mate choice. No fighting between females was observed.

Mating occurred 'at a distance' with the male extending his hectocotylised arm to the female (Fig. 6.21e). Mating duration averaged 4.5 min by guarding males, and 8.5 min by sneaker males. No behavioural significance could be attributed to the longer matings by sneakers, although some octopus species have a ligula that engorges (Thompson & Voight, 2003; Huffard & Godfrey-Smith, 2010), and there has been speculation that this may be a mechanism of sperm removal (Cigliano, 1995).

Sneaker males obtained copulations by visually hiding themselves from guarding males to avoid aggression, while mating with the guarded female. The sneakers were the smallest males known to inhabit the vicinity of the female, and they would approach with great stealth and camouflage to avoid visual detection by the guarding males. This is the first confirmed and quantified report of sneaker male tactics in octopuses.

Sex identification at a distance seemed to be occurring in these field observations. When large males spotted another individual, they frequently displayed the 'Black and White Stripe' pattern (Fig. 6.21f). This pattern was speculated by Huffard, Caldwell and Boneka (2008) to be a signal of sex identity, yet this is questionable since females also are capable of showing this display (Huffard, 2007; Huffard, Caldwell & Boneka 2008). Most mating encounters (206 out of 223 insertions) began with males touching the female with their hectocotylised arm, and this would provide another suite of sensory cues for sex identification. By contrast, males initiated physical aggression with other males using other arms.

Direct female choice was very low: males were rejected in only 7% of cases, and females mated with males of different sizes and using different tactics. Females used what appeared to be a proceptive display of raising the dorsal arms and curling them only at the distal portions; this display was associated with a decreased delay to mating (Huffard, 2007). Although nearly 1/3 of females occasionally rejected a mating attempt, overall they accepted nearly every opportunity to copulate; thus females of this species are apparently non-selective with respect to mate choice.

Huffard, Caldwell and Boneka (2008) and Huffard and Godfrey-Smith (2010) drew attention to anecdotes published on several other shallow-water benthic octopuses indicating that some of the more complex features of the mating system of *A. aculeatus* may not be so unusual among octopods. More detailed field studies are encouraged.

6.4.3 Sexual Cannibalism

Sexual cannibalism by a large female *Octopus cyanea* was documented in the field by Hanlon and Forsythe (2008). The male in this case was approximately 500 g, roughly half the size of the female. He followed her for more than 3 hours as she foraged more than 70 m, during which time he mated with her 'at a distance' 13 times. The male was extremely cautious at all times, and often flinched backwards whenever the female made the slightest movement toward him. During the 13th mating, she turned towards him, swiftly engulfed him and swam off with him despite a huge struggle and copious inking by the male, who was completely wrapped within her arms and web. She returned to her den and consumed the large meal for the next 2 days, presumably benefiting from reduced foraging time as well as his possible genetic contribution to her sperm stores. Curiously, as this female was consuming the cannibalised male the next day, another small male cautiously approached and mated once with her for 3 hours, which was in stark contrast to the very short mating durations (mean 6 min, range 0.5–23) of the 13 matings by the first small male. Female receptivity remains enigmatic, since she could not even see the male in this last mating.

This report of cannibalism in the wild perhaps adds perspective to the curious laboratory and field observations in other species in which mating occurs out in the open. For example, Hanlon and Wolterding (1989) suggested that it may be advantageous for the smaller partner to mate in the open, despite the risk of predation, because of known cannibalistic tendencies in octopuses. Once, in the sea, Hanlon (1983b) observed a small male *Octopus briareus* (53 mm ML) swim across the substrate, mount a larger female (85 mm ML) and mate with her for 58 minutes, both of them completely exposed in the open area. A remarkable feature of this encounter was that the same female had only minutes before eaten another male of 53 mm ML.

6.4.4 Reproductive Behaviour in Other Octopods

Table 6.3 provides references to reproduction in various shallow-water octopods, and many of them share behaviours similar to those of *O. vulgaris* and *O. cyanea* described above. For example, the newer observations of *O. tetricus*, *Amphioctopus marginatus* and the blue-ringed octopus *Hapalochlaena lunulata* (Fig. 6.21c; Cheng & Caldwell, 2000) are not very different from the known suite of reproductive behaviours seen in *O. vulgaris* and *O. cyanea*.

Two striking exceptions to these schemes have been reported. In two small Panamanian species of Pacific striped octopus, Rodaniche (1984, 1991) and Hofmeister *et al.* (2011) briefly described widely different behaviours that included a previously unseen

'beak-to-beak mating position', possible clustering of individuals and 'pair bonding', the use by males of inking and arm twirling as courtship signals, a permanent white line down the hectocotylised arm, multiple clutches of eggs by individual females, and feeding by females during egg brooding. The second exception is that reported above for *Abdopus aculeatus*, which demonstrated previously unseen male agonistic contests over females, and two forms of mate guarding of females (Huffard, Caldwell & Boneka, 2008, 2010). It is probable that many other variations in octopod mating systems will be discovered when the other 250+ species come to be studied.

Deep-sea and pelagic octopods are rarely studied, and to our knowledge there are no observations of mating-related behaviour except the report by Lutz and Voight (1994) of attempted copulation between two males of different species at 2512 m. The authors surmise that octopuses in these low-density environments attempt copulation with any octopus encountered 'rather than to leave opportunities for reproduction unexplored'.

6.5 Fertilisation, Egg Deposition, Egg Care and Parental Death

In cuttlefish, fertilisation is external, occurring within the arm bundle as sperm are released from the seminal receptacle on the buccal region or in the mantle cavity (see Table 6.1) as eggs are extruded from the oviducts. *Sepia* eggs are often exceptionally large (up to 2 cm diameter; Corner & Moore, 1980; Arnold, 1984; Hall & Hanlon, 2002), and each is held amidst the arms while spermatozoa are released from the seminal receptacle for fertilisation. In other sepioids, the seminal receptacle – the pharetra – is in the mantle cavity near the oviduct, and the method of fertilisation is unknown. Female *Sepia* wrap the stalk of each fertilised egg around some elongated structure that, in nature, is usually hidden somewhat. Eggs of *Sepia officinalis* are usually blackened with ink, which obscures the developing embryo and may reduce predation. In *Sepia esculenta* (Natsukari & Tashiro, 1991), the eggs are deposited with a sticky clear exterior that accumulates sand or other adventitious material for cryptic coloration (see Fig. 5.4). *Euprymna scolopes* females incorporate sand grains or bottom rubble on the egg capsules, which are laid on the undersides of coral ledges (Singley, 1983; Fig. 6.12b). Females of *Sepia latimanus* search for coral crevices of the exact dimensions to protect eggs from predation by chaetodontid fish (Corner & Moore, 1980). *Sepia pharaonis* in the Arabian Sea also protect their eggs from predation by finding a suitable crevice; fish of the families Chaetodontidae, Balistidae, Monacanthidae and Zanclidae have been observed preying on eggs (Gutsal, 1989). It is common in *Sepia* to find communal egg masses in which different females have deposited eggs (Fig. 6.4h), and in *Sepiola* and *Sepietta* eggs of both species have been found together in the Mediterranean (Boletzky, 1983a). Males in some species accompany females during the several hours needed for egg laying (e.g. Corner & Moore, 1980; Moynihan, 1983a, b; Hall & Hanlon, 2002); this form of temporary mate guarding is thought to be important because of sperm competition (§6.6.4). So far as is known, females usually die shortly after spawning, although spawning can extend over several weeks or even months in the laboratory (cf. Boletzky 1983a, b, 1987a, 1988). No form of egg care has ever been reported.

In squid, fertilisation is also external, but there are two methods, according to which mating style occurred. In myopsids, each finger-sized capsule may contain three to seven large eggs (*Sepioteuthis* spp.) or hundreds of small eggs (*Loligo* and *Doryteuthis* spp.), and these are fertilised either as the capsules pass out of the oviduct (when spermatophores have been placed in the mantle cavity) or amidst the arms (when spermatozoa are released from the seminal receptacle near the mouth). Eggs may be deposited in a variety of ways: as large communal masses on the substrate (Figs. 6.16b, 6.18b), or small groups laid 'up and in' amidst rocks, seagrasses or corals. The behavioural cues initiating egg laying are not understood. Although Arnold (1962) showed that female *Loligo pealeii* will lay eggs when artificial or real eggs are placed in their tank, it is not clear what causes the first female to commence laying on a spawning ground. Female loliginid squid such as *L. pealeii* or *L. opalescens* are spent after producing large numbers of eggs (up to 55000 eggs per individual; Maxwell & Hanlon, 2000). Some oegopsid (oceanic) squid produce neutrally buoyant, gelatinous egg masses with small eggs, and little is known about these (§6.3.5). However, the fertilisation process appears to be similar, and there are indications that oegopsids also die shortly after spawning (see reviews in Boyle, 1983b). Parental care of eggs or young is unknown among loliginid squid, although some oegopsid squid care for their eggs (§6.3.5).

In octopods, fertilisation is internal, occurring in the oviducal gland (where sperm are stored) as eggs pass down the oviduct. The behavioural cues for initiating egg laying are unknown. Eggs are deposited by several methods: cemented individually to a hard substrate in the den; individual eggs intertwined with other eggs in clusters of many tens, then cemented to a substrate; or carried in the arms (next section). Females guard and care for the eggs (Fig. 6.22a, b) partly because in some species the males will eat the eggs if not guarded, and other benthic invertebrates such as chitons will remove eggs that are unattended (Narvarte *et al.*, 2013). Moreover, there is a vital behavioural and physiological change in female octopuses associated with this phase of the life cycle. Basic metabolic changes occur in *Octopus vulgaris* females as they approach full maturity, with somatic growth stopping (O'Dor & Wells, 1978), and the ovary becoming full of ripe eggs; in some octopods, the other internal organs become squeezed in the mantle cavity. Feeding behaviour becomes erratic. After egg laying, the female never leaves the den, does not feed, and continuously cleans and aerates the eggs while she broods them, which can be for as long as 1 to 3 months, perhaps a quarter of her life span. Arnold (1984) reported that when he replaced brooding females with non-brooders, they too groomed and aerated the eggs, which were apparently the releasing stimulus for that behaviour. Females sometimes expel water forcibly over the eggs, which indirectly helps to eject hatchlings out of the den; this usually occurs at night, which may enhance initial survival of the hatchlings. Females become emaciated and die soon after hatching occurs (with the exception of *O. chierchiae*; Rodaniche, 1984; Hofmeister *et al.*, 2011). Males also die at about the same age, owing to the same physical degradation as females: hormonally induced muscle catabolism, high amino acid levels in the blood, and consequent high metabolic rate (O'Dor & Wells, 1978).

Female octopods in habitats without obvious dens or shelters carry their eggs with them during development. For example, the following benthic species, which generally live on sand or mud plains, carry developing embryos in the web of the arms: *Octopus burryi*

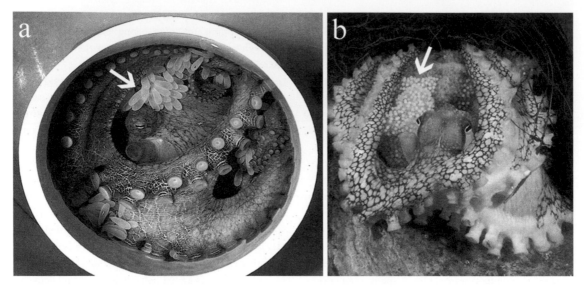

Figure 6.22 Parental care of eggs illustrated in female *Octopus bimaculoides* with large eggs (**a**), and a field image of *Octopus marginatus* with small eggs (**b**) that become planktonic. The eggs (white, shaped like tear drops) are continually aerated and cleaned by the female for several months of development, but she does not care for the young (photograph **a** by J.W. Forsythe and **b** © Fred Bavendam).

(Forsythe & Hanlon, 1985), *Octopus defilippi* (Hanlon, Forsythe & Boletzky, 1985), *Hapalochlaena maculosa* (Tranter & Augustine, 1973), *H. fasciata, H. lunulata* (Norman, 2000), one unidentified *Scaeurgus* species from Hawaii (W.F. Van Heukelem, unpublished data), *Wunderpus photogenicus* (Miske & Kirchhauser, 2006), and *Amphioctopus* spp. including *marginatus* (Huffard & Hochberg, 2005). One *Octopus burryi* was described carrying an egg mass 45% the weight of its own body (total 36000 eggs), despite which it managed to feed on crabs on at least three occasions (Forsythe & Hanlon, 1985). The pelagic octopod *Argonauta* attaches its eggs to the papery 'shell' (Arnold, 1984; Norman, 2000), *Tremoctopus* attaches its embryos to a film of organic material secreted by the web (Naef, 1923 in Young, 1972a), *Vitreledonella* retains its embryos within the mantle cavity (Joubin, 1929 in Young, 1972a), and the bathypelagic *Bolitaena microcotyla* and *Japetella diaphana* apparently brood their embryos and young in chambers formed by the arms (Young, 1972a). *Ocythoe* retains the embryos in its long and convoluted oviducts, the eggs moving slowly down them as development proceeds; this is apparently the only ovoviviparous cephalopod (Naef, 1923 in Young, 1972a).

6.6 Reproductive Strategies

It is a characteristic of most animals that, to promote their individual reproductive fitness, there is variation built into the reproductive strategy adopted by individuals of each species (Lott, 1991; Oliveira, Taborsky & Brockmann, 2008). This variation occurs in the form of alternative tactics that give individuals the flexibility needed to achieve their overall reproductive strategy. Flexibility is necessary because all animals, and particularly highly mobile ones like cephalopods, face changing ecological and demographic influences during their lifetime. The degree of flexibility is so great in some animals that they change tactics within their mating system yearly, monthly, daily and sometimes even hourly. It is not known whether cephalopods change tactics or overall strategy on yearly bases, but experiments on loliginid squid and cuttlefish show that they change tactics by the minute and even second in some circumstances. Given the very different life styles and body forms of decapods and octopods, we should expect the mating systems and reproductive strategies to be diverse.

Sexual selection is the evolutionary process that favours the increase in frequency of genes that confer a reproductive advantage (Birkhead & Pizzari, 2009). It was recognised and described by Darwin (1871), but it is only in recent decades that it has become a primary focus in evolutionary biology (Harvey & Bradbury, 1991; Andersson & Simmons, 2006; Clutton-Brock, 2009). Sexual selection of a trait can be viewed as a shorthand phrase for 'differences in reproductive success, caused by competition over males, and related to the expression of the trait' (Andersson, 1994). Sexual selection strategies augment the numbers of mates and/or effective matings. Many of the characters selected for attracting mates are recognised as being costly (either energetically or in terms of risk of predation) or a hindrance to survival (because of conspicuousness). Yet sexual selection and natural selection share some commonalities insofar as the end product is the production of as many viable offspring as possible. Males can be expected to compete for females because (1) females produce fewer and larger gametes than males, and thus female fitness does not necessarily increase with the number of mates she has, and (2) male fitness generally rises with each additional female he fertilises. Sexual selection comprises intrasexual selection (usually male competition), intersexual selection (e.g. female choice) and postcopulatory selection (male–male competition in the form of sperm competition, and cryptic female choice).

One of the most interesting aspects of reproduction in cephalopods is the facultative nature of sexual selection behaviours. For example, small unpaired males (in *Loligo pealeii* or *Sepia apama*) seek extra-pair copulations using various 'sneaking' tactics, but these are usually non-aggressive tactics that actively avoid confrontations with the paired males (for an unusual case involving sexual mimicry, see Hanlon *et al.*, 2005). However, if the large consort male leaves or is displaced, the small males recognise the new behavioural context and immediately switch tactics by becoming paired consorts to the female; they will then use agonistic displays to ward off other small males. This transition between sneaker/non-aggressive and consort/highly aggressive is quite remarkable for its speed and fluidity, and testifies to the cognitive abilities of these marine invertebrates.

6.6.1 Mating Systems

The term *mating system* of a population refers to the general behavioural strategy used in obtaining mates (Emlen & Oring, 1977). A mating system – sometimes called a breeding system – includes such features as: the

number of mates acquired; forms of courtship, coercion and competition; the presence and characteristics of any pair bonds; and the form and duration of parental care provided by each sex (*op. cit.*; Reynolds, 1996). The variety of mating systems among animals appears bewildering at first sight, not only between species but also within populations of the same species (Shuster & Wade, 2003; Alcock, 2013). One fruitful approach has been to view mating systems as outcomes of the behaviour of individuals competing to maximise their reproductive fitness; therefore, because of male competition, the dispersion of females in a population and the investment in parental care by each sex are strong influences on the nature of any mating system (Davies, 1991). Mating systems are often described mainly according to the number of individuals that each member of each sex mates with. Lott (1991) pointed out that the conceptual beginning point of mating systems is promiscuity (i.e. all pairings are random and multiple) and that the ultimate non-random system is monogamy.

To our current knowledge, there are no monogamous cephalopods, nor is there any parental care of the young (Table 6.4). Males of studied species are polygynous (mate with multiple females), and most females that have been studied are polyandrous (mating with multiple males). In other taxa, there are many tactics available within each sex's mating system, but cephalopods have not been studied in sufficient detail to assign such specific tactics. For examples of other tactics, males in some taxa practise female defence polygyny by monopolising females, or resource defence polygyny by defending territories that are prime egg-laying sites, or lek polygyny in which territories are reserved solely for displaying to females. Certainly variations of these and other tactics are practised by cephalopods, and these deserve future experimental study.

Operational sex ratio is the number of sexually available males to females; specifically, it is the average ratio of males to females who are ready to mate at a given time and place, thus forming the 'mating pool' (Emlen & Oring, 1977; Kokko, Klug & Jennions, 2012). This varies greatly in some cephalopod species and has a strong impact on the nature of the mating system. For example, in the giant Australian cuttlefish, *Sepia apama*, the OSR is 11:1 at the peak of the spawning aggregation, and averages 4:1 over the course of the 3-month event (Hall & Hanlon, 2002). Such high ratios engender strong

Table 6.4 Mating systems in cephalopods (based upon coastal species only)

	Cuttlefish	Squid	Octopuses
I. Male–male competition	Agonistic behaviour strong (Zebra patterns)	Agonistic behaviour strong (Zebra patterns/Lateral Displays)	Little if any
	Contest escalation (little-to-moderate physical contact)	Contest escalation (little-to-moderate physical contact)	None
	Dominance behaviour	Dominance behaviour	None
II. Courtship	Brief–moderate courtship	Brief–moderate courtship	Little or no courtship
	♂-initiated	♂-initiated	?
	Very brief pairing	Very brief pairing	None or little
	No 'nuptial gifts'	No 'nuptial gifts'	No 'nuptial gifts'
	♂ ornamentation	♂ ornamentation	None
	(Zebra bands)	(Zebra bands/Lateral flame marks)	(only 1/16 species)
	Moderate sexual dimorphism of body patterns	Strong or moderate sexual dimorphism of body patterns	None
	None	None	♂ sucker display/ligula display
III. Mating	♂ & ♀ promiscuous	♂ & ♀ promiscuous	♂ & ♀ promiscuous
	Extra-pair copulations	Extra-pair copulations	Not applicable
	Head-to-head position	Head-to-head position	♂ distant from ♀
	or ♂ holds mantle	or ♂ parallel	or ♂ mounts ♀
	0.5–80 minutes	1–20 seconds	40–60 minutes
	External fertilisation	External fertilisation	Internal fertilisation
	Temporary mate guarding	Temporary mate guarding	None
IV. Sperm competition	Yes	Yes	Probable
V. Parental care			
Eggs	No	No (except a few deep sea)	Yes (by ♀ only)
Young	No	No	No

competition among males and lead to multiple tactics including sexual mimicry by some small males. Only a few OSRs have been determined in field studies of cephalopods. Among loliginid squid, the OSRs are in the range of 1.6–3M:1F, and in the only reported case for octopus it is 1.8M:1F in *Abdopus aculeatus*. Future studies of sexual selection in cephalopods should consider that the impact of OSR on sexual selection is also interdependent on other factors such as the Bateman gradient (Kokko, Klug & Jennions, 2012).

Pair formation and the nature of mate acquisition are not well understood. Octopuses seem generally not to form pairs, and mate acquisition in most species seems to depend upon mutual consent, with the female generally allowing most, or in some species all, males to mate with her. This would appear to be a useful tactic since most octopus species are solitary and may seldom encounter potential mates. *Abdopus aculeatus* is the exception among octopods: it lives in somewhat dense aggregations, and male/female pairs mated over successive days (§6.4.2). In the few decapods that we know about, pairs are maintained only briefly (perhaps only hours or a day or two) and can hardly be said to have 'bonded' in any way. Female decapods generally seem to mate both consort and sneaker males, yet females are choosy and do reject mating attempts (see §6.2, 6.3). Perhaps the overall strategy for decapod females is to produce different offspring from multiple mating tactics; DNA fingerprinting of coastal squid egg capsules (each containing hundreds of embryos) and cuttlefish egg clutches indicates high levels of multiple paternity (§6.6.4).

A source of variation in cephalopod mating systems is that individuals in many loliginid squid and octopus species have two or three methods of copulation. Female choice may be involved in the decision of which mating position is used; we do not know how mating position is determined, or by which sex. What determines the technique to be used by a particular pair of animals at any given mating encounter is also unknown. In squid such as *Loligo*, it is worthwhile to point out that in the 'head-to-head' position the spermatophores appear to be passed only to the seminal receptacle near the mouth for storage by the female, whereas in the 'male-parallel' position the spermatophores are placed inside the mantle cavity near the opening of the oviduct. In the latter case, eggs are generally extruded and fertilised within minutes or hours to take advantage of the sperm seeping out of the extruded spermatangia, because many spermatophores are partially or wholly broken as they are deposited there. *Lolliguncula*, with its spermatophoric 'pad' on the inner mantle wall near the oviduct, and some oceanic squid, which 'plant' spermatophores at various positions on the mantle, are exceptions to this. Another variation in *Loligo plei*, *L. pealeii*, *L. bleekeri* and *L. vulgaris reynaudii* is that extra-pair copulations (with 'sneaker' males) are known to occur; thus, these females may have up to three sources of sperm to fertilise their eggs. Figure 6.16a shows a schema for multiple mating methods of *L. pealeii* on spawning grounds in Massachusetts, USA. Some octopuses mate in two positions (Fig. 6.21), but the significance of this is unknown because all spermatophores are deposited in the same place, the oviducal glands in the paired distal oviducts. All of these important behaviours require future study.

Moynihan and Rodaniche (1982) interpreted the mating system data of Corner and Moore (1980) on *Sepia*

latimanus (see §6.2.3) as possibly fulfilling some of the requirements of a lek (i.e. 'an area used consistently for communal courtship displays'; Wilson, 1975). Certainly, on those reefs in Guam, males seemed temporarily aggregated on 'hotspots', the term used to describe places where encounters with females are higher (Davies, 1991; Alcock, 2013). However, in leks, the 'females visit males solely for mating and males provide no parental care' (Davies, 1991). The latter is true for all cephalopods: there is no parental care by males, and the only known form in females is egg care, but not care of the young. However, the former requirement may not be met for *S. latimanus*, because females seem to be coming to the corals to lay eggs. Sauer *et al.* (1997) suggested that *Loligo reynaudii* mating aggregations may exhibit some features of leks, yet there are substantial differences from the traditionally described leks as in prairie fowl. Nevertheless, the possibility of a lek deserves consideration by cephalopod field workers.

Time investment allocated to reproductive behaviour in cephalopods has been estimated by Hanlon and Forsythe (unpublished data). In their direct, quantified field observations they found that mature and reproductively active squid, *Sepioteuthis sepioidea*, in the Cayman Islands, spent about 8% of each day in agonistic, courtship or mating activity (Fig. 9.9). This reproductive behaviour occurred throughout the daylight hours with a noticeable peak 2 hours before dark; this activity cycle was noted in spring and summer, but was lower in the autumn when smaller individuals were present. Moynihan and Rodaniche (1982) indicated a similar time investment for the same species off Panama. Thus, on tropical coral reefs, mature *S. sepioidea* engages in reproductive activity most months of the year at a level near 3% of its time each day. This aspect of cephalopod reproduction requires more research.

6.6.2 Intrasexual Selection: Male Fighting Tactics

Fighting among males is a common feature in many animals, and competitive aggression is a means by which males gain access to preferred females for mating; this is known as sexual competition.

A typical decision during an aggressive contest involves, firstly, whether to escalate or terminate the contest (i.e. is the resource worth fighting over?), and secondly, whether this particular opponent is worth fighting. Archer (1988) pointed out that during conspecific fights for mates, there is scope for adopting a set of tactics of 'living to fight another day' by retreating, or perhaps by adopting an alternative mating tactic such as becoming a 'sneaker male' to obtain extra-pair copulations. The decision to continue or withdraw from a contest is usually influenced by the fighting ability of a contestant, termed resource holding potential (RHP) (Maynard Smith 1974).

Game theory is an analytical tool used to understand behavioural relationships between animals as they attempt to obtain resources (e.g. mates). It became a central part of ethological thinking owing to the influential work of Maynard Smith (1974, 1982). Game theory examines the evolutionary consequences of biological interactions. It considers the fitness pay-off for an individual based not only on its own decision-making strategy, but also on those of other individuals involved in the interaction. In the context of animal conflict, game theory specifies that decision-making by each individual must take into account the dynamic aspects of the interaction.

Game theory models can be applied to animal contests to determine the assessment tactics used for decision-making. Currently, there are three major game theory models, which can be divided into self-assessment strategies and mutual-assessment strategies. Self-assessment is modelled through the energetic war of attrition model (E-WOA; Payne & Pagel 1996, 1997) and the cumulative assessment model (CAM: Payne 1998). Both models assume that contestants evaluate their own fighting ability, but fail to assess their opponent's fighting ability. Mutual assessment is modelled through the sequential assessment model (SAM), which assumes that contestants evaluate their own fighting ability relative to their opponent's (Enquist & Leimar 1983). Predictions for all three models are established on estimates of fighting ability and vary in their assumptions about how opponents gather information on fighting ability. Mutual assessment is assumed to be a more efficient strategy than self-assessment because animals can reduce costly and futile persistence by gathering information about relative fighting ability (Enquist & Leimar, 1983). The following two examples in cephalopods provide support for the concept of a mutual assessment strategy, specifically the SAM.

Incidentally, there is no evidence of intrasexual selection among female cephalopods.

Tropical arrow squid *Loligo plei*. In the first experimental study in cephalopods, DiMarco and Hanlon (1997) found five features of squid contests that were consistent with the predictions of game theory: (1) the influence of size asymmetry on contest outcome; (2) the effects of resource value (i.e. a female mate), (3) effects of role asymmetry (owner/intruder) on contest duration; (4) the sequential assessment of opponents; and (5) alternative fighting tactics. By testing squid in six different combinations, DiMarco and Hanlon found that: males never fought females, but only other males; and when a female was present, the nature of the fight between males changed depending on the combination of squid in each test.

Two general features of the fights were noteworthy: they were mainly visual, with only minor physical contact, and the visual signalling was complex. Over 21 dynamically controlled visual signals were observed and scored, and sometimes six of these visual signals were shown simultaneously to form a striking Lateral Display (Figs. 6.23a, 6.24b, c; Box 7.1).

There was step-wise escalation of contests (Fig. 6.23a): (1) short sequences of miscellaneous visual signalling followed by Lateral Display and physical Fin beating (i.e. clashing of the fins), and ending with Chase and Flee, or (2) long sequences of miscellaneous visual signalling followed by Chase and Flee. Escalation was defined as the progression from miscellaneous visual signalling to Lateral Display, then through physical contact (i.e. Fin beating) to Chase and Flee (i.e. one male vigorously chased the other male around the tank). Escalation was measured quantitatively in two ways. First, as shown in Fig. 6.23b, there was a sharp escalation in the frequency of signalling during the contest duration. Second, the average frequency of Lateral Displays was plotted (Fig. 6.23c) and the same degree of escalation was evident. Note that the contest was considered finished after the last Fin beating and the beginning of Chase (Fig. 6.23a).

Larger males won 16 of 24 contests, which supports the SAM assumption that relative size can predict contest outcome. Yet it is noteworthy that squid could signal dishonestly, making themselves look bigger by

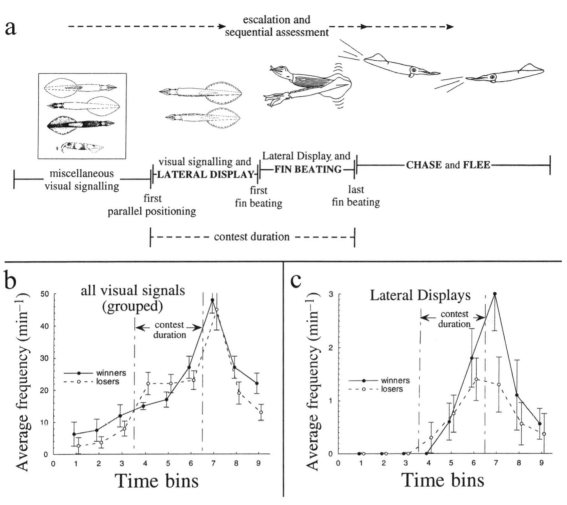

Figure 6.23 Escalation and sequential assessment of fighting ability by male *Loligo plei*. (**a**) Summary of the typical sequence of escalating fights. (**b**) Increasing frequency of all visual signals. (**c**) Increasing Lateral Displaying. Because the contest duration varied, the abscissa was plotted in nine time bins, the first three representing 60 seconds before the contest began, the next three the variable contest duration, and the final three the 80 seconds after the contest (from DiMarco & Hanlon, 1997, by permission of Wiley).

using their neurally controlled chromatophores (§7.1.1). Most contests escalated to Fin beating and Chase and Flee, and these behaviours may have provided a more realistic physical assessment of size and fighting ability.

Contest duration and fighting tactics were influenced by the presence of a female (the 'resource value') and whether temporary pairing of the female had been established. As shown in Fig. 6.24d, contests were very long when no female was present (♂ ♂; mean 13.7 minutes), but became progressively shorter when females were present in different combinations. In

particular, when a male and female were together for 10 minutes and an intruder male was added (♂ ♀ + ♂), the mean contest duration was only 30 seconds. As a control, when females were added to male/female pairs, there were no agonistic interactions. Presumably, when an owner/intruder asymmetry is perceived and one male has gained temporary pairing of the female (a resource value), then both males escalate quickly through all stages of fighting. These stages are all costly in terms of energy expended and risk of injury. Fighting tactics were different: only in the short contests were Lateral

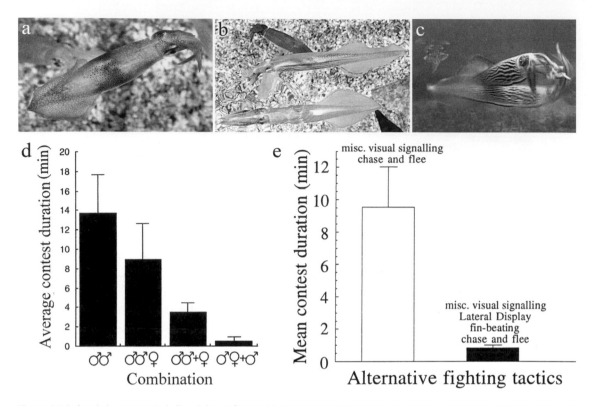

Figure 6.24 Agonistic contests in *Loligo plei* are influenced by 'resource value' such as the presence and status of female mates. (**a**) Female (93 mm ML) showing Band with Downward curling arms; these two components of the body pattern act as a signal that stimulates courtship and agonistic competition by males. (**b**) A consort male (middle) defending two potential mates against a rival male (bottom); note the unilateral expression of flame markings in the Lateral Display. (**c**) Field shot in Venezuela of two males in an escalated fight in which Lateral Displays have escalated to strong Fin beating; the female is in the background. (**d**) Contest duration decreased significantly when a female 'resource' was present, especially when temporary pairing had occurred (♂♀ + ♂). (**e**) Different fighting tactics by males: shorter contests included escalation to Fin beating. (**d**, **e** from DiMarco & Hanlon, 1997, by permission of Wiley; photographs by R. Hanlon.)

Display and Fin beating used (Fig. 6.24e) (see also van Staaden, Searcy & Hanlon, 2011).

Thus, there were two distinct tactics used as part of the strategy for fighting in male *L. plei*, and the general sequence of activity followed the predicted escalation of the sequential assessment model of game theory.

Giant Australian cuttlefish *Sepia apama*. In the field, large male consorts pair temporarily with females and guard them with complex agonistic behaviours towards challenging unpaired males (Hall & Hanlon, 2002). A detailed field study showed that males (*n* = 116) used progressive visual signalling to convey increasing levels of threat (Fig. 6.25a) (Schnell *et al.*, 2015). The consort male usually matched the Frontal Display (82%) of the unpaired challenging male, but some retreated (18%). The consort male would then progress to the Lateral Display (69%) or the Shovel Display (31%), which in turn led to physical actions (pushing). Corroborative laboratory experiments (*op. cit.*) demonstrated a similar pattern, showing that *S. apama* used a hierarchical sequence of agonistic signals. That is, the likelihood of an escalation to physical combat was greater immediately following a Shovel or Lateral Display than a Frontal Display (Fig. 6.25b, left). Escalations were significantly greater in response to the threatening stimuli (Frontal and Lateral Displays) than to the Control and the non-threatening HNS stimulus (Hovering and Not Signalling; Fig. 6.25b, right). These

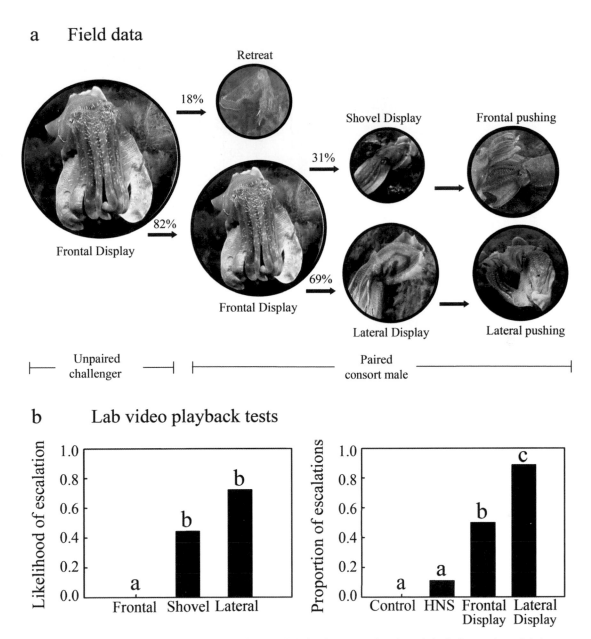

Figure 6.25 Agonistic contests in *Sepia apama*. (**a**) Schema depicting the progression of agonistic displays performed during contests between 116 males. Diameter of circles is proportional to the frequency of occurrence of each display. Each step in the series is contingent upon the behaviours of the opponents. See text for details. (**b**) Escalation of agonistic bouts during video playback experiments in the laboratory. Left: proportion of test males that escalated following a specific aggressive display; Shovel and Lateral Displays evoked escalation strongly. Right: proportion of males that escalated in response to different video playback stimuli: HNS, hovering and non-signalling males; $n = 18$ test males. Bars that do not share a common upper-case letter were significantly different ($p < 0.05$). From Schnell *et al.*, 2015: Animal Behaviour **107**, 31–40, Giant Australian cuttlefish use mutual assessment to resolve male–male contests, Schnell, A. K., Smith, C. L., Hanlon, R. T. & Harcourt, R., © (2015), with permission of Elsevier.

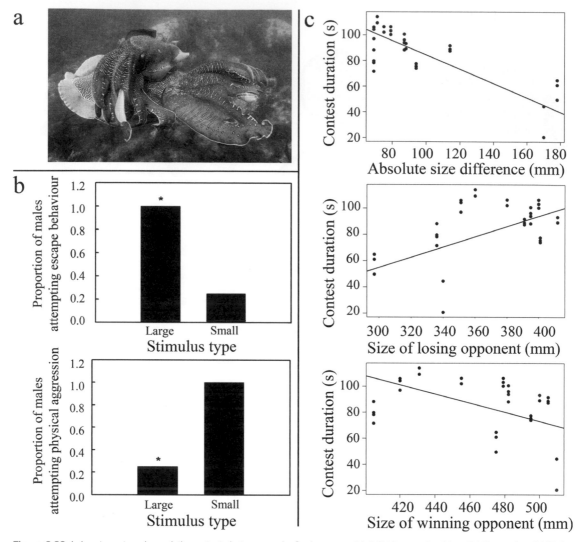

Figure 6.26 Laboratory-staged agonistic contests between male *Sepia apama*. (**a**) A field example of two fighting males. (**b**) Males (*n* = 8) exhibited escape behaviour from larger males but not smaller ones (top), and attempted physical aggression towards smaller males but not larger ones (bottom). The asterisk represents a significant difference in bars (*p* < 0.05). (**c**) Contests lasted longer when opponents (*n* = 20) were more similar in size (top) or loser males were larger (middle), but were shorter when the winner male was significantly larger (bottom). Modified from Schnell *et al.*, 2015: *Animal Behaviour* **107**, 31–40, Giant Australian cuttlefish use mutual assessment to resolve male–male contests, Schnell, A. K., Smith, C. L., Hanlon, R. T. & Harcourt, R., © (2015), with permission of Elsevier.

results suggest that the Frontal Display by a challenging (unpaired) male acts as a low-level threat signal to assess the aggressive responsiveness of the consort male, whereas the Shovel and Lateral displays convey greater aggression and motivation to escalate to physical aggression.

In a separate laboratory study, Schnell *et al.* (2015) determined the fighting strategy used to resolve male contests (Fig. 6.26a) by investigating (1) decision-making and (2) contest duration. In 22 staged laboratory trials between live males, 75 contests lasted on average for 1 min 46.0 s ± 30.7 s but ranged greatly (13.3 s to 18 min

22.0 s). Cuttlefish decision-making during agonistic interactions was based on size-related information: males were more aggressive towards smaller rivals and attempted to flee from large ones (Fig. 6.26b). This shows that males perceive size-related information during contests and adjust their behaviour as a function of the realised size of their opponent relative to their own. This study also found that contest duration was correlated negatively with relative fighting ability; contests were longer in duration when opponents were more similar in size, but shorter in duration when opponents varied in size (Fig. 6.26c). This pattern demonstrates that persistence to continue a contest is based upon fighting ability asymmetry rather than loser and/or winner fighting ability. This suggests that competing males are able to gather information concerning fighting ability asymmetries, all of which are characteristics of mutual assessment. Collectively, these results demonstrate that male cuttlefish assess the fighting ability of their opponent relative to their own, providing evidence that supports mutual assessment.

Collectively, the results of these experiments provide evidence that supports mutual assessment via the sequential assessment model and refutes the self-assessment strategies of the energetic war of attrition and cumulative assessment models.

The presumed outcome of these contests of intrasexual selection is that it leads to access to preferred females for mating. This appears to be the case in *Sepia apama* based upon field studies (Hall & Hanlon, 2002; Naud *et al.*, 2004; Schnell, 2014) and paternity testing (next section). Adamo *et al.* (2000) found in the cuttlefish *Sepia officinalis* that consort males who won fights obtained more copulations in a very large indoor tank system. Although DiMarco and Hanlon (1997) did not carry out a longitudinal laboratory study or a field study that would have included matings, they did measure how frequently each male was 'with the female' throughout the staged contests, and the winner was with her substantially more at the conclusion of the contest. In filmed contests of the squid *Loligo reynaudii* in South Africa (Hanlon, Smale & Sauer, 2002), paired mates won 16 and intruders 9. Obviously then, intruders can win often in some species; when this did occur, the winner stayed with the female most of the time, and in a few observed cases the winner male mated the female and then accompanied her to the communal egg mass while she deposited eggs that she fertilised immediately after their mating. Paternity testing indicated that such consorts also obtained most fertilisations (Shaw & Sauer, 2004). Similar behaviours and fertilisations by consorts that have won fights have been reported in *Loligo pealeii* (Buresch *et al.*, 2009; Shashar & Hanlon, 2013) and *Loligo bleekeri* (Iwata, Munehara & Sakurai, 2005).

6.6.3 Intersexual Selection: Female Choice, Male Ornamentation and Sexual Mimicry

The concept underlying intersexual selection is that females mate selectively with males, and that female choice has led to the evolution of male traits (some unusual, some bizarre) that enhance the opportunities of mating for those males. Often no other function can be attributed to these male traits, and some of them are thought to impair survival (cf. Andersson, 1982; Davies, 1991). In other animal groups, female choice sometimes involves material benefits such as 'nuptial gifts' from males, resources such as shelter or food monopolised in male territories, or differing parental investment by males, but these are unknown in cephalopods.

Female choice may involve simply the attractiveness of extravagant or even arbitrary traits in males. Such traits or male ornamentation in the traditional sense are unknown in the cephalopods. That is, there are no permanent morphological features on male cephalopods that seem to correspond to the extravagant (or epigamic) traits common in vertebrates (e.g. the peacock's feathers). In the loliginid squid *Alloteuthis subulata* and *A. africana*, the males have longer tails than females (Rodhouse, Swinfen & Murray, 1988; Jereb & Roper, 2010) and it is possible that this character is influential in female choice; tail size in birds has been shown to be important in this regard (Harvey & Arnold, 1982).

However, because body patterns can change through neural control (Chapter 3), males can show some sexually dimorphic body patterns to females, and it is possible that some of these could function as ornamentation. For example, the Zebra displays of *Sepia officinalis*, *Sepia latimanus*, *Sepioteuthis sepioidea* and *Loligo plei* (see previous sections) towards other males are often seen by nearby females, who may interpret them as a sign of health and vigour. However, the experiments on *Sepia officinalis* by Boal (1997) and field observations of *Sepia apama* (Hall & Hanlon, 2002) cast doubt on this function since females appear to be ignoring these traits or choosing mates on other factors. The shimmering chromatophores and iridophores (called 'Blushing' by Moynihan & Rodaniche, 1982) shown by male squid *Sepioteuthis sepioidea* during intensive courtship and just prior to mating may provide an attractive appearance and a sign of vigour to females.

In octopuses, ornamentation by display of enlarged suckers on the arms of males (Packard, 1961) and presentation of the ligula (the terminal part of the hectocotylus) (Voight, 1991a) have been suggested as instruments of mate choice (Fig. 6.3). In both cases, they could be indicators of 'maleness' and status of maturity, since both can be correlated with sexual maturity among males of that species. Enlarged suckers are also reported in males of the sepioids *Euprymna scolopes* and *Sepietta oweniana* (Bergstrom & Summers, 1983; Singley, 1983), and in various oegopsid squid, but behavioural observations of their use are lacking. In some mid- and deep-water cephalopods, the photophores are distributed differently in each sex (Fig. 6.3; §6.1), and it is possible that some photophore displays may be used in mate choice.

Female choice in the form of rejecting male mating attempts has been demonstrated during several field studies. For example, in the mass spawning aggregation of the giant Australian cuttlefish, *Sepia apama*, females rejected 70% of 122 mating attempts (Hall & Hanlon, 2002; §6.2.1). In a follow-on study, Naud *et al.* (2004) found that 43% of 49 mating attempts were rejected by females. Schnell (2014; Schnell *et al.* 2015) compared field observations with lab experiments. In the field ($n = 116$ matings), there was only a 22% probability of mating when the female signalled non-receptivity with the White lateral stripe display, whereas when she did not display the lateral stripe there was 80% probability that a mating would ensue (Fig. 6.27a, b). In lab experiments ($n = 14$), female receptivity varied substantially. Females rarely displayed the non-receptive White lateral stripe to unfamiliar unmated males, and this resulted in a high proportion of matings with unfamiliar males; conversely, females often displayed to familiar males that had been flushed of spermatophores (Fig. 6.27c), resulting in significantly fewer matings, all of which suggests some form of cryptic female choice. In squid, Shashar and Hanlon (2013) found that *Loligo pealeii* females rejected 55% of 73 mating attempts in the field (§6.3.3), although the vast majority of those were from sneaker males, not consorts. *Loligo reynaudii* females also mated with both consorts ($n = 9$) and sneakers ($n = 6$) but rejected one sneaker male (Hanlon, Smale & Sauer, 2002). These forms of overt female choice require more detailed study, but it seems as though various squid and cuttlefish exert such intersexual selection.

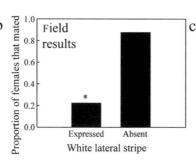
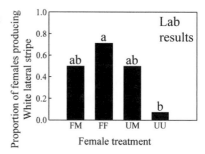

Figure 6.27 Sexual signalling of non-receptivity by female giant Australian cuttlefish, *Sepia apama*. (**a**) Female (left) unilaterally signalling to a male suitor by displaying the White lateral stripe along the base of her fin (frame grab from field video by R. Hanlon). (**b**) Proportion of females ($n = 116$) in the field that mated following the expression or absence of the White lateral stripe; asterisk represents a significant difference ($p < 0.05$). (**c**) Proportion of females ($n = 14$) in the laboratory displaying the White lateral stripe during different female treatment groups: FM = Familiar Mated – mated previously with the same male; FF = Familiar Flushed – previously mated with the same male but any spermatophores accumulated from the previous mating were removed; UM = Unfamiliar Mated – previously mated with a different male; and UU = Unfamiliar Unmated – female had not come in contact with the test male and had not mated for at least 48 hours. Female treatments that do not share a common lower case letter were significantly different. Modified from Schnell, 2014, with kind permission.

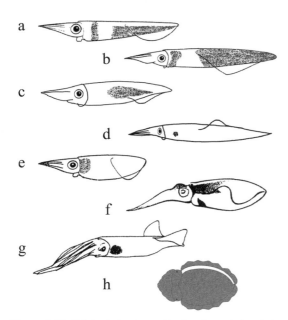

Figure 6.28 Unilateral signals used by female cephalopods to repel courting males. (**a**) *Loligo plei* (Hanlon, 1982: *Malacologia*, **23**, 89–119, with permission); (**b**) *Loligo reynaudii* (Hanlon, Smale & Sauer, 1994: *Biological Bulletin* **187**, 363–372, © (1994), by permission of the University of Chicago Press); (**c**) *Loligo pealeii* (Hanlon et al., 1999: *Biological Bulletin* **197**, 49–62, © (1999), by permission of the University of Chicago Press); (**d**) *Loligo bleekeri* (based on data from Iwata, Ito & Sakurai, 2008); (**e**) *Lolliguncula brevis* (Dubas et al., 1986, by permission of the Company of Biologists); (**f**) *Sepioteuthis sepioidea* (Moynihan, 1985a); (**g**) *Todarodes pacificus* (Y. Sakurai, unpublished data); (**h**) *Sepia apama* (based on data from Hall & Hanlon, 2002).

Some female squid (Family Loliginidae) have visual signals seemingly analogous to the White lateral stripe of *Sepia apama* – various sorts of spot, blush or bar on the lateral mantle – that they use as a mild repellent towards males, and these may function in some way to signal their choice (Fig. 6.28). We still know rather little about overt female choice in cephalopods, and these observations are difficult to understand since in many species females are highly polyandrous and sometimes seem promiscuous. Clearly, these ideas require investigation.

Sexual mimicry has clearly been demonstrated in *Sepia apama* (see §6.2.1 and Fig. 6.6) and is an example of how rapid neural polyphenism can implement facultative mimicry as part of a conditional mating

strategy (Hanlon et al., 2005). A key unanswered question is: how and why do females accept these sneakers as mates? Females generally reject 70% of mating attempts in this mating system but only rejected ca. 35% of the female mimics (although the numbers to support this last metric are small). There is much yet to learn about the perceptual criteria that females use for such sexual selection processes.

Finally, we should not exclude the rarer possibility of male choice, which may be one explanation for the circumoral photophores in females of some deep-sea octopods (Robison & Young, 1981).

6.6.4 Postcopulatory Sexual Selection: Sperm Competition and Cryptic Female Choice

Female promiscuity, or polyandry, has key biological implications: it signifies that sexual selection persists after copulation up to the point of fertilisation, and in some cases beyond (Birkhead & Pizzari, 2002). Postcopulatory sexual selection involves two components: male–male competition in the form of sperm competition, and cryptic female choice.

Sperm competition is the competition within a single female between the sperm from two or more males for the fertilisation of the ova (Parker, 1970). It occurs whenever a female mates with more than one male in one breeding cycle (cf. Smith, 1984; Birkhead & Parker, 1997) and is now recognised to be almost ubiquitous across the animal kingdom (Birkhead & Moller, 1998). Sperm competition simultaneously favours both the ability to usurp any sperm previously inseminated by other males, and the ability to prevent any female that they inseminate from being inseminated by other males. Because sperm competition impinges on so many aspects of biology (e.g. behaviour, evolution and reproductive physiology), its meaning has expanded to encompass all the behaviours and morphologies associated with copulation, including multiple mating, paternity guards, male and female genitalia and many others. It is a behavioural phenomenon that involves mating tactics used to achieve the reproductive evolutionary stable strategy of individuals, both males and females.

Cryptic female choice is the ability of a female to bias the fertilisation success of the males that copulate with and inseminate them (Eberhard, 1996), and criteria for demonstrating it have been continually developed (Eberhard, 2000, 2009). In decapod cephalopods, females can choose to use sperm from recently deposited spermatangia or from stored sperm in seminal receptacles, and this is 'cryptic' because the male is not involved or is not present when this is accomplished. DNA fingerprinting has revolutionised the study of these two processes of sexual selection by enabling genetic paternity assessment rather than merely mating success.

The following features support the evolution of sperm competition in cephalopods: the large testis is capable of producing large quantities of sperm, the sperm are packaged in spermatophores, sperm have different morphologies in consorts and sneakers, sperm are stored and sometimes selectively chosen by females, spermatophores are often placed by males in two locations on the female, the morphologies of the oviduct and spermatheca are appropriate, mating systems are 'polygamous', there are multiple styles of mating, mate guarding by males is common in teuthoids, and there are delays between mating and egg laying.

Strong evidence for sperm competition in cephalopods is available through field studies of the giant Australian cuttlefish, *Sepia apama* (Naud *et al.*, 2004, 2005; Hanlon *et al.*, 2005; Naud & Havenhand, 2006). Females have access to sperm from (1) packages (spermatangia) deposited in the female's buccal area from recent matings, and (2) internal seminal receptacles located just below the beak (Fig. 6.29a, c), each containing sperm from previous matings. In a study of 21 mating pairs, sperm from multiple males was found in both locations: spermatangia had a mean of 2.8 (range 2–5) male genotypes; seminal receptacles had a mean of 1.9 (range 1–3). In a separate study, 18 females were videotaped as they mated with up to 6 males per hour. Thirty-nine eggs were sampled during 17 observational sequences. One-third of females mated with multiple males. One female had five mates before laying an egg; of the six eggs subsequently collected from her, three sires were identified. In nine females for which multiple eggs were collected, six of them had multiple paternity. Most of the mating duration was spent by the male flushing water into the female's buccal area, presumably to flush out sperm from previous matings by other males; however, longer flushing did not result in more fertilisations (Fig. 6.29b). Matings that occurred

20–40 min before egg laying were significantly more successful in obtaining fertilisations than those with shorter or longer periods (Fig. 6.29b). Fertilisation success did not differ significantly between paired consorts (36%) and unpaired males (31%), or between large (33%) or small males (29%). Although last-male precedence was expected from the extensive flushing times as well as the intensive mate guarding by consorts, the paternity results do not support this mechanism: only four of eight resulted in last-male precedence. Females used sperm from paired, unpaired and sneaker males as well as from previous matings to fertilise eggs. Thus, there was a high level of multiple matings and multiple paternity, and males of any size or status obtained successful fertilisations. Overt female choice was exerted by females who can deter harassing advances of males by displaying the unilateral White mantle stripe or by direct rejection via moving away or attacking the male. In a third study, small unpaired males used the sneaker tactic of female mimicry to acquire copulations that led to immediate fertilisations even though the consort male had been guarding and mating the female (Fig. 6.6). Clearly there was cryptic female choice occurring during all of these study examples.

In the Japanese cuttlefish *Sepia lycidas*, males increased their duration of sperm removal and the number of ejaculations per mating when another male had recently mated the female (Wada *et al.*, 2010). Different tactics were used by males of different sizes: larger males prolonged sperm removal, whereas smaller males with relatively larger testes ejaculated more times during a mating. These behaviours begin to address the issues of spermatophore depletion and tactical decisions to increase fertilisation success.

Sperm competition and cryptic female choice have been demonstrated behaviourally in the Japanese pygmy squid *Idiosepius paradoxus*. Five males mated a female serially in one case, and in another three males were seen mating a single female simultaneously; thus there were ample sperm from many potential sires (Fig. 6.11; Sato *et al.*, 2010). Sato, Kasugai and Munehara (2014) found that females selectively removed the spermatangia attached to their body by large males (or males that mated for longer periods) by elongating their buccal mass and either blowing them away or eating them (Sato, Kasugai & Munehara, 2013). Conversely,

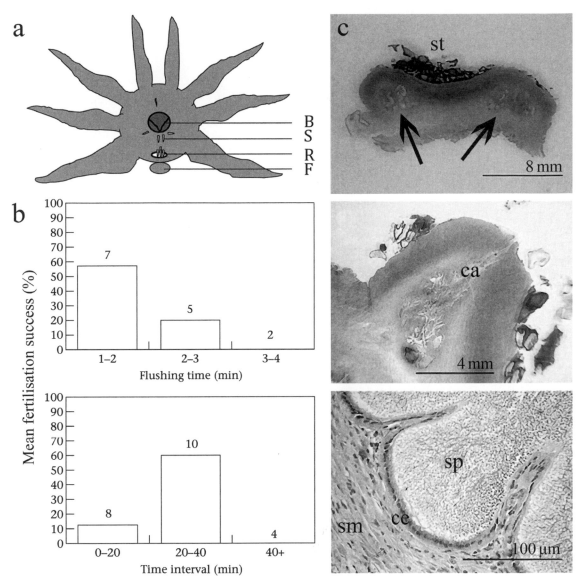

Figure 6.29 Aspects of sperm competition in the giant Australian cuttlefish *Sepia apama*. (**a**) Location of sperm sources in females. Spermatangia (S) are attached externally around the beak (B), and sperm are stored internally in the receptacles (R). F is the funnel. (**b**) Fertilisation success of males in relation to time spent flushing water towards the female's buccal area during the first phase of mating, and (bottom) the time between sperm transfer and egg laying. Sample sizes are given above bars. (**c**) Cross section of the two sperm receptacles (arrows); note the externally attached spermatangia (st). Middle: single receptacle illustrating the canal (ca) opening to the buccal area. Bottom: bulb of the sperm receptacle showing stored sperm (sp) and simple columnar epithelium cells (ce) and smooth muscle cells (sm). (**a, c** from Naud *et al.*, 2005, by permission of Royal Society Publishing; **b** reprinted from Naud *et al.*, 2004: *Animal Behaviour* **67**, 1043–1050, Behavioral and genetic assessment of mating success in a natural spawning aggregation of the giant cuttlefish (*Sepia apama*) in southern Australia, Naud, M. J. *et al.*, © (2004), with permission from Elsevier.)

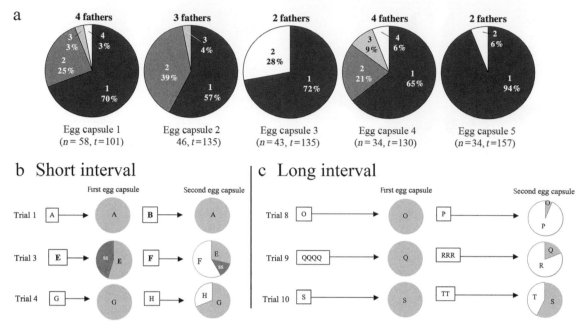

Figure 6.30 Sperm competition manifest through demonstration of multiple paternity in egg capsules of the squid *Loligo pealeii* both in the field and laboratory. (**a**) Percentage of offspring fathered by different males from field-collected egg capsules. *n* is number of hatchlings sampled; *t* is total number of hatchlings from that capsule (modified from Buresch *et al.*, 2009, © (2009), with permission from Inter-Research). (**b, c**) Controlled mating trials in the laboratory. The first male to mate garnered most fertilisations, but when a second male then mated the female, the resulting paternity was very different, depending on the interval between mating and egg laying. See text (modified from Buresch *et al.*, 2001, © (2001), with permission from Inter-Research). Each letter represents one male squid; ss = surreptitious sneaker. QQQQ means that male squid Q mated four times.

spermatangia from small males or from males that mated for a shorter time were left alone, and these sperm were shown to end up in the seminal receptacle within about 2 hours. The authors conclude that this behaviour by females is an example of cryptic female choice.

Among squid, *Loligo pealeii* is most thoroughly studied and illustrates many features of sperm competition (Buresch *et al.*, 2001, 2009; Shashar & Hanlon, 2013). In loliginid squid such as *Loligo*, females lay egg capsules (or 'fingers') each containing 100–300 small eggs. Figure 6.30a shows multiple paternity in field-collected egg capsules: there were two to four sires in every capsule, and in each there was one male who sired the majority of hatchlings (57–94%). These results demonstrate that sperm competition is taking place (1) year to year, (2) among different egg capsules within the communal egg mops, and (3) within individual egg

capsules. In the laboratory, ten mating trials were conducted in which a single female mated with two males over the course of a few hours while laying successive egg capsules. Relative paternity of the first egg capsule was typically in favour of the first male to mate (Fig. 6.30b, c). When the female mated with an additional male before the second egg capsule was laid, the first male to mate still achieved high relative paternity when the interval between first mating and second egg laying was brief (< 40 min). However, very different relative paternities occurred when the interval was longer than 140 min (Fig. 6.30c). These results argue against routine last-male sperm precedence in this species, and suggest other influences on paternity, such as the interval between insemination and egg laying.

Last-male precedence as a mechanism of sperm competition could not be excluded as an explanation for

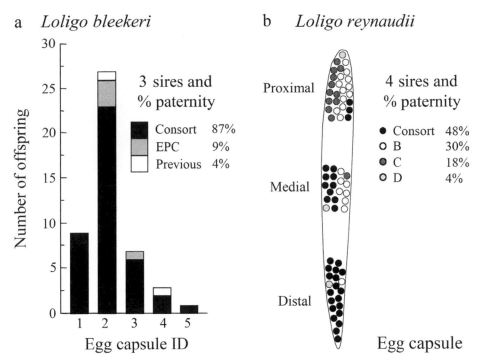

Figure 6.31 Multiple paternity in egg capsules of other loliginid squid. (**a**) In controlled laboratory experiments with *Loligo bleekeri* (Japan), the last male to mate (e.g. the consort in this case) received most of the fertilisations even for the second spawning. EPC = extra-pair copulation (from Iwata, Munehara & Sakurai, 2005, © (2005), with permission from Inter-Research). (**b**) In *Loligo reynaudii* (S. Africa), field-collected egg capsules had multiple sires. In this example with four sires, the consort had 48% of fertilisations but distribution within the capsule was different: he had 90% of distal and only 11% of proximal (from Shaw & Sauer, 2004, © (2004), with permission from Inter-Research).

laboratory mating trials in *Loligo bleekeri* because the male to mate last with the female before she spawned sired 87–100% of the eggs (Fig. 6.31a). However, those males had also mated with the female more frequently. One possible explanation for this difference from *Loligo pealeii* and *Loligo reynaudii* is that *L. bleekeri* mates for much longer than those species (300 versus 25 seconds) and actually transfers spermatophores into the oviduct of the female (Iwata et al., 2011), whereas the other species place spermatophores 'externally' near the outside of the oviduct. These differences could explain the advantage to the last male to mate. Shaw and Sauer (2004) also found multiple paternity in field samples of *Loligo reynaudii* in South Africa, with up to five sires in a single egg capsule. With precise sampling of eggs within each capsule, they found a non-random

distribution of eggs sired by different males (Fig. 6.31b). That is, consorts sired 90% of eggs at the distal end of the egg but only 11% at the proximal end, which may correspond to the different placement of spermatophores by consorts versus sneakers. The interplay of sperm competition with both overt and cryptic female choice is behaviourally complex in squid.

In a revealing study, Iwata et al. (2011) discovered sperm polymorphism between consort and sneaker males of *Loligo bleekeri*. Large consort males have larger spermatophores with shorter sperm that they transfer into the female's oviduct, whereas smaller sneaker males transfer smaller spermatophores with 50% longer sperm to the seminal receptacle; consorts have fivefold greater number of spermatozoa in their spermatophores than sneakers (Fig. 6.32a–d). However, in a fascinating

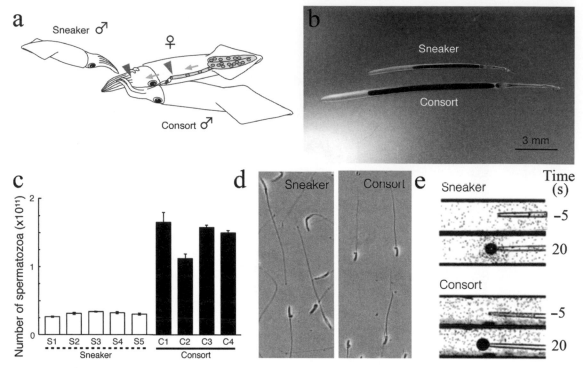

Figure 6.32 Dichotomies in sperm, spermatophores and mating positions between consort and sneaker males in the squid *Loligo bleekeri*. (**a**) Two mating positions, each with different sperm placement. The externally located seminal receptacle and the internal oviduct opening are the targets for sperm transfer by smaller sneaker and large consort males, respectively. (**b**) Consort spermatophores are larger than those of sneakers. (**c**) Consorts have more sperm per spermatophore than sneakers. (**d**) Sneaker sperm have a longer head and flagellum. (**e**) Sneaker sperm swarm to ova more strongly than those of consort males (**a–d** from Iwata, *et al.*, 2011: © Iwata *et al.*; licensee BioMed Central Ltd., 2011; **e** modified from Hirohashi *et al.*, 2013: *Current Biology* **23**, 775–781, Sperm from sneaker male squids exhibit chemotactic swarming to CO_2, Hirohashi, N. *et al.*, © (2013), with permission from Elsevier).

discovery, Hirohashi *et al.* (2013) found that sneaker sperm release a self-attracting molecule that causes their sperm to swarm by way of a signalling pathway that elicits Ca^{2+}-dependent turning behaviour (Fig. 6.32e), thus presumably enhancing fertilisation success since more sneaker sperm than consort sperm would be aggregated around each ovum. These researchers submit that 'sperm competition is also a strong selective agent in the evolution of sperm and ejaculate characteristics in *L. bleekeri*, perhaps in optimising these characteristics for each insemination/fertilisation site'. It will be intriguing to study why only sneaker sperm acquired the swarming trait in light of postcopulatory sexual selection and natural selection.

Cryptic female choice by the Caribbean reef squid *Sepioteuthis sepioidea* was observed during video

recording of 42 matings (Hanlon & Forsythe, unpublished). Males typically place spermatophores anywhere on the head or arms, then the female determines where, how and perhaps even if they are deposited in her seminal receptacle (Fig. 6.13j). She may discard the spermatophores, but none are placed in the mantle near the oviduct as in other loliginid squid.

Octopods are also likely to show sperm competition because females are polyandrous, sperm are stored in the oviducal gland, fertilisation occurs as eggs pass through the oviducal gland, and mating and egg laying are not linked temporally. Mating lasts a long time (Table 6.3) and each spermatophore is deposited carefully by the male's hectocotylus as it is inserted into the oviduct and near the spermatheca within the oviducal gland (Fig. 6.1). The spermatheca is an elongated 'cul de sac'

(Froesch & Marthy, 1975), which in the absence of sperm mixing would place the sperm of the last male in a position to fertilise the eggs. The tip of the male's hectocotylus – the ligula – is spoon-shaped and may be designed to remove sperm from the spermatheca in the female's oviducal gland. Cigliano (1995) conducted controlled laboratory matings with pygmy octopus and found that mating duration increased greatly for males that were second to mate (from 103 s for first male to 263 s for second male to mate). Since the number of spermatophores transferred did not increase, this extra time was attributed to the possibility of sperm removal or displacement but no evidence was obtained. Voight and Feldheim (2009) used microsatellite markers to demonstrate at least two sires in a clutch of the deep sea *Graneledone boreopacifica* at 1600 m depth. Quinteiro *et al.* (2011) sampled an egg clutch from each of four *Octopus vulgaris* and found three to four sires, thus providing genetic evidence for sperm competition.

In some octopuses, there appears to be strong male–male competition for females and ample opportunity for sperm competition. There are several reports of a single female octopus engaging in reproductive behaviour with several males simultaneously. Wood (1963), observing *Octopus vulgaris* in captivity, stated that 'two males may mate simultaneously with the same female'. Voight (1991b) noted three field observations of *Octopus digueti* in which two males were beside a female; she could verify that at least one male was actively mating but could not tell if both males actually mated with the female; recall, however, that female octopuses have two oviducts. In the blue-ringed octopus, *Hapalochlaena maculosa*, Tranter and Augustine (1973) once noted a male that intruded upon a mating couple and supplanted the resident male. The case of sneaker mating in *O. cyanea* observed by Tsuchiya and Uzu (1997) involved male–male aggression. In an extraordinary underwater observation of *Octopus bimaculatus*, R.F. Ambrose (unpublished data) saw six males attempting to mate simultaneously with one female (Fig. 6.21g). The idea of six hectocotyli searching for two oviducts brings a new dimension to one aspect of sperm competition (§6.6.4). These seven octopuses were still together 24 hours later. In other observations, Ambrose noted sequential mating with different males, and that males searched actively for females. Norman (2000) reported that female *Octopus kaurna* attract groups of males that

swarm over them, suggesting a chemical attractant, but the phenomenon has yet to be studied. The mating behaviours of *Abdopus aculeatus* (§6.4.2) suggest sperm competition as well. Overall, these preliminary studies in octopods suggest some interesting and different mechanisms of sperm competition and cryptic female choice, and will repay future efforts for more thorough research.

6.6.5 Semelparity, Iteroparity and Reproductive Success

Reproductive success measures fitness in terms of the number of progeny that are produced in succeeding generations (Clutton-Brock, 1988). This is a complex set of measures bound up in life span, mating system, competition, population fluctuations, spatial patterns, etc. (cf. Bradbury & Vehrencamp, 2011). DNA fingerprinting combined with laboratory experiments has revolutionised the assessment of reproductive success in individuals of many animal groups, including squid and cuttlefish (§6.2, 6.3, 6.6.4). Yet many questions of broader scope remain for populations of any species. Here we briefly consider life span and relative frequency of reproductive behaviour and spawning in a few representative cephalopods.

Cephalopods appear to be mostly short–lived (usually 1 to 2 years) and previously were thought to breed only once before dying (Boyle, 1983b, 1987). This sort of semelparous life style has been referred to as 'Big Bang' reproduction (Gadgil & Bossert, 1970). As experimental studies and field observations began to accumulate in the 1980s, disagreement began to arise about how strictly 'semelparous' cephalopods are (cf. Mangold, Young & Nixon, 1993). Since then, there have been numerous studies demonstrating that there is substantial variation in reproductive strategies among cephalopods (most summarised in this chapter; see also Boyle & Rodhouse, 2005, and Rosa, O'Dor & Pierce, 2013). Thus, it is simplistic to characterise cephalopods as semelparous. A newer framework was proposed by Rocha, Guerra and Gonzalez (2001) based solely upon spawning patterns of cephalopods. Specifically, they focused on three factors: the type of ovulation, spawning pattern, and growth between egg batches or spawning periods. They arrived at five overall strategies that effectively replace the terms semelparous and iteroparous. It is noteworthy to acknowledge (as the authors did) that their classification does not fully address reproductive strategies, which also include intra- and intersexual selection, mating systems and sperm competition. Nevertheless, their classification has some merit, as it addresses aspects of physiology and behaviour of different species. The five spawning strategies are: simultaneous terminal spawning (formerly semelparity), polycyclic spawning, multiple spawning, intermittent terminal spawning, and continual spawning (the last four are variations of former iteroparity). The future task for the ethologist is to consider these factors in the broader scope of behavioural factors as presented in this chapter.

A suitable example of simultaneous terminal spawning (i.e. semelparity) is *Octopus vulgaris*. However, a previous 'classic' example of semelparity among squid was the California market squid *Loligo opalescens*, yet by deploying ROVs to spawning grounds and conducting extensive video monitoring of reproductive behaviour day and night, it was discovered that spawning dynamics were flexible and that females were not spawning once and dying (Forsythe, Kangas & Hanlon, 2004; Zeidberg *et al.*, 2012; §6.3.4). Direct *in situ* observations and video recording by acceptable ethological methods (Martin & Bateson, 2007) are invaluable for determining accurate reproductive strategies.

The oft-cited example of iteroparity in cephalopods is that of the Panamanian *Octopus chierchiae* (Rodaniche, 1984), and Hofmeister *et al.* (2011) corroborated this multiple spawning strategy through laboratory husbandry. Other examples of the various cephalopod strategies are provided by Rocha, Guerra and Gonzalez (2001), and the interested reader can use those as a starting point for future field work and laboratory experimentation to better understand cephalopod reproduction and how it relates to life span.

6.7 Summary and Future Research Directions

Some of the most diverse behaviours of cephalopods are related to reproduction, and the complexity of some of their mating systems rivals that of many vertebrates. There has been considerable progress in the past two decades on several behavioural aspects of cephalopod reproduction: there have been nearly 150 papers since 1996 on mating systems, alternative mating tactics, sperm competition, female choice and fertilisation success.

The advent of DNA fingerprinting to assess paternity has revolutionised studies of sexual selection, and cephalopod biologists have taken advantage of this relatively new technology to study sperm competition and cryptic female choice. Detailed field studies and laboratory experiments on several cephalopod species have transformed our view of their reproductive tactics and strategies. Perhaps the most noteworthy finding is the behavioural flexibility demonstrated by males and females during agonistic contests, mating and mate guarding.

Both sexes of all species appear to have multiple mates and, as far as we know, there are no monogamous cephalopods. They reproduce once or several times during a short period (usually weeks) and then die. There is no parental care of the young, but female octopods (including the pelagic *Argonauta* and *Tremoctopus*) and some oceanic and deep-sea squid carry the eggs. There are at least two different mating systems: some cuttlefish and squid show elaborate agonistic behaviour, extensive courtship and brief copulation, but no protection of the eggs, whereas in most octopuses there is little agonistic or courtship behaviour, copulation is relatively lengthy and females care for the eggs. Coleoids mature early in their short life. Males are reproductively capable for the greater part of their life cycle; females can receive and store sperm for most of their lives, so that fertilisation and egg laying are temporally independent of mating.

Aspects of agonistic behaviour among males of the squid *Loligo plei* and the cuttlefish *Sepia apama* have been analysed in terms of game theory. The evidence suggests that the evolutionarily stable strategy includes some sort of mutual assessment (rather than self-assessment), where contests begin with various visual signals before escalating to elaborate displays. Some contests continue to escalate to physical fin beating and finally to chasing and occasionally biting. These behaviours can be correlated with the squids' assessment of each other's fighting ability or the resource value present (e.g. a mature and receptive female). Similar behaviour has been described in other decapods.

Courtship, usually initiated by males, is elaborate in some decapod species. However, we know little about the influence of female choice during courtship, except that in cephalopods there are no obvious material benefits to the female such as nuptial gifts from males, shelter, food resources or differing parental investment by males.

During the past two decades, there have been reports of detailed observations and experimental studies of alternative mating tactics, sperm competition, cryptic female choice and reproductive success. Cuttlefish and squid form only temporary mating pairs, lasting a day or two at most. Many species have two methods of copulation, some even three, but it is not known how they choose which method, nor how many mates individual animals have.

We now know that sperm competition and cryptic female choice are common features of several cephalopod mating systems. Cephalopods, like birds and insects, have elaborate pre- and postcopulatory mechanisms for storing, removing or manipulating sperm before fertilisation. One of the most elaborate is the female mimicry used by small sneaker males of *Sepia apama*. Why do females accept the female mimics as mates, and even give them fertilisation priority? What are the tactics of sperm precedence by males? The discovery of specialised sperm in sneaker males of the squid *Loligo bleekeri* requires us to re-think sperm precedence mechanisms. How exactly does cryptic female choice operate? Why is copulation so prolonged in cuttlefish and octopuses if not to displace previous sperm? Lott (1991) provides a useful and provocative review of mating systems in vertebrates that could guide future research on cephalopods, and many papers and books explore behavioural and methodological aspects of sperm competition (e.g. Birkhead, 1989; Parker, 1990; Parker, Simmons & Kirk, 1990; Birkhead & Möller, 1998).

The ultimate evaluation of reproductive success is measured in terms of numbers of progeny produced in succeeding generations. Much has been learned about a few cephalopod species in the past 10 years, but further studies are required, in particular cross-sectional and longitudinal studies similar to those reported by Clutton-Brock (1988) and Reynolds (1996) for many animal groups.

It is not clear why these short-lived animals have evolved such varied and complex mating systems and reproductive strategies. Compared with fish and mammals, cephalopods grow fast, breed once or a few times but all within a single yearly period, die relatively soon after egg laying and have a short life span. Despite considerable progress in understanding many behavioural aspects of reproduction, there has not been commensurate progress on the physiology and endocrinology of the reproductive system; these subjects are not well understood, even in the common octopus, and there is scope here for future experimentation. Questions to be investigated include: are microseminoproteins, which act as potent contact pheromones in loliginid squid, present in other cephalopods? How important is olfaction in sexual selection? Finally, we have to acknowledge that we know almost nothing about reproduction in oceanic and deep-sea cephalopods.

CHAPTER SEVEN

Communication

We now consider the ways in which individual cephalopods influence the behaviour of other animals, both conspecifics and others. In previous chapters, we saw that cephalopods often change pattern and colour during prey capture, during secondary defence and during agonistic encounters and courtship. In this chapter, we examine the nature of their communication systems more closely, and attempt to provide a framework for future studies.

We begin by characterising signals and communication before considering how this applies to cephalopods with their neurally controlled chromatophores. We then review the communication channels available in cephalopods and the particular features of the visual channel that they mainly exploit. After reviewing some limitations of visual signals, we consider the vexed question of 'language'.

7.1 Signals and Displays

7.1.1 Definitions and Concepts

There are innumerable definitions of animal communication in the ethological literature. The meaning of the terms communication and signal may seem straightforward but their definitions have been the subject of much debate. For example, Wilson (1975) defined communication as an 'action on the part of one organism (or cell) that alters the probability pattern of behaviour in another organism (or cell) in a fashion adaptive to either one or both of the participants'. Dawkins and Krebs (1978) and Slater (1983) stressed the benefit to the sender while Philips and Austad (1992) and Smith (1997) emphasised the sharing of information and interaction between sender and receiver. Most of the current concepts of communication and signalling have been summarised by Bradbury & Vehrencamp

(2011) and by Stevens (2013), the latter defining communication as 'A process involving signalling between a sender and receiver, resulting in a perceptual response in the receiver having extracted information from the signal, potentially influencing the receiver's behaviour.'

Organisms communicate with each other via signals, and one function of signals is to provide information that receivers can use in making decisions (Bradbury & Vehrencamp, 2011). A signal is 'Any act or structure which influences the behaviour of other organisms (receivers), and which evolved specifically because of that effect' (Stevens, 2013). A signal is different from a 'cue', which has not evolved specifically to modify the behaviour of the receiver. Following the work of Guilford and Dawkins (1991) and Endler (1993), signals can be thought of as having (1) a 'strategic aspect' referring to the actual content or message of the signal, and (2) 'signal efficacy' that refers to the form or structure of the signal, which did evolve to influence the response of the receiver. Efficacy is usually considered as the most relevant to sensory ecologists because they want to know how the signal actually transmits through the environment and stimulates the sensory system of the receiver (Stevens, 2013). For example, signals that are more intense and contrasting (with the background environment) should be better at eliciting responses.

Signals in many animal groups tend to evolve from incomplete locomotor or postural movements, and also from autonomic movements and conflict movements (Hinde, 1970; Zahavi, 1980; Harper, 1991). Simple signals are often discrete behaviours that are more stereotyped than non-signal movements, so that they are relatively easy to identify and classify (Pearce, 2008). Signals may be either honest or dishonest. Honest signals provide accurate information on the underlying quality of the signaller; dishonest ones do not (Zahavi, 1975, 1987).

Dishonest signals may have advantages for some signallers, for example when lures are used to attract prey, when fighting males signal greater apparent size, fighting ability or endurance than they actually possess, or when palatable prey mimic poisonous animals (Dawkins & Krebs, 1978; Krebs & Dawkins, 1984; Bond, 1989a; Dawkins & Guilford, 1991). For current debates and reviews of benefits, costs and interpretations of signals, and on defining and measuring 'information', the reader is referred to Maynard Smith and Harper (2003), Seyfarth and Cheney (2003), Bradbury and Vehrencamp (2011) and Stevens (2013).

7.1.2 Signalling Behaviour in Cephalopods

In cephalopods, fine neural control of body patterns means that most individual signals can be varied in their visual expression (termed 'multivariate signals' by some researchers; Bradbury & Veherencamp, 2011). Furthermore, signals are sometimes combined together into 'signal sets' (*op. cit.*). In cephalopods, certain types of these signal sets, shown simultaneously or sequentially, are what we define as *displays*. Cephalopod displays tend to be conspicuous, stereotyped, exaggerated or otherwise specialised expressly to facilitate the transmission of information, and they tend to be long-lasting (often tens of seconds) as opposed to most individual signals, which when shown alone are most often transient. Thus, we distinguish a 'display' from a 'signal' not only by a display's complexity, but also by how long it lasts and by its greater degree of ritualisation. Ritualised displays have the following features: (1) redundancy (repetition of the same signal set), (2) conspicuousness, to provide high intensity or strong contrast, (3) stereotypy, to help the receiver classify the signal it detects, and (4) alerting components, to give the receiver a warning that the sender is about to signal (Wiley, 1983).

The terminology used to describe the hierarchical chromatophore system of cephalopods sometimes conflicts with that commonly used by ethologists. For example, ethologists often refer to 'components' of behaviour or to ritualised behaviour 'patterns'. As we saw in Chapter 3, the words 'component' and 'body pattern' have very specific meanings when used to describe the appearance of a cephalopod. In discussing communication in cephalopods, we suggest that it might be useful to define the terms 'signal' and 'display' in

Table 7.1 Terminology: relationship between body patterns and displays

Crypsis		Communication
Body pattern	=	Display
Comprises:		**Comprises:**
Postural components	=	Postural signals
Locomotor components	=	Locomotor signals
Textural components	=	Textural signals
Chromatic components	=	Chromatic signals

such a way that they can be related to the terminology used by cephalopod biologists when describing body patterns. Thus, although the skin or body of the animal may look the same, we shall speak about postural, locomotor, textural, and chromatic *components* of body patterns for crypsis, but postural, locomotor, textural and chromatic *signals* for displays. This emphasises that cephalopods produce body patterns for two quite different reasons: crypsis and communication. In crypsis, a number of components contribute to a body pattern; in communication, a number of signals contribute to a display. Table 7.1 clarifies this terminology.

Agonistic displays of cephalopods comprise a dozen or so signals with rather complex stripes and bands (Box 7.1), whereas deimatic displays (startle or threat) are characterised by large areas of bright whiteness complemented with high-contrast dark spots and outlines (Box 7.2). Sometimes, cephalopods use one or two signals by themselves, for example, only one or two dark mantle spots in *Sepia* surrounded by general paling: we do not consider that these constitute a display, and we refer to them as signals.

Finally, we wish to make it clear that we do not regard crypsis to be a form of communication as some authors have suggested: it is essentially *non-communication*. Moynihan (1975, 1985a), who actually uses the term non-communication, even went so far as to call cryptic body patterns 'anti-displays'. This does not seem particularly useful; apart from anything else, the body patterns used for crypsis (Chapter 5) are not highly ritualised but quite variable. Alternatively, we can think of certain cryptic body patterns as signalling false, or misleading, information, such as 'I am not an octopus but a random sample of the background.' However, signals and displays are attention-getting (Philips &

Box 7.1 Visual signals used during intraspecific encounters: decapod agonistic displays

Sepia officinalis   *Loligo plei*   *Sepioteuthis sepioidea*

Intense Zebra Display	Lateral Display	Zebra Spread Display
Postural signals	*Postural signals*	*Postural signals*
antiparallel/parallel	parallel positioning	over/under
fourth arm extended	arms compressed dorso-ventrally	arms spread
	tentacles extended/dark	tentacles flared
	ventral midline ridge extended	flattened body
Locomotor signals	*Locomotor signals*	*Locomotor signals*
circling	fin beating	touching mantle tips
pushing	chase/flee/forward rush	chase/flee
chase/flee		
Textural signals	*Textural signals*	*Textural signals*
smooth	smooth	smooth
Chromatic signals:	*Chromatic signals:*	*Chromatic signals:*
zebra bands	lateral flame markings	zebra bands
dark eye ring	dark eye ring	dark eye ring
dark anterior head	arm spots/stripes	arm and belly spots
white arm spots	dark stitchwork fins	white fin line
white fin line	iridescent arms	iridescent eye sclera

COLOUR

Colour seems unimportant in these intraspecific displays between males. Instead, the message is probably framed in terms of brightness.

BRIGHTNESS/INTENSITY

Complex displays using many signals. High contrast is achieved by maximal chromatophore expansion in dark areas and maximal retraction of chromatophores in light areas (combined with enhancement by underlying reflecting cells to achieve bright whites). Conspicuousness is attained by: transverse zebra bands on *Sepia officinalis* and on the lower individual of the two *Sepioteuthis sepioidea*; lateral flame markings on the mantle of *Loligo plei*; and dark eye rings.

APPARENT SIZE

Greater apparent size is achieved with arm and tentacle postures as well as with the body. This can be flattened in *Sepioteuthis*; in *Loligo plei*, there is a protrusible flap of skin along the ventral midline that can form a conspicuous mid-ventral ridge.

See Table 7.2 and text for further details.

Box 7.2 Visual signals used during interspecific encounters: deimatic displays

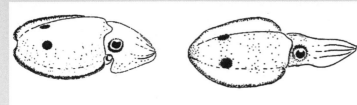

Sepia officinalis   *Sepioteuthis sepioidea*   *Octopus vulgaris*

Postural signals orient to receiver spreading (fins or arms & web) flattening	***Textural signals*** smooth ***Chromatic signals*** general paling dark eye rings dilated pupil false eyespots dark-edged suckers dark fin line
Locomotor signals stationary blowing water (octopuses)	

COLOUR

None. Essentially black on white.

BRIGHTNESS/INTENSITY

Simple displays of high contrast: general paling with only three to four large black signals.

APPARENT SIZE

Large apparent increase due to extensive white areas and spread body.

See Table 7.2 and text for further details. Based on Packard (1972), with permission from John Wiley & Sons, Ltd.

Austad, 1992), and most forms of crypsis are the antithesis of this: they avoid attention. If crypsis is thought of as non-communication, then cephalopods have to be considered as the masters of non-communication!

7.2 Communication Modalities in Cephalopods

After reading Chapter 2, it should come as no surprise that all the current evidence suggests that cephalopods, and not only the shallow-water forms, communicate mainly with visual signals, and these are explored in detail in Section 7.3. Nevertheless, other communication channels may sometimes be employed by cephalopods,

and future researchers should be aware of this possibility.

As we mentioned, cephalopods seem to have no electric sense, so that electric signalling is precluded; and although cephalopods are clearly not deaf (Chapter 2), there is no evidence that they communicate by sounds. There are only two reports in the literature that squid produce sounds (Nishimura, 1961; Iversen, Perkins & Dionne, 1963), but the 'faint poppings' described (vibrations at 1000–4000 Hz) are thought to result from accidental fluttering of the external lips of the funnel as water is expelled through them. In any case, these frequencies are considerably higher than those to which squid are sensitive, the so-called 'infrasounds' (Chapter 2). Moynihan and Rodaniche (1977) emphasise that cephalopods lack hard parts to vibrate or with which to stridulate, and we agree with them that it is most unlikely that cephalopods communicate by sound.

It is possible that squid can detect mechanical disturbances created by conspecifics (using the lateral line analogues on the head and arms: Chapter 2) and use such information to facilitate shoaling at night or in murky water, as in fish (Pitcher, 1993), but to date there is no evidence for this.

Tactile communication, so important in other groups, seems to be relatively uncommon in cephalopods, which are either solitary or live in shoals. However, in *Sepia latimanus*, Corner and Moore (1980) described the males as 'stroking' the head of the female during courtship (§6.2.3). In loliginid squid such as *Loligo pealeii*, *L. plei*, *L. vulgaris reynaudii*, *Sepioteuthis sepioidea* and *S. lessoniana*, fin beating occurs between competing males during agonistic encounters (§6.3); this involves intermittent physical contact of the fins along their length. During such encounters in *S. sepioidea*, the upper male may touch its mantle tip to that of its rival beneath quite precisely (Box 7.1). In *Loligo plei*, the fin beating is very rigorous (DiMarco & Hanlon, 1997) and may function as a signal of strength and aggression (van Staaden, Searcy & Hanlon, 2011). In the mimic octopus of Indonesia, freely ranging conspecifics have been filmed approaching each other and performing a delicate, brief touching of suckers on the distal arm, during which they could exchange chemosensory information. This information could then be used to identify the other animal's species or sex (Fig. 7.1; Hanlon, Conroy & Forsythe, 2008). Collectively, these may be some kind of tactile signal.

Chemoreception in cephalopods is still not well understood (Chapter 2), but there is now incontrovertible evidence for chemical communication in one species, the long-finned squid *Loligo pealeii* (King, Adamo & Hanlon, 2003; Buresch *et al.*, 2003; Cummins *et al.*, 2011). Females of this species lay large masses of fertilised eggs ('egg mops') on the sea floor, and during spawning the males of this species actively compete for females. After pairing with winning males, the females lay eggs on the sea floor in large masses of egg mops. Males are attracted to these mops visually, but if they touch the capsules their agonistic behaviour increases markedly (Fig. 7.2). Buresch *et al.* (2003) found that the factor responsible for the abrupt change in behaviour resides in the egg capsule, and recently Cummins *et al.* (2011) have established that the pheromone is produced in the nidamental glands in the female's reproductive tract before being embedded in the outer tunic of the egg capsules. They also showed it to be a protein closely related to a protein found in the mammalian reproductive system (β-microseminoprotein).

King, Adamo and Hanlon (2003) also showed that the chemical cue is ineffective when water-borne, emphasising the chemotactile nature of this signalling. They suggest that the function of this pheromone might be to signal to males that mature receptive females in the vicinity are about to lay eggs. Whatever the function, this is a beautiful example of the interplay of two communication channels, visual and chemical, in these advanced invertebrates (Wells, 1978).

We should recall that the familiar cephalopod ink might function as a chemical signal. Lucero, Farrington and Gilly (1994) showed that cells in the olfactory organ of *Loligo opalescens* respond to ink (and to the L-DOPA and dopamine that ink contains: Chapter 2). Moreover, behavioural experiments with another species of

Figure 7.1 Two mimic octopuses (*Thaumoctopus mimicus*) approaching each other and briefly touching their arms, presumably to sense the other; subsequently they swam away from each other (from Hanlon, Conroy & Forsythe, 2008: *Biological Journal of the Linnean Society* **93**, 23–38, Mimicry and foraging behaviour of two tropical sand-flat octopus species off North Sulawesi, Indonesia, Hanlon, R. T., Conroy, L. A. & Forsythe, J. W., © 2008, by permission of the Linnean Society).

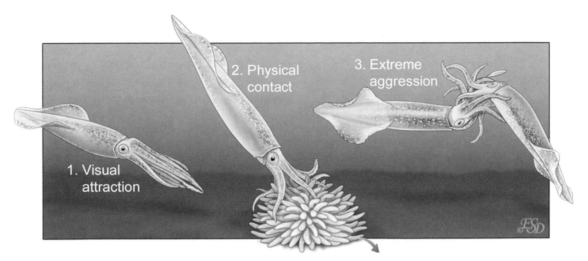

Figure 7.2 Male squid (*Loligo pealeii*) sees egg mops on the sea floor (1), approaches and touches them (2) and immediately switches from calm swimming to extreme fighting (3), owing to the presence of a contact pheromone in the egg tunics (based on Cummins *et al.*, 2011: *Current Biology* **21**, 322–327, Extreme aggression in male squid induced by a beta-MSP-like pheromone, Cummins, S. F. *et al.*, © (2011), with permission from Elsevier).

squid (*Sepioteuthis sepioidea*) have shown that conspecific ink consistently elicits deimatic, cryptic or even protean responses from members of a shoal (Wood, Pennoyer & Derby, 2008). Thus, the ink, which presumably evolved to aid escape from predators, can also function as an alarm substance. Whether or not chemical cues are involved in such signalling has yet to be established. Wood and his collaborators found that the squid responded positively to ink released into an adjacent glass tank (visual but no chemical stimulus), whereas melanin-free ink released into the animals' own tank (chemical but no visual stimulus) elicited no response. Yet the authors recognise that, in the procedures to remove melanin from the ink and make it colourless, they may have removed some other chemical cue, so that at present chemical communication cannot be ruled out. In this species, however, ink undoubtedly functions as an intraspecific visual signal as it does in a number of mesopelagic squid (Bush & Robison, 2007). Ink can also function as an interspecific signal: Caldwell (2005) describes how the emission of ink pseudomorphs by *Octopus bocki* can deter young green turtles from attacking.

In other cephalopods, the evidence for possible chemical communication is much weaker, and this is an area in which experiments could be very rewarding. Male and female *Octopus vulgaris*, visually isolated in adjacent tanks in the laboratory, often move close to one another on either side of the partition between them and may copulate if the partition has a gap large enough to admit the extended third right arm of the male. It is not known how the hectocotylus finds the oviduct subsequently: this could also involve chemical cues (§6.4). In the deep sea, many mesopelagic squid have their olfactory organs raised on stalks: might this enhance their chance of detecting pheromones produced by conspecifics? There is also evidence that female cuttlefish, *Sepia officinalis*, seem to prefer mating with recently mated males (Boal, 1997), and they apparently hyperventilate after exposure to odours from ovary extract (Boal *et al.*, 2010).

However, while chemical communication may be used prior to courtship in *Sepia*, there is again evidence that visual signals are involved in courtship itself (Tinbergen, 1939). Both male–male and male–female interactions occur when individuals are separated by glass, and males show an Intense Zebra Display to their reflection in a mirror. Furthermore, a unilaterally blinded male will be seized for copulation by another male approaching on the blind side, because it is unable to see the Extended Fourth Arm and Intense Zebra Display of the 'courting' male and return the identical display, which communicates, presumably, 'I am a male.' If approached on its intact side, it returns the signal and no attempted male–male copulations are made (Messenger, 1970). This suggests again that most cephalopods, most of the time, communicate with one another via the visual channel, which we consider in detail below.

7.3 Visual Signalling in Cephalopods

Cephalopods use postural, locomotor, textural and, above all, chromatic signals to communicate with their conspecifics and others: these signals are listed in Table 7.2. In this section, we discuss their salient features and then explore the types of information conveyed during intraspecific and interspecific communication.

7.3.1 The Nature and Uses of Visual Signals

Postural, locomotor, textural and chromatic signals may all contribute to a display (Table 7.1). The postural signals may involve the entire body, which is often oriented towards the receiver so as to maximise the effect of the chromatic signals, and which is sometimes flattened and spread to create an impression of greater size. The arms are particularly important in postural signals: Jantzen and Havenhand (2003a) describe no fewer than 15 postures in the squid *Sepioteuthis australis* that mainly differ because of arm position (Fig. 7.3). Such signals are often very obvious (Box 7.1), but there are also subtle arm signals that are only now being documented (Mather, Griebel & Byrne, 2010). Mather and her collaborators emphasise the stereotyped nature of the position of the arms relative to the body: this may

Table 7.2 Visual signals available to cephalopods

A. Postural signals	
Whole body:	Orientation to receiver; upward or downward pointing; spreading and flattening
Arms only:	Singly, in pairs or all together; raised, lowered, splayed, split, V-curled or contorted; male ligula presentation in octopuses
B. Locomotor signals	Chase, flee, forward rush, (anti)parallel positioning
C. Textural signals	Papillate or smooth
D. Chromatic signals	
Whole body:	1. General paling
	2. Intense whitening
	3. General darkening
	4. Flashing (pulsating)
	5. Passing cloud (dark moving waves)
	6. Conflict mottle
Part of body only (often unilateral):	7. Dark stripes or streaks (longitudinal)
	8. Dark bars, bands or rings (transverse)
	9. Dark spots (large or small)
	10. Bright white spots (large or small)
	11. Dark eye rings
	12. Dilated pupil
	13. False eyespots
	14. Dark waving arms
	15. Suckers (white or dark-edged)
	16. Zebra bands or flame markings
	17. Lateral mantle blush
	18. Fin lines (dark or light)
	19. Accentuated white gonad
	20. Red accessory nidamental glands
	21. Iridescent rings or stripes
	22. Polarised light from arms
E. Inking	Pseudomorphs
F. Photophores	Mainly in mid-water cephalopods; signalling on the basis of anatomical arrangement, or active flashing (Chapters 6 and 9)

See Boxes 7.1 and 7.2 for examples.

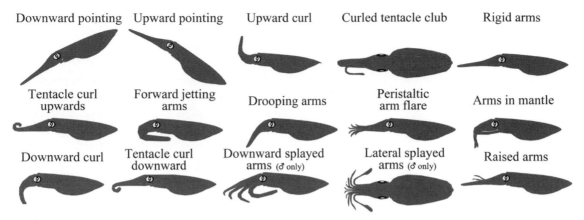

Downward pointing   Upward pointing   Upward curl   Curled tentacle club   Rigid arms

Tentacle curl upwards   Forward jetting arms   Drooping arms   Peristaltic arm flare   Arms in mantle

Downward curl   Tentacle curl downward   Downward splayed arms (♂ only)   Lateral splayed arms (♂ only)   Raised arms

Figure 7.3 Different postures in the squid *Sepioteuthis australis*, each representing different signals to conspecifics or predators (modified from Jantzen & Havenhand, 2003a, *Biological Bulletin* **204**, 290–304, Reproductive behavior in the squid *Sepioteuthis australis* from South Australia: ethogram of reproductive body patterns, Jantzen, T. M. & Havenhand, J. N., © (2003), by permission of the University of Chicago Press).

be particularly important in signalling. Locomotor signals can also be numerous. Jantzen and Havenhand (*op. cit.*) list 15 locomotor signals in *S. australis*. Textural signals, less pronounced in squid, are important in cuttlefish and octopuses. In the Flamboyant Displays that *Sepia officinalis* shows against predatory teleosts, the papillae are prominent (Hanlon & Messenger, 1988; Langridge, 2009; see also §5.2.3).

Chromatic signals are many and varied (Table 7.2). For the most part they involve high-contrast black-and-white signals, although in some interspecific signals the colour content may be important. These are achieved by expansion and retraction of specific motor fields of chromatophores (physiological units: Chapter 3); whiteness is often augmented by the reflecting elements (Box 7.1). Most of the signals are line stimuli: stripes, bars, bands, lines or circles (spots or annuli, e.g. eye rings), which may be black or white (Boxes 7.1 and 7.2). There is abundant evidence that cephalopods can discriminate black from white, horizontal from vertical, and circles from other figures with the same area (Fig. 2.4), so it should be no surprise to find that they use these kinds of signal for intraspecific communication. Contrast seems especially important. For example, Adamo and Hanlon (1996) noted that the chromatic component 'dark face' of the Intense Zebra Display is very variable and showed that males that do not escalate a fight have paler faces than their opponents in the early stages of an encounter. Palmer, Calvé and

Adamo (2006) identified a body pattern ('Splotch') in females that is shown to other females and to their mirror image; this pattern too depends on intensity differences. Recently, Zylinski *et al.* (2011) have quantified the visual characteristics of body patterns in signalling *Sepia apama* and confirmed the importance of contrast from the background for efficient signalling. High-contrast signals also characterise the moving 'Passing Cloud' Display shown by *Octopus* and by *Sepia* (Packard & Sanders, 1971; Mather & Mather, 2004; Fig. 7.4; see also Fig. 4.6). In this, waves of dark chromatophore expansion pass across the body, head and arms.

The use of false 'eye-spots' for interspecific signalling (as in Box 7.2) may exploit the existence of cells in the retinae of their vertebrate predators with 'centre–surround' receptor fields well adapted for detecting spots (Packard, 1972, 1988a). An advantage of using brightness contrast for signalling is that it is effective in many conditions of illumination. The functions and efficiency of eyespot signals in animals are reviewed by Stevens (2005) and further explained in Stevens *et al.* (2008) and in Stevens and Ruxton (2014).

The ability to perceive polarised light has been shown to be important in the life of several species of cephalopod (see Chapter 2), and, ever since Shashar, Rutledge and Cronin (1996) demonstrated that cuttlefish have a conspicuous polarisation pattern around the eyes and down the arms (Fig. 7.5), there has been speculation

Figure 7.4 Male Giant Australian cuttlefish *Sepia apama* displaying the Passing Cloud pattern to a rival male during an agonistic bout (photo: R. Hanlon).

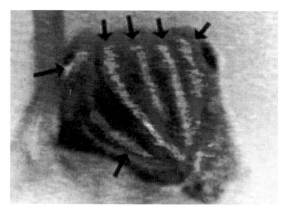

Figure 7.5 *Sepia officinalis* showing the polarisation signals (arrows) emanating from the iridescent arm stripes; photograph from an imaging polarimeter video camera (from Boal *et al.*, 2004: figure reproduced with permission from N. Shashar).

that *Sepia*, and possibly other genera, could communicate with polarised signals. Mäthger and Denton (2001) studied the iridescent signals, which are also polarised, made by two loliginid squid (*Alloteuthis subulata* and *Loligo vulgaris*). Some iridophores in these squid create 'red', 'green' and 'blue' reflective stripes running along the mantle (as seen to a human observer in white light). Theoretical considerations suggest that in shallow seawater these would be highly conspicuous and permit signalling between neighbours in a shoal. Experimental confirmation of this would be technically very difficult, however (Mäthger, Shashar & Hanlon, 2009).

Mäthger and Hanlon (2006) subsequently investigated the anatomical basis for polarisation signalling in the squid, *Loligo pealeii*. They demonstrated not only that the polarisation derives from the iridophores but also that this polarisation remains unchanged after it has passed through the overlying chromatophore layer. Furthermore, when Chiou *et al.* (2007) investigated the polarisation characteristics of the light reflected from the iridophore layer in the arms of *L. pealeii* and *Sepia officinalis*, they showed that it does not alter with changes in arm orientation, obviously a great potential advantage for signalling.

Yet the evidence that cephalopods actually use polarisation signalling is less impressive. Shashar, Rutledge and Cronin (1996) reported that the polarisation pattern on the head and arms of the cuttlefish disappears during extreme aggression, attacks on prey, copulation and egg laying; they also found that male cuttlefish retreated from their reflection in a mirror unless the polarisation of that image was deliberately distorted by a special filter. Boal *et al.* (2004), in an extensive set of experiments, only found 'limited evidence that . . . polarized patterns varied in the presence of conspecifics'. They found females were more likely to show polarisation patterns than males, but that they did not modify these patterns in the presence of conspecifics. Mäthger, Shashar and Hanlon (2009) provide a very useful review of all this work. Clearly, new experiments would be most welcome to establish whether some cephalopods can communicate with conspecifics by using a visual channel inaccessible to most of their (vertebrate) predators (Land & Nilsson, 2012).

We have seen that colour does not seem to be so important in visual signalling, certainly not for intraspecific communication, which agrees with the findings from discrimination training and other evidence suggesting that most cephalopods appear to be colour blind (Chapter 2). Nevertheless, we must be cautious about generalising from the experimental evidence available from *Octopus vulgaris* (Chapter 2), particularly as Moynihan and Rodaniche (1982) report what they term 'pastel coloration' in *Sepioteuthis sepioidea* during sexual interactions. Especially in the males, much of the upper surface can turn 'pinkish-tan', pink or lilac, and a rapid succession of yellow, pink or lilac waves may pass along the body. Of course, such signals might be distinguishable in terms of contrast, or perhaps they could involve ultraviolet reflection: the possibility that squid can see into the UV has never been explored, and experiments are required to clarify this.

The use of colour in interspecific signalling has also been reported in *Sepioteuthis*: the same authors, and Hanlon and Forsythe (unpublished data), also report the use of bright yellow coloration towards goatfish and other large herbivorous fish. Several species of *Octopus* have iridescent blue rings in the eye-spots, and the venomous *Hapalochlaena* can show iridescent blue rings and lines on the black patches of the body and arms. At

normal incidence, the rings reflect at approximately 500 nm, which is not only within the range of all potential vertebrate predators (cetaceans, pinnipeds, birds and teleosts) and other cephalopods, but is also well tuned to the ambient underwater light field (Mäthger *et al.*, 2012; see also Chapter 2). In this latter genus, the body can also be boldly striped black and yellow, giving the animal a wasp-like appearance that could constitute aposematic coloration (see also *Metasepia*, Fig. 5.21e). Unfortunately, little is known about the use of the displays by these blue-ringed octopuses.

In Box 7.1, we illustrate one type of intraspecific encounter and list the range of visual signals from which agonistic displays are constructed by male decapods. Particular signals carried by postures, movements, skin texture and skin patterning (i.e. chromatic signals) are used to convey information about apparent size and, presumably, fighting ability. Note that the arms are an important part of making the animal appear larger, and that the arms usually have light or dark spots that may carry the message of 'hostility' in this context. The zebra bands or flame markings on the mantle are also highly conspicuous visual signals contributing to these displays, at least to a human observer. We are ignorant of the exact information content of the many individual signals, or even of the details of the behavioural contests themselves (§6.6.2), but generally these agonistic displays are complex and comprise 8–12 signals.

Interspecific Deimatic Displays (§5.2.3) also comprise a dozen or so individual signals (Box 7.2), but differ in one important manner. Here, there is widespread general paling combined with high-contrast circular black signals (including dilated pupil, eye ring, or body outline as delineated by fin lines or dark suckers) and body flattening to create the illusion of large size. Note how conserved these pattern displays are in cephalopods as different as octopus, squid and cuttlefish (Box 7.2). The information conveyed here seems to be about apparent size and 'fierceness'. Other displays, such as Flamboyant Displays, use different signals, such as textured skin, contorted arms and facing the predator, to convey a message of threat. There are many more examples in the figures in Chapters 5 and 6.

In Table 7.3 we list some of the presumed functions of the individual signals used by cephalopods. Of the various categories of signal function listed by Halliday

Table 7.3 Functions of some cephalopod signalling behaviour

	Communication context	Directed toward	Display	Signal	Presumed message	Presumed function
Intraspecific signals	Courtship	Potential mates	♀ Pied	Vertical bar Bright white Arms drooped	'Court me'	Initiate mating
	Courtship	Rivals and mate	♂ Lateral Silver	Ipsilateral bright white Contralateral dark	'Males, keep away; female, stay near'	Repel rival Maintain courtship
	Agonistic contest	Rival male	♂ Lateral Display	See Box 7.1	'I am stronger, fitter'	Repel rival
	♀ Non-receptivity	Male suitor	♀ White lateral stripe	Unilateral white stripe	'Don't want to mate'	Reject male
Interspecific signals	Feeding	Prey	Passing Cloud	Dark moving waves	'Stop and watch me'	'Mesmerise' prey
	Feeding	Prey	n/a	Dark arm waving	'Stop and watch this'	'Mesmerise' prey
	Defence	Predators	Flamboyant	Arm posture Heavy texture Dark mottle	'See my weapons'	Threaten predator
	Defence	Predators	n/a	Ink pseudomorph	'Attack me'	Deceive predator
	Defence	Predators	Deimatic	See Box 7.2	'I am large and fierce'	Startle predator

n/a: not applicable; these signals act alone

Many other examples are discussed in Chapters 5 and 6.

(1983) and Harper (1991), cephalopods certainly transmit signals about identity, ability and motivation in the sexual domain, indicating competence to mate (i.e. courtship), and fighting ability (or resource holding power, §6.6.1). We have no data showing that cephalopods can recognise individuals of their species, such as kin, former rivals or former sexual partners (Boal, 1996). Nor do we know whether they can communicate the position of food sources or other information that might strengthen group adhesion, both of which are found in other social animals. During the early stages of predation, some cephalopods use signals that may aid in prey capture (§4.4.2). As far as is known, cephalopods do not signal about their environment. They are not territorial (see §9.2.1), but they sometimes use ink as an alarm signal. There is, however, circumstantial evidence that shoals of *Sepioteuthis sepioidea* have sentinels that may transmit subtle alarm signals (Moynihan & Rodaniche, 1982).

During defence, many cephalopods attempt to foil the attack sequence of predators after crypsis has failed, by using signals that seem to threaten, startle, frighten or bluff predators, or otherwise confuse them or misdirect their attacks (see §5.2.3). It is known that signalling to predators can be expressed asymmetrically towards the threat (Moynihan, 1985a; Hanlon & Messenger, 1988; Langridge, 2006). More interestingly, cuttlefish signalling for defence is highly selective in two respects. First, they show different signals to herbivorous fish than to predator fish (Hanlon & Messenger, 1988). Second, they use the 'eyespot' Deimatic Display to visual predators (teleost fish, *op. cit.*; bird models, Adamo *et al.*, 2006) but not to chemosensory or electrosensory predators (crabs or dogfish) (Langridge, Broom & Osorio, 2007; Langridge, 2009).

A surprising fact that has emerged recently is that several mid-water cephalopods have been recorded communicating visually, using a whole variety of chromatic and postural signals even in deep water where light levels are presumably quite low (Bush, Robison & Caldwell, 2009). Of course, many mesopelagic squid possess light organs, or photophores, some of which are almost certainly used for visual signalling (Chapter 9). There is often a characteristic arrangement of the light organs that might serve to identify species. Moreover, in several species there is a clear sexual dimorphism in the

arrangement or in the type of photophores, suggesting that they could be used to identify sex. It is also possible that the light organs could be used for interspecific signalling: the bright dorsal patch of *Ommastrephes pteropus*, for example (Herring, 1988), or the arm tip flashes of *Watasenia scintillans* or *Taningia danae* (Roper & Vecchione, 1993) could be used to startle teleost, bird or mammalian predators. Some of the light organs on the tentacles of *Chiroteuthis* or *Mastigoteuthis* might function as lures (§4.4.2). If true, the latter would be an example of the kind of signal that some commentators call manipulatory or dishonest (Krebs & Dawkins, 1984). The other functions postulated above for the photophores would involve honest signalling.

7.3.2 Advantages of Signalling with Chromatophores or Reflectors

There are at least five advantages arising from having chromatophores under direct neural control from the brain. The first is rapidity of signalling. Several different signals can be made in a few seconds or one signal can be repeated (at variable frequency). This flexibility can set the characteristic fast tempo of interactions seen in squid such as *Sepioteuthis sepioidea* (Moynihan, 1985a). Of course, it is also possible to 'hold' a given signal for a long time.

The second is that signals can be graded in intensity of expression. This is impossible to convey in photographs but well known to anyone who has worked with living cephalopods or viewed video sequences. For example, low contrast between pale and dark zebra bands in *Sepia officinalis* can be instantly increased to produce a signal that is a key part of the Intense Zebra Display (Box 7.1). Such a finely graded neural system permits signalling of small changes in motivational state. Moynihan (1985a) has gone so far as to state explicitly that 'the visual patterns of coleoids often indicate moods, i.e. probabilities of future acts' and this might be worth investigating (cf. Bond, 1989b).

Third, neural control of chromatophores enables cephalopods to switch quickly between camouflage and signalling, whereas most other animals can only ever do one or the other. In this respect, cephalopods may provide a good system in which to test when a particular strategy is favoured.

Fourth, because chromatophores are independent of the other body muscles, the cephalopod can signal without interfering with its other activities. A moth revealing its coloured hind wings, or a primate presenting its rump, has to interrupt other motor actions to signal, but cephalopods can continue to swim, eat or copulate while signalling.

Fifth, and uniquely, neurally controlled chromatophores permit unilateral signalling. During sexual encounters, male squid such as *Loligo plei* (Box 7.1 and Fig. 6.24) and *Sepioteuthis sepioidea* (Fig. 6.14) can show a Lateral Display on one side of the body to repel an approaching rival male, while simultaneously showing Clear or Beige (a signal of calmness) on the other side to maintain its attraction for the female. Because the actors change places frequently during the agonistic encounter, as the intruder male tries to get near the female, the defending male must frequently change the side on which he shows Lateral flame so that he does not repel his mate (DiMarco & Hanlon, 1997).

Iridophores also contribute to signalling: they produce structural colours, and the colours they produce can be altered slowly by acetylcholine (Chapter 3). Most iridescence, however, is passive. All iridescence is highly directional, depending on the position of the light source and the viewing angle. The advantage of this is that squid could theoretically target a receiver, for example a conspecific in a school, so that the signal is transmitted horizontally but remains invisible to a predator situated above or below (Mäthger & Denton, 2001). Moreover, since iridescence is polarised, such a signal has the potential to function as a 'private' communication channel, invisible to most predators (Mäthger & Hanlon, 2006; Mäthger, Shashar & Hanlon, 2009; Mäthger *et al.*, 2009).

Leucophores are also structural elements important for camouflage and signalling. They act as almost perfect diffusers, appearing equally bright from all angles of view, and they reflect light at very high levels (up to 70%) at wavelengths from 300 to 900 nm (Mäthger *et al.*, 2009). They are critical for producing the bright whites in high-contrast signals that constitute the Intense Zebra Display of cuttlefish (Box 7.1, Fig. 7.3), where stripes of leucophores alternate with stripes of expanded black chromatophores. This reminds us of the importance of the precise arrangement in the skin of the various reflectors and pigments.

7.3.3 Limitations of Visual Signalling

The different communication modalities of animals all have their advantages and disadvantages (Marler & Hamilton, 1966; Krebs & Davies, 1993; Bradbury & Vehrencamp, 2011; Stevens, 2013), and it is interesting that cephalopods seem to be so reliant on a single channel, vision, to communicate with each other (Moynihan & Rodaniche, 1977). Obviously, the eyes are highly developed in these animals, so they can profit from the fact that visual signals spread in all directions, convey a lot of information quickly, and can be made to appear and disappear instantly. For those cephalopods that shoal, visual signals constitute a highly suitable means of exchanging information quickly.

On the other hand, visual signalling is only effective over short distances, in clear water and in the daytime (unless there are photophores). Another disadvantage of most visual signals, of course, is that they give away the position of the signaller to any onlookers in the vicinity: signals are perceived by animals for which they are not intended. This would be one advantage of polarised light signals (§7.3.1), for these would be invisible to most predators of cephalopods. Another way of overcoming this is to limit the signal temporally, to as little as 1 to 2 seconds if necessary (as many other animals do). The neurally controlled chromatophores are, of course, ideally suited for this.

7.3.4 Language?

The Caribbean reef squid, *Sepioteuthis sepioidea*, is remarkable for the richness of its chromatic signals and the complexity of its displays: this has led Moynihan and Rodaniche (1982) and Moynihan (1985a) to speculate that the visual signalling of this species amounts to a 'language', with its own rules, syntax and grammar. Their view stems from their observations that particular components of body patterns occur in certain combinations. They report that this squid has 'at least 10 to 11 components with a corresponding, almost geometrical increase of permutations. Unritualised movements and intention movements must add hundreds more components. And all of these occur in sequential as well as synchronous combinations' (Moynihan & Rodaniche, 1982, p. 124).

Yet by themselves these numbers can tell us nothing about whether squid are communicating by using a language. For example, in a series of experiments on language acquisition by apes, Nim, a chimpanzee trained by Terrace *et al.* (1979), learned 125 different signs and, over 18 months, combined them, by making signs, into no fewer than 19 000 multi-word utterances of over 5000 different types. Despite this, the experimenters found no evidence that Nim's utterances were structured, with rules, syntax and grammar. Thus, his apparent fluent use of signs did not constitute language as we understand it.

In *Sepioteuthis*, Moynihan himself (1985a, p. 95) notes the apparent absence of phrases or sentences: 'the components of the utterances of known coleoids are not, to human eyes, grouped in sections of stereotyped intermediate lengths. They do not seem to be marked by internal starts or restarts, pauses, or full stops'. Moreover, it is not immediately obvious why cephalopods, without an elaborate social structure and with no parent–offspring relationships, would need to transmit the large amounts of information that a true language permits (Hockett, 1960; Anderson, 2005). There have been many attempts to show that animals other than humans may use language, with the result that there is a considerable literature on what is still a contentious issue: Pearce (2008) offers a critical discussion of communication and language in animals that is particularly useful for biologists. In the present context, it seems safe to conclude that although *Sepioteuthis* undoubtedly signals to conspecifics, there is insufficient evidence to claim that it uses a 'language'.

7.4 Summary and Future Research Directions

It is obvious that cephalopods signal to other individuals. Although evidence for chemical communication is now emerging, cephalopods probably signal mainly via the visual channel. The iridophores may contribute to signalling in some species, but the chromatophores are of prime importance here, especially in creating messages couched in high-contrast intensity differences, and postures can also be important, even in the deep sea, where signalling with photophores also occurs. The chromatophores and photophores probably both evolved for crypsis rather than communication, but the fact that both systems are under neural control has meant that they have proved particularly well suited for visual signalling. There are many advantages for signalling in having neurally controlled chromatophores.

In discussing communication in cephalopods, we found it useful to define the terms 'signal' and 'display' in a way that relates them to the terminology used when describing body patterns. Cephalopods sometimes communicate with simple signals. However, they also combine postural, locomotor, textural and chromatic signals into complex displays, both to conspecifics (during a wide range of agonistic and reproductive behaviour) and to predators (for defence).

Some of the highly ritualised intraspecific displays are complex even by the standards of the higher vertebrates (with as many as 12 signals per display) and there are almost certainly other, more subtle, signals used in intraspecific communication that are inconspicuous to human observers. Interspecific displays (e.g. the Deimatic Displays) tend to be much simpler, and the individual signals are often strikingly similar in members of all three major groups of present-day coleoids. These displays, with their highly conserved signals, could be considered 'dishonest' in that they tend to make the sender look much larger and, sometimes, misdirect a predator to false eyes. On the other hand, the displays used in agonistic contests between males may be 'honestly' signalling fighting ability or motivation.

Little is known about the functions of signalling in cephalopods: structured field observations and laboratory experiments, even simple ones, could be most rewarding. For example, signals about danger, which may be widespread among shoaling squid, have only been described cursorily, partly because they are so subtle. Nor are the precise functions of the various signals and displays made during sexual encounters understood. These might be amenable to analysis since both sender and receiver can be easily identified. With the improved methods of captive rearing now available, some of these questions could be answered in the laboratory without elaborate apparatus.

There has been a suggestion that in one social cephalopod, *Sepioteuthis sepioidea*, where visual signalling is particularly well developed, the signals constitute a 'language' with a vocabulary and a syntax. There are at least 35 different displays in this species, but the many signals can be combined, simultaneously or successively, in different ways, and not all individuals always show exactly the same displays in the same situation. It would be highly informative to know whether recombinations of signals mean different things to conspecific receivers; that is, do they evoke different responses? Despite this complexity, the evidence that this is a language seems untenable at present, and more rigorous analysis of better-recorded data will be necessary before such a claim can be seriously entertained.

However, this should in no way deter a younger generation of biologists from studying the signalling of these beautiful animals. There are so many questions to investigate. For example, does the great variety of signals serve as a measure of the signaller's fitness, as in bird song (Catchpole, 1980)? How does cephalopod signalling develop? Do different populations of *Sepioteuthis* always use exactly similar signals to make exactly the same displays? Could some features of the signals be learned? Not least, might the study of sophisticated signalling in animals unrelated to vertebrates help to establish some general principles of visual communication?

CHAPTER EIGHT

The Development of Behaviour, Learning and Cognition

It is obvious that most animals, even apparently 'simple' ones, do not always respond to stimuli in the same way; in particular, there are often striking differences in the way that young and old animals respond to events in the external world. Some of these are linked to normal growth and development, including maturation of the gonads, and to development within the CNS (Bateson, 1991); others involve the relatively abrupt change in behaviour that we call learning. Learning in many 'higher' animals often involves quite complex behaviour, often referred to as cognition (Shettleworth, 2010), and undoubtedly shown by some cephalopods. There may also be individual differences in behaviour that appear to be genetically guided, sometimes referred to as behavioural syndromes or 'personalities'.

8.1 Development

The development of three types of behaviour in cephalopods has been studied in detail: the escape response in squid, body patterning in cuttlefish and prey capture by cuttlefish.

There are also several other reports of developmental change in young cephalopods. In three species of loliginid squid, Hanlon *et al.* (1987) have shown that shoaling only commences when the animals have reached about 10 mm ML, which may be at 3 to 9 weeks according to species. In *Octopus joubini*, Mather (1984) has provided good evidence that the diel rhythm is 'tuned' or sharpened over the first 3 to 4 weeks of life (Fig. 8.1), and there is complementary evidence in newly settled *O. cyanea* (Wells & Wells, 1970).

Juvenile cuttlefish conceal themselves in sand by digging, burying and blowing sand particles over their body with the funnel, in what Mather (1986b) regarded

as a 'fixed' behaviour pattern. However, Poirier, Chichery and Dickel (2004) found that the proportion of cuttlefish that bury themselves increases steadily in the first 2 weeks after hatching. They also carried out an experiment that compared two groups of hatchlings, one reared in a black tank with a sand substrate, and one in a black tank without sand. At 3-day intervals, animals from each group were introduced to a novel grey tank with sand and their burying behaviour recorded. With both groups, the proportion of partially buried animals increased during the first 9 days of life, but those reared with sand showed a greater increase in the first 6 days, and significantly more of them covered themselves completely at days 12 to 15. It seems clear that digging behaviour in cuttlefish is not completely pre-programmed.

8.1.1 Development of the Escape Response in Squid

The giant fibre system of squid, discovered in the 1930s, is important in mediating escape by jetting (Chapter 2: Box 2.4). It used to be considered as operating in a relatively straightforward reflex manner, but in the 1990s, evidence began to emerge showing how it develops during ontogeny.

Using embryos and hatchlings of the squid *Loligo opalescens*, Gilly, Hopkins and Mackie (1991) measured changes in the timing and nature of the response and correlated these with anatomical evidence about the morphology of the stellar nerves, which in adults each carry a third-order giant axon (Box 2.4). Escape jetting can be elicited from embryos before hatching (at stage 26 according to the staging criteria of Segawa *et al.*, 1988), even though these embryos lack giant axons in the stellar nerves. Jetting improves greatly between

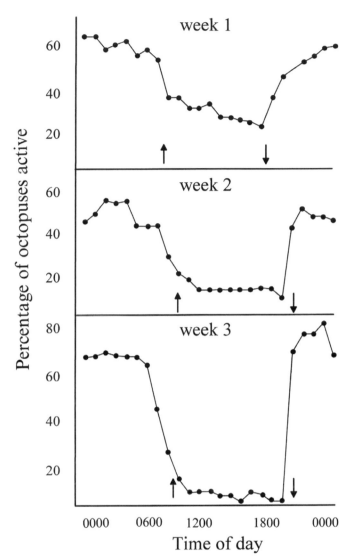

Figure 8.1 Tuning of the diel activity pattern of young *Octopus joubini* during the first weeks of life. They are active at night but return to their shelters in daylight. Arrows indicate dawn and dusk. Each dot represents the percentage of animals (*N* = 70) active at a given hour (from Mather, 1984).

stage 28 (when a larger axon is identifiable) and stage 30 (hatching, when there is a well-developed giant axon), but even then it is not as fast as in adults. In all embryos, the escape responses are 'fast-start', with the giant axon response always preceding activity in the small motor fibres. In adults, however, the escape response of the giant fibres can be delayed so that they produce a boost to the jetting mechanism at the critical time to achieve maximum acceleration (Otis & Gilly, 1990). Preliminary results with hatchling

Sepioteuthis lessoniana had suggested that the adult style of escape response might appear in the first week of post-embryonic life, if the animals had been actively feeding (Gilly, Hopkins & Mackie, 1991).

Subsequently, Preuss and Gilly (2000) showed unequivocally how early experience of prey capture by young *L. opalescens* could influence the escape response to a short electrical stimulus. Juveniles fed only slow-moving prey (*Artemia*) developed deficits in coordinating the activity of the giant fibres with that of the non-giant fibres; such deficits did not appear when juveniles were offered copepods, their preferred, fast-moving prey. The influence of prey-capture experience is especially marked during the first 1–3 weeks post-hatching. Thus, motor skills acquired during one unrelated activity (prey capture) can influence coordination of the systems utilised in another activity: escape jetting.

8.1.2 Development of Body Patterning

This has been studied mainly in *Sepia officinalis*. Hanlon and Messenger (1988) found that hatchling cuttlefish already possess most of the components, including textural components, used for body patterning and can produce 10 out of the total adult repertoire of 13 body patterns. They can change body patterns as quickly and effectively as adults, too, and experiments suggest that, from the earliest moments of life, visual cues are of prime importance in controlling body patterning. In *S. latimanus*, Corner and Moore (1980) observed a very late embryo showing the Splayed arm posture and papillate skin while still in the egg.

Nevertheless, there are differences between the body patterns shown by hatchling and adult *Sepia officinalis*. For example, the Weak Zebra body pattern, which is an effective form of camouflage for a larger animal, does not appear until the animals are about 45 mm or more in mantle length. On a background of gravel, hatchlings show a Disruptive pattern; early juveniles a less Disruptive pattern with Mottle components; and late juveniles a Mottle pattern (Fig. 8.2). To a human observer, these patterns appear equally effective. A late juvenile, or an adult, showing a Disruptive pattern on this substrate would be very conspicuous: the Disruptive is reserved for a background containing relatively large, pale stones on a dark substrate. Recent experiments that tested cuttlefish of different sizes on black and white chequerboards of various sizes showed that Disruptive is shown when the chequer size is between 40% and 120% of the White square *irrespective of animal size* (Fig. 8.2). This suggests that the cuttlefish may employ a single 'visual sampling rule' throughout life, which would seem to be an economical way of

maintaining camouflage with growth (Barbosa *et al.*, 2007). How the animal can achieve this without employing visual feedback to monitor the appearance of its White square (see Chapter 3) remains a mystery, however.

Further evidence that early experience can influence body patterning has been provided by Poirier, Chichery and Dickel (2005). These authors reared large numbers of cuttlefish (*Sepia officinalis*) under two quite different conditions and followed their body patterning behaviour from hatching to two months. One group comprised isolated individuals ($n = 120$) kept in black plastic tanks, conditions described as '*uniform-solitary*'. The other comprised three batches of animals ($n = 40$), each kept in tanks with grey walls and a substrate of fine yellow sand on top of which lay stones of different colour, shells and plastic 'seaweed', conditions described as '*varied-social*'. The cuttlefish were tested at 15, 30 and 60 days post-hatching on a background of either small, variegated stones or a uniform pale grey plastic. In the course of development, it became clear that the cuttlefish in the varied-social group concealed themselves differently from those from the uniform-solitary group. In particular, the varied social cuttlefish used significantly more chromatic components in their Disruptive pattern on a heterogeneous background than the uniform-solitary cuttlefish.

The most striking change in body patterning behaviour between hatchling and adult cuttlefish is the shift in emphasis from crypsis to communication. Of course, mature cuttlefish sometimes have to hide, but they also have to indulge in reproduction, which involves signalling. The Weak Zebra (cryptic) develops into the Intense Zebra (signal), which is expressed fully in mature males (Box 7.1), although females also show the white and black zebra bands on the mantle. This highly contrasting pattern in the adult relies for its full expression on the development in the skin of iridophores and leucophores, which are absent in hatchlings (Chapter 3).

The Deimatic Display is another that matures as the cuttlefish grows. In its final and most familiar form it is a simple bold display (Box 7.2). Its most conspicuous feature is a pair of black spots, but these, which do not appear until the cuttlefish is about

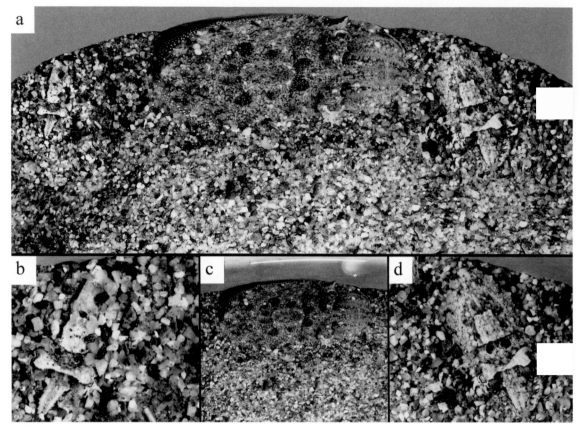

Figure 8.2 *Sepia officinalis* of different ages and sizes use different body patterns for concealment on the same substrate. (**a**) Montage of a hatchling (left, 10 mm), late juvenile (centre, 35 mm) and early juvenile (right, 18 mm). (**b**, **c**, **d**) Photographs of the same three individuals, printed so that their mantle lengths are equal (from Hanlon & Messenger, 1988, with permission of Royal Society Publishing).

35 mm long, are the remnants of four pairs of spots present in hatchlings, three of which drop out of the repertoire with maturity. Young cuttlefish do not show the Deimatic Display: they respond to threat with a different display, the Flamboyant (Chapter 5).

A corresponding change has been shown in *Octopus vulgaris* by Packard and Sanders (1971). Newly settled juveniles (less than 20 days post-settling) respond to a threatening object (a large rubber bung on a wire rod) with the highly stereotyped Flamboyant Display. Gradually, the Flamboyant Display drops out of the repertoire and adult octopuses respond by showing the Deimatic Display (Fig. 8.3).

In *O. vulgaris*, there is evidence that the Deimatic Display may develop out of the Flamboyant Display by way of the Conflict Mottle Display. Deimatic Display is shown regularly to disturbing stimuli by animals of all age groups, and the arm-spreading of the Conflict Mottle can be seen as an incomplete version of the Deimatic arm-spreading; there are also some chromatic signals shared among Conflict Mottle, late Flamboyant and early Deimatic (Packard & Sanders, 1971).

In *Octopus briareus*, which hatches from a large egg as a replica of the adult (Messenger, 1963), it has been found that there is no Flamboyant Display; there is a Deimatic Display, but this is not fully developed until about 3 months (Hanlon & Wolterding, 1989).

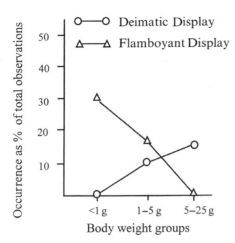

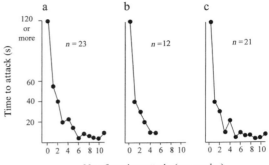

Figure 8.4 On the first occasion that a hatchling cuttlefish is shown a *Mysis*, it may wait for more than 2 minutes before attacking it. The delay is greatly reduced with repeated presentations, but this change is independent of whether the animal has been (**a**) rewarded for attacking, (**b**) not rewarded or (**c**) starved for 5 days before the first trial (from Wells, 1958: *Behaviour* **13**, 96–111, Factors affecting reactions to *Mysis* by newly hatched *Sepia*, Wells, M. J., © (1958), by permission of Koninklijke Brill NV).

Figure 8.3 Very young *Octopus vulgaris* respond to disturbance by showing the Flamboyant Display, but this is soon replaced by the Deimatic Display, which continues into adult life (reprinted from Packard & Sanders, 1971: *Animal Behaviour* 19, 780–790, Body patterns of *Octopus vulgaris* and maturation of the response to disturbance, Packard, A. & Sanders, G. D., © (1971), with permission from Elsevier).

8.1.3 Development of Prey Capture in *Sepia officinalis*

The eggs of the common cuttlefish, which are untended and receive no parental care, look like small black grapes; they are laid in clusters in shallow water that abounds with small fish and crustaceans, notably *Mysis*. The hatchlings are miniature replicas of the adults and at once take up life on the seabed, sitting concealed in the substrate. They are relatively easy to maintain in the laboratory, and their behaviour immediately after hatching was first studied by Wells (1958).

Although newly hatched cuttlefish look like adults, they are often less active, and at first may not respond to the prey species, *Mysis*. Wells found that they begin to attack mysids after 24–48 hours, with the precise sequence of motor acts shown by mature animals. The exact age at which they start attacking seems to depend on the rate at which they absorb the yolk reserve. The latency to the first attack made on a live *Mysis* may be more than a minute, but subsequently this delay gets shorter and shorter (Fig. 8.4): Wells (1958) showed that

this change depends entirely on the number of previous attacks the baby cuttlefish has made, not on its age, nor on the time since its last meal, nor on whether it has been rewarded for attacking. The accuracy of the attack does not improve with age, and other experiments confirm that the binocular depth-detecting system is operative the very first time a cuttlefish attacks a mysid (Messenger, 1977b). The long latency to attack a mysid on first encounter seems unlikely to result from major defects in either the visual system or the motor system (hatchlings can match the background, swim, use their tentacles and change body patterning). It is as if there is some 'defect' in the CNS at this time, but unfortunately we know nothing about the nature of such a defect.

By using models and parts of mysids, Wells also found that naive hatchling *Sepia* respond only to moving objects, and only to a *Mysis* shape when it is moved along its long axis (Table 8.1). This preference for an elongate figure is similar to that shown by octopuses for rectangles moved along their long axis (Chapter 2) and suggests there may be some pre-programming for the detection of prey shaped like fish or mysids. (In the absence of live food, hatchling cuttlefish can be reared on the gill

Table 8.1 Attacks by young *Sepia* on moving shapes

Rotated shapes	Naive	5 previous attacks	10 previous attacks
[horizontal dotted/bead shape]	18%	47%	66%
[horizontal bar shape]	0	26%	35%
[vertical bar shape]	0	6%	31%
[asterisk/star shape]	0	0	0
Shapes moved up and down			
[horizontal dotted/bead shape]	0	0	–
[horizontal bar shape]	0	0	–

Hatchling cuttlefish clearly respond much more strongly to elongate shapes moving along their long axis. Shapes were 10 mm long, painted in black on a transparent plastic cylinder, rotated or moved up and down (modified from Wells 1962b).

filaments of crabs or on pieces of shrimp if these are made to move in the water.) Cuttlefish become less selective once they have attacked mysids five to ten times, and Wells (1958) showed that they then respond to a far greater variety of models. It appears that during development the rather restrictive feeding programme that keeps them attacking *Mysis* gives way to a 'looser' programme, running for a month or two, that allows them to sample different prey items. Wells suggested that gradually, perhaps as a result of unpleasant reinforcement, hatchlings would begin to learn to restrict themselves to prey that are safe as well as tasty; certainly, adult cuttlefish are selective in what they attack.

More recently, in a very interesting series of experiments, Darmaillacq and her colleagues have extended these findings to reveal the importance of experience on the feeding behaviour of young cuttlefish. Without previous feeding experience, 3-day-old cuttlefish prefer shrimps to crabs, but a group given a single crab on Day 3 and tested on Day 7 showed a significant preference for crabs (Darmaillacq *et al.*, 2004a). Moreover, cuttlefish exposed to crabs for 5 hours at hatching and tested on Day 3 significantly preferred crabs, as compared with cuttlefish that had no exposure to crabs, which preferred shrimps. The preference for crabs was greater when both visual and

chemical cues were available to the cuttlefish, but visual exposure alone was sufficient to change preference (Darmaillacq *et al.*, 2006). These results reveal that – as far as prey preference is concerned – the behaviour of newly hatched cuttlefish is plastic and modifiable by experience. Indeed, further experiments strongly suggest that there may be a brief window, 2 to 4 hours after hatching, when exposure to crabs (without reinforcement) is most effective in switching preference from shrimp to crab. This effect persists at least until 7 days after hatching, and Darmaillacq, Chichery and Dickel (2006) consider that the existence of such a *sensitive period* implies that 'food imprinting' occurs shortly after hatching, which could explain the prey preferences exhibited by adult cuttlefish. Darmaillacq, Lesimple and Dickel (2008) also have evidence that the window may be open before hatching, in the final week of embryonic life. It is becoming clear that food imprinting may occur in the embryo (Guibé *et al.*, 2012); that the chemosensory visual and tactile systems are already functional in the embryo (Romagny *et al.*, 2012); and that odours and visual stimuli can interact very early in determining food preferences of young *Sepia* (Guibé, Boal & Dickel, 2010; Guibé & Dickel, 2011; Guibé *et al.*, 2012).

Warnke (1994) found that cuttlefish reared in groups (of four) attacked mysids or shrimps faster than cuttlefish reared individually; they also captured more than twice as many prey animals during the tests, suggesting that there was some kind of social interaction (or simply stimulus enhancement?) during feeding. This finding has never been followed up.

Adult cuttlefish generally attack crabs from behind, thus avoiding the chelae (Boycott, 1958; Messenger, 1968; Duval, Chichery & Chichery, 1984). Younger, smaller cuttlefish do not show such behaviour, but this change in feeding tactics has not been followed in any detail. Boal, Wittenberg and Hanlon (2000) found that adult cuttlefish that had been reared without experience of crabs improved their prey capture technique (to avoid being nipped) over only five trials. They also found that exposure to crab odour led to enhanced performance, suggesting that chemical cues can influence attacking behaviour, perhaps by triggering food arousal. Incidentally, cuttlefish can capture crabs with the tentacles or with the arms alone; Chichery and Chichery (1992) have shown that it is the size ratio between predator and prey that determines which attack tactic is used, and obviously this changes with growth and development.

8.2 Learning

Learning is 'that process which manifests itself by adaptive changes in individual behaviour as a result of experience' (Thorpe, 1963). It involves the storage in the nervous system of a record – a memory – of the outcome of recent actions, in order to increase the chances of survival. Because cephalopods are very adept at learning, at least in the laboratory, J.Z. Young chose them in the 1950s as a model for the study of learning and memory. His choice led to three extended series of experiments on different aspects of learning in *Octopus vulgaris*, most of which were carried out at the Zoological Station in Naples during the 1950s and 1960s.

The first of these, by Boycott and Young, sought to establish a general model for the physical basis of memory, using a visual discrimination paradigm. The second, by Wells and Wells and by Wells and Young, used tactile discriminations with a similar aim. Both of these approaches involved brain lesioning experiments, designed to establish the sites of the memory stores in the brain. The third series of experiments, by Sutherland and his colleagues, explored the visual capacities of octopuses, as these could be revealed by discrimination learning. As a result, there is an extensive literature on learning in octopuses that it would be inappropriate to summarise here, first because there already exist several comprehensive reviews (Boycott & Young, 1950; Young, 1965b; Wells, 1966, 1978; Sanders, 1975), and second because the thrust of much of the work is too neurological and speculative for this book (but see Chapter 2).

Instead, we attempt to survey the different kinds of learning shown by cephalopods in the laboratory and in the sea. In so doing, we partly follow the classification of learning types originally proposed by Thorpe (1963), although we have not hesitated to depart from it where appropriate.

8.2.1 Habituation and Sensitisation

Habituation, which is often regarded as the simplest kind of learning, is the relatively persistent waning of a response as a result of repeated stimulation without any kind of reinforcement (Hinde, 1970). A particularly clear example was reported by Wells and Wells (1956): if a Perspex cylinder is placed on the arm of a blind octopus, the animal will pass it under the web to the mouth, where it may be examined for as long as 20 minutes before being rejected. If the same object is repeatedly presented to the arm at intervals of 2 minutes, the octopus stops passing it under the web to the mouth after two to four trials, stops pulling it towards the mouth after another three to five trials, and after another four to ten trials spends only 4 seconds examining it. Finally, the arm is withdrawn from the object after the briefest of examinations.

The habituation of a visual response has been demonstrated in the bay squid, *Lolliguncula brevis*. Long *et al.* (1989) exposed individual squid to models of fish predators and showed that the number of escape jets and durations of rings shown on the mantle (shown to novel stimuli) declined with repeated presentation (Fig. 8.5); the individuals in these experiments also showed signs of dishabituation to the same model after a threat stimulus (hand waving near the animal). Recently, visual habituation has been demonstrated in the late-stage embryo of the cuttlefish, *Sepia officinalis*, suggesting that habituation must be very important, even in the early stages of life, in eliminating responses to irrelevant stimuli (Romagny *et al.*, 2012).

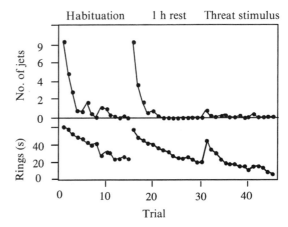

Figure 8.5 Habituation, recovery after an hour's rest and dishabituation of a visual response in a squid (*Lolliguncula*) shown a model predator. See text (reprinted from Long *et al.*, 1989, with permission from Taylor & Francis Group).

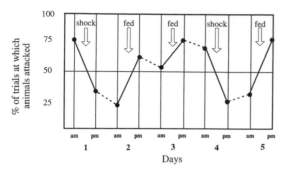

Figure 8.6 Sensitisation. Dots represent the mean test scores of eight animals that had been previously trained to attack a vertical rectangle. For 5 days, they were then shown a rectangle without reward five times during the morning, and five times in the afternoon. In the middle of the day, each animal was given either a shock or food in the home, following a single showing of the rectangle whether it attacked or not. It is clear that shocks depress and food elevates the number of attacks (modified from Wells, 1967a, based on Young, 1960c: *Proceedings of the Royal Society of London B: Biological Sciences* **153**, 1–17, Unit processes in the formation of representations in the memory of *Octopus*, Young, J. Z., © (1960), by permission of the Royal Society).

Sensitisation, or the increased likelihood of an animal responding to a stimulus, has also been demonstrated clearly in *Octopus vulgaris*. Thus, feeding an animal prior to showing it a plastic shape increases the probability that it will attack the shape, just as shocking it decreases this likelihood (Fig. 8.6). That is, the reward or punishment takes place *before* the presentation of the test shape. Sensitisation is an adaptive response, at least in the very short term, which makes it likely that food or pain will both be responded to appropriately; it can occur during tactile discriminations (Wells & Wells, 1958) and with olfactory stimuli (Chase & Wells, 1986). The importance of sensitisation usually diminishes during discrimination training, but it can still exert a powerful influence on behaviour; Wells and Wells (1958) deliberately manipulated sensitisation to reveal a hidden ability to make a tactile discrimination (see discussion in Wells, 1967a).

8.2.2 Associative Learning

This important class of learning (Pearce, 2008) is generally taken to include *classical* conditioning (including discrimination learning), and *instrumental* or operant conditioning (including trial-and-error learning). Both of these involve long-term changes in behaviour as a consequence of an association between particular sets of contingencies, which can be manipulated by the experimenter in the laboratory.

In cephalopods, there have been few studies specifically focused on either classical or instrumental conditioning (but see Papini & Bitterman, 1991). Some early papers successfully employed classical conditioning techniques, as did the study of 'infra-sound' detection by Packard, Karlsen and Sand (1990). Early attempts to demonstrate instrumental conditioning in *Octopus*, both of which required animals to pull a lever for a food reward, were unsuccessful (Dews, 1959; Crancher *et al.*, 1972). More recently, however, in a series of experiments investigating arm preference (Chapter 2), Byrne *et al.* (2006a) showed that it was possible to train an octopus to insert an arm into a T-maze to retrieve a food reward. Using a similar operant task, Gutnick *et al.* (2011) obtained important evidence that octopuses can combine positional information from the arms with visual information (see Chapter 2).

The most extensive evidence for classical conditioning in cephalopods comes from the numerous examples of discrimination learning by *Octopus vulgaris*, several of which have been referred to already in Chapter 2. Octopuses readily learn to attack one of a pair of shapes or objects (Fig. 8.7), when an attack on one leads to a food reward and on the other to a small electric shock or to no food. Learning can be very rapid, especially with easy discriminations, although performance tends to remain erratic (Fig. 8.8) and never seems to reach the near-perfect levels sometimes achieved by mammals in discrimination training.

In *visual discrimination* training, the shapes can be presented successively or simultaneously: higher scores can be achieved with simultaneous presentation (Sutherland & Muntz, 1959). Octopuses can also learn more than one visual discrimination at the same time (Boycott & Young, 1957; Messenger, 1977a). Objections have been raised to the methodology of some of the visual training techniques employed with *Octopus vulgaris* at Naples, and it is certainly true that manual

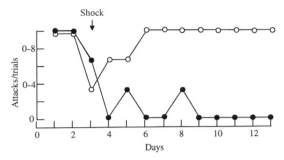

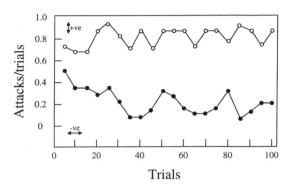

Figure 8.7 Visual discrimination learning in *Octopus*. Attacks persist on the positive (crab alone; open circle), but decline on the negative (crab plus white square giving shock from the third day; filled circle). Results from a single individual (redrawn from Boycott & Young, 1955a: *Proceedings of the Royal Society B: Biological Sciences* **143**, 449, A memory system in *Octopus vulgaris*, Lamarck, Boycott, B. B. & Young, J. Z., © (1955), by permission of the Royal Society).

Figure 8.8 Octopuses learning to discriminate between the different planes of polarised light (as shown) never achieve a perfect score even after 100 trials. Attacks on positive shown by open circles; attacks on negative by filled circles. (*n* = 4; redrawn from Moody & Parriss, 1961, with permission from Springer.)

presentation of shapes has its disadvantages (Bitterman, 1975). However, although it is to be hoped that future workers will continue to develop automated training procedures and such novelties as computer-generated stimuli projected onto screens (Papini & Bitterman, 1991; Pronk, Wilson & Harcourt, 2010), the rigour of some of the experimental designs employed in Naples in the 1950s, particularly those of Sutherland and his collaborators, should leave no one in any doubt that octopuses can make complex visual discriminations (e.g. Sutherland, 1957b, 1958, 1960, 1962; Chapter 2).

Visual discrimination training has also been successfully carried out with *O. apollyon*, *O. maya*, *O. bimaculatus* and *O. bimaculoides* (Roffe, 1975; Hanlon, Forsythe & Messenger, 1984; Allen, Michels & Young, 1986; Boal, 1991), and with the decapods *Sepia officinalis* (Messenger, 1977b), *Lolliguncula brevis* (Allen, Michels & Young, 1985) and *Todarodes pacificus* (Flores, 1983).

As shown in Chapter 2, octopuses can also be trained to make successive *tactile discriminations*, accepting and passing to the mouth one object, but rejecting and pushing away another object that differs in texture or in taste (for a summary see Wells, 1978). Because the accept-or-reject criterion is so clear, and because the touch-learning centres in the brain are relatively small and localised, tactile, rather than

visual, discrimination learning in *Octopus* could still be a potential model for studying the mechanism of memory formation in *Octopus* and perhaps more generally (Robertson, 1994; Robertson, Bonaventura & Kohm, 1994; see §8.4).

There is some evidence that in *Octopus* the visual and tactile memories may interact (Messenger, 1983; Allen, Michels & Young, 1986). Allen *et al.* trained octopuses (*O. bimaculatus*, *O. maya*, *O. vulgaris*) first with a tactile and then a visual discrimination, using plastic balls as discriminanda (black, white or clear; rough or smooth). They found that a negatively associated visual memory was sometimes able to interfere with a previously learned, positively associated tactile memory.

Avoidance learning by octopuses has been known since the time of von Uexküll (1905); Polimanti (1910) described how *Eledone moschata*, after being stung, will avoid hermit crabs that bear sea anemones. Boycott (1954) reported that *Octopus vulgaris*, too, can learn to avoid the crab *Eupagurus bernhardus* carrying *Calliactis parasitica* but persist with attacks on *E. prideauxii* carrying *Adamsia palliata*, perhaps because *Adamsia* is less noxious. He also invited an octopus to attack a crab presented between two *Anemonia sulcata* and describes the various kinds of 'cautious' behaviour it began to adopt after it had got stung, such as a slow approach or the extension of a single arm between the two anemones. Ross (1971) also demonstrated how effective a deterrent *Calliactis* could be to *O. vulgaris* (§4.2); there are comparable findings for *O. maorum* (Hand, 1975) and *O. joubini* (McClean, 1983; Brooks, 1988).

It is not difficult to envisage behaviours of this kind being employed in the sea, but octopuses are equally able to learn avoidance in other, less natural, circumstances. Maldonado (1968, 1969) trained octopuses to leave a dark box for the light to avoid an electric shock.

More recently, Darmaillacq *et al.* (2004b) have obtained striking evidence for *taste aversion* learning in *Sepia*. They first established (with over 60 mature animals) that cuttlefish caught in the same feeding ground in the English Channel show a marked preference for either shrimps or crabs when they enter the laboratory. They then showed that cuttlefish would attack their preferred prey after it had been painted with a solution of quinine dissolved in nail polish, although the cuttlefish quickly released it as soon as they touched it. In careful experiments with matched controls, Darmaillacq *et al.* followed the time of attack of experimental cuttlefish over 15 trials and obtained clear evidence not only of a significant increase in latency to attack but an increase as early as the second trial. Moreover, 26 out of 32 cuttlefish attacked a different prey from the preferred one in a choice test 24 or 72 hours after completing training. Interestingly, the authors report that in the later stages of the experiment some cuttlefish change their behaviour in quite subtle ways, approaching their prey slowly and extending their arms or tentacles to touch it, or blowing jets of water at the prey.

Little attention has been paid to *latent learning*, or 'irrelevant incentive learning' (Mackintosh, 1974); that is, the association of indifferent stimuli or situations without patent reward (Thorpe, 1963). However, the fact that octopuses in the sea have been shown to learn about their surroundings (see below) suggests that such learning almost certainly occurs. Moreover, in the laboratory, Boal *et al.* (2000) have provided good evidence for exploratory behaviour in *Octopus bimaculoides*; and Cartron *et al.* (2012) have established that cuttlefish can utilise redundant cues in learning a spatial task.

8.2.3 Spatial Learning

Over the years, there have been several attempts to train octopuses to run mazes. Boycott (1954) reported his own and others' failure to train octopuses in a maze. Wells (1964), however, using a rather elaborate maze design (developed by Schiller, 1949), did eventually succeed, although the octopuses learned extremely slowly. A more convincing experiment to provide evidence of maze learning in *Octopus* was carried out by Walker, Longo and Bitterman (1970), who trained *O. maya* to turn left or right in a simple T-maze to regain entry to water: they reported zero errors after 27 days of training at three trials a day and subsequently obtained good scores in reversal training.

Many years later, in a field study in Bermuda, Mather (1991b) provided more impressive evidence for spatial memory in young *Octopus vulgaris*. Analysing 60 trips made by four individuals, Mather found that the octopuses at this particular site spent, on average, 55 minutes foraging away from their dens, travelling about 9 m in distance. She showed that octopuses did not go back to their den the same way: rather, they returned along different routes (Fig. 8.9), sometimes along conspicuous features of the landscape, and usually

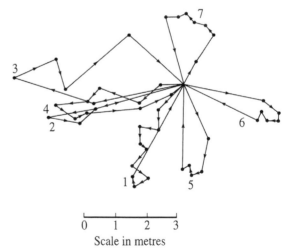

Scale in metres

Figure 8.9 Foraging trips of a single juvenile *Octopus vulgaris* recorded in the sea in Bermuda (from Mather, 1991b: *Journal of Comparative Physiology A* **168**, 491–497, Navigation by spatial memory and use of visual landmarks in octopuses, Mather, J. A., © (1991), with permission of Springer). See Fig. 4.12 for similar foraging paths in the Pacific species *O. cyanea*.

without reference to the outgoing route. They generally swam rather than walked around their foraging area, thus eliminating the possibility they were using chemical cues from the substrate. When disturbed, they went swiftly and directly back to the den, suggesting that they had formed some kind of map of the area. There was also some evidence that none of them was delayed in regaining its den by the experimenter moving artificial landmarks in the foraging area, but the sample size was small and the landmarks had only been placed there recently.

A comparable field study of natural foraging by adult *Octopus cyanea* on a Polynesian coral reef also strongly supports the notion that octopuses use spatial learning in very complex visual habitats (Forsythe & Hanlon, 1997; see Fig. 4.12). Remarkably, on 6 of 24 forages, octopuses abruptly stopped foraging and made a swift swimming 'bee line' directly back to their dens, indicating exceptionally accurate knowledge of spatial features of the diverse seascape.

In the laboratory, Boal *et al.* (2000) performed three experiments with *Octopus bimaculoides* and demonstrated clearly that octopuses could learn the location of an open burrow (among five others that were not open), and that they could remember its location for at least a week. Hvorecny *et al.* (2007) trained *O. bimaculoides* (and two cuttlefish species) in two different maze configurations: about half the animals tested selected the correct escape route in each maze, thus demonstrating an undoubted ability for spatial learning.

At this juncture, brief mention must be made of jar-opening behaviour in octopuses, which has been investigated by Piéron (1911), Boycott (1954), Cousteau and Diolé (1973) and more recently Fiorito, von Planta and Scotto (1990). An octopus presented with a crab in a glass jar with a conspicuous bung swims to it and envelops it. After a while the bung comes out, and eventually any arms that are inside the jar locate the crab, seize it and pass it to the mouth. The time needed for the octopus to get the crab is remarkably long, however. In the experiments of Fiorito and his colleagues, not only did the time spent enveloping the jar fail to decrease significantly in 20 trials (over 10 days), but it also remained at between 30–50 seconds. The time taken to open the jar did decline significantly, but even after 10 days it was well over a minute, and the

time taken subsequently to seize the crab remained over 50 seconds. Such a slow performance is unimpressive to anyone who has ever watched an octopus in a visual discrimination training experiment swoop out in two or three seconds to seize a shape: jar-opening is patently a poor paradigm for studying learning in these animals.

Although the life style of octopuses cannot normally involve maze-running (or jar-opening), it seems highly likely that they learn about spatial relationships in their natural environment as they forage from, and return to, a den. The same would not, on the face of it, seem to be true for cuttlefish, yet a series of recent experiments have shown unequivocally that *Sepia officinalis*, too, can learn a simple T-maze.

Karson, Boal and Hanlon (2003) were the first to suggest that cuttlefish might be capable of spatial learning, a finding elaborated on by Alves *et al.* (2007), who trained *Sepia* to enter a T-maze to reach a dark compartment, where they could bury themselves in sand. The cuttlefish, which could see out of the maze to various objects in the testing laboratory ('distal visual cues'), rapidly learned this task. The day after successful training, the animals were tested by being placed in a new start position in the maze such that, had the animals previously learned on the basis of a left or right turn ('response tactic'), they would be expected to turn in a direction opposite to that taken previously. As can be seen in Fig. 8.10, nine out of ten cuttlefish did just this; only one appeared to have used a 'place tactic', using external visual cues. In another experiment, cuttlefish were trained in a T-maze with black-and-white stripes and spots ('proximal visual cues') placed a few centimetres above the water surface, but with the laboratory surrounds curtained off. Again, the animals learned very quickly, but when tested with the left/right positions of the visual cues reversed, five out of seven cuttlefish oriented using a place tactic while two used a response tactic, a difference that was not significant. Cuttlefish seem to be able to use one of two tactics for maze learning, although curiously there is evidence that adult males may rely more on place learning than females (Jozet-Alves, Modéran & Dickel, 2008). More recently, Cartron *et al.* (2012) successfully trained cuttlefish to solve a Y-maze with black-and-white visual 'landmarks' and polarised light as cues; they also showed that the animals were able to use either cue when the other was missing.

a Experiment 1: T-maze with distal cues

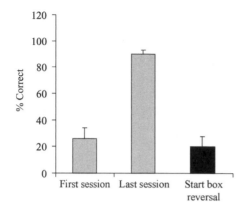

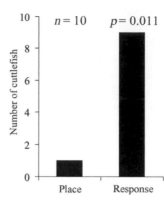

b Experiment 2: T-maze with proximal cues

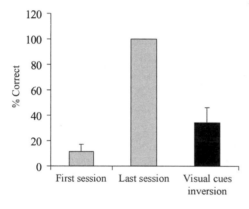

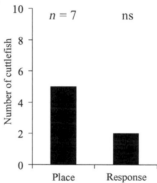

Figure 8.10 Spatial learning in *Sepia*. (**a**) T-maze with distal visual cues. Mean correct responses in the maze improved from the first to the last training session, but fell when the start box was reversed; nine cuttlefish used a response strategy to orient, and one a place strategy. (**b**) T-maze with proxImal visual cues. Mean correct responses improved during training but fell after inversion; five animals used a place strategy to orient, two a response strategy (based on Alves *et al.*, 2007: *Animal Cognition* **10**, 29–36, Orientation in the cuttlefish *Sepia officinalis*: response versus place learning, Alves, C., Chichery, R., Boal, J. G. & Dickel, L., © (2007), with permission of Springer).

8.2.4 Social Learning

Thorpe (1963) identified another category of learning that he termed *insight-learning*: insight was defined as 'the apprehension of relations' and as 'the solution of a problem by the sudden adaptive reorganisation of experience'. There are difficulties with these definitions, and probably few ethologists or experimental psychologists would now consider that this was a particularly useful learning category.

However, one category of behaviour that Thorpe considered to be a type of insight-learning is *imitation*, which he defined as 'the copying of a novel or otherwise improbable act or utterance, or some act for which there is no instinctive tendency'. This category embraces social learning (or social facilitation) as well as

local enhancement, and there has been a challenging demonstration that octopuses may be able to learn from conspecifics. Fiorito and Scotto (1992) trained two groups of 'demonstrator' octopuses to discriminate between red and white spheres (presumably on the basis of brightness differences; see Chapter 2) until they reached criterion (no errors in five consecutive trials). Each animal was then tested without reward four times in the field of view of a naive 'observer' octopus in an adjacent tank. None of the demonstrators made any errors during testing, so each observer octopus saw its demonstrator attack one of a pair of objects four times (at 5-minute intervals). When the observers were themselves tested without any kind of reward, they attacked the positive shape significantly more than the negative one, and their performance was significantly better after four trials than that of the demonstrators at that stage of their own training. It has long been known that a brightness discrimination is a simple task for an octopus (Boycott & Young, 1957) but the rapidity claimed for 'observational learning' in this experiment is remarkable, especially as the observers never saw a food reward being given to the demonstrators for an attack on a sphere. Fiorito and Chichery (1995) again obtained positive results using this same paradigm in an experiment in which they were studying the effects of lesions to the vertical lobe (see §8.2.5).

Several important questions remain about Fiorito and Scotto's results, particularly whether 'observational learning' is different from imitation (Biedermann & Davey, 1993), or whether this is imitation at all. It could be an example of 'stimulus enhancement', whereby the observer's attention is directed towards a particular object in the environment (Pearce, 2008). Whatever the theoretical basis for this type of learning (Suboski, Muir & Hall, 1993), it does seem remarkable in animals such as octopuses. Not surprisingly, Fiorito and Scotto (1992) themselves questioned the function of such learning in the life of octopuses, which are solitary animals whose generations do not overlap (see Chapter 11). We might expect to find social learning in shoaling squid, especially in *Sepioteuthis sepioidea*, where young and old individuals are often found together, but this has never been investigated.

There have been two inconclusive attempts to show observational learning in cuttlefish. First, Boal, Wittenberg and Hanlon (2000) sought to establish whether naive observer *Sepia officinalis* could learn to avoid being nipped while attacking crabs by watching demonstrator crabs. The authors found no evidence that cuttlefish could learn from conspecifics, but they concede that, under the conditions of their experiments, odour cues were confounding their results. Second, Huang and Chiao (2011), using *S. pharaonis*, had the observer animals watch other individuals (the 'experiencer group') being threatened and displaced from their preferred resting place by a toy submarine (*sic*!). Although one individual (out of ten) may have learned by observation, the experiencer group as a whole did not show evidence of observational learning.

8.2.5 Short- and Long-Term Memories

Learning involves memory storage and in cephalopods, as in other animals, there is good evidence for a distinct short-term memory (STM), lasting little more than an hour or so, and a long-term memory (LTM) lasting for at least several months in an octopus. Typically the STM is 'labile', and thus susceptible to interference, by electro-convulsive shock (ECS) or by drugs; the LTM is more stable and resistant to interference.

The existence of STM in octopuses is evident from various experiments on delayed rewards (Schiller, 1949; Dilly, 1963; Wells, 1964, 1967b; Sanders, 1970). They can easily learn visual discriminations with delays of up to 2 minutes between responding to a food stimulus and gaining reinforcement; perhaps octopuses in the sea often experience a delay between first seeing a prey animal and subsequently being able to capture it. Delays of much more than 30 seconds seriously impair the acquisition of tactile discriminations (Wells & Young, 1968). In *Octopus*, Maldonado (1968, 1969) also demonstrated that ECS administered at the appropriate time could interfere with a learned avoidance response. In addition, there have been innumerable brain-lesion experiments demonstrating the existence of an STM in *Octopus*. The idea that the vertical lobe might be the STM store, as well as being involved in the establishment of the LTM, dates back to the experiments of Boycott and Young (1955b), Young (1958, 1960c, d) and Wells and Wells (1957b, c). Sanders (1975) gives a particularly useful summary of this early work. More recently, Fiorito and Chichery (1995) also demonstrated STM in *Octopus*, using the observational learning paradigm (§8.2.4). Interestingly, they showed that lesions to the vertical lobe impaired STM, but not LTM; and Fiorito *et al.* (1998) found that scopolamine interfered significantly with STM and LTM in octopuses presented with the crab-in-jar problem and in an observational learning situation.

In *Sepia officinalis*, too, there is evidence for short- and long-term memories. Messenger (1973b), using a simple technique developed by Wells (1958), tested a large number of adult cuttlefish with a transparent tube containing two prawns. The tube was left in the individual's tank for a period of 20 minutes, during which time the animals gradually stopped attacking the inaccessible prawns. The tube with prawns was removed and then re-presented for 5 minutes after an interval that varied from 2 minutes to 24 hours. Not surprisingly, cuttlefish tested after a few minutes made no or very few strikes; those tested after 24 hours made more (although considerably fewer than on first presentation). The recovery of the striking response is not smooth, however (Fig. 8.11), and Messenger (1977b) suggested that *Sepia* might have separate STM and LTM memory stores, with parallel entries, which had been revealed by the conditions of the experiment.

Support for the existence of a specific STM in *Sepia* has been provided more recently by the careful experiments of Agin *et al.* (1998) and Dickel, Chichery and Chichery (1998), again using the same 'prawn-in-tube' paradigm. Moreover, cycloheximide, a protein synthesis inhibitor, has been shown to interfere with LTM if administered between 1 and 4 hours after training (Agin *et al.*, 2003). This important finding is the only evidence from cephalopods that *de novo* protein synthesis is required for the formation of LTM, although this has, of course, been shown in many other animal groups. Agin and her collaborators have also raised the possibility of there being intermediate-term memory (ITM) in cuttlefish (see §8.2.6; Agin *et al.*, 2006). In a series of neat experiments, Jozet-Alves, Bertin and Clayton (2013) provided evidence of episodic-like memory in *Sepia* by requiring them to recall the 'what, where and when' components of a foraging and feeding event. This approach deserves future experimentation with octopus and other cuttlefish species; it appears to be the first behavioural evidence of such complex memory in an invertebrate.

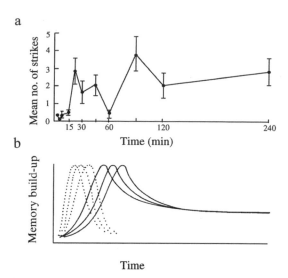

Figure 8.11 Are there separate short- and long-term memory systems in *Sepia*? (**a**) Animals that had learned not to strike at prawns in a glass tube were shown the prawns again after varying intervals. Note the very low strike-level of animals ($n = 19$) tested 60 minutes after first presentation: these animals appear to have retained more than those tested after 20 minutes (reprinted from Messenger, 1973b: *Animal Behavior* **21**, 801–826, Learning in the cuttlefish, *Sepia*, Messenger, J. B., © 1973). (**b**) Such a result could be explained if *Sepia* has separate short-term and long-term memory systems with parallel entries. The parameters adopted in these particular experiments could have allowed the STM (dotted curve) to decay before the LTM (solid curve) had built up (from Messenger, 1977b: *Symposia of the Zoological Society of London* **38**, 347–376, Prey-capture and learning in the cuttlefish, *Sepia*, Messenger, J. B., © (1977), with permission from Elsevier).

8.2.6 Changes in Learning Capacity with Age

Over 70 years ago, Sanders and Young (1940) showed that adult cuttlefish could learn *not* to attack prawns presented behind glass. They also showed that surgical lesions to a particular region of the brain (comprising the vertical, subvertical and superior frontal lobes, hereafter the 'vertical lobe system') interfered in some way with this learning. Wells (1962a) examined the way that hatchling cuttlefish reacted to *Mysis* in a glass tube: he found that, like the adults, they attacked in the usual way but struck their tentacles against the glass, and obtained no food reward. A remarkable feature of this behaviour, however, is that whereas adults gradually

ceased attacking inaccessible prey, the young continued to attack for up to 6 hours with only a slight reduction in attack level, which could have been due to fatigue. Wells drew attention to the important finding of Wirz (1954), who had shown that the vertical lobe of young cuttlefish is small and that during development it nearly doubles in size relative to the rest of the brain. The vertical lobe has long been known to be important for the establishment of memories in *Octopus*, as well as in *Sepia* (Chapter 2).

Messenger (1973b) subsequently extended these findings using the same experimental paradigm as

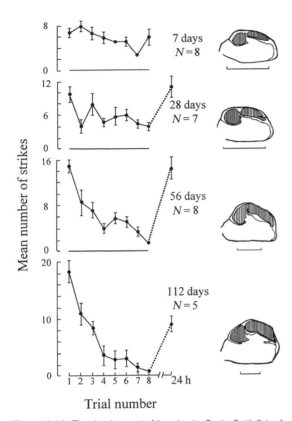

Figure 8.12 The development of learning in *Sepia*. Cuttlefish of different ages were shown prawns in a glass tube for a series of 3-minute trials at 30-minute intervals. Midline sagittal sections of the brain (right-hand side) chart the development of the vertical lobe system (from Messenger, 1973a: *Brain Research*, **58**, 519–523, Learning performance and brain structure: a study in development, Messenger, J. B. © (1973), with permission from Elsevier).

Wells (1962a). Groups of *Sepia officinalis* were reared from hatching and maintained for 1, 4, 8 and 16 weeks before being tested with prawns in a glass tube. The results are shown in Fig. 8.12, which shows the gradual improvement, with age, in learning not to attack prawns in a glass tube. It turns out that 1-week-old animals behave like hatchlings and show no significant decline in attack level over a series of repeated presentations. Animals that are 4 weeks old attack less on the last of a series of eight trials than on the first, but on a retention test one day later revert to the initial level. Not until cuttlefish are 8 weeks old is there a steep decline in the rate of attack over a series of repeated presentations; but even then the animals show

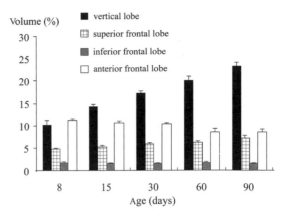

Figure 8.13 Increase in the relative volume of the vertical lobe in *Sepia* during the first 3 months of life (from Dickel, Chichery & Chichery, 2001: *Developmental Psychobiology* 39, 92–98, Dickel, L., Chichery, M. P. & Chichery, R., © (2001), with permission of the American Psychological Association).

no retention when tested a day later. By the age of 16 weeks, however, naive animals not only learn but also show retention.

Histological examination of sections of the brains of the cuttlefish used in these experiments (Fig. 8.12) provided evidence of the increasing size of the vertical lobe as learning performance improved. Dickel, Chichery and Chichery (2001) extended and refined these experiments and, broadly, confirmed that learning in *Sepia* improves over the first 3 months of life, while the vertical lobe doubles in size (Fig. 8.13). Agin, Chichery and Chichery (2001), using cytochrome oxidase as a marker for neuronal activity, reported localised changes in activity in two regions of the superior frontal lobe as a result of training adult cuttlefish with the prawns-in-tube paradigm. All these findings clearly implicate the vertical lobe system in the establishment of visual memories in *Sepia* as in *Octopus*.

Another series of experiments by Agin *et al.* (2006), using the prawn-in-tube paradigm, is noteworthy, because, while broadly confirming the conclusions of Messenger, it highlights a marked improvement in learning performance between 15 and 21 days of age. The authors speculate that there may be a sensitive period around that time for the establishment of what may be an intermediate-term memory; further experiments are required to clarify this issue.

Finally, it is worth noting another type of behaviour that shows evidence of change with age. Adult cuttlefish will pursue ('hunt') prawns that move out of the visual field, a response that is abolished by lesions to the vertical lobe (Sanders & Young, 1940). In contrast, hatchlings, 1-week-old and 4-week-old animals do not hunt (Messenger, 1977b); not until they are 2 months old do some cuttlefish occasionally show hunting. Again, it seems as if a fully functional vertical lobe system may be necessary for this kind of behaviour. Confirmation of this idea has been provided by Chichery and Chichery (1992), who found that senescent cuttlefish, with marked degeneration in the vertical lobe system, no longer hunt. Dickel, Chichery and Chichery (1997), working with newly hatched cuttlefish, found that a crucial feature for hunting behaviour was the development of the tracts between the vertical and the subvertical lobe. Collectively, these experiments suggest that learned behaviour such as hunting, which presumably involves STM, appears rather late in an animal that does not live for much more than 12–18 months.

8.3 Cognition

Most cognitive scientists now see cognition as an elaborate structure of evolved traits, and the central questions are ones of how those traits evolved, how they are interrelated, and how they develop (e.g. Balda, Pepperberg & Kamil, 1998; Byrne, 2000; Byrne & Bates, 2006). These issues mirror Tinbergen's (1963) famous 'four whys of ethology'. Definitions of cognition abound. In Balda, Pepperberg and Kamil's edited volume (1998), a general consensus is that an ethological approach suggests that what we call 'cognition' is not a tightly unified system, but a collection of skills and aptitudes, each of which might have a different evolutionary story. According to Shettleworth (2010) 'cognition refers to the mechanisms by which animals acquire, process, store and act on information from the environment. These include perception, learning, memory, and decision-making.' These are all topics addressed in the present book. These subjects are also considered in several recent cephalopod publications (Mather & Kuba, 2013; Darmaillacq, Dickel & Mather, 2014). What we consider in this section are complex behaviours of cephalopods that have been observed in the laboratory or in the sea.

We consider cognition under two headings: first the 'hard' evidence, from a series of well designed, rigorous experiments with octopuses, for higher-order processing in visual discrimination learning; second, the more preliminary evidence that suggests cephalopods may use 'tools', explore and play, which has led some authors to claim that octopuses may even exhibit 'consciousness'.

8.3.1 Perceptual Processes in Visual Learning

The experiments of Young, Sutherland and others on visual learning in *Octopus vulgaris* have established beyond doubt that this species can make visual discriminations of some complexity (see the form discrimination shown in Fig. 2.4). Such findings could be regarded as trivial, but there are complementary experiments of more general interest, notably by Mackintosh, showing that some of the subsequent stages of visual processing in octopuses are similar to those made by mammals (Box 8.1).

For example, octopuses show *stimulus generalisation*. This can be shown by means of a simple 'transfer test' interposed at a particular stage of training. After having been trained, octopuses are tested by allowing them to respond to a series of novel shapes shown without any kind of reinforcement. For example, octopuses that had learned to attack a small (solid) square, but not attack a large one, 'transferred to' (i.e. attacked) small outline squares significantly more than large outline squares even though attacks on these squares were not rewarded in any way (Sutherland, 1960, 1969). Octopuses show other kinds of stimulus generalisation: in the visual system, size invariance and brightness invariance (Messenger, 1981); and in the tactile system, degrees of roughness (Wells & Young, 1970). Not surprisingly, octopuses also show *receptor generalisation:* after being trained to make a discrimination using part of the retina only, they perform significantly better than chance when the discriminanda are now presented to a different part of the retina (Muntz, 1962). Interocular transfer has also been demonstrated: if, after monocular training on discrimination, the 'untrained eye' is presented with the discriminanda, the octopus continues to attack the correct positive shape and avoid the negative (Muntz, 1961).

The same is true of tactile discriminations: Wells (1959b) trained octopuses to reject an object that was always presented to one arm and then tested the response of a different arm to the same object. He found that adjacent arms, and arms on the other side of the body, could also make the correct response to the object as long as sufficient time had elapsed after training, in practice about 20 minutes.

Cue-additivity has also been demonstrated in octopuses: when two or more cues are made relevant in a visual discrimination, subjects learn significantly faster than when only a single cue is present (Sutherland & Mackintosh, 1971; Mackintosh, 1974). Combinations of size and shape (Sutherland, Mackintosh & Mackintosh, 1965) and of brightness and orientation (Messenger & Sanders, 1972) have been tested. Although all the octopuses learn something about both cues, they do not attend equally to all the features of the stimulus (Box 8.1). When both cues are made relevant in a training situation, some octopuses learn the discrimination in terms of one cue, while some learn it in terms of the other.

Reversal learning by octopuses has been investigated in some detail by Mackintosh and his collaborators (Mackintosh, 1965; Box 8.1). Animals trained to criterion on a discrimination between a positively associated white rectangle and a negatively associated black rectangle were given a discrimination task in which the black rectangle was positive and the white negative. Once criterion was reached, the task was reversed again, and so on. The octopuses made fewer and fewer errors to criterion with progressive reversals (Fig. 8.14); there is evidence that the performance of octopuses in reversal learning depends on earlier training regimes, as in rats. Such findings suggest that some features of concept learning in octopuses are comparable to those of mammals (Mackintosh, 1965; Box 8.1). Reversal learning has also been demonstrated in cuttlefish, in the context of a spatial learning task (Karson, Boal & Hanlon, 2003).

Reversal learning experiments like these suggest that octopuses might be able to form other *learning sets* (Harlow, 1949; Pearce, 2008), but this has not been explored in depth. Boal (1991) began to investigate whether octopuses could form the relative class concept of oddity, but unfortunately it transpired that the apparently positive results obtained at first could have been due to cueing.

8.3.2 Tool Use, Play and Exploration

In a field study conducted over an extended period, Mather (1994) examined the 'homes', or dens, of juvenile *Octopus vulgaris* in Bermuda. She found that octopuses that had occupied a site potentially suitable for a den sometimes modified it in two ways: by removing rocks or sand (using the arms or jets of water from the funnel), and by bringing in fresh rocks or shells, specifically to block the entrance to the den. Such modification often continued over many days of occupancy. The ability to acquire a protective shelter from which to forage is clearly adaptive for any animal, perhaps especially for a vulnerable, soft-bodied one, but whether the kind of modification observed can be designated 'tool use' depends on the definition chosen, as Mather makes clear.

In another octopod, Finn, Tregenza and Norman (2009a) also observed some interesting behaviour that might be termed 'tool use'. The veined octopus, *Amphioctopus marginatus*, occurs in shallow water in Indonesia, and the authors report observations over a 10-year period of more than 20 individual octopuses, which were found occupying empty gastropod shells as well as discarded coconut shell halves. When flushed from a coconut shell by the observing diver, individuals would swiftly re-occupy it. It is perhaps not surprising to find octopuses occupying coconut shells if these are locally abundant because of nearby human habitation. However, the authors observed four animals carrying one or two shell halves for distances of up to 20 m for future use as a den. To do this, the animals were constrained to adopt an unusual and awkward type of locomotion ('stilt walking') that presumably made them more vulnerable to predation. Yet the increased risk seems to be outweighed by the potential future benefit of the shells as shelter. It is also noteworthy that these octopuses on at least some occasions assembled the two shell halves to make the shelter (Fig. 8.15). The authors also observed two octopuses in the act of extracting buried coconut shells from the muddy substrate by using jets from the funnel.

Once again, these activities may or may not constitute 'tool use', but this kind of object manipulation is certainly remarkable and it calls for experimental investigation by workers lucky enough to have access to this species.

Box 8.1 Octopuses and rats: similarities in visual processing

1. NEITHER ATTENDS EQUALLY TO ALL FEATURES OF THE STIMULUS INPUT

Two groups of octopuses were trained on a two-cue discrimination for which size and shape were relevant. One group had been pre-trained on a (different) shape discrimination; the other group had been pre-trained on a size discrimination. Animals in the second group learned significantly less about shape in the two-cue training than animals in the first group (they attended more to the cue they had already been trained to attend to, i.e. size): Sutherland, Mackintosh and Mackintosh (1965).

In another experiment, octopuses that had learned a two-cue discrimination in which brightness and orientation were relevant were given transfer tests. Twenty-two out of 29 animals were found to be attending mainly to brightness; six attending mainly to orientation; one to both: Messenger and Sanders (1972).

2. BOTH SHOW THE PHENOMENON OF 'TRANSFER ALONG A CONTINUUM'

(a)

(b)

Octopuses were trained simultaneously to discriminate between a square and a parallelogram (P5). In discrimination (**a**), which was straightforward, animals scored 54% correct after 17 days' training. In discrimination (**b**) animals were trained for 10 days on the easy discrimination (square versus P1) then trained for one day each on a square versus P2, P3 and P4, before being trained for three days on square versus P5. This group achieved a score of 73% correct. Based upon Sutherland, Mackintosh and Mackintosh (1963), with kind permission from APA.

3. BOTH LEARN REVERSALS FASTER AFTER OVERTRAINING ON THE ORIGINAL TASK

Three groups of octopuses were trained on a brightness discrimination and its reversal; half the animals in each group were overtrained and half were not. One group was given a straightforward discrimination; one group had position as an additional irrelevant cue; one group had orientation as an additional irrelevant cue. The overtrained animals in the two latter groups learned the reversal significantly faster than the non-overtrained ones: Mackintosh and Mackintosh (1963).

4. BOTH LEARN LATER REVERSALS FASTER THAN EARLIER REVERSALS IN A SERIES

Two groups of octopuses were trained to make a brightness discrimination to a criterion of eight out of ten correct responses; the following day the former negative stimulus became the positive and so on (cf. Fig. 8.14): Mackintosh and Mackintosh (1964a).

5. BOTH LEARN A NON-REVERSAL SHIFT MORE SLOWLY AFTER PRIOR OVERTRAINING

A group of octopuses were trained on an orientation discrimination: half were overtrained. When they were then shifted to a shape discrimination, the overtrained animals learned significantly slower than the controls. This (and the other findings described here) could be explained in terms of the switching-in of analysers in the visual system: Mackintosh and Mackintosh (1964b).

For a discussion of these experiments see Mackintosh (1965).

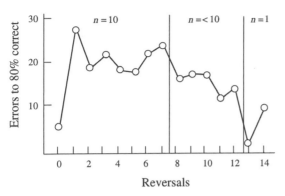

Figure 8.14 Serial reversal learning in *Octopus*. Black and white rectangles were presented simultaneously and animals were trained to a criterion of 80% correct; the following day, the former negative shape was made positive, and so on. Note the improvement as reversal training proceeded (from Mackintosh & Mackintosh, 1964a: *Animal Behavior*, **12**, 321–324 (1964); redrawn by Wells, 1978: *Octopus: Physiology and Behaviour of an Advanced Invertebrate*, Wells, M. J., © (1978), with permission from Elsevier).

Figure 8.15 *Amphioctopus marginatus* occupying coconut shell halves assembled as a shelter. Photographs © Fred Bavendam.

There are many anecdotes about octopuses in aquaria showing exploratory behaviour, and Mather and Anderson (1999) built on such observations to investigate 'play' or 'exploratory play'. In their experiments they presented, without reward, eight octopuses (*O. dofleini*) with a series of novel floating objects (plastic pill bottles). At first, the animals began to palpate the objects with the arms, but later, unsurprisingly, they showed varying degrees of habituation to them. However, two of the octopuses tested repeatedly aimed jets of water at the bottles while viewing them, with the result that the bottles moved around the tank because of the circulatory system, until they returned, to be jetted at again.

Kuba *et al.* (2003) presented *O. vulgaris* with two different objects and attempted to relate the subsequent behaviours to a five-level set of criteria formulated by Burghardt (1999, 2005) for research on animal play behaviour. Kuba *et al.* (2006) continued this approach with *O. vulgaris* and added two different food items for comparison with two Lego objects. The results were mixed, but the authors surmise that 9 of the 14 octopuses showed Level 3 play-like behaviour towards the Lego objects. There is

inherent subjectivity in both the methodology of the five-criteria 'test' and the interpretation of how cephalopod behaviours might fit into those criteria; these factors complicate such 'difficult to prove' sorts of trials.

The authors of these three papers consider these to be examples of play, and compare these behaviours to play behaviour in mammals and even humans; yet without further evidence it seems safer to exercise caution and resist the temptation to attribute too much to these admittedly remarkable molluscs (see Chapter 11).

The need for caution applies even more to claims that octopuses might possess consciousness. Curiously, such claims come not only from psychologists (Mather, 2008a, b) but also from neuroscientists (Edelman & Seth, 2009). As Pearce (2008) puts it, 'it is not possible to observe directly the mental state of an animal', so that, for the moment, it would seem better to avoid speculation about the unobservable.

8.4 Individual Differences in Behaviour: 'Personalities'

In our first edition we noted that little attention had been paid to differences in individual behaviour, and wrote that 'anyone who has ever trained a group of octopuses on a visual discrimination task will know how varied the behaviour of each can be'. Since then, Mather, Sinn and their colleagues have specifically addressed this issue, in two octopuses and the charmingly named southern dumpling squid (*Euprymna tasmanica*).

Mather and Anderson (1993) were the first to refer to 'personalities' in their study of *Octopus rubescens*. They observed over 40 subjects in three situations (described as 'alerting', 'threat' and 'feeding') and used principal component analysis (PCA) to isolate three 'dimensions' of variability that they designated 'activity', 'reactivity' and 'avoidance', which together accounted for 45% of the variance in behaviour.

Sinn *et al.* (2001) went on to study 'temperamental traits' in *O. bimaculoides* during the third week of life. Again using PCA, they identified four components – 'active engagement', 'arousal/readiness', 'aggression' and 'avoidance/disinterest' – accounting for 53% of the variance. They also found changes in the expression of these traits from week 3 to week 6. In both these studies, the authors cautiously attempted to relate their findings with octopuses to those reported in a variety of vertebrates, even in humans.

Sinn and his colleagues have continued with this approach in their studies of *Euprymna*. Sinn and Moltschaniwskyj (2005) claimed that *Euprymna* showed four 'personality traits', whose expression was context-specific, and whose variation was partially dependent on an individual's sexual maturity and size. Sinn, Apiolaza and Moltschaniwskyj (2006) shifted the emphasis onto 'shy–bold axes', as did Sinn, Gosling and Moltschaniwskyj (2008), who analysed behaviours in terms of risk taking. Sinn *et al.* (2010) continued to pursue the study of shy/bold behaviour, which they refer to as a 'behavioural syndrome' in keeping with other literature. Pronk, Wilson and Harcourt (2010), studying *Octopus tetricus*, found that individuals' responses within a given context were consistent within a given day but inconsistent across three test days, a result they termed 'episodic personality'. Clearly there is much yet to be discovered regarding differences in behaviour among individual cephalopods. All these investigations are characterised by sophisticated statistical analyses, and they undoubtedly confirm that there are marked differences in individual behaviour. So far, however, these studies have told us little new about learning in cephalopods, although it may transpire that the dumpling squid will prove a useful model for studying this aspect of behavioural ecology.

8.5 Summary and Future Research Directions

Several examples are now known of changes in cephalopod behaviour that must be attributed to development rather than learning: for example, the development of escape jetting in squid, the development of body patterns and the development of prey capture in *Sepia*.

Newly hatched cuttlefish in the sea feed preferentially on small shrimps or crabs, but in the laboratory such preferences can easily be reversed, sometimes by exposure to chemical cues alone, even in late embryonic stages. Learning not to attack prey in a glass tube is difficult for hatchling cuttlefish, but they become progressively better at this task as they grow older; this ability is correlated with the gradual development in the brain of the vertical, superior frontal and subvertical lobes, and their connecting tracts.

Learning has been studied most extensively in *Octopus*, but the cuttlefish, *Sepia*, is becoming an increasingly useful model for studying learning in cephalopods. Octopuses have been shown to exhibit habituation, associative learning (notably classical conditioning) and possibly social learning. They also show spatial learning, in the sea and in the laboratory, which is understandable in terms of their life style. Recently, spatial learning in *Sepia* has attracted some attention from researchers, although it is not obvious how relevant such learning is to life in the sea.

For those interested in studying the physical basis of learning and memory in animals, the octopus brain continues to attract investigators (see Chapter 2). Currently, the emphasis is still on the vertical/superior frontal lobe system, which is important in establishing and maintaining visual memories. However, the *Octopus*

Box 8.2 Relative sizes of memory areas in some cephalopods

A crude measure of brain function is size. Wirz (1959) and Maddock and Young (1987) have measured the volumes of different lobes of the brain in over 60 species of cephalopods, revealing differences that can often be related to function and habit but may also reflect phylogeny.

The parts of the cephalopod brain involved in establishing and storing memories are, for visual learning, the optic lobes and the vertical/superior frontal lobe system; and, for tactile learning (in octopods only), the inferior frontal/subfrontal system (Chapter 2).

The figures below (based on Maddock & Young, 1987; Nixon & Young, 2003) are the volumes of these areas expressed as percentages of the total brain volume.

Lobes involved in visual learning (all orders)

Genus	Optic lobes (%)	Vertical lobe system (%)
Octopus vulgaris	85	13
O. bimaculatus	115	18
Sepia	142	24
Loligo	336	21
Illex	287	14
Taonius	449	14
Cranchia	610	16
Vampyroteuthis	378	28

Lobes involved in tactile learning (octopods only)

	Inferior frontal system
O. vulgaris	8.5 %
O. bimaculatus	9.6 %
Benthoctopus	10.7 %

These data show that, on grounds of size alone, there is no reason to believe that other cephalopods are less 'intelligent' than *Octopus vulgaris*; quite the reverse, indeed. It is interesting that *Vampyroteuthis*, which is restricted to life in deep water, has surprisingly large optic lobes and a well-developed vertical lobe system.

brain also contains a separate, compact touch learning system, first investigated by Wells and Wells in the 1950s, and the subject of some elegant experiments in the 1990s by Robertson and his colleagues (Chapter 2). This seems to have been overlooked by physiologists despite its great potential.

Given the compelling experimental evidence that *Octopus*, *Sepia* and two or three other species of cephalopod can learn quickly in the laboratory, it seems obvious to ask: can all the other 750 species of cephalopods learn? We only know that the parts of the brain that are involved in storing and setting up memories in *Octopus* are present and well developed in all of the 62 species of coleoid cephalopods that were examined by Maddock and Young (1987; Box 8.2). Indeed, the measurements of these authors reveal that the 'memory lobes' of many cephalopods often constitute a larger proportion of the brain than they do in *Octopus*. The inference must be that learning is likely to be important to all coleoids, emphasising the need for field observations and experiments (however difficult) on learning in oceanic and deep-water species.

This is not to say that laboratory studies have nothing more to tell us about learning in these animals. Looking back at the elegant psychological experiments of Mackintosh in the 1960s on *Octopus*, we can see that, for anyone interested in cognition, these studies stopped just when they were becoming interesting! The evidence that octopuses appear capable of some form of social learning is also remarkable; attempts to show this in cuttlefish have been less clear so far and further experiments with other cephalopods – especially reef squid – would be most welcome.

Finally, there are a number of observations in the literature of octopuses doing apparently complex things that could be interpreted as exploration, play or even tool use. Fascinating though these findings are, it may be useful to remind future students of cephalopod behaviour of the importance of exercising restraint when interpreting the behaviour of these molluscs in terms that have to be used with caution, even with higher vertebrates.

CHAPTER NINE

Ecological Aspects of Behaviour

This chapter focuses on life cycles, shoaling, migration and other distinctive cephalopod behaviours not covered elsewhere.

9.1 Life Cycles

As far as we know, all living coleoids have a single breeding season and a short life span: they 'live fast and die young' according to the aphorism that so vividly describes their life cycle (O'Dor & Webber, 1986). Current evidence indicates that most cephalopods complete their life cycles in 1–2 years (Boyle, 1983b; Boyle & Rodhouse, 2005). The exceptions are species that live in very cold water (near the poles or in deep water) or grow very large. For example, the octopus *Bathypolypus arcticus*, living in 4 °C water, apparently completes its cycle in 4 years (O'Dor & Macalaster, 1983). The giant squid *Architeuthis* (Roper & Boss, 1982; Jackson, Lu & Dunning, 1991) and the very large *Enteroctopus dofleini*, also seem to survive up to 5 years (Hartwick, 1983).

A single terminal breeding season is characteristic of coleoids, though not of *Nautilus* (Chapter 10). This is unusual for large animals, but apparently few cephalopods consistently exhibit features of r- or K-selected life cycle strategies (Boletzky, 1981). However, doubts have been expressed about the usefulness of the concept of r/K selection to understanding life histories (Stearns, 1992), and Boyle and Rodhouse (2005) consider that a 'theoretical framework for the cephalopod life cycle is perhaps premature'. These authors describe several cephalopod life cycles, and they offer a useful discussion of the concept of semelparity and how it applies to reproduction and short life cycles in coleoids.

There are many variations of life cycles among cephalopods, making generalisations difficult, even within Orders. For example, in certain species there is plasticity in the onset and duration of maturity and spawning, resulting in some individuals being in a mature condition throughout the year. However, with annual death occurring in most species, it is difficult to assign a standard life cycle descriptor to this scenario when strict seasonality of reproduction is not occurring. Nevertheless, the seemingly universal characteristic of endogenously determined death relatively soon after spawning is a basic tenet of cephalopod biology, albeit one in need of greater attention by experimental biologists, ecologists and life history theorists.

9.2 Shoaling and Social Organisation

Social organisation (defined and reviewed in Hinde, 1970; Wilson, 1975) is weakly developed or non-existent in cephalopods, except for that related to reproduction (Chapter 6). Cuttlefish and octopuses are basically solitary animals. Most squid species shoal during much of their life, but there is no cooperative hunting for prey and no clear division of labour among individuals for activities such as defence, much less home building or care of the young. In short, there are no signs of any of the intricate long-lasting social relationships such as those found in birds, mammals or some insects.

9.2.1 Octopuses and Cuttlefish

Most field studies suggest that octopuses are solitary in the wild (Aronson, 1986; Forsythe & Hanlon, 1988). In the laboratory, too, individuals usually avoid each other, although species differ: for example *Octopus briareus* is highly intolerant of conspecifics and often cannibalises them, whereas *O. bimaculoides* is tolerant of crowding and is seldom cannibalistic even under crowded laboratory conditions (Hanlon & Forsythe, 1985). The octopus *Eledone moschata* has been called 'social' (Mather, 1985), but this reflects only a relatively higher degree of tolerance among individuals in a laboratory environment. A few laboratory studies have claimed that octopuses form linear, size-based dominance relationships (Van Heukelem, 1977; Boyle, 1980; Mather, 1980; Cigliano, 1993), but such behaviour could be an artefact of the unnatural conditions. In most of these studies (as well as in some field observations mentioned below) the larger individual retained or gained use of the den, a situation common in many solitary animals. Intraspecific fights are described in §9.4. However, Huffard and Caldwell (2008) have discovered natural aggregations of the octopus *Abdopus aculeatus* in shallow habitats in the tropical Indo-Pacific, and Godfrey-Smith and Lawrence (2012) have reported high-density occupation of a field site by *Octopus tetricus* in southern Australia. Thus, there may be a few octopus species that are not strictly solitary for all of their life cycle.

Territoriality is unproven. In field studies in the Bahamas, *O. briareus* has been observed engaging in 'den defence', but the area defended extended only a few centimetres from the lair, leading Aronson (1986) to conclude that they are not territorial. In large outdoor ponds in Hawaii, the smaller *O. cyanea* generally retreated from larger conspecifics whether in dens or in the open, yet there was no evidence of territoriality (Yarnall, 1969). Field studies of *O. vulgaris* in the Mediterranean and in Bermuda have indicated no signs of territoriality or dominance relationships (Altman, 1967; Kayes, 1974; Mather & O'Dor, 1991), and field studies of *O. joubini* (Butterworth, 1982; Mather, 1982) and *O. bimaculatus* (Ambrose, 1988) have also failed to demonstrate such evidence. Field studies of *O. dofleini* indicate that this species may show intraspecific aggression related to den utilisation; no experiments have been performed, but these observations appear to indicate den defence rather than territoriality (Hartwick, Breen & Tulloch, 1978; Kyte & Courtney, 1978).

Although most octopuses are asocial, a few species have been seen to occur in clumped aggregations in the sea. It is not always clear whether the clumping is due to habitat restrictions, abundant prey, den availability or some form of social organisation. In *Vulcanoctopus*, an aggregation caused by swarming crustaceans has been

observed (Voight, 2005; see Chapter 4). In some locations, *O. vulgaris*, *O. joubini* and *O. bimaculoides* show non-random distributions that seem correlated with clumping of available dens (e.g. Kayes, 1974; Guerra, 1981; Butterworth, 1982; Mather, 1982; Forsythe & Hanlon, 1988). In Panama, the Larger Pacific Striped Octopus (as yet unnamed) has been observed living in 'colonies' of 30–40 individuals at a depth of 9 m; holes or lairs were spaced about 1 metre apart, and three holes were occupied by two octopuses each (Moynihan & Rodaniche, 1982; Rodaniche, 1991). In California, *O. rubescens* has been observed living in clumped dens in the mud, with animals about 1 m apart. Another highly unusual behaviour of *O. rubescens* is that the young sometimes form large shoals that move throughout the water column, appearing and behaving like a shoal of fish or squid. This phenomenon has been observed regularly by submersibles in Monterey Bay, California, but its function is unclear.

The cuttlefish and sepiolids appear to tolerate conspecifics rather than show any attraction for them (except, again, at reproduction). Many are solitary for much or all of their life cycle. In the laboratory, cuttlefish and sepiolids (e.g. *Sepia*, *Sepiola*, *Sepietta*, *Rossia*, *Euprymna*) seem tolerant of one another except under extreme food deprivation (Boletzky *et al.*, 1971; Boletzky & Hanlon, 1983). Field observations are few, but they generally corroborate this (Mangold-Wirz, 1963; Moynihan, 1983a; Hanlon & Messenger, 1988). In one laboratory study of social interactions in *Sepia officinalis*, Mather (1986a) failed to demonstrate dominance relationships among animals. However, Boal (1996) found a male dominance hierarchy in this species, apparently related to size. She obtained no clear evidence for social recognition; although there is mate guarding, there is no recognition of individual mates. Boal and her colleagues also studied the effects of crowding on laboratory-cultured *S. officinalis*, in the course of which they described how the animals spaced themselves to avoid one another (Boal *et al.*, 1999). This would agree with a field study of *S. latimanus* by Corner and Moore (1980), who also suggest that these cuttlefish are likely to be solitary.

9.2.2 Squid

Following Pitcher's descriptions of fish behaviour, we term aggregations of squid that remain together for social reasons *shoals*, much as the term flock is used for birds. Synchronised and polarised swimming groups are termed *schools*. Schooling is thus only one of the behaviours shown by squid in shoals (Kennedy & Pitcher, 1975; Pitcher, 1983; Pitcher & Parrish, 1993). Figure 9.1, based on Pitcher (1983), illustrates some distinctions between these terms.

Shoaling is a common behaviour of many squid species and it is extremely common amongst fish, about 10 000 species of which are estimated to shoal at some time in their lives (Shaw, 1978). Schooling seems to be far more restricted in both groups: in squid, it is restricted mainly to various loliginids and to some neritic oegopsids. The California market squid, *Loligo opalescens*, begins shoaling as soon as it is large enough to swim against a current (about 8–11 mm ML, or 6–7 weeks old; Yang *et al.*, 1986) and apparently shoals day and night for the duration of its life (Fields, 1965; Hurley, 1976, 1978). Investigation of the structure of *schools* of this species revealed that squid maintain a small mean angular deviation relative to one another (0–32°: random deviation would be 69°). The schools are usually composed of individuals of the same size, and vision seems to be important in forming them, at least in the laboratory (Hurley, 1978). The same is true of the neritic Northern shortfin squid *Illex illecebrosus* (Mather & O'Dor, 1984). Both species show many similarities to fish in their schooling: there is a specific spatial organisation that (1) can be described by distance and angular deviations to other individuals, (2) is affected by group size, and (3) is largely a result of visual attraction (Pitcher & Partridge, 1979). Blinded fish with intact lateral lines can still school, however (Pitcher, Partridge & Wardle, 1976; Partridge & Pitcher, 1980), and the discovery of a lateral line analogue in cephalopods suggests that it might be profitable to re-examine the mechanisms by which squid maintain their spatial relationships when schooling (Budelmann & Bleckmann, 1988; Budelmann, Riese & Bleckmann, 1991). Shoal size varies greatly, but can be huge, into the thousands or tens of thousands in *Loligo vulgaris reynaudii* in South Africa and *Illex illecebrosus* off northeast North America (Squires, 1957; Sauer, Smale & Lipinski, 1992).

The shoaling behaviour of the Caribbean reef squid, *Sepioteuthis sepioidea*, has received much attention, notably by Moynihan and Rodaniche (1982). Unlike most squid, this species commonly swims near the bottom among corals and seagrasses in shallow water. Its shoals are highly variable in size and shape (from linear to spherical; Fig. 5.24); large and small squid of both sexes are present in the same shoals, unlike any other cephalopod; the squid occasionally shoal with another loliginid species, *L. plei*; 'vigilance' is thought to be maintained by sentinels placed throughout the shoal; and information about predators appears to be passed among members of the shoal.

The hatchlings of *Sepioteuthis* appear to be asocial and are often solitary, but they quickly become gregarious, beginning to shoal within 2 weeks of age at a mantle length of 10 mm (LaRoe, 1971; Lee *et al.*, 1994). In the sea, well-formed shoals of about 100 very small individuals have been observed over the seagrass *Thalassia* (Moynihan & Rodaniche, 1982); these individuals are probably only a few weeks old. Generally, a shoal comprises 20–40 squid, although it is not uncommon to see as few as 2–5 squid or the occasional large shoal of 100–200 individuals (Hanlon & Forsythe, unpublished data). This species apparently forms polarised schools only when approached closely or attacked by predators, or when the young are travelling high in the water column.

There are some obvious social interactions within shoals of *S. sepioidea* (Moynihan & Rodaniche, 1982). Large individuals are joined and followed by smaller individuals more frequently than the reverse, and small individuals commonly give way to larger individuals. Thus, there seems to be a subtle relationship based upon size. There is no evidence of any sort of individual leadership. Many complex social interactions have been noted among *Sepioteuthis*, including cases where small subgroups dispersed by predators have been guided back to larger shoals of squid (Hanlon & Forsythe, unpublished data). Social fidelity is maintained only loosely, probably for days rather than weeks or months. Agonistic and courtship behaviour among mature individuals within the shoals do not appreciably change the dynamics of the shoal; these sexual activities occur throughout the day, with an intense peak one hour before dark (see Fig. 9.9), and it is noteworthy that courtship often occurs immediately after the shoal has had an interaction with a predator. It may be that group vigilance is increased at this time, and the few sexually active individuals take advantage of this situation to engage in courtship for 1 to 2 minutes when 'the coast

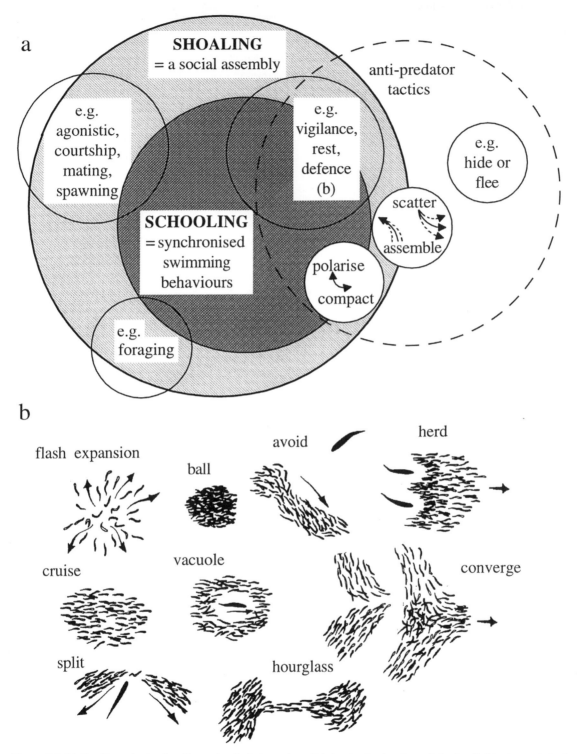

Figure 9.1 (**a**) Shoaling and schooling. The main criteria for each are listed: schooling is a form of shoaling that involves polarised, parallel swimming in the same direction at the same speed. Shoals may be diffuse or compact and have various shapes; circles describe examples of behaviour. Modified from Pitcher (1983): *Animal Behaviour*, **31**, 611–612, Heuristic definitions of fish shoaling behaviour, Pitcher, T. J. © (1983), with permission from Elsevier. (**b**) Examples of anti-predator tactics shown by sand eels which could apply equally well to squid. Modified from Pitcher & Wyche, 1983: *Predators and Prey in Fishes* (ed. D. L. G. Noakes, D. G. Linquist, G. S. Helfman & J. A. Ward), 193–204, © (1983), with permission of Springer.

seems clear'; this may be significant since these shoals have, on average, seven interactions per hour with predators in the site investigated (based upon nearly 1000 observations at Little Cayman Island; Hanlon & Forsythe, unpublished data). Mather (2010) noted a similar level of shoal interruption in Bonaire. No cooperative hunting has been noted, but this species hunts individually and almost exclusively at night. As in many fish (Radakov, 1973), shoals of *Sepioteuthis* generally disband at dusk and re-form at dawn; many species of *Loligo*, in contrast, shoal both by day and by night.

'Vigilance' (§5.1.6) has been described in *Sepioteuthis sepioidea* by Moynihan and Rodaniche (1982), Hanlon and Forsythe (extensive unpublished field data) and Mather (2010). There are reports of what appear to be 'sentinel' squid that may be large or small and appear to take turns in this role. Sentinels are usually found at the ends of linear 'picket lines' and near the middle as well, and collectively they face in multiple directions (Fig. 5.24). Video-taped sequences of reactions to predators suggest that squid react within a fraction of a second to the responses of a squid that responds to the approaching predator. The nature of the alarm signal is unknown (Chapter 7). This aspect of predator detection and transmission of predator information appears to be highly effective: in one field study, Hanlon and Forsythe (unpublished data) saw no successful attacks by fish during the nearly 1000 interactions. However, Mather (2010) questions the presence of sentinels in her study at Bonaire; future quantified field or laboratory studies could help to clarify whether *S. sepioidea* possess this defensive shoaling behaviour.

It is curious that a related species, *S. lessoniana*, from coastal waters in the Indo-West Pacific Ocean, shows little sign of complex social behaviour, at least when it is reared and observed in captivity (Boal & Gonzalez, 1998).

Recently, the shoaling behaviour of the jumbo squid, *Dosidicus gigas*, has been the subject of a detailed field study, using sonar localisation (Benoit-Bird & Gilly, 2012). This has revealed that, in this species at least, shoaling behaviour is not fixed or stereotyped. The authors observed two types of aggregation: 'incredibly dense groups in which individual squid could not be acoustically localized' and 'looser, highly coordinated groups of several individuals (up to 40) following parallel paths upward from the dense groups'. The function of the former is not clear, but the looser,

'breakaway' aggregations are probably for foraging. It is noteworthy, however, that this important study has revealed no evidence that shoaling by jumbo squid conferred any benefit in foraging (Chapter 4).

This brings us to the possible functions of shoaling. Presumably it confers some of the same advantages as in fish (Pitcher, 1986; Pitcher & Parrish, 1993; Krause, 1994; Hoare & Krause, 2003), notably in reducing predation, but there have been no experimental studies on the functions of shoaling in cephalopods. Moynihan and Rodaniche (1982) suggest that *Sepioteuthis* may have vigilant 'sentinels' (see §5.1.6), and evidence for or against this would be welcome, as would investigation of the way the various shoal shapes (Fig. 9.1b) are employed for secondary defence against predators. Shoaling must facilitate finding and selecting mates (Chapter 6), and schooling may also offer opportunities for observational learning (§8.2.4), especially in species in which old and young individuals school together. Moynihan and Rodaniche (1982) even suggest that shoaling and observational learning in *Sepioteuthis sepioidea* could foster the development of group or local traditions, which could be useful in a short-lived species. Lastly, it is possible that shoaling confers an advantage by providing conspecifics to be cannibalised during long horizontal migrations (see §9.5).

9.3 Interspecific Associations

There are only two known examples of symbiotic relationships involving a cephalopod. First, there is cleaning by fish, known in *Octopus cyanea* (Fig. 9.2a), and in *O. vulgaris*, which Johnson and Chase (1982) saw in the Caribbean being cleaned by the goby *Gobiosoma*. These gobies are obligate cleaners on coral reefs; the goby entered the funnel of the octopus twice, probably to remove ectoparasites as it does on many reef fish. Other invertebrates are not known to take advantage of such a cleaning service, and it may be a learned behaviour on the part of the octopus. Second, there are reports of small octopuses of the genera *Ocythoe* and *Argonauta* sometimes occurring inside the gelatinous bodies of pelagic salps (Jatta, 1896; Banas, Smith & Biggs, 1982; Okutani & Osuga, 1986). The occurrences have simply been noted, but no behavioural evidence or experimentation is available to substantiate the nature of this interaction, although Banas, Smith and Biggs (1982)

Figure 9.2 (**a**) *Octopus cyanea* adult being cleaned by the wrasse, *Labroides dimidiatus*, on a coral reef in Palau, Western Micronesia (photo R.T. Hanlon). (**b**) A foraging *O. cyanea* on an Indo-Pacific coral reef, being followed by scavenging fish (photo © Fred Bavendam).

noted that the benefits might include food (there are commensal amphipods inside the salps) as well as transport and shelter.

On Caribbean reefs, Moynihan and Rodaniche (1982) described a very common association between the squid *Sepioteuthis sepioidea* and two species of goatfish. The goatfish and their occasional associates (parrotfish, surgeonfish and wrasses) forage on the substrate for small invertebrates and plants, and the squid hover near them (usually within 0.5 to 3 m) for up to several hours. The nature of the relationship is unclear, but the authors suggest that goatfish use the squid as sentinels while they grub into the substrate for food. The squid do not react to the fish and do not feed upon them or upon the small organisms that they often stir up, so that the benefit to squid is unclear.

Different squid species occasionally mix within the same shoal, at least for brief periods. *Sepioteuthis sepioidea* have been seen mixed with *Loligo plei* in Panama, in the Bahamas and in the Cayman Islands (Moynihan & Rodaniche, 1982). Also in the Caribbean, the small *Lolliguncula brevis* has been seen with *L. plei* and *L. pealeii* around night-lights, and all three have been captured in the same trawl tow (Cohen, 1976; Hanlon, Hixon & Hulet, 1983).

Foraging octopuses are often followed by scavenging fish (Fig. 9.2b). *Octopus vulgaris* in the Mediterranean is often followed by the grouper *Serranus scriba;* Kayes (1974) noted that this made it easier for researchers and fishermen to find octopuses and speculated on whether it could help other predators too. In Bermuda,

O. vulgaris juveniles are often followed by a variety of small fish that feed upon prey stirred up by, or escaping from, the octopuses (Mather, 1991b). On coral reefs in the Red Sea, no fewer than five grouper species, two goatfish species, two scorpionfish species and one coronetfish species have been observed following foraging *O. cyanea* and *O. macropus* and feeding upon escaping prey (Ormond, 1980; Diamant & Shpigel, 1985). In one observation, four grouper species were recorded following a single octopus simultaneously, and apparently competing for prey flushed by the octopus. On coral reefs in Polynesia, *O. cyanea* is also followed by groupers and wrasses (Forsythe & Hanlon, 1997); on one occasion, an octopus sitting at its den feeding on a large crab was seen being approached by several fish that darted in to take pieces of the crab. On tropical rock reefs in the Sea of Cortez, Strand (1988) recorded several species of grouper wrasse and hawkfish following an unidentified octopus on 31 foraging occasions. In colder temperate waters, Hartwick and Thorarinsson (1978) reported groupers and sculpins scavenging remains in middens of *O. dofleini*. R.F. Ambrose (pers. comm., 1993) has observed fish following *O. bimaculatus* in a kelp habitat in California. All these incidents suggest that the presence of conspicuous fish might sometimes adversely affect an octopus by drawing attention to it.

9.4 Competition for Resources

Competition arises when resources are too scarce to support all the progeny of any given species. An individual's competitive success is determined not only by its physical abilities but also by the behavioural tactics it uses, and those tactics often depend upon what its competitors do. One of the concerns of behavioural ecology is to define and measure the interference that competitors impose on an individual's search for such resources as shelter and food. The term 'interference' has been used in a variety of contexts to describe how the density of competitors affects the rate at which resources are found and exploited; it is often defined as the reduction of a given individual's success rate as a result of competition. The mechanisms of interference include: exploitation competition, scramble competition and contest competition (Milinski & Parker, 1991).

Temporal spacing and niche partitioning are ways in which sympatric cephalopods reduce interference from one another. Partitioning of activity periods, microhabitat preferences, foraging methods and food choices help to separate them on a daily basis; such resource partitioning is a widespread phenomenon among animals (Schoener, 1974). On Hawaiian coral reefs, *Octopus cyanea*, *O. ornatus* and the crescent octopus (unnamed) maintain species-specific temporal spacing (Fig. 9.3) and microhabitat preferences (i.e. coral

heads, rubble/sand and tide pools, respectively; Houck, 1982). On Caribbean coral reefs, *O. vulgaris*, *O. briareus*, *O. macropus*, *O. burryi* and *O. defilippi* can co-exist on the same reef; *O. vulgaris*, *O. briareus* and *O. macropus* may occasionally all be seen on a single SCUBA dive. Their interactions are apparently infrequent, presumably because they forage in different ways and at different times, have different prey, and reside in different types of den (Hochberg & Couch, 1971; Hanlon, 1988). It is possible that temporal spacing also reduces the chances of predation among cephalopods. These observations merely suggest that competition is reduced by such mechanisms, but more data and experiments are needed before anything more definite can be stated. In the Caribbean Sea, as pointed out by Aronson (1986, 1989, 1991), it appears that predation keeps octopus populations so low that niche partitioning may not be so important in reducing interference between sympatric octopuses.

Intraspecific competition for *dens* (or *shelters*: both terms are used) is obviously important for octopuses , and there have been several studies of den selection, the most detailed of which are those on *Octopus joubini* (Butterworth, 1982; Mather, 1982), *O. briareus* (Aronson, 1986) and *O. tehuelchus* (Iribarne, 1990) in the western Atlantic; and *O. bimaculatus* (Ambrose, 1982), *O. dofleini* (Hartwick, Ambrose & Robinson, 1984) and *O. bimaculoides* (Cigliano, 1993) in the eastern Pacific. Some species have rather specific requirements for den structure (usually a constricted entrance, and a volume similar to that of the occupier; see Fig. 5.16), and octopus abundance is partly restricted by the availability of such dens (Butterworth, 1982; Mather, 1982; Aronson, 1986; Iribarne, 1990). Anderson *et al.* (1999), studying the use by juvenile *O. rubescens* of beer-bottle dens in Puget Sound, found that they preferred bottles that were dark brown or covered with marine growth, which enabled the octopuses to inhabit a sandy or muddy habitat lacking in shelter. Katsanevakis and Verriopoulos (2004) studied the different kinds of dens made by *O. vulgaris* in soft sediment in the Mediterranean, where again paucity of materials limits the distribution of octopuses. The shelters used by *O. tetricus* in New Zealand have been studied by T.J. Anderson (1997), who observed that brooding females were more likely to modify their den than non-brooding females or males, always

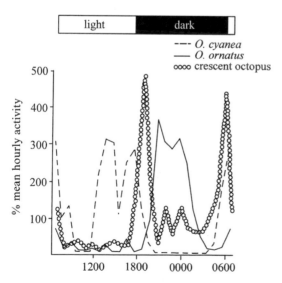

Figure 9.3 Temporal spacing in three sympatric species of shallow-water Hawaiian octopods: the diurnal *O. cyanea*, the nocturnal *O. ornatus* and the crepuscular crescent octopus (modified from Houck, 1982: image from the Biodiversity Heritage Library).

barricaded the den more completely and were always located deep within it, never exposed at the entrance: again, evidence that behaviour in cephalopods is rarely simple. In a 2-year study of the same species in Australia, Godfrey-Smith and Lawrence (2012) found evidence that long-term occupation might have actually modified the site studied so that it came to support a denser population of octopuses. Originally there had been a single solid object in the area suitable for a den, but, as large numbers of scallop shells were amassed by foraging octopuses, the area suitable for providing shelter increased: the shell bed permitted burrowing, and some animals were found in shafts as deep as 20 cm, lined with shells.

Fighting among octopuses mainly involves arm wrestling, although it has not been studied in detail. Yarnall (1969), who observed 19 encounters by *O. cyanea*, noted that, after visual contact, one or both advanced slowly with an arm (usually one of the first pair) extended. With brief contact of the arms, or slightly before, one octopus retreated, indicating that some visual assessment might have occurred. Packard and Sanders (1971) illustrated an encounter of *O. vulgaris* in which the arms and suckers of each animal were aligned, and Boyle (1980) described agonistic bouts in the laboratory in which they engaged each other in this position 'trying to force the other backwards'. Boyle (1980) speculated that this 'arm alignment' may represent a method of assessing relative size of the opponent, but this has yet to be tested. Hanlon and Wolterding (1989) observed that *O. briareus* often showed the Passing Cloud pattern after visual contact, and then the initiator extended an arm to touch the other. The second octopus often withdrew and moved into the 'protective posture', but sometimes they both entangled arms in an apparent attempt to manoeuvre into top position. The eventual winner would wrap its arms around the mantle of the octopus underneath and restrict its breathing. Cigliano (1993) described different fights in which *O. bimaculoides* in tanks engaged in full contact, or web-to-web fighting, or would 'leap' at another octopus, 'whip' an arm rapidly at an opponent, 'move' rapidly at an opponent by jetting, or 'chase' an opponent that might actively 'retreat'. No specific body patterns have been mentioned in these reports, but *O. cyanea* shows a bold striped display during intraspecific encounters both in the laboratory (Van Heukelem, 1966;

Wells & Wells, 1972) and in the field (Forsythe & Hanlon, 1997).

Interspecific competition with other benthic species may influence the abundance of octopuses. In the laboratory, *O. joubini* prefer to make their dens in gastropod rather than bivalve shells; yet in the field they were found more often in bivalve shells, partly because these were more abundant, but also because gastropod shells were often inhabited by competitors such as fish or hermit crabs (Mather, 1982). The stomatopod shrimp *Gonodactylus* also competes vigorously with *O. joubini* for shelters in the laboratory, and possibly in the field too (Caldwell & Lamp, 1981).

Octopus briareus also has rather stringent den requirements, but it is probably fish predation that limits its abundance and distribution on Caribbean coral reefs. Aronson (1986, 1989, 1991) concluded that release from fish predation in a Bahamian salt lake is the probable cause for the high density of *O. briareus* compared with that on 'normal' reefs on the island of Eleuthera. In den enrichment experiments on normal reef environments, octopus density did not improve; instead, small fish (*Paraclinus*, *Coryphopterus*, *Astrapogon*) occupied most of the artificial dens (Aronson, 1986, 1991). There is a good deal of interspecific competition for available dens (Butler & Lear, 2009).

Elsewhere, there is also evidence that teleosts limit octopus populations. Heavy fishing pressure on sparids and other demersal fish off the Saharan Bank (West Africa) is correlated with increases in octopus populations. Bas (1979) attributes this to a reduction in competition for food between sparids and octopuses, but it could partly be a reduction in competition for dens; Caddy (1983) attributes the increase to reduced predation by fish. It is also possible that it is because of reduced predation on paralarvae by sparids (Gulland & Garcia, 1984).

Competition in cuttlefish and squid has not been studied. Squid in all habitats certainly have interactions with fish, but the nature of these competitions is unknown. In the northern Gulf of Mexico, the loliginid squid *Loligo pealeii*, *L. plei* and *Lolliguncula brevis* are sympatric, but differences in peak activity time, feeding preference and shoal structure appear to segregate them, although at times all three species are sometimes caught in the same trawl (Hixon, 1980; Hixon *et al.*, 1980; Hanlon, Hixon & Hulet, 1983).

9.5 Dispersal and Horizontal Migration

Most adult molluscs, like many other marine invertebrates, are relatively sluggish or sedentary creatures, so they rely on *larvae* for dispersal. Such larvae, which are usually very different in form from the adult, feed and travel within the plankton, before ultimately settling to the bottom. Such a dispersal phase is unnecessary in some mobile coleoids that undertake long migrations as adults. Other coleoids have developed paralarvae that facilitate dispersal early in the life cycle. Yet other species only produce very large hatchlings, and dispersal is limited to latter stages of the life cycle and to the locomotory capabilities of the adult forms.

9.5.1 Paralarval Dispersal

Cephalopods do not have a true larva; that is, there is not a distinct metamorphosis. Some cephalopods, however, pass through a 'paralarval' stage. Paralarva is a term defined by behavioural and ecological characteristics: some hatchlings are pelagic in near-surface waters during the day and have a distinctively different mode of life from older conspecifics (Young & Harman, 1988). 'Day' is stipulated in the definition because near-surface waters are often occupied by older conspecifics during the night (because of vertical migrations, §9.6). By this definition, paralarvae do not necessarily have major morphological differences from older conspecifics, although some do (Sweeney *et al.*, 1992). *Octopus vulgaris* hatchlings, for example, look more like tiny squid than octopuses, so they might be considered a larva.

A benthic octopus that hatches as a planktonic paralarva first behaves as a squid: it swims in the plankton, then begins to move daily between the plankton and the seabed, until finally it settles on the bottom, only swimming occasionally. *In situ* observations of this transition have been made only in *O. defilippi* (Hanlon, Forsythe & Boletzky, 1985). Many open-ocean squid also exhibit different behavioural and morphological adaptations during their paralarval stage: they inhabit near-surface waters during the day and descend at night, the opposite of the adult habit (§9.6;

also Young & Harman, 1988). These represent diverse behavioural adaptations to changing life styles. Paralarval squid often have a different chromatophore distribution from the adults, different body proportions, and modified arms or tentacles.

Conversely, other cephalopods produce fewer, large progeny that are developmentally more advanced and hatch out as 'miniature adults'. For example, among shallow-water octopuses, *O. briareus* lays only 200–500 large eggs, which grow slowly and develop into miniature adults that immediately assume the benthic life style of the adults (reviewed in Boyle, 1983b). Off the California coast, two sympatric and nearly identical species *O. bimaculatus* (small eggs) and *O. bimaculoides* (large eggs) share many habitats but have very different life history strategies (Pickford & McConnaughey, 1949; Ambrose, 1988). How these large-egged species with benthic hatchlings achieve dispersal remains an unstudied subject.

9.5.2 Migration Patterns

Cephalopods are widely distributed in the seas of the world although, for reasons that are not clear, they have never succeeded in colonising fresh water or land-locked seas of low salinity. Only a few can tolerate even brackish or estuarine water: the bay squid, *Lolliguncula brevis*, is the best known (Hendrix, Hulet & Greenberg, 1981; Bartol, Mann & Vecchione, 2002).

Their distribution and zoogeography are beginning to be understood more fully as data become available (e.g. Jefferts, 1988). Mangold (1989) gives a useful summary account: according to her, about 53% of species are exclusively Indo-Pacific, 29% are confined to Atlantic, 6% to South African, and 4% to Antarctic Australian waters. About 8% of species are circumtropical or found in several oceans. The areas with the richest cephalopod fauna in the world include the West Pacific, the seas around Japan and Indonesia, where there are myopsids, sepioids and octopods in abundance although apparently fewer oegopsids, the North Pacific, with numerous oegopsids and octopods, and the Gulf of California, with many endemic species of *Octopus* (Mangold, 1989). Areas of the world where some cephalopods are lacking include the Americas (no sepioids on either coast), and the Antarctic (no myopsids, including loliginids). We do not know why, especially since the energy-budget studies of O'Dor (1988) suggest that large squid can make lengthy migrations very quickly. According to Clarke (1966), most oceanic squid seem able to tolerate a relatively wide range of temperatures and even some small variations in salinities, at least for brief periods.

Mangold (1989) points out that the most widely distributed cephalopods are either pelagic, or benthic with planktonic young (Octopodinae). The young stages, which move in the major currents of the oceans, must be important for the dispersal of many cephalopods, including strongly swimming pelagic forms, many of which have paralarvae that live near the surface. For example, in the North Atlantic, young *Illex illecebrosus* are thought to drift north for thousands of kilometres in the Gulf Stream to feeding grounds off Newfoundland (Fig. 9.4a). The adults of this species swim in the opposite direction in the autumn to spawn south of Cape Hatteras. Thus, as in several other squid, the adults swim against the currents that are taking their young in the opposite direction (Amaratunga, 1987).

Following Kennedy (1986), we can define migration as 'persistent and straightened-out movement effected by the animal's own locomotory exertions or by its active embarkation on a vehicle'. The vehicle for a cephalopod is, of course, a horizontal current, and, for many oegopsids where data are available, migrations have been found to be associated closely with movements of water masses (Clarke, 1966). This is illustrated in Fig. 9.4. Such movements may be associated with reproduction (e.g. *Ommastrephes*, Clarke, 1966) or with following prey. Sometimes they are inshore/offshore; sometimes they occur in the open oceans, and then over distances of hundreds or even thousands of kilometres. Some of the octopods, however, especially those laying large eggs (Iribarne, 1991) can be termed 'sedentary' (Mangold, 1987). Yet tag-and-release studies make it clear that many cephalopods regularly travel medium to long distances (relative to body length), for example, *Illex* around Newfoundland (Hurley & Dawe, 1980).

Sustained, large-scale migrations, for which many squid seem well fitted both physiologically and biochemically (O'Dor & Webber, 1991), have been documented in varying degrees of detail in several squid: here we restrict our detailed discussion to three

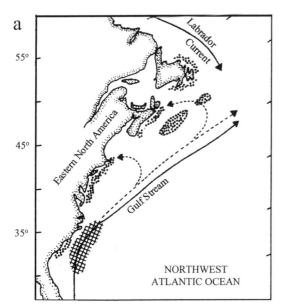
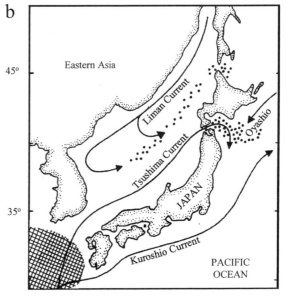

Figure 9.4 Relationship of squid migrations to major currents (solid lines). (**a**) *Illex illecebrosus*, (**b**) *Todarodes pacificus* (winter spawning population). Cross hatching indicates spawning area; dots, fishing areas; dotted lines, possible migration route (from Coelho, 1985: Review of the influence of oceanographic factors on cephalopod distribution and life cycles. NAFO Scientific Council Studies 9, 47–57).

species, *Illex illecebrosus*, *Todarodes pacificus* and *Dosidicus gigas*. The north–south migration of different age-classes of *Illex* has already been touched on. Because of the numerous data available for this species, especially from the fisheries statistics of the United States and Canadian governments, the events shown in Fig. 9.4a seem reasonably well established even if the precise timings are not (O'Dor, 1983). A most interesting feature of this migration is that it is apparently associated with cannibalism of the males during the long journey south. O'Dor and Wells (1987) have suggested that only in this way could a population of migrating squid acquire enough energy to fuel such a long trip and at the same time develop gonads. *Illex* travels up to 2000 km in 2–3 months to reach the warm waters necessary for egg development, but whereas the females are immature at the outset of the journey their gonads constitute a third of the body weight by the time they arrive on the spawning grounds (*ibid.*). However, this hypothesis remains to be substantiated.

In the waters off Japan, the movements of *Todarodes pacificus* have been documented even more fully, partly because of its importance to the country's commercial fisheries, whose statistical records go back to 1910, but also because of a whole series of tag-and-release experiments from Soeda (1965) onwards; Okutani (1983) gives a brief summary of this work, and other references can be found in Clarke (1966), Mangold (1989), and Nagasawa, Takayanagi and Takami (1993). Again, we find movements related to the prevailing currents (Fig. 9.4b) and a pattern of a northward migration to feed and a southward one for reproduction (Takami & Suzu-Uchi, 1993). The situation is more complex than this, however: there are 'winter', 'summer' and 'autumn' populations, which may occasionally intermingle, perhaps at the extremes of the ranges. Some of these populations are large enough to support major fisheries. Coelho (1985) discusses the way in which local hydrography in Japan can influence the distribution and population structure of *T. pacificus*, and compares this with the way in which local conditions in the North Atlantic affect *Illex* populations.

The jumbo or Humboldt squid, *Dosidicus gigas*, occurs in the eastern Pacific where its range extends from Chile to Baja California (Nigmatullin, Nesis & Arkhipkin, 2001). The migration pattern along the South American coast is the reverse of that of *Illex* and

Todarodes in the northern hemisphere. There is a main migration in the southern summer that brings the squid down to southern Peru and Chile, where they feed; in the southern winter and spring they move north towards the Equator, maturing as they near their spawning grounds. The young are dispersed westward by currents but with age swim southeast (Nesis, 1983). Individuals are thought to migrate over thousands of kilometres, and Nesis (1970) has described large schools of *Dosidicus* that may have been migrating; individuals were moving in schools, side-by-side, closely spaced, and were not feeding.

On a smaller scale, there is now a great deal of information about the migrations of *Dosidicus* into and within the Gulf of California, Mexico. Here this species supports the largest cephalopod fishery in the world. In 1979, about 5000 metric tonnes per year of these squid were caught commercially; in 1980, 22 000 metric tonnes; in 1981, only 9000 metric tonnes (Ehrhardt, 1991). Yet in 2004, 800 000 metric tonnes were landed (FAO data, cited in Gilly *et al.*, 2006). *Dosidicus* is ecologically as well as economically important in this region: it is a major predator (of fish, crustaceans and small squid) and falls prey to fish, birds and mammals, especially sperm whales. The important study of Gilly and his collaborators (*op. cit.*) has revealed much about its horizontal and vertical migrations and set new standards for the investigation of squid movements in the sea. The authors attached either pop-up archival transmitting (PAT) tags to the underside of the fins or archival tags to the mantle of nearly 100 large squid (mean ML of about 70 cm). The PAT tags transmitted data (lasting for 842 hours) to a satellite system; the archival tags had to be physically recovered (and yielded a total of 780 hours of data). Previous estimates of the rate and distance of migrations of *Dosidicus* within the Gulf of California had suggested a rate of 8 km per day (Markaida, Rosenthal & Gilly, 2005) but the PAT tags revealed much higher velocities, on average 30 km per day, over distances of 100 to 200 km. Stewart *et al.* (2012) recorded one individual swimming at 34 km a day for over 17 days, the longest horizontal migration known for this species. The PAT data provided new evidence about migration routes within the Gulf, which probably match the seasonal currents; the data also suggest that this species does not migrate *en masse* but in fairly small groups (Gilly *et al.*, 2006). Another PAT study of *Dosidicus* migration has been made in the Pacific Ocean, outside the Gulf of California (Bazzino *et al.*, 2010). This suggests that in the spring the squid move in a southerly direction, again in small groups, and enter the Gulf in the summer, where they are fished commercially; this long-distance migration may be related to reproduction and spawning.

Sepioids and octopods also undertake horizontal migrations. Thus, there are migrations of *Sepia officinalis hierredda* off the West African coast (Bakhayokho, 1983) and of *S. officinalis* in the English Channel (Boucaud-Camou & Boismery, 1991). The giant Australian cuttlefish (*S. apama*) may also undertake long-distance migrations, but a careful study of tagged individuals by Aitken, O'Dor and Jackson (2005) revealed the interesting fact that this cuttlefish barely moves at all once it has reached the inshore spawning ground in South Australia. Over the 19 days of the study, the maximum distance travelled by tagged individuals varied from as little as 90 m to 550 m.

In Japan, Itami (1964) reported movements of marked *O. vulgaris* individuals for up to 50 km over 40 days. *Enteroctopus (Octopus) dofleini* in Japan apparently makes two seasonal migrations per year, travelling up to 4 km per day (Kanamuru, 1964, quoted in Hartwick, 1983). In British Columbia, there are no mass migrations (onshore/offshore) but frequent movements in shallow-water populations (Hartwick, Ambrose & Robinson, 1984), although the maximum distance travelled in one study was less than 5 km (Scheel & Bisson, 2012). In the Mediterranean (Catalan Sea), there is evidence that both *Eledone cirrhosa* and *E. moschata* move inshore to spawn, although over comparatively short distances (Mangold-Wirz, 1963). However, because of the nature of the Catalan coast, the migrations of both these species also involve a vertical component, and many cephalopod migrations involve changes of depth as well as horizontal position (§9.6).

The control and regulation of migration are not well understood. Light, temperature, water density and salinity have been suggested as abiotic factors influencing migration; and predation, productivity and competition as biotic factors. However, there is little direct evidence for any of these and none concerning the way they might interact.

The effect of light on daily movements has been seen during field observations of *Sepioteuthis sepioidea*

(Hanlon and Forsythe, unpublished data) in the Cayman Islands. There was a clear daily movement on and off the reef associated with daylight; at dusk, the squid dispersed into shallow sandy areas with seagrass, where they fed throughout the night, before convening again at dawn to shoal all day.

The migration of the common cuttlefish *Sepia officinalis* in the English Channel is related to temperature, the animals moving inshore as spring advances and offshore in winter to depths where the minimum temperature exceeds 9 °C (Boucaud-Camou & Boismery, 1991).

We can only speculate about the mechanisms that cephalopods such as squid use to navigate during very long-distance migrations. Do they utilise mechanical information about currents or current boundaries? Or do they rely on inertial clues, as has been suggested for fish by Harden Jones (1984)? Do they use chemical clues? Have they got a geomagnetic sense? None of these can be excluded but, as always with cephalopods, vision seems to be the most likely sensory channel to be involved: we saw in Chapter 2 that cephalopods have polarised-light sensitivity, which could potentially allow them to navigate using solar positional information.

The function of migration in cephalopods is also poorly understood. Looking at the long-distance migrations described above for Japan, the East Atlantic and the Pacific Coast of South America, we saw that there was a pattern involving movement of populations to cold waters (high latitudes) for feeding and perhaps to initiate sexual maturation using the longer day length (see Chapter 6); this was followed by movement to warm waters (low latitudes) for spawning. Yet it would be premature and simplistic to generalise at this stage. Clearly, some of the inshore migrations of ommastrephids into shallow (and therefore warmer) water are for spawning: but others are for feeding (Clarke, 1966).

9.6 Vertical Migration and Activity Patterns

Cephalopods are distributed vertically in the sea as well as horizontally (Fig. 1.2), but as long ago as 1841, d'Orbigny recognised that some species were distributed differently by night and by day (d'Orbigny,

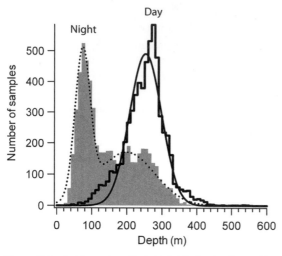

Figure 9.5 *Dosidicus gigas.* Vertical distribution by day and by night, based on over 15 000 data points. Daytime data show a normal distribution, peaking at a mean of 253 m; night-time data are bimodal (dotted line), peaking at 73 m and 198 m (modified from Gilly *et al.*, 2006: *Marine Ecology Progress Series* **324**, 1–17, with permission from Inter-Research).

1841). Subsequently, particularly with the advent of closing nets (Clarke, 1969), and, more recently, implanted telemetering devices (e.g. Ward *et al.*, 1984; Nakamura, 1991; O'Dor et al., 1991, 1995; Gilly *et al.*, 2006; Benoit-Bird & Gilly, 2012), it has become established beyond doubt that many species of cephalopod, both adult and young, make vertical movements on a daily basis. The pattern is generally the same: ascent at dusk, descent at dawn (Fig. 9.5). Such a rhythmical pattern of movement, termed *diel*, is shown by many marine animals, and has been the subject of numerous theories and speculation (Marshall, 1954; Hardy, 1956). At present, the most widely accepted theory seems to be Hardy's: because water masses move horizontally at different speeds and in different directions at different depths, a weak swimmer, by moving a short distance up or down, can achieve many kilometres of horizontal dispersion for little expenditure of energy and a short travel time (Barnes, 1991).

In pelagic cephalopods, R.E. Young (1975a) has recognised no fewer than eight categories of vertical distribution. These range from the non-migrators (whether living at the top or at the bottom of the water column), through 'shifters' (genera like *Gonatus* that

only move slightly so that there is considerable overlap: 400–800 m by day, 300–500 m by night) and 'spreaders' (e.g. *Octopoteuthis deletron*, whose range expands from 200–400 m by day to 0–500 m by night), to true diel migrators, which make considerable vertical excursions. Two examples given by Roper and Young (1975) are *Chtenopteryx*, which may migrate from a day depth of 800 m to less than 150 m, and *Brachioteuthis*, which may come up from 1000 m to less than 200 m. Moiseev (1991) reported seeing a 500-mm *Todarodes sagittatus* from a submersible at a depth of about 1950 m during a day dive; at night this species occurs close to the surface.

More recently, in the Gulf of California, a revealing study of *Dosidicus* using electronic tagging (Gilly *et al.*, 2006; see also §9.5) has yielded detailed data showing that:

(1) most daylight hours are spent at depths of about 250 m; during darkness at depths of between 73 and 198 m (Fig. 9.5); this diel change is similar to that of its main prey (myctophid fish);

(2) the dynamics of its vertical movements are much more complex than a simple diel up–down migration, with considerable individual variation (Fig. 9.6);

(3) these squid regularly penetrate a hypoxic zone in the water column (the oxygen minimum zone, OMZ) below 300 m;

(4) there are lunar effects on vertical distribution;

(5) there are changes in vertical distribution during horizontal migrations.

Several of these points are shown in Figure 9.7, which, in particular, shows how this individual squid often made high-frequency dives, sometimes within the hypoxic zone. The reason for this is not clear, but Gilly and his collaborators suggest that it might be to escape from predators: fascinating data are now emerging about the complex interaction of sperm whales and their *Dosidicus* prey (Davis *et al.*, 2007). The change in behaviour during different lunar phases may relate to changes in prey distribution, but an active predator such as *Dosidicus* is likely to spend time searching within a zone rich in prey. As so often with cephalopods, we find that the behaviour of a particular species in a given situation is far from simple.

The physiological implications of *Dosidicus* foraging in the OMZ – perhaps for as long as 6 hours – have

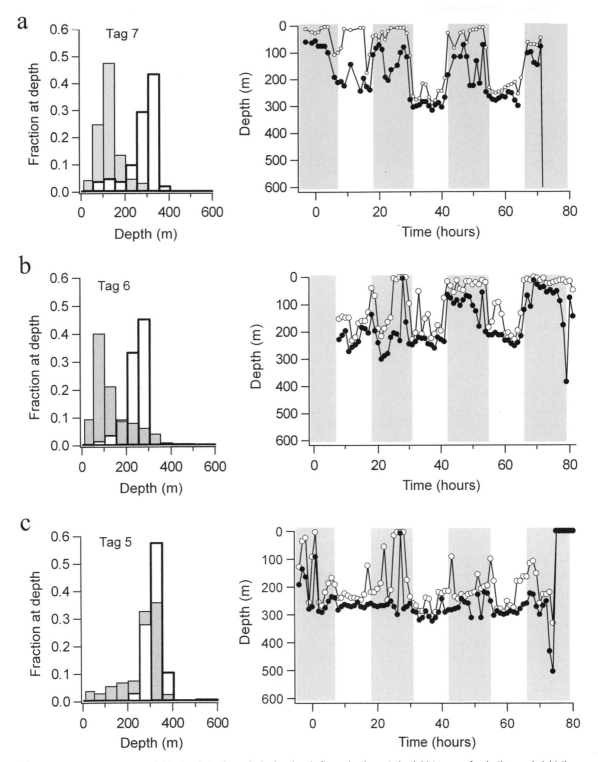

Figure 9.6 *Dosidicus gigas*. Individual variation in vertical migration. Left panels: time-at-depth histograms for daytime and night-time for three individual squid (a, b, c). Right panels: records of their maximum (open circle) and minimum (closed circle) depths over time (night-time shaded). Note that squid C (Tag 5) showed almost no diel change in vertical distribution, staying in the 200–300 m zone for most of its horizontal migration. (From Gilly *et al.*, 2006: *Marine Ecology Progress Series* **324**, 1–17, with permission from Inter-Research.)

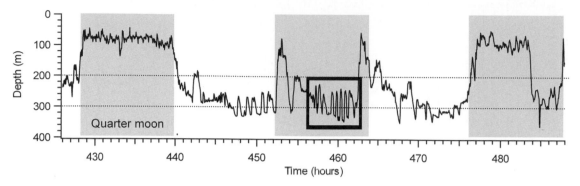

Figure 9.7 *Dosidicus gigas.* Frequent dives by an individual squid (box) during the night in the hypoxic zone (dotted lines: 200–300 m). See text. (From Gilly *et al.*, 2006: *Marine Ecology Progress Series* **324**, 1–17, with permission from Inter-Research.)

attracted much interest (e.g. Seibel, 2011; Gilly *et al.*, 2012; Stewart *et al.*, 2013). Respirometry studies in the laboratory by Gilly *et al.* (2006) have shown that *Dosidicus* can suppress its normally high rate of oxygen consumption when exposed to conditions like those encountered in the OMZ, but this is in marked contrast to the hypoxia intolerance shown by many ommastrephids (Pörtner, 2002).

It is clear from the literature not only that many cephalopods make vertical migrations but also that there is considerable variation in depth distribution and vertical migration pattern among cephalopods of different species and different ages. Even within species, there can be geographical variation: Nesis (1982) notes that *Illex coindetii* occurs at 60–400 m in the Mediterranean, 150–300 m in the East Atlantic, 180–450 m in the Caribbean and 200–600 m in the West Atlantic. It is not easy, therefore, to make general statements about vertical movements, and as more data accumulate it may become less so. For example, the measurements by Arkhipkin and Fedulov (1986) have revealed unexpected complexities in the vertical organisation of the water column in the North Atlantic and corresponding differences in the distribution of juvenile *Illex illecebrosus*. They found at least five vertically distributed layers in their study area, differing in salinity, temperature and oxygen content, and in the vertical gradients for these factors. Vertical distribution and diel migrations of the young animals were closely associated with the layers, but there were individual differences relating to size. The distribution

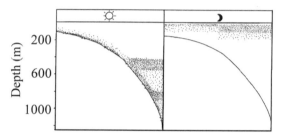

Figure 9.8 Vertical distribution of *Illex argentinus* above the continental slope near Patagonia during the day and by night. Density of dots reflects population density (from Moiseev, 1991: *Bulletin of Marine Science* **49**, 446–456, Observation of the vertical distribution and behavior of nektonic squids using manned submersibles, Moiseev, S. I., © (1991), by permission of Rosenstiel School of Marine and Atmospheric Science/Allen Press).

and vertical migration seem to be related not simply to depth but to the 'vertical structure of the water column'.

Direct observation from manned submersibles has also revealed complex patterns of distribution and migration (Moiseev, 1991). While confirming the basic down-at-dawn/up-at-dusk pattern, these studies have shown that in some squid the population may separate into distinct layers, either during the day or at night (Fig. 9.8). Incidentally, these observations provide direct evidence that some squid (e.g. *Sthenoteuthis pteropus*) are most concentrated at depths where prey is densest.

Mention must also be made of the phenomenon known as 'ontogenetic descent' (a better term might be ontogenetic vertical migration). In many pelagic cephalopods, the youngest stages inhabit the top layers of the sea and then progressively descend the water column as they mature. According to Mangold (1989) this phenomenon is especially common in cranchiids and in some pelagic octopods, but it is known in other teuthoid families and in sepioids (e.g. *Sepiella japonica*, Ueda, 1985). However the phenomenon is not simple: some young cephalopods move upwards as they grow and some, for example *Gonatus fabricii* (Lu & Clarke, 1975), seem to show an early ascent followed by a descent (see summary by Vecchione, 1987). Submersible observations on *G. onyx* in California reveal that the vertical distribution is bimodal, peaking at 400 and 800 m depth in the day and 300 and 500 m at night (Hunt & Seibel, 2000), which reflects younger (shallower) and older (deeper) age classes. The young squid actively shoal for about 3 months before beginning their descent to deeper waters, where they become solitary hunters making long vertical migrations to feed at night. There has been a claim for an upward ontogenetic migration in two benthic cephalopods, *Bathypolypus sponsalis* and *Neorossia caroli* (Villanueva, 1992), and more data might complicate the picture further.

We know little about how light regulates the behaviour of cephalopods, but there is little doubt that both daylight and moonlight (Fig. 9.7) are factors influencing their vertical distribution. In one study, fewer specimens of *Octopus digueti* were present in the intertidal zone during moonlit nights (Voight, 1992) and many squid (Baker, 1957; Hamabe, 1964; Wormuth, 1976; Gilly *et al.*, 2006) and *Nautilus* (Chapter 10) are known to go deeper when there is moonlight. The response of the squid to the moon seems paradoxical, in that some of the species that avoid moonlight will approach bright lights, as Mediterranean fishermen have known since antiquity.

The striking influence of light on cuttlefish behaviour was demonstrated unequivocally by Denton and Gilpin-Brown (1961): in bright light, cuttlefish bury themselves in the gravel at the bottom of their tank; after twilight they emerge and swim around their tank until dawn. If kept in total darkness for one or two days they become extremely buoyant, unable to stay on the bottom. This is due to changes in the volume of the gas space within the cuttlebone; there is little evidence of any inherent diurnal rhythm of density change. The daily activity of

several species of octopus is also known to be strongly influenced by light: for example, in the Mediterranean off Malta, Kayes (1974) obtained good evidence that *Octopus vulgaris* tend to be in dens during the day, but out at night, and Mather (1984) showed the same for *O. joubini* (Fig. 8.1). In Bermuda, Mather (1988) also found differences in daytime activity of *O. vulgaris* that might be related to feeding and, directly or indirectly, to tidal changes. She points out that the cycles of rest and activity shown by these octopuses are not unlike those of their mammalian predators, reminding us that the whole question of rhythmic behaviour in cephalopods has not yet been seriously addressed. Palmer and O'Dor (1978) kept *Illex* in continuous light and found that some up–down movements persisted, suggesting the operation of some kind of internal 'clock', but in *O. vulgaris* Bradley and Young (1975) obtained no evidence of a circadian rhythm affecting learning performance (Chapter 8). In the lesser octopus, *Eledone cirrhosa*, there is also little convincing evidence of a circadian rhythm (Cobb, Pope & Williamson, 1995). This species shows a nocturnal activity pattern, but when maintained in continuous dark or light this rhythm does not persist; the main factor determining locomotor activity was apparently 'exogenous changes in illumination . . . [and] . . . although there is the suspicion of an endogenous factor at work . . . the evidence is far too weak to be convincing' (*ibid.*, page 54).

It may be relevant to note here that there is now evidence for the presence of melatonin in *Octopus vulgaris* (Muñoz *et al.*, 2011); moreover, in animals maintained in a natural day/night cycle, the levels of melatonin in the blood, retina and optic lobe fluctuate widely, peaking regularly at night. This obviously suggests that this molecule may have a role in transducing the day/night cycle information to regulate rhythmic behaviour in cephalopods as in other animals.

Apart from the influence of light, we know very little about daily activity patterns in cephalopods. There are only three field studies documenting the types of behaviour exhibited throughout the day by shallow-water octopuses and squid; all three studies were carried out by teams of divers who could watch animals continuously.

In Bermuda, four juvenile *Octopus vulgaris* (100–200 g) were observed throughout daylight hours

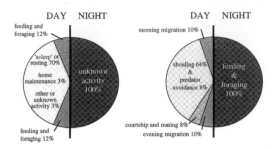

Figure 9.9 Daily activity patterns. (**a**) Four juvenile *Octopus vulgaris*, observed every 10 minutes from 0630 to 1830 hours daily for 5 weeks in Bermuda (modified from Mather, 1988: *Malacologia*, **29**, 69–76, with permission). (**b**) Shoals of *Sepioteuthis sepioidea* (5–80 individuals) recorded continuously from 0630 to 1930 hours (and sporadically at night) at Little Cayman Island (based on data from Hanlon & Forsythe, unpublished).

(Mather, 1988; Fig. 9.9a). Over a 5-week period, the octopuses spent about 70% of their time in dens during the day, and 24% of the day foraging and feeding, with peak activity between 0600 and 0800 h. The inference that the juveniles in this intertidal zone may have been essentially nocturnal is borne out by the pioneering studies of adults of the same species in the Mediterranean, where good evidence exists that there is a nocturnal peak of activity (Altman, 1967; Kayes, 1974).

Adults of the Indo-Pacific coral reef octopus, *Octopus cyanea*, have been observed continuously during daylight for 6 days on an atoll near Tahiti (Forsythe & Hanlon, 1997). These animals generally foraged once in the early morning and once in the late afternoon (representing approximately 20–30% of their daytime activity) and retired to their dens in the middle of the day; checks at night found the octopuses to be in their dens. This study corroborates the indirect observations of Yarnall (1969) and Houck (1982) in Hawaii.

In Little Cayman (Fig. 9.9b), Caribbean reef squid (*Sepioteuthis sepioidea*) disperse from their shoals to begin feeding at dusk, feeding throughout the night, until they reassemble at dawn for their daytime 'resting' period. They spend about 64% of the daytime resting in shoals, 8% actively avoiding predators, 8% in courtship and mating (mostly the hour before dark), and the

remaining 20% moving to and from feeding areas. This daily pattern appears to hold throughout the year (Hanlon & Forsythe, unpublished data).

There have been several laboratory studies of rest–activity cycles in octopuses (e.g. Meisel *et al.*, 2006, 2011) and in *Sepia* (Frank *et al.*, 2012), but the most thorough analysis of such behaviour derives from the study of Brown and his colleagues, working with *Octopus vulgaris* (Fig. 9.10; Brown *et al.*, 2006). They first found clear evidence that during an artificial 12/12 h light/dark cycle the animals are more active in the dark and quieter in the light, which agrees with most field studies. They identified four distinct behaviours – 'active', 'alert', 'quiet', 'eyes closed' (a sub-state of 'quiet' behaviour) – and recorded how these changed after a light period in which the subject was deprived of rest for 12 hours (by gentle tactile stimulation with a net every time the animal appeared to rest). In the 12 hours following deprivation, there was little change in active behaviour, a marked decrease in alert behaviour, and an increase in quiet behaviour. The authors noted that there was a rebound of quiet behaviour soon after the deprivation period, which extended into the ensuing dark and light periods; however, a similar but stronger rebound of the 'eyes closed' behaviour was delayed until the post-deprivation light period. Brown *et al.* also showed that octopuses indulging in quiet behaviour were barely responsive to external stimuli (light flashes or tank tapping) and slower to attack a white ball. More importantly, Brown *et al.* monitored brain activity while recording the animals' behaviour by means of electrodes implanted close to the vertical lobe (§2.3). The activity they recorded in the lobe was similar to that reported previously (Bullock & Basar, 1988), but during behavioural rest electrical activity in the vertical lobe intensified. During sleep-like states in other invertebrates the rule appears to be 'body off/brain off'; in vertebrates, it is 'body off/brain on'. As the authors put it, with admirable caution, 'our data provide the first evidence that something like the well known off-line activity associated with vertebrate sleep and memory processing when sensory input is closed down is also occurring in an invertebrate. This study underlines the special status of cephalopods amongst invertebrates as "honorary vertebrates" (Brown *et al.*, 2006, page 359).

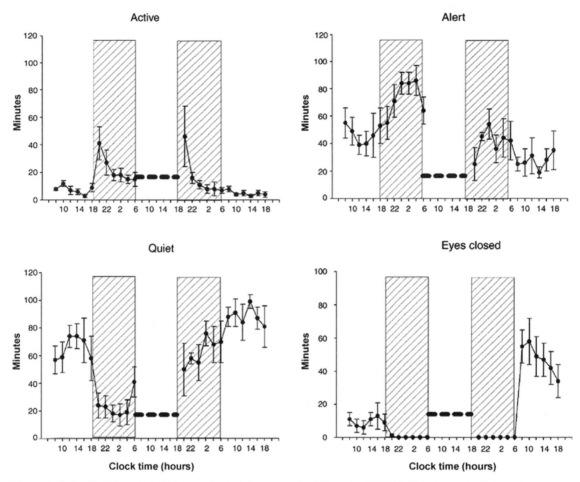

Figure 9.10 Rest/activity cycle in *Octopus vulgaris* during a period of 60 h under 12 h light, 12 h dark cycle. Data points are mean duration (min) of each behaviour state during 2-h intervals in six animals. Heavy horizontal dashed line indicates time during which animals were deprived of rest (see text) (reprinted from Brown *et al.*, 2006: *Behavioural Brain Research* **172**, 355–359, Brain and behavioural evidence for rest–activity cycles in *Octopus vulgaris*, Brown, E. R., Piscopo, S., De Stefano, R. & Giuditta, A., © (2006), with permission from Elsevier).

9.7 Life in the Deep Sea

Inevitably, much of this book has been concerned with shallow-water inshore forms, but according to Clarke (1966) as many as 66% of oegopsid squid species live below 500 m during the day and 33% live below 1500 m. *Vampyroteuthis* and many of the octopods also live in the depths: one cirrate octopod captured at 7279 m holds the depth record for a cephalopod (Voss, cited in Aldred, Nixon & Young, 1983).

Deep-sea cephalopods often have gelatinous bodies and most are quite small, a fact that may disappoint readers of science fiction but will not surprise students of deep-sea fish (Marshall, 1954). Clarke (1966) estimates that about 60% of oegopsids species are less than 100 mm ML. In most deep-sea cephalopods, the chromatophores are far less important than in epipelagic forms and, as far as we know, body patterning is much simpler or absent. In cranchiid squid, for example, the number of chromatophores is greatly reduced and there are correspondingly much smaller chromatophore lobes in the brain (Maddock & Young, 1987). In *Vampyroteuthis* there are no chromatophores except those associated with the photophores, and in *Cirrothauma* they are lacking totally.

Another striking feature of some deep-sea forms is the absence of an ink sac. There is none in *Vampyroteuthis*, nor in the cirrate octopods, nor in several members of the incirrate sub-family Eledoninae that live in deep water, for example *Thaumeledone*, *Graneledone* and *Bentheledone* (Mangold, 1989). However, all deep-sea squid retain the ink sac (R.E. Young, pers. comm., 1993), and an important recent paper has revealed the unexpected fact that many mesopelagic squid release ink at depths down to 1800 m or more (Bush & Robison, 2007). Video-recordings, taken from ROVs in Monterey Bay, California, have revealed complex inking behaviour by many individuals of as many as 20 species of squid, including smokescreens, pseudomorphs and ink-ropes (see §5.2.5). The function of this inking is presumably often the same as it would be in shallow water (camouflage or other forms of defence: Chapter 5), but the significance of ink-ropes and ink-filled mantle cavities is obscure. The fact that inking is so common in low-light conditions is slightly less surprising now we know that in at least one squid species, *Octopoteuthis deletron*, chromatic, postural and locomotor body patterns may also be shown at similar depths (Bush, Robison & Caldwell, 2009).

Clearly the visual system must be fully functional even at these depths: it is certainly well developed in all the meso- and bathypelagic cephalopods so far examined. The eyes may be very large, an advantage for gathering light and for high resolution; some genera (*Histioteuthis*, *Amphitretus*, *Sandalops*) have developed tubular eyes, to optimise vision upwards in the only direction where there is a little light (Land, 1990). In some genera, such as *Bathyteuthis* (formerly *Benthoteuthis*), there is a ventral 'fovea' (Chun, 1914). Only in one deep-sea form, the finned octopod, *Cirrothauma*, is the visual system markedly reduced. This genus is mostly taken below 3000 m: the eyes lack a lens, there is no optic chiasm, and the optic lobes are smaller and simpler in their organisation than in other coleoids (Aldred, Nixon & Young, 1983). Nevertheless, it is not a blind octopod, as originally thought (Chun, 1914), and the large-diameter eye (14 mm) must certainly be adequate to detect bioluminescence. It is curious that only in this genus is there such a reduction of the visual

system with depth. The eyes of the squid *Histioteuthis* are especially remarkable in that the two eyes are quite different. One eye is 'normal' in size and in being spherical; the other eye is unusual in that it is twice as big, tubular, with a yellow lens. The gaze of the ordinary eye is ventro-lateral, while that of the large eye is upwards. The explanation of this bizarre arrangement seems to be that the small eye is concerned with events around and below it: it is surrounded by light organs that may act as searchlights (Young, 1975b). The large eye, however, is concerned with the world above and, specifically, with the light produced by the ventral light organs of fish, squid and other organisms for crypsis (see next section). The yellow lens would attenuate the intensity of daylight reaching the retina much more strongly than bioluminescent light, so that an eye like this might be able to 'break' the camouflaging effect of potential predators (Muntz, 1976).

Light organs are among the most remarkable adaptations to life in the deep sea, and they are so important in behaviour that they merit a separate section.

9.7.1 Light Organs and Bioluminescence

Bioluminescence, or biologically generated light, is a widespread phenomenon in marine organisms, and in 'higher' animals, such as crustaceans, cephalopods and teleost fish it is produced in special light organs, or photophores (§2.2.4).

In cephalopods, light organs are often found in species living between 400 and 1200 m during the day (Young, 1983). They are especially common in oceanic squid (oegopsids), but are found also in *Spirula*, in many sepiolids, in *Loligo*, in *Vampyroteuthis*, and in two octopods, *Eledonella* and *Japetella*. By all accounts they are remarkably beautiful: 'lucky is he who has seen the brilliance of a *Thaumatolampas*' [now *Lycoteuthis*], wrote Chun (1910) in his famous report on the *Valdivia* deep-sea expedition.

The first clue about the possible function of light organs comes from their anatomical distribution: this is not uniform, and there are clearly a number of 'preferred' positions on the body (Herring, 1977, 1988). There are nearly always more ventrally than dorsally; they are especially common below the eye; they occur on the mantle and arms, often at the tips, and sometimes

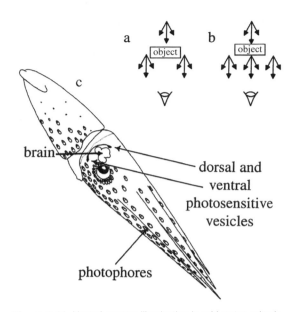

brain

c

dorsal and
ventral
photosensitive
vesicles

photophores

Figure 9.11 Ventral counter-illumination in mid-water animals. In (**a**), an observer in the deep sea views an object that is conspicuous against a background of submarine daylight. In (**b**), the object is now camouflaged because it emits light from its lower surface to replace that which its upper surface absorbs or reflects. (**c**) Distribution of photosensitive vesicles and photophores in *Histioteuthis dofleini* (**a, b** from Denton, 1990: *Light and Life in the Sea*, ed. P. J. Herring *et al.*, 127–148, © (1990), with permission from Cambridge University Press); **c** from R.E. Young, 1977: *Symposia of the Zoological Society of London* **38**, 377–434, Brain, behaviour and evolution of cephalopods, © (1977), with permission from Elsevier).

on the tentacles; and they may occur at the posterior tip of the body. This distribution becomes understandable if we realise that the two main functions of the photophores are *ventral counter-illumination* (i.e. crypsis) and *signalling*.

The principle behind ventral counter-illumination in mid-water animals, first proposed by Dahlgren in 1916, is shown in Fig. 9.11. A fish or a squid can easily conceal itself from predators located above or at the side by making the upper surface darker and having reflecting sides, but concealment from a predator below is more difficult. One solution, available to teleosts because of their method of locomotion, is to reduce the ventral surface area, and indeed many fish are wedge-shaped in cross section. However, others employ a different

technique: they carry a series of 'lanterns' on the ventral side of the body that produce a highly directional light that matches the light downwelling from the surface both in intensity and wavelength. This is the solution adopted by many cephalopods (Young, 1977, 1983), notably the histioteuthids and enoploteuthids, but also the bobtail squid, *Euprymna scolopes* (Jones & Nishiguchi, 2004).

These 'lanterns' are complex. They contain photogenic tissue and filters enabling them to produce a weak blue light, whose maximum intensity lies close to the peak value for downwelling light in the open ocean (about 475–480 nm) (Herring, 1988). There are many hundreds of such organs, each measuring only $400 \times 200\,\mu m$, on the ventral surface of *Histioteuthis* (Young, 1977), and on theoretical grounds alone we would expect such a bank of blue-emitting photophores to make the squid invisible to a predator beneath it.

In fact, R.E. Young and his collaborators have provided direct evidence about the functioning of the ventral photophores in squid in a series of important papers. They have shown, first, that deep-sea squid tested in a ship-borne aquarium mimicking natural conditions as far as possible can be made to turn on their photophores by switching on an overhead light and to turn them off by plunging them into darkness (Young & Roper, 1976, 1977). Second, the intensity of light produced by the squid (*Abralia* sp., *Abraliopsis*) matches that of the downwelling light over the range encountered in the sea at the depths at which these species are taken (Young *et al.*, 1980; Fig. 9.12). Third, at high levels of illumination, corresponding to shallow depths where they do not normally occur, the photophores are unable to match the intensity of downwelling light. Fourth, these species can alter the colour of their bioluminescent light to conceal themselves better under the different colours of downwelling light encountered by day and by night (perhaps as a result of pressure cues; R.E. Young, pers. comm., 1994). Lastly, the organs responsible for detecting and monitoring the light produced by the photophores are the ventral photosensitive vesicles (PSV) (Young, Roper & Walters, 1979).

These 'extraocular' photoreceptors were mentioned briefly in Chapter 2. R.E. Young (1973, 1977) had already implicated them in the control of bioluminescence and

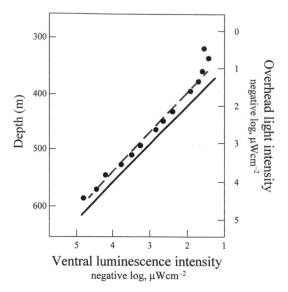

Depth (m) — left ordinate: 300, 400, 500, 600

Overhead light intensity
negative log, μWcm^{-2} — right ordinate: 0, 1, 2, 3, 4, 5

Ventral luminescence intensity
negative log, μWcm^{-2} — 5, 4, 3, 2, 1

Figure 9.12 The intensity of counter-illumination produced by young *Abraliopsis* (dots) increases as the overhead light intensity increases (right ordinate) or the depth decreases (left ordinate). Note how the matching breaks down in bright, shallow water. Thick line: calibration. Dashed line: regression line for first 12 points (modified from R.E. Young *et al.*,1980: *Deep-Sea Research A* **27**, 671–691, Counterillumination and the upper depth limits of midwater animals, Young, R. E. *et al.*, © (1980), with permission from Elsevier).

suggested that the dorsal PSV monitored downwelling light, while the ventral PSV measured the output of the photophores (Fig. 9.11). By covering the eyes and/or the dorsal PSV in *Abraliopsis*, Young, Roper and Walters (1979) found good evidence that it is indeed the PSVs that are the prime regulators of light production. These elegant experiments go a long way to establishing the functioning and importance of ventral countershading in mesopelagic squid, and stress once again the elaborate morphological and physiological adaptations that have evolved in cephalopods.

In contrast, our understanding of the role of the photophores in *signalling* is rudimentary. Starting again with the morphology, it is clear that all the photophores cannot have the same (counter-illumination) function, since they are organised differently. For example, in *Histioteuthis macrohista* the dorsal and arm-tip photophores lack the pigmented filter of the ventral ones, so that the light they produce cannot be of the

same wavelength. The arm-tip photophores contain considerably more photogenic material than the small photophores of the ventral mantle, so that they must generate a stronger light (Dilly & Herring, 1981). This suggests such photophores might be used for communication.

Figure 9.13 shows the kinds of array of photophores that could be used for signalling in several cephalopod species. Apart from signalling species-identity, the photophores could be used to identify sex, and many cranchiids, enoploteuthids and histioteuthids exhibit sexual dimorphism of the photophores (Herring, 1977, 1988). For example, in *Leachia pacifica*, only the female has photophores on the third arms; these seem to develop just before sexual maturation (Young, 1975a). In *Lycoteuthis*, the males have more posterior photophores than females, more dorsal integumental photophores (Chapter 6, Fig. 6.3) and brachial organs on the second and third arms. Indeed, until recently, the males were even thought to be another genus, *Oregoniateuthis* (Toll, 1983). In *Chtenopteryx*, only sexually mature males develop a posterior visceral light organ, whose emission, peaking at 415 nm, is unlikely to be involved in countershading (Young, 1983). Among the octopods, *Japetella* and *Eledonella* also show sexual dimorphism in their photophores: only mature females possess the circumoral organ (Fig. 6.3).

It is also possible that some signals produced by the photophores have evolved to influence animals of another species. The bright, arm-tip flashes of *Taningia danae* have been described as startling by Roper and Vecchione (1993), and possibly serve to distract or confuse a predator. Kubodera, Koyama and Mori (2007) also reported flashes in *Taningia*, but their remarkable videos, taken at depths of between 240 m and 940 m, revealed more complex behaviour. They found that the animals emitted flashes of at least three different lengths (lasting from about 1 to 8 seconds); they suggest that the shortest of these may be for blinding prey. The squid *Ommastrephes pteropus* displays a very large bright dorsal patch when attacked or when caught on a line (Herring, 1988). Could this act as secondary defence against birds or other predators attacking from above (§5.2.3)? Might the luminescent 'cloud' produced by *Heteroteuthis* (Herring, 1988) have a similar function? And could the suggestion of Voss (1956) that the photophores on the tentacles of *Chiroteuthis* or

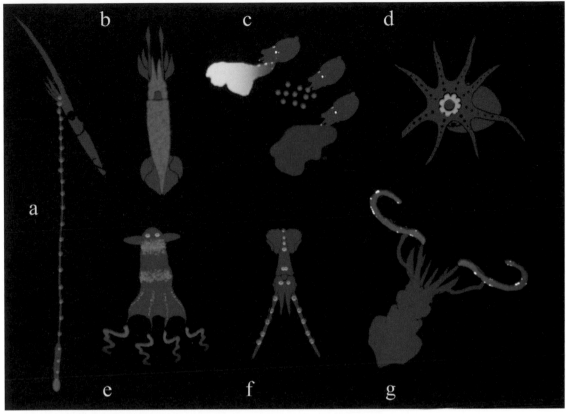

Figure 9.13 Bioluminescence in oceanic cephalopods.
Bioluminescent patterns for signalling: (**a**) *Chiroteuthis imperator*: flashing photophores on elongate tentacles; (**b**) *Sthenoteuthis oualaniensis*: glowing head and ventral mantle; (**c**) *Heteroteuthis hawaiiensis*: three types of luminous clouds produced by the visceral photophore; (**d**) female *Eledonella pygmaea*: circumoral photophores; (**e**) *Vampyroteuthis infernalis*: Passing Cloud of luminescence; (**f**) *Pyroteuthis addolux*: long flashes making a Y-display; (**g**) *Abraliopsis pacificus*: large flashing photophores on waving ventral arms (unpublished data from R.E. Young, with permission).

Megalocranchia may function as lures be correct and an example of 'manipulation' by the sender (§7.3.1)?

Much of this is speculation, but direct observations of photophore responses in living animals have been made. R.E. Young *et al.* (1982) observed the behaviour of two species of *Pterygioteuthis* in shipboard aquaria. These animals have a pair of photophores under the eyes ('ocular flashers') and another pair within the mantle cavity at the base of the gills; it is easy to elicit bright flashes from these by administering an electric shock. There are four kinds of responses, involving long (2 to 6 s) or short (1 to 2 s) flashes. Although the flashes may 'have a generally defensive function by discouraging or disrupting an attack by a potential predator' (*ibid.*), the

authors recognised that they might also be used as offensive weapons to confuse prey in some way.

In *Selenoteuthis*, there is a large spherical tail organ, larger in males than in females, that produces a bright blue light (Fig. 9.14). Its surface is covered with numerous red chromatophores that have been seen to expand and retract regularly once every second, thus enabling a steady luminescence to be converted into a series of light pulses (Herring, Dilly & Cope, 1985). This direct evidence of modulation of light from a photophore is strong presumptive evidence that this organ could serve for signalling. *Selenoteuthis* is a fascinating animal because of the morphological diversity of the numerous other photophores scattered

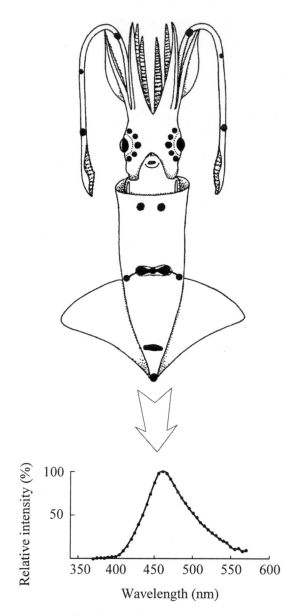

Relative intensity (%)

Wavelength (nm)

Figure 9.14 The tail organ of the female squid *Selenoteuthis*, which emits blue light (peak wavelength 460 nm), may provide a sexual signal (modified from Herring *et al.*, 1985).

over the body (Fig. 9.14). There are three distinct types of organ, and some of the ocular organs are double structures. Such complexity is bewildering, and although it is comforting to read that 'lycoteuthid photophores are among the most elaborate in structure and dramatic in appearance of any cephalopod

photophores' (*ibid.*), one is still left wondering what they are all for. Recent work continues to confirm the extraordinary variety of light-emitting organs in cephalopods not only among species but also within certain species (*Octopoteuthis danae*: Herring, Dilly & Cope, 1992). New methods of light production are also being discovered: in *Vampyroteuthis*, for example, there are unusual light-organs at the tips of all eight arms that produce flashing blue light; the arm tips can also release a viscous fluid containing microscopic luminous particles to create a glowing cloud around the animal (Robison *et al.*, 2003). Both types of light production are thought to be anti-predatory devices: the authors note that 'visual trickery is common within the depth range and light regime that *Vampyroteuthis* occupies' and speculate that the luminous fluid might stick to a potential predator and mark it out for a secondary predator.

It seems curious that we know so little about the behaviour of what is perhaps the most famous bioluminescent squid of all: the firefly squid, *Watasenia scintillans* (Fig. 9.15), which is unusual among cephalopods in having three visual pigments (§2.1.1) and therefore presumably colour vision. B. Skerry (personal communication) writes that 'following spawning the females die . . . typically in deeper water. But for a few nights each year, when there is a new moon during the spawning season, the animals rise up towards the surface. After spawning the females get caught in surface currents and wash ashore.' This leads to remarkable scenes on the shoreline, such as that shown in Fig. 9.15a.

9.7.2 Large Squid and the Giant Squid, *Architeuthis*

It is a commonplace that some squid are extremely large, the largest of all invertebrates. They live at great depths and are rarely seen alive, so that their habits and life style are a mystery and open to speculation. This fact has attracted novelists for some 150 years, and fictional accounts of animals 'as large as a locomotive' (Fleming, 1958) and malevolently attacking humans and sinking their boats continue to appear (Benchley, 1991).

Strictly, the term 'giant squid' refers to animals belonging to the genus *Architeuthis*, a usage adhered to

Figure 9.15 Firefly squid *Watasenia scintillans*. (**a**) Dying female squid washed onshore in Japan during a new moon; this common occurrence is thought to coincide with peak spawning activity in deeper water close to the shore. (**b**) Bioluminescent patterns of photophores on two moribund squid; this image required an exposure time of 90 seconds, emphasising that the natural bioluminescent signal is weak. (Photographs by Brian Skerry.)

in this book, but before discussing *Architeuthis* we must briefly mention several other squid genera that if not 'giant' are as big as or larger than a human, i.e. approaching or exceeding 2 m ML (Fig. 9.17).

There are at least four genera of oegopsids that meet this criterion: *Dosidicus*, *Mesonychoteuthis*, *Taningia* and *Moroteuthis* (now *Onykia*); five, if we include the problematical *Magnapinna*. The best known of these is *Dosidicus gigas*, the Humboldt or jumbo squid. This attains sizes up to about 1.2 m ML (50 kg weight) according to Nesis (1983), and Clarke (1966) reports claims of a total length of 4 m. *Dosidicus* is common in the eastern Pacific, from Chile north to California (Nesis, 1983), supporting large commercial fisheries. It is swift and agile, sometimes leaping out of the water to 'fly' (Cole & Gilbert, 1970); its complex migration behaviour is discussed in §9.5.2. It is a social species, and huge shoals have been observed from on board ship (Nesis, 1970). It is also one of the most voracious cephalopods known, attacking virtually anything (Chapter 4; Nesis, 1983). There is even an account of an attack on a SCUBA diver (Hall, 1990).

Mesonychoteuthis hamiltoni is a cranchiid squid, first described in 1925 and recently given the name of 'colossal squid' (Fig. 9.16d). It is distributed widely in the Southern Ocean around the Antarctic, where it may provide a major food source for sperm whales

(Clarke, 1980). Freshly caught specimens recovered in 2003 and 2007 have prompted much discussion about whether it is larger than *Architeuthis*: it seems that it may have a greater mass, with a longer mantle but shorter tentacles and therefore shorter overall length. Almost nothing is known about its behaviour. It has the largest beak of any cephalopod, very large eyes (between 270 and 280 mm diameter in one specimen: Nilsson *et al.*, 2012), and remarkable hooks on the arms and tentacles. Rosa and Seibel (2010a) have suggested that the eyes are principally to detect predators, and the hooks merely serve to 'ensnare prey that unwittingly approach', suggesting that it is an ambush predator. Such an interpretation must surely be treated with caution given the remarkable videos of living *Taningia* and *Architeuthis* obtained by Kubodera and his colleagues.

Taningia danae is a large deep-sea squid, very abundant in tropical and sub-tropical oceans worldwide (Clarke, 1980). It can grow to at least 2.3 m overall (Roper & Vecchione, 1993), and the feeding tentacles are absent in adults. The mantle musculature contains many vacuoles containing ammonium chloride for neutral buoyancy (Chapter 2: Table 2.1), and this has led several workers to assume that this species is sluggish and slow-moving. Such a notion has been dispelled recently by video recordings of its vigorous attacks on bait – rapid

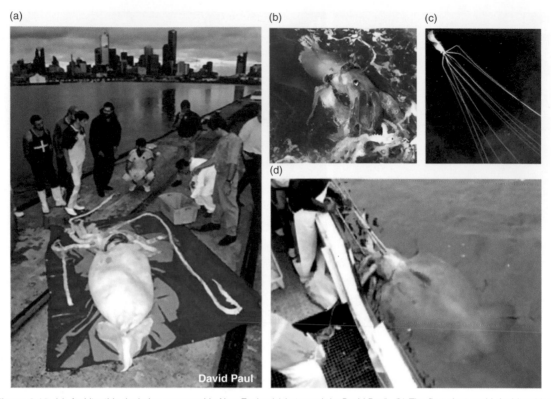

Figure 9.16 (**a**) *Architeuthis dux* being measured in New Zealand (photograph by David Paul). (**b**) The first giant squid *Architeuthis* to be landed alive southeast of Japan. Photograph courtesy of Tsunemi Kubodera of the National Museum of Nature and Science of Japan/AP. (**c**) The bigfin squid *Magnapinna* sp. photographed from a submersible in the Gulf of Mexico at nearly 2000 m; total length about 7 m (from Vecchione *et al.* 2001: *Science* **294**, 2505–2506, Worldwide observations of remarkable deep-sea squids, Vecchione, M., Young, R. E., Guerra, A., *et al.*, © (2001), reprinted with permission from AAAS). (**d**) Colossal squid *Mesonychoteuthis hamiltoni* being landed alive (with permission from Getty).

backwards and forward swimming, brought about by the large fins, and extremely fast changes of direction (Kubodera, Koyama & Mori, 2007).

Moroteuthis (*Onykia*) *robusta*, the 'robust clubhook squid', has a mantle length of at least 1.6 m (Kubodera *et al.*, 1998), although there are claims that it may reach 2 m. Little seems to be known about its behaviour.

Magnapinna, the 'bigfin squid', is a curious animal, perhaps more remarkable for the length of its arm and tentacles than for the size of its fins. The extensible arms may be ten times the length of the mantle, and are sometimes bent, making the animal look as if it had 'elbows' (Fig. 9.16c). Because of the great length of the arms, the overall length is considerable, perhaps as much as 7 m. Once again almost nothing is known about its behaviour.

The true giant squid (Fig. 9.16a, b) belongs to the genus *Architeuthis*, which we now know, on the basis of mitochondrial DNA studies (Winkelmann *et al.*, 2013), to contain only a single species, *Architeuthis dux* Steenstrup, 1857. This is widely distributed in all the world's oceans and, on the basis of beaks found in sperm whale stomachs, it seems to be an especially important constituent of the cephalopod fauna off Madeira, Spain and South Africa (Clarke, 1977). There used to be some speculation about its depth distribution. Data from bottom trawls (see Roeleveld & Lipinski, 1991) suggested that *Architeuthis* normally occurs between 200 and 500 m. However, Clarke (pers. comm.) always maintained that the sperm whales that feed on them regularly dive to 1000 m, and we now have a record, in a photograph taken by Kubodera and Mori (2005), of an individual squid

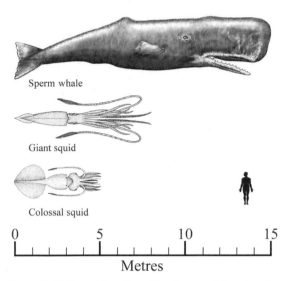

Sperm whale

Giant squid

Colossal squid

```
0          5         10         15
├───┬───┬───┬───┬───┬───┬───┬───┬───┤
```

Metres

Figure 9.17 Size relationships of two large cephalopods and a sperm whale. The *Architeuthis* giant squid drawing is from Jereb and Roper (2010): Food and Agriculture Organization of the United Nations, *Cephalopods of the World: An Annotated and Illustrated Catalogue of Cephalopod Species Known to Date*, Vol. 2: *Myopsid and Oegopsid Squids*, Jereb, P. & Roper, C. F. E., © (2010), reproduced with permission. The *Mesonychoteuthis* colossal squid drawing is from Voss (1980): *Bulletin of Marine Science* **30**, 365–412, A generic revision of the Cranchiidae (Cephalopoda, Oegopsida), Voss, N.A., © (1980), by permission of Rosenstiel School of Marine Science/Allen Press.

(later confirmed as *A. dux*) caught on camera at a depth of 900 m.

The paper of Kubodera and Mori (2005) was entitled the 'first-ever observations on a live giant squid in the wild', and since it is still the only paper describing a living *Architeuthis*, it is worth reporting in some detail. The authors suspended a vertical long-line (400–1000 m) with a digital camera facing downwards and below it a hook and squid jig baited (with fresh squid, *Todarodes pacificus*), as well as an odour lure (freshly mashed euphausiids). The camera was configured to take images every 30 s for 4–5 hours; the system also carried a depth sensor, strobe and various data recorders.

Kubodera and Mori write:

> At 9.15 on 30 September 2004, an individual giant squid attacked the lower squid bait of one of our camera systems at 900 m over a seafloor depth of 1200 m [in the

North Pacific]. The squid's initial attack was captured on camera ... and shows the two long tentacles characteristic of giant squid wrapped in a ball round the bait. The giant squid became snagged on the squid jig by the club of one of these long tentacles.

There followed a 4-hour struggle between the squid and the recording system, until eventually the squid broke free, leaving a severed tentacle, 5.5 m long, attached to the camera system. This was subsequently retrieved – 'still functioning, with the large suckers of the tentacle club repeatedly gripping the boat deck and any offered fingers' – and subjected to measurement and DNA analysis.

The recovered tentacle club measured 720 mm; the largest sucker diameter was 28 mm. Estimates based on these figures give a mantle length for the squid of between 1615 and 1709 mm; an approximate length, from tip of the fin to tips of normal arms, of 4.7 m; and a total length including the long tentacles, of over 8 m.

Clarke (1966) states that the longest giant squid on record is 18 m total length, of which 12 m was made up of the tentacles, but Kubodera and Mori (2005) point out that these can stretch as they decay, so that upper size claims for this species may be exaggerated.

However, for students of behaviour, size is not everything, and the most interesting behavioural observations in this important paper are those revealing that *Architeuthis* is a much more active predator than previously thought. The individual filmed attacked the lure vigorously and horizontally, and the tentacles are 'clearly not weak fishing lines dangled below the body' (*ibid.*). As for *Taningia*, mentioned above, one reason for assuming that the animals were sluggish is that the musculature of the mantle and arms contains many spaces filled with ammonium chloride to confer neutral buoyancy (Clarke, Denton & Gilpin-Brown, 1979). Other reasons include the facts the fins are not especially large and the 'giant' nerve fibres are much smaller than those of the common squid, *Loligo*: only 250 μm in diameter instead of 800–1000 μm (Young, 1977a). However, the beak is large and heavily muscled, the suckered arms are massive, and in the brain the brachial lobe, which controls the arms, is especially well developed (Nixon & Young, 2003). The tentacles are remarkable (Kubodera & Mori, 2005): apart from their great length they bear suckers with corresponding lugs along their entire length, so that the shafts can be 'zipped' together to form a single structure with paired clubs at the tip to seize prey. It would seem that the giant squid is indeed a formidable predator. Like *Mesonychoteuthis* it has extremely large eyes, although these may be an adaptation for predator avoidance rather than prey capture (§2.1.3).

Little is known about reproduction and development. Males apparently mature at a smaller size than females (Roeleveld & Lipinski, 1991). The only evidence of sperm transfer is embedded spermatophores in skin of a female's ventral arms (Norman & Lu, 1997). In one ripe female, there were about 10^7 eggs, which were surprisingly small, less than 2 mm (Boyle, 1986c). Practically no data on growth are available. Jackson, Lu and Dunning (1991) counted 153 growth rings in the statocyst of an immature female of 422 mm ML, giving an estimated growth rate of 2.76 mm per day. Gauldie, West and Förch (1994) counted 393 rings in a specimen of 1610 mm ML, which translates to 4.09 mm per day. These figures suggest that it might take nearly 4 or 5 years for an *Architeuthis* to reach 5 m ML or more. In a different approach, Landman *et al.* (2004) used isotopic analyses of statoliths and calculated 'an age of 14 years or less' for three female *Architeuthis* collected off Tasmania, but they warned that more refined estimates await better understanding of the variations of temperature and depth.

There is obviously much that we still need to know about the giant squid, but we agree with the appeal of Roeleveld and Lipinski (1991) for a more measured, less sensational, response to *Architeuthis* data. It is sad to reflect that the truth as it emerges will almost certainly be less exciting than the fiction, at least to some people.

9.8 Food Webs and Community Structure

Trophic relationships depend upon community structure but are also dependent, in large part, on the behaviour of the species involved. Community structure results from complex interactions of many factors, including the physical structure of the habitat, numerous other abiotic influences, and the types and numbers of species present. Cephalopods inhabit nearly all known types of marine community, including coral reefs, seagrass beds, rock reefs, sand plains, open ocean

(all depths, including hydrothermal vents), kelp forests, intertidal zones and polar environments. Yet we understand relatively little about the roles played by cephalopods in these communities.

Food webs describe functional relationships among species in a community (Ricklefs & Miller, 1999), but their analysis, as far as cephalopods are concerned, is still in the early descriptive phase, and only a few studies exist. Recent reviews indicate that the apparent complexity of food webs is amenable to analysis and useful in understanding the dynamics of communities (Pimm, Lawton & Cohen, 1991).

The place of squid in food webs has been reported in three studies. The California market squid *Loligo opalescens* (Fig. 9.18) plays a central role in the Monterey Bay ecosystem. Nineteen species of fish, eight species of birds and at least two marine mammals feed on *L. opalescens*, while the squid feed primarily on krill and anchovies (Morejohn, Harvey & Krasnow, 1978). Vovk (1977) described a similar food web for another loliginid, *L. pealeii*, in the northwest Atlantic, and concluded that this squid plays a part at three of the four trophic levels they describe for that community. Very young squid (2–10 cm ML) fed upon zooplankton (second trophic level), squid 11–18 cm ML fed upon planktophages (third trophic level) and large squid (more than 18 cm ML) fed upon larger fish and other squid (fourth trophic level). Furthermore, there was clear evidence that squid of all size groups were affecting the community at three of the four levels throughout the year. A similar food web has been described for squid in Antarctic waters. Roles are often reversed as individuals grow. For example, squid are consumed by many fish when young, but eat those same predator species as they grow (*op. cit.*). Amaratunga (1983, 1987), Vecchione (1987), and Boyle and Rodhouse (2005) review some of these aspects of the ecology of cephalopods.

Fotheringham (1974) described the role of *Octopus bimaculatus* in a homogeneous littoral boulder field of low trophic complexity off California. Four other invertebrate carnivores were studied for their collective impact on shelled molluscs and barnacles. *Octopus bimaculatus* was an important grazer of these shelled molluscs; this has been confirmed by Ambrose (1984, 1986) and Schmitt (1982, 1987). These studies did not consider the predators of octopus, but Taylor and Chen

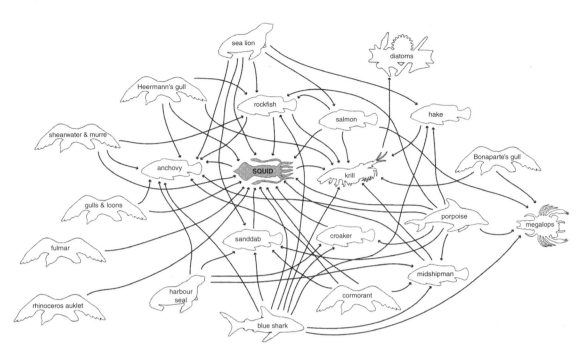

Figure 9.18 Food web of the California market squid *Loligo opalescens* (modified from Morejohn, Harvey & Krasnow, 1978, courtesy of California Department of Fish and Wildlife Office of Communications, Education and Outreach; and Boyle & Rodhouse, 2005: *Cephalopods. Ecology and Fisheries*, © Blackwell Science (2005), with permission from Wiley).

(1969) had considered the changing predator/prey roles between rockfish and *O. bimaculatus* as the animals grew; however, these were laboratory studies and require field corroboration.

Octopus vulgaris off the coast of West Africa occur in large numbers and are the subject of a major fishery (Caddy, 1983; Amaratunga, 1987). In this community, they may play an important role. Octopus biomass increased greatly in the 1970s. Part of the reason for this may have been fishing pressure on sparid fish; sparid stocks were depleted heavily during this period as octopus stocks increased. Caddy (1983) attributes this to the opportunistic behaviour of octopuses, particularly their flexible catholic diet (Chapter 4), their numerous ways of avoiding predation (Chapter 5) and their ability to reproduce quickly (Chapter 6).

Only one study has considered the possible behavioural decisions of a cephalopod in a specific community. On the basis of several years' study of *Octopus vulgaris* juveniles in the intertidal zone in Bermuda, Mather and O'Dor (1991) have identified the various (and often conflicting) paths of selective pressure

on these juveniles (Fig. 9.19). These small octopuses spent a surprisingly small amount of time foraging: only 12% of the daylight period. They made the same sorts of trade-offs that many other animals do when making the compromise between eating food and escaping predators. Enough is known about energetics in this species to conclude that they could grow much faster, given the available food, but were opting to eat less because of predation pressure.

9.9 Summary and Future Research Directions

Cephalopods generally 'live fast and die young', few surviving more than 2 years. Most are semelparous. Some produce a large number of small vulnerable progeny; others produce fewer large progeny that are developmentally more advanced, but there are intermediate conditions. There are no true larvae, but 'paralarvae' (i.e. hatchlings that are pelagic in near-surface waters during the day and have a distinctively different mode of life from older

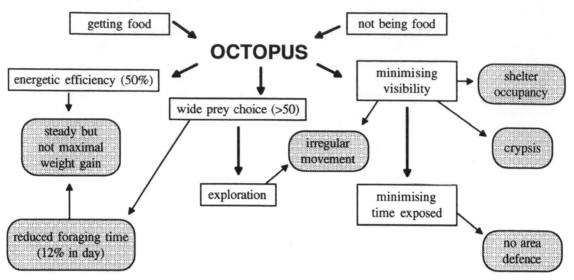

Figure 9.19 Hypothesised paths of selection pressure on juvenile *Octopus vulgaris* (from Mather & O'Dor, 1991: *Bulletin of Marine Science* **49**, 256–269, Foraging strategies and predation risk shape the natural history of juvenile *Octopus vulgaris*, Mather, J. A. & O'Dor, R. K., © (1991), by permission of Rosenstiel School of Marine Science/Allen Press). Shaded panels indicate the resultant behaviour and energetic results.

conspecifics) occur in many squid and octopods. Studies of life history strategies of cephalopods are still hampered by the lack of accurate ageing techniques.

Most cephalopods are asocial, apart from squid. This is associated with the promiscuous mating systems, lack of parental care, and the great mobility of most cephalopods, which seldom occupy an area for more than a few days. Among the social squid, many of which shoal, the costs and benefits of sociality have never been analysed (Alcock, 2013). Discovery and selection of mates is an obvious benefit of social organisation, and perhaps young squid benefit by observing and learning from adults. There may be behavioural plasticity conferred by shoaling, and it may be useful to use elective group size to measure squid perception of the current balance of costs and benefits of shoaling, as fish biologists have done (Krause *et al.*, 2000; Hoare *et al.*, 2004; Ward *et al.*, 2011).

Symbioses in cephalopods are rare, but some squid and octopus species have other interspecific associations. Foraging octopuses often have attendant fish that follow them and prey upon small invertebrates that escape their webs and arms. This is clearly to the advantage of the fish, but could draw attention to the octopus.

Competition for resources can be keen, especially for octopuses that may have to compete (intra- and interspecifically) for dens and food in the same habitat. Intraspecific competition aggression among octopuses for resources has been observed, but the nature of these fights and their relation to resource acquisition are not known in any detail. Temporal spacing and niche partitioning have been described briefly for coral reef habitats in Hawaii and the Caribbean, but these sorts of analyses have been rare, and are lacking for cuttlefish and nearly all squid.

There is abundant evidence that many cephalopods make migrations over distances to be measured in thousands of kilometres. For these long-distance migrations, there seems to be a pattern, in both hemispheres, of movement between higher latitudes, where feeding predominates, and lower latitudes, where spawning takes place. The functions of migration are not understood, nor are the factors that trigger it; in some situations, abiotic factors such as temperature seem to be a major factor, but there may be biological reasons for large-scale displacements, such as movements of prey. Nor is it clear how cephalopods orient during migrations.

In common with other marine animals many cephalopods, adults and paralarvae, undertake diel vertical migrations at dawn and dusk. There is great variation in the extent of these excursions, but some species travel 700–800 m twice daily. Recent data from tagging studies has shown that the vertical distribution of some cephalopod species in the sea is much more variable and complex than previously thought.

Little is known about the daily activity patterns of cephalopods, and SCUBA divers could make a valuable contribution here, at least for shallow-water forms. In the deep sea, this is not so easy, of course. Yet this world is beginning to be opened up now, and our understanding of the sense organs and light organs of deep-sea cephalopods continues to advance. It has been experimentally established in cephalopods that some of the ventral light organs function in counter-illumination, and emit light of the appropriate intensity to match the downwelling light at a given depth. On the other hand, our understanding of the life and habits of giant, colossal and big-fin squid remains tantalisingly incomplete, despite some exciting new data.

Few examples of food webs and community structure are available from the literature. It is obvious that squid occupy mid-to-high trophic levels in some ecosystems, but few generalisations can be made about the food webs of octopuses.

CHAPTER TEN

Nautilus

The Pearly Nautilus, renowned for its beautiful shell, is so different from all other cephalopods that it must be considered separately, especially as the differences extend to its behaviour (Table 10.1). Unlike the coleoids, these animals live primarily in an olfactory world, 'smelling and groping' rather than 'visually spotting and attacking' (Saunders, 1985). Thus they are probably far more like their ancient molluscan ancestors than present-day coleoids (Ward, 1987; Shigeno *et al.*, 2008; Saunders & Landman, 2009).

10.1 Major Body Features

Among living cephalopods, only *Nautilus* has an external chambered shell (Fig. 10.1). This serves not only for protection but also as a buoyancy device. There is a series of chambers in the shell, connected by a tube called the siphuncle. Most of the chambers contain gas, although in the newest (youngest) chambers there may be large amounts of liquid, which can be partially removed to alter the density of the animal and so maintain neutral buoyancy (Denton & Gilpin-Brown, 1966; Ward, 1987). The shell bears irregular brown stripes, making a pattern rather like the Weak Zebra pattern of *Sepia officinalis* (Hanlon & Messenger, 1988). It needs emphasising that this is a fixed pattern, for there are no chromatophores in these animals. The stripes are probably disruptive (Cowen, Gertman & Wiggett, 1973; Packard, 1988a), but in mature animals they are restricted to the upper part of the shell and, together

Table 10.1 Some differences between *Nautilus* and other living cephalopods

	Nautilus	*Coleoids*
Shell	External	Internal or absent
Chromatophores	None	Generally conspicuous
Ink sac	None	Usually present
Eye	Pinhole	Lens and iris
Olfactory organ	Large	Very small
Appendages	About 90 very thin tentacles	Eight arms (plus two tentacles in decapodiformes)
Suckers	None (although tentacles are adhesive)	Generally conspicuous
Salivary toxins	Absent	Present (in several genera, but see Chapter 4)
Feeding strategy	'Smelling and groping'	'Visual spotting and attacking'
Growth	Relatively slow (egg incubation 1 year)	Extremely fast
Longevity	More than 20 years	1–2 years (rarely 4 or 5)
Reproduction	Iteroparous	Mostly semelparous (but see Chapter 6)
Eggs	Large (27 mm), laid singly or in small groups over many months	Usually small, laid in batches of 10^2–10^6 generally over a relatively short period

with the pigment on the hood and upper parts of the tentacles, may contribute to countershading (Cott, 1940). In young animals, the shell is striped all over (Fig. 10.2a).

The body nestles in the last chamber of the shell, into which the animal can partially retract, leaving the very tough 'hood' to seal the opening rather like the operculum of a gastropod (Fig. 10.2b). Withdrawal into the shell is the animal's only known defence tactic, and it is noteworthy that *Nautilus* can survive for long periods without oxygen (Wells, Wells & O'Dor, 1992). Indeed, the low metabolic rate induced by hypoxia (Boutilier *et al.*, 1996; Boutilier *et al.*, 2000) may have been important in the evolution of these animals. There is a conspicuous, mobile funnel, formed from two separable flaps of muscle rather than being a single tube as in coleoids. Water in the mantle cavity is expelled through this funnel, which can be pointed in any

 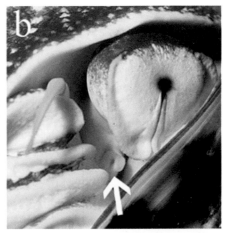

Figure 10.1 (**a**) An adult *Nautilus pompilius* swimming at night on a deep reef at Milne Bay, Papua New Guinea (photo © Fred Bavendam). (**b**) Live nautilus in the laboratory, illustrating the pinhole-camera-type eye and, just below it, the rhinophore (arrow) that is used for distance odour detection (from Basil *et al.*, 2000).

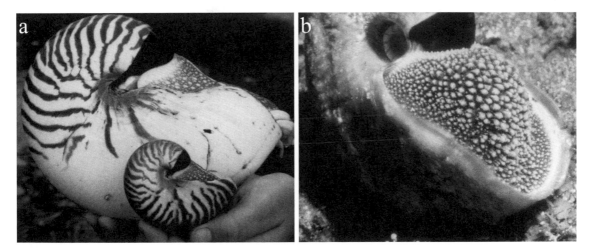

Figure 10.2 (**a**) Adult and young shell markings (original W.B. Saunders). (**b**) Defensive behaviour of a *Nautilus* is snail-like. *Nautilus scrobiculatus* retracts into its shell, leaving the tough hood to block the shell aperture (from Saunders & Landman, 2009: *Nautilus: The Biology and Paleobiology of a Living Fossil*, Plate III. New York: Plenum Press. © Plenum Press, with permission from Springer).

direction, and swimming in *Nautilus* is entirely by jet-propulsion. Pressure for the jet comes from either the wings in the funnel (for slow swimming) or by head retractions (for rapid swimming) (Packard, Bone & Hignette, 1980). The locomotion of *Nautilus* has been studied extensively by Chamberlain (1987).

The head bears large eyes and about 90 slender, delicate tentacles, which are substantially different from those of coleoids, although their musculature is organised in a similar manner (Kier, 1987). There are several groups of tentacles (Fig. 10.3): (1) a pair of post-ocular tentacles, whose function is still unclear, but which may bear mechanoreceptors (Barber & Wright, 1969), and a pair of chemosensory pre-ocular tentacles (Bidder, 1962; Fukuda, 1980); (2) 19 pairs of digital tentacles, some of which are termed lateral digital tentacles, and all of which bear chemoreceptors (Ruth *et al.*, 2002); and (3) numerous labial (or buccal) tentacles internal to these. The ocular and digital tentacles are composed of an extendable inner cirrus and an outer sheath into which the cirrus can be withdrawn. Encircling the cirrus are numerous fine circular grooves, and on the oral side of these there is a thick epithelium with many cells containing mucopolysaccharide (Muntz & Wentworth, 1995; von Byerna *et al.*, 2012) that apparently make the tentacle adhesive and enable the animal to grasp and manipulate food.

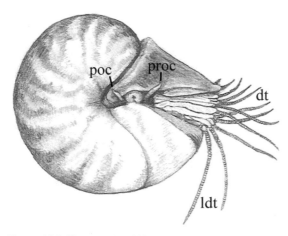

Figure 10.3 The tentacles of *Nautilus*: the post-ocular (poc) and pre-ocular tentacles (proc); the digital tentacles (dt); and the lateral digital tentacles (ldt). The labial (or buccal) tentacles are internal to the digital tentacles and are not shown (original by R.J. Crook).

The tentacles are supplied with numerous sensory cells, the ultrastructure of which has been studied by Ruth *et al.* (2002), who recognised no fewer than nine types of ciliated cells, four of which appear to be chemosensory, three mechanosensory and two of unknown function. Chemosensory cells were found on the pre- and post-ocular tentacles and on the lateral digital tentacles.

Bidder (1962) was the first to describe how the tentacles respond in 'alert' animals. The ocular tentacles and the most lateral digital tentacles are held out sideways, but if food is located they extend still farther (up to 10 cm), as do numerous other median digital tentacles, so that they form a 'cone of search' reaching towards the direction of putative food while the animal circles to locate it. Another set of inner digitals is used to manipulate any food into the mouth. Some of the labial tentacles are modified into sexual organs, including the male's large spadix, used to transfer spermatophores during copulation.

The functions of the various tentacles have been clarified by Basil *et al.* (2005). She and her colleagues recorded the responses of nautiluses in a flume to fish-extract pipetted unilaterally onto the pre-ocular, post-ocular and digital tentacles, and onto the rhinophore (see below). These careful experiments showed that stimulation of the digital tentacles resulted in the animal's immediately releasing its lateral digital tentacles from their sheaths and swimming towards the substrate with the medial digital tentacles making contact with the odour source ('touch bottom'), whereas stimulation of the pre-ocular tentacle mainly elicited the 'cone of search' response, although also a weak 'touch bottom' response. Stimulation of the post-ocular tentacles gave inconclusive results, and Basil *et al.* (2005) speculate that they may have a 'hydrodynamic function during orientation'.

These experiments also confirmed the importance of the paired *rhinophores* in distance chemoreception. These large organs, projecting from below each eye (Fig. 10.1b), have long been held to be olfactory organs on the basis of their ultrastructure (Barber & Wright, 1969); each projects to the brain in a large 'olfactory' nerve (Young, 1965a). In the experimental flume apparatus, fish-extract directed at the rhinophore consistently elicited the distinctive 'far field' cone-of-search behaviour, usually instantly (and always within

10 seconds): that is, the animal swam towards the odour source with digital (and sometimes ocular) tentacles extended. Significantly, fish-extract directed at the rhinophore rarely elicited 'near-field' behaviours, such as 'touch bottom', so that the *Nautilus* brain clearly possesses separate sensory-motor pathways for dealing with distant and close odours.

However, Basil and her colleagues had already demonstrated the great importance of the rhinophores in the life of *Nautilus* in an earlier paper (Basil *et al.*, 2000). They showed that under dark conditions in a flume, nautiluses detect and follow, in three dimensions, turbulent odour plumes to the source even when the source is as much as 10 m distant. This ability is abolished by unilateral or bilateral blocking of the rhinophores with plugs of petroleum jelly: such animals detect odours but cannot track the plume and locate the source (Fig. 10.4).

Yet the same individuals tested after they had lost the plugs were able to locate the odour source. This unequivocal demonstration that *Nautilus* uses the rhinophores for distance chemoreception is in agreement with the important histological evidence of Ruth *et al.* (2002) and the considerable anecdotal evidence that *Nautilus* locates its prey using chemical information. Fishermen and biologists capture them by lowering baited traps, often at night; and all the evidence suggests that the animals forage on the reef face, eating prey that they have located by olfactory rather than visual clues, often in conditions of very low light (see below). Finally, we should note that there is now good evidence that nautiluses may use olfactory information to locate a mate (§10.3).

The conspicuous eye (Fig 10.1b) is not only very different from the coleoid eye, it is unique in the animal kingdom because it lacks a lens and is constructed like a pinhole camera. This must have profound consequences for sensitivity as well as resolution. Land and Nilsson (2012) have calculated that with a wide-open pupil the image must be 400 times dimmer than in the *Octopus* eye. Visual acuity is also much poorer than in octopuses, fish or seals: according to Muntz (1991), it is at least 60 times worse than in any of these other marine animals. Nevertheless, the pupil changes diameter with light intensity, and there is also a marked optomotor response (Muntz & Raj, 1984). O'Dor, Wells and Wells (1990) suggest that this may help nautiluses to maintain

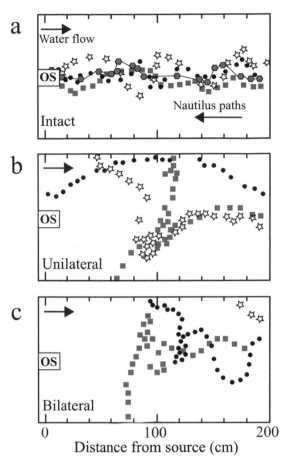

Figure 10.4 Paths of *Nautilus pompilius* locating an odour source (OS) in a laboratory flume, illustrating the effects of temporarily blocking one or both rhinophores with a petroleum jelly plug. (**a**) Intact animals (*n* = 3, each a different symbol) find the odour source, as does a nautilus tested after it had ejected both petroleum jelly plugs (hexagonal symbols connected by line). (**b**) Unilaterally blocked animals were unable to find the source. (**c**) Bilaterally blocked animals showed highly erratic paths and did not get near the odour source. (Modified from Basil *et al.*, 2000, with permission from the Company of Biologists.)

their station on a reef-front in a strong current. A visual fix, even one made by a poor eye operating at very low light levels in the depths, could serve to tell an animal out of physical contact with the bottom whether or not it is being moved (by a current) and help establish the direction of a chemical source (M.J. Wells, pers. comm., 1994). The eyes, which are equipped with extraocular muscles, also show compensatory movements under the influence of receptors in the statocyst (Hartline, Hurley & Lange, 1979). Moreover, each eye contains four million receptors, far more than occur in the eye of any other non-cephalopod mollusc (Messenger, 1981). And although the receptors are not organised with their microvilli in a strict two-plane system, as in coleoids (Muntz & Wentworth, 1987), we should be cautious before discounting visual influences on the life of *Nautilus*. For example, it has been shown experimentally that *Nautilus* is positively phototactic: they move towards the light or towards the brighter of two lights when given the choice (Muntz, 1987, 1994a). Muntz (1994b) also found that illuminated traps increase the catch of *Nautilus*, suggesting that the eyes may respond to the various kinds of bioluminescence common in the deep sea. It is clear from behavioural and biochemical studies that the spectral sensitivity of the *Nautilus* eye is well adapted to its deep-water habitat: the λ_{max} of *Nautilus* rhodopsin is very short indeed (465–467 nm), but this would also enable it to respond to the bioluminescent output of many deep-sea animals (Muntz, 1986, 1987; Hara *et al.*, 1995).

In the arrangement of the internal organs, too, there are profound morphological differences between *Nautilus* and the coleoids; for example, there are four (rather than two) gills and four 'kidneys'. There is no ink sac. The well-differentiated gut has no posterior salivary glands secreting toxins to subdue large, active prey. The buccal mass is massive, with large and powerful beaks bearing heavily calcified tips, presumably for grinding and crunching hard coral and crustacean moults; there is also a curious, and very large, radular appendage, part secretory, whose function remains obscure but which may keep the radular teeth clean (Messenger & Young, 1999). The gut is chiefly remarkable for its relatively enormous crop, which can enlarge to four times its original size (Westermann *et al.*, 2002). This suggests that opportunistic feeding may prevail; yet although there is experimental evidence (Mangold, cited in Ward, 1987) that food may remain in the crop for at least 2 weeks, an X-ray analytical study by Westermann and her colleagues showed that in *Nautilus pompilius* food enters the stomach in 20 minutes, chyme reaches the midgut gland in 3 hours and the rectal loop in only 5 hours (Westermann *et al.*, 2002). The total time between food intake and

elimination is 12 hours, approximately the same as in benthic octopods and sepioids.

The brain is also different from that of coleoids, although it is organised along a similar plan (Young, 1965a), with supra- and sub-oesophageal regions, peri-oesophageal (magnocellular) lobes and, laterally, the optic lobes. The magnocellular lobes are not well developed, nor are there basal or peduncle lobes in the supra-oesophageal brain, perhaps because the animal does not make the fast, precisely timed ballistic movements that characterise coleoids (Chapter 2). The statocysts (§2.1.2) are simpler than in coleoids, open to the exterior and with no division into crista and macula. There are numerous free statoconia, many receptor cells (primary hair cells with kinocilia) and a large volume of endolymph (Neumeister & Budelmann, 1997). The receptor cells are of two types: type A, which occur in the ventral half of the statocyst, where the statoconia will lie when the animal is in the upright position; and type B, which are in the dorsal half. These features alone suggest that as well as detecting gravity, nautiluses might be able to detect small positional changes and rotatory movements, and this has been confirmed by a series of careful experiments (*ibid.*). When a restrained nautilus is rotated about its vertical body axis, there are horizontal, phase-locked movements of the funnel: either in the opposite sense (compensatory) or same sense ('funnel-follow') as the imposed rotation. The statocysts of *Nautilus* can therefore respond to angular as well as linear acceleration (gravity). This must be important in its life next to the coral reef face (see below), enabling it to compensate for passive displacement in the yaw plane caused by currents; the animal is stable in the roll and pitch plane. It is noteworthy that the positioning of the funnel is also under visual control and there is a marked optokinetic response (*ibid.*): the funnel turns in the opposite direction to the visual stimulus so as to reduce retinal slip. This finding is further evidence that, although nautiluses are primarily 'macrosmatic', their eyes may be more important than is usually accepted. The optic lobes are much less developed than in coleoids. They constitute about one-sixth of the brain in volume, as opposed to *Sepia officinalis* in which they are nearly three times larger. The outer cortical region shows none of the precise organisation found in coleoids, and there is no dorso-ventral chiasm of the optic nerves

(Chapter 2), which suggests that visual processing in *Nautilus* may not include detailed form perception, although they can use visual stimuli of different shapes to locate a previously learned spatial location (Crook, Hanlon & Basil, 2009; Crook & Basil, 2013). The olfactory lobes, however, are larger and more complex than in coleoids, and the olfactory nerve from the well-developed rhinophore is large. There are numerous nerves from the digital tentacles and conspicuous nerves from the ocular tentacles, emphasising that chemosensory input must play a larger part in the life of *Nautilus* than it does in most coleoids.

10.2 Behaviour and Ecology

There are from three to ten species of *Nautilus* (a new genus, *Allonautilus*, has recently been established: Ward & Saunders, 1997) all living in a vast warm-water area extending from the Indian Ocean to Australia, Samoa and the Philippine Islands. Wherever the animals have been studied in detail, such as in New Caledonia, Palau, New Guinea, Fiji, and Osprey Island, Australia, they occur on the steep, fore reef slopes of hermatypic coral reefs (never in the lagoons) and occur from the surface down to about 600 m. Ever since Willey's (1899) pioneering studies, it had been supposed that *Nautilus* underwent a daily vertical migration, but it was only fairly recently that direct confirmation of this was obtained. Two groups of investigators working in Palau (Carlson, McKibben & DeGruy, 1984; Ward *et al.*, 1984) attached sonic transmitters to the shells of *N. belauensis*, released them and tracked them from the surface in small boats. They obtained good evidence for vertical movements from several individuals, and favourable weather conditions enabled one particular animal to be followed for six consecutive days and nights (Fig. 10.5a). A clear diel rhythm was shown: upward (at about 120 m per hour) prior to sunset to depths around 150–100 m, downward (at about 180 m per hour) at sunrise to about 250–350 m. The suggestion that it is change in light intensity that triggers vertical movement is supported by the fact that the overriding rhythm can be interrupted under certain circumstances. Thus, on June 28 (Fig. 10.5a), Ward and his colleagues recorded a sudden descent, beginning at about 7:30 pm when a full moon rose, and lasting for 2 hours. Such beautiful data provide striking evidence

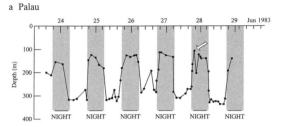

a Palau

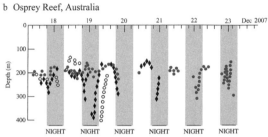

b Osprey Reef, Australia

Figure 10.5 (a) Vertical movements of a single tagged *Nautilus belauensis* in Palau. Note the sudden descent (arrow) one night at the onset of a full moon (reprinted from Ward *et al.*, 1984, by permission from Macmillan Publishers Ltd: *Nature* **309**, 248–252, Remote telemetry of daily vertical and horizontal movement of *Nautilus* in Palau, Ward, P. D., Carlson, B., Weekly, M. & Brumbaugh, B., © 1984). **(b)** Vertical movements of three tagged *N. pompilius* (each animal has a different symbol) at Osprey Reef, Australia (from Dunstan, Ward & Marshall, 2011a: *PLoS ONE* **6**, e16312, *Nautilus pompilius* life history and demographics at the Osprey Reef Seamount, Coral Sea, Australia, Dunstan, A. J., Ward, P. D. & Marshall, N. J.). Note the differing day/night activities of these two species.

for a diel rhythm in this species, and subsequent records confirm this (Ward, 1987; O'Dor *et al.*, 1993). The findings agree with other evidence for vertical migration such as capture records and oxygen isotope measurements on shell formation (Ward, 1987; Zakharov *et al.*, 2006).

However, recent extensive field work on *N. pompilius* at Osprey Island, based on data obtained by tracking and from ROVs, has revealed a more complex pattern of vertical movements (Dunstan, Ward & Marshall, 2011a; Fig. 10.5b). The animals in this population showed continuous nightly movements between 130 and 700 m; in the day, they either stayed between 160 and 225 m or were found between 490 and 700 m, exhibiting a good deal of active foraging. The authors suggest that vertical distribution may be influenced by the search for optimal feeding grounds, the avoidance of daytime visual predators, the need to rest to regain neutral buoyancy, an upper temperature limit of 25 °C and an implosion depth of 800 m, or any combination of these factors. Food availability may be the major factor in depth distribution.

Because *Nautilus* never moves far from the reef face, many 'vertical' migrations can be better thought of as onshore/offshore migrations, and there is no doubt that vertical migrations are superimposed on horizontal migrations, as tag–release experiments over a 14-month period in Palau have shown (Saunders & Spinosa, 1979). Animals of both sexes may remain at or return to the same site, or may move independently of currents for long distances along the reef front (Fig. 10.6). One animal was observed to travel 30 km in only 10 days; the longest journey recorded was 150 km in 332 days. These rates of travel (more than 0.5 km per day) are easily within the capability of *Nautilus* based on the swimming speeds calculated by Packard, Bone and Hignette (1980), by Chamberlain (1987) and by O'Dor, Wells and Wells (1990): around 0.16 to 0.33 m per second or about 0.5 to 1 km per hour. Recent tracking data in Osprey Island has shown that *N. pompilius* are also quite mobile, with maximum long-term movements of 5–29 km over 9 to 52 days (Dunstan, Bradshaw & Marshall, 2011).

At one time, there was a debate about whether *Nautilus* is nocturnal. Zann (1984) obtained evidence, in the aquarium, that *N. pompilius* is more active at dusk and also reported that *Nautilus* exhibits characteristic 'activity–rest' cycles throughout the day and night: brief bouts of swimming alternating with long rests. However, Saunders (1984) found that *N. belauensis* in the field was equally active in the day and, indeed, discovered bait more quickly in the daytime. A high-resolution telemetric tracking study of a single individual showed that it was more active at night, generally making crepuscular vertical excursions (O'Dor et al., 1993). The recent longitudinal study of *N. pompilius* at Osprey Island (Dunstan, Ward & Marshall, 2011a) puts these data into perspective, however: in this species, at this site, there is considerable variation in individual migratory behaviour (Fig. 10.5b), so that we must exercise caution before making simplistic generalisations about 'nocturnal/diurnal' activity.

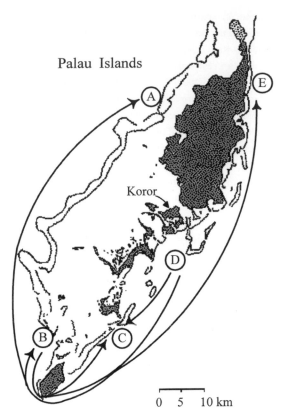

Figure 10.6 Long-term horizontal movements of *N. belauensis* around the reefs of Palau. Four tagged animals released at B and D were recaptured 10–12 months later at A, B, C and E, 40–150 km away (from Saunders & Spinosa, 1979: *Science* **204**, 1199–1201, *Nautilus* movement and distribution in Palau, Western Caroline Islands, Saunders, W. B. & Spinosa, C., © (1979), with permission from AAAS).

The animals, adults and juveniles, appear to feed on any food item they can find, and the general belief seems to be that they locate food by sampling lateral currents sweeping across the reef face for chemical trails (O'Dor et al., 1993). The cast-off exoskeletons or moults of lobster, various kinds of crabs (especially hermit crabs; Ward & Wicksten, 1980) and other crustaceans, nematodes, deep-water echinoids, fish, and even pieces of chicken put down as bait, have been recovered from gut contents (Saunders & Ward, 1987). Apparently, it has long been known that fresh chicken is the best bait for them (Haven, 1977). M.J. Wells (pers. comm., 1991) has found small bivalves and, very often, annelid

chaetae in the crop and suggests that *Nautilus* may
sweep the bottom and dig in fine mud with its tentacles.
Certainly Ward and Hewitt (pers. comm., 1991) were
able to elicit directed digging behaviour from a *Nautilus*
by burying lobster carapaces filled with fish meat under
the substrate. The beaks of coleoids and fragments of
Nautilus tentacles have also been found in the gut, and
according to Carlson (1987) cannibalism is not
uncommon, at least under aquarium conditions.

Fish, including sharks, recovered from traps with
Nautilus have been found to bear their beak marks
(Saunders & Ward, 1987), but this is almost certainly the
result of confinement within a trap and, although
Nautilus may be a predator as well as a scavenger, there
are no observations under natural conditions to support
this idea. It is surely not agile or swift enough to stalk
and capture a healthy fish or prawn in mid-water even if
its vision were adequate to the task, which seems
unlikely. Some kinds of food may be detected visually,
for *Nautilus* may be able to detect bioluminescent
organisms, the bioluminescent clouds produced by
caridian shrimps (King, cited in Zann, 1984), or
luminescence produced as a result of decay. Although
its spectral sensitivity curve is typical of a deep-sea
animal, it also matches the bioluminescence of many
deep-sea organisms (Lythgoe, 1972).

The predators of *Nautilus* include other *Nautilus*,
octopuses, teleosts, sharks and, in aquaria at least,
turtles (Ward, 1987). Predation by octopuses is inferred
from the frequent presence, in the shells of living
animals and in empty shells (Tucker & Mapes, 1978), of
appropriately shaped bore-holes (Nixon, 1979). It is
interesting, incidentally, that there is some evidence
that octopuses drill holes predominantly in the posterior
part of the body chamber, in the region of the retractor
muscles and viscera (Saunders, Spinosa & Davis, 1987)
(Fig. 10.7 and Fig. 4.13). Direct evidence of predation by
octopuses has been provided by Arnold (1985), who
found an *Octopus cyanea* in a trap with six *N. pompilius*,
one of which had been bored and partially eaten. It now
appears that octopuses are major predators whose
impact on *N. pompilius* and *N. scrobiculatus* populations
may be limiting (Saunders, Knight & Bond, 1991).

Direct evidence of predation has been obtained in
Palau (for *N. belauensis*) and in Papua, New Guinea (for
N. pompilius) (Saunders, Spinosa & Davies, 1987). At the
former site, triggerfish (*Balistoides viridescens*) were

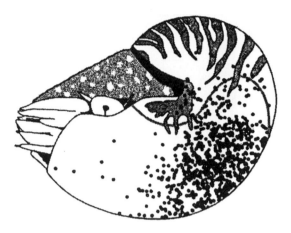

Figure 10.7 Predation by *Octopus*. Composite diagram showing
positions of 779 bore-holes (black spots) found in live
N. pompilius shells. Note that the borings are concentrated over
the body chamber, the site of the retractor muscles and viscera
(from Saunders *et al.*, 1991: *Bulletin of Marine Science* **49**,
280–287, *Octopus* predation on *Nautilus*: evidence from Papua
New Guinea, Saunders, W. B., Knight, R. L. & Bond, P. N., ©
(1991), by permission of Rosenstiel School of Marine Science/
Allen Press).

recorded attacking animals at 20 m; at the latter site, a
grouper (*Epinephalus* sp.) was seen attacking an
individual against the reef-front at 27 m. In Papua New
Guinea, deep-water remote photography has recorded a
snapper (*Etelis carbunculatus*) attracted to a baited trap
at 270 m containing *N. pompilius* and *N. scrobiculatus*.
At 300 m off Osprey Island, large groupers (ca. 2 m total
length) have been filmed 'inhaling nautiluses, biting
them head on and spitting them out' (P.D. Ward, pers.
comm., 2012). These and other observations confirm that
Nautilus is vulnerable to attack by large teleosts with
powerful jaws (Weekley & Ward, cited in Ward, 1987).
Teleosts dominate the upper 75 m of the reef, but in
deeper water Saunders (1984) regularly observed
elasmobranchs in his photographic sequences, especially
the large sharks *Hexanchus griseus* and *Enchinorhinus
cookei*, and these should be considered potential
predators too. In some areas humans, too, are seriously
depleting nautilus populations by unsustainable fishing,
mainly to serve the ornamental shell trade (Dunstan,
Alanis & Marshall, 2010).

Taking these results together, a picture begins
to emerge of the life of *N. belauensis* in Palau,

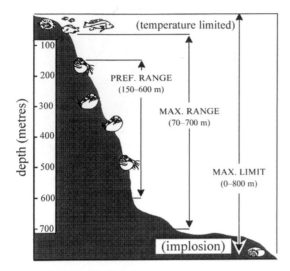

Figure 10.8 Vertical distribution, range and depth limits of *Nautilus belauensis* in Palau (from Saunders, 1984, by kind permission of the Paleontological Society) combined with recent data from *N. pompilius* in Australia showing that this species can dive to at least 700 m (Dunstan, Ward & Marshall, 2011a).

Figure 10.9 The deepest record of a nautilus: *N. pompilius* feeding on bait during daytime at 703 m at Osprey Reef in Australia (from Dunstan, Ward & Marshall, 2011a: *PLoS ONE* **6**, e16312, *Nautilus pompilius* life history and demographics at the Osprey Reef Seamount, Coral Sea, Australia, Dunstan, A. J., Ward, P. D. & Marshall, N. J., © 2011).

N. macromphalus in New Caledonia (Saunders & Ward, 1987) and *N. pompilius* in Papua, New Guinea (O'Dor *et al.*, 1993; Dunstan, Ward & Marshall, 2011a, b). Foraging on the fore reef slope and never straying far from it, they scavenge and feed opportunistically on whatever food is available: they probably detect it largely by following patchy, turbulent chemical plumes. In so doing, they can move fairly long distances horizontally and will probably change depth considerably. In Palau, they live mostly at 150–300 m, but they can descend much deeper and rise to 70 m, or even to the surface in New Caledonia (Fig. 10.8). In Australia, *N. pompilius* commonly forage to 400 m (Fig. 10.5b) and appear to feed normally at 700 m (Fig. 10.9). The lower limit will probably be set by pressure constraints: high pressures can cause siphuncle rupture and ultimately shell implosion (Denton & Gilpin-Brown, 1966; Jordan, Chamberlain & Chamberlain, 1988). It seems certain that the lower limit is not set by oxygen requirements, for Wells, Wells and O'Dor (1992) and Staples, Webb and Boutilier (2003) have shown that these animals have a remarkable tolerance of low oxygen tensions. The upper limit will

be set by temperature, for values in excess of 25 °C are probably adverse for *Nautilus*, and also by the presence of teleosts and other predators. Perhaps it is for these reasons that diel movements take them away from the surface during the daytime.

As we have seen, the triggering mechanisms for the downward and upward movements are not understood. They may include light, but Jordan, Chamberlain and Chamberlain (1988) have experimental evidence that *Nautilus* is quite sensitive to increases in pressure, and responds by swimming upwards.

10.3 Reproduction and Growth

The sexes are separate, with males slightly larger and, in all the populations sampled, males far outnumber females (Saunders & Ward, 1987). In a population of *Nautilus pompilius* on Osprey Island, there were 83% males to 17% females (Dunstan, Bradshaw & Marshall, 2011). Juveniles represent less than 10% of the population, evidence of low fecundity. It is not clear how sexual recognition takes place; there seems to be nothing resembling courtship as this is known in coleoids (Chapter 6). According to Arnold (1987), males will approach and initiate copulation 'with anything of the general shape and size of another *Nautilus*'. A new *Nautilus* introduced into a tank will be grasped irrespective of sex or species. Perhaps some kind of olfactory recognition takes place, because when a *Nautilus* is removed from a tank containing several animals, its shell wiped with a paper towel and then replaced in the tank, it will immediately be subject to copulation attempts from any males present (Carlson, quoted in Arnold, 1987). Males grasped in this way will eventually be released but females will be forced into copulation: the hoods will be brought together, the shell apertures apposed and the tentacles grasped. At the end of this preliminary phase, during which the female appears passive or even retracts into the shell, the animals appear locked together at a slight angle (Fig. 10.10). This is to accommodate the male's spadix (Fig. 10.10b), a large modified labial tentacle whose function may be to push aside the female's buccal tentacles so that the spermatophore can reach the organ of Valenciennes, where the female stores it. The details of how the spermatophore is passed to the female are not understood. Some authors describe copulation as lasting

for many hours (Mikami & Okutani, 1977; Arnold, 1987), but in *N. belauensis* Kakinuma *et al.* (1995) found it only lasted about an hour. The male often bites the female's shell or body during copulation. Nothing is known about multiple matings or sperm competition. Because of the relative ease with which *Nautilus* can be kept in aquaria, it is to be hoped that someone will address this problem soon.

It has become apparent that nautiluses can be attracted to conspecifics by chemical cues, at least in the laboratory. Basil *et al.* (2002), testing animals in a simple choice chamber in the dark, showed that female *Nautilus* chose the compartment with male odour significantly more than the branch with no odour; they also spent more time in the 'male compartment' once they had chosen it. Moreover females, but not males, extended their tentacles fully when tracking conspecific odour. No experiments were made to determine whether the rhinophores were involved in this tracking behaviour. In another study, Westermann and Beuerlein (2005), who had earlier described a secretory rectal gland in the *Nautilus* gut (Westermann & Schipp, 1998), investigated

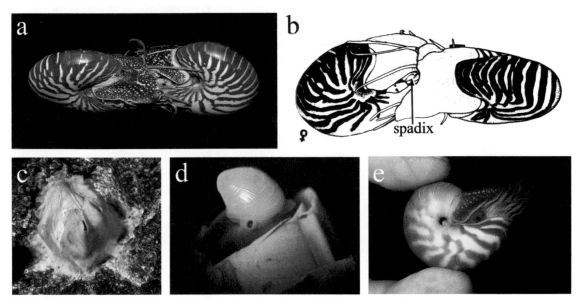

Figure 10.10 *Nautilus pompilius* reproduction. (**a**) Mating in a laboratory tank (photo R. Hanlon). (**b**) Top view illustrating the male's spadix to the right of the female's eye (from Arnold, 1987: *Nautilus. The Biology and Paleobiology of a Living Fossil*, ed. W. B. Saunders & N. H. Landman, 353–372, © Plenum Press, with permission of Springer). (**c**) A single egg case that is ca. 3 cm wide. (**d**) Developing embryo in an egg case that has been cut away; shell diameter ca. 1 cm. Note the dark eyespot and the overall resemblance to a snail. (**e**) A hatchling nautilus ca. 2.5 cm long. (Photos **c** and **d** by B. Carlson; photo **e** courtesy of Waikiki Aquarium, University of Hawaii.)

whether extracts of this gland attracted conspecifics. They used a Y-maze and compared the responses of animals to frozen shrimp, to homogenates of gill, mantle or rectal gland. The animals did not respond to gill or mantle homogenates but responded positively to shrimp with lateral extension of the ocular and some digital tentacles. A similar positive response was made to female rectal gland homogenates, but only by *mature males*, not by immature male or females. The authors suggest that the female rectal gland may be producing pheromones that are released in the faeces, which on the reef face might act as scent marks to bring solitary males and females together.

The eggs are remarkable in that they are among the largest of invertebrate eggs: up to 40 mm in *N. belauensis* (Kakinuma *et al.*, 1995). They are enclosed in a flexible, tough milky-white case (Fig. 10.10c, d) and are laid singly or in small batches often over an extended period (Martin, Catala-Stucki & Ward, 1978; Carlson, Awai & Arnold, 1992). Fecundity is low, only 10–20 eggs per year in the aquarium (Uchiyama & Tanabe, 1999). Although it was once thought that the eggs were laid in deep, cool water (Ward & Martin, 1980), recent experiments in captivity have shown that, for *N. belauensis* at least, warm temperatures are necessary for successful development (Carlson, Awai & Arnold, 1992). Such temperatures (21–25 °C) correspond to depths of 80–100 m in Palau, and this suggests that the eggs may be laid at these depths during the night ascent (Carlson, McKibben & DeGruy, 1984; Ward *et al.*, 1984). Landman *et al.* (1994) demonstrated that hatching of *N. belauensis* in a variety of sites is likely to occur at between 100 and 200 m, and for *N. pompilius* in Fiji, Landman, Jones and Davis (2001) have evidence that hatching occurs at 160–210 m, followed by a descent to 300–370 m (cf. Zakahrov *et al.* (2006). However, the recent detailed study of *N. pompilius* in Osprey Island, extending over 10 years, yielded no data about what the authors term 'the mysteries of egg location . . . or behavior of post hatching juveniles' (Dunstan, Ward & Marshall, 2011b). The incubation time is surprisingly long: at least 9 months, sometimes nearly a year (Uchiyama & Tanabe, 1999). This need for warm water year-round may explain why *Nautilus* is limited to the tropics (B.A. Carlson, pers. comm., 1991). *Nautilus* hatchlings have only been observed and studied in an aquarium (Waikiki, in Hawaii: Carlson, Awai & Arnold, 1992). They are neutrally buoyant at hatching and feed at once on chopped shrimp; they are quiescent during the day and active at night. In an aquarium at 22 °C, their growth is rapid, perhaps because of superabundance of food or perhaps because at atmospheric pressure it is much easier to pump the shell chambers empty. Depth is undoubtedly an important regulator of growth; indeed at depths greater than 250 m it becomes impossible to pump liquid from the shell chambers (Ward, 1987).

The growth and development of *Nautilus* has been the subject of great interest and, despite exceedingly different estimates of age and growth rates, there is a consensus that these animals may not become sexually mature for many years in the wild (Landman & Cochran, 1987), although they can reach maturity in 3 years in aquaria. In the laboratory *Nautilus pompilius* may reach sexual maturity in 8 years, and Westermann *et al.* (2004) suggest that the total age could exceed 12 years. Data from the same species collected over many years in Osprey Island, however, suggest that in this location they take over 15 years to reach maturity, and must live to 20 years or more (Dunstan, Ward & Marshall, 2011b). This, of course, contrasts markedly with the coleoid condition, where sexual maturity is generally achieved in well under a year (Chapter 6). Saunders (1983) also had evidence of long survival in males after the attainment of maturity, and it is known that egg laying is repeated from year to year (Martin, Catala-Stucki & Ward, 1978; Arnold, 1987). The ovaries always contain oocytes of several sizes, with 6–10 oocytes in any one size class. The exact number of oocyte size classes has not yet been determined, but it is clear that *Nautilus* is not semelparous.

It is worth emphasising that the life history of *Nautilus* differs markedly from that of coleoids; so much so that for some authors its 'low fecundity, late maturity, long gestation and long life span', together with the lack of a larva for dispersal, render it vulnerable to over-fishing (Dunstan, Bradshaw & Marshall, 2011).

10.4 The Development of Behaviour and Learning

There are few data on the development of behaviour. Wells (1962a) cites an interesting unpublished

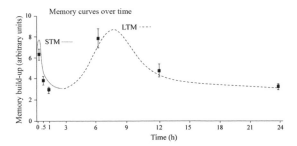

Figure 10.11 *Nautilus pompilius* showing biphasic memory of an associative learning task. Dotted and dashed lines indicate hypothetical short-term and long-term memory curves, STM and LTM (see text; modified from Basil *et al.*, 2011, with permission of Elsevier Masson, based on data in Crook & Basil, 2008).

experiment by Anna Bidder, who found that over a 10-day period the average time taken by a *Nautilus* to extract a piece of fish from a glass tube fell from over a minute to fewer than 20 seconds. This was the first, and until recently the only, hint of learning in these animals. In recent years, however, Basil and her collaborators have provided unequivocal evidence of learning by *Nautilus*. First, in a carefully controlled experiment, Crook and Basil (2008) showed it is possible to condition nautiluses to respond to a pale blue light when this is paired with an olfactory stimulus (a weak solution of frozen fish-head juice): such a solution elicits clear and measurable excitatory responses that include rapid ventilation and tentacle extension. Interestingly, their results suggest that *Nautilus* may have separate short-term and long-term memory stores (Fig. 10.11) whose time courses are similar to those of the cuttlefish *Sepia officinalis* (Messenger, 1971). Second, Crook, Hanlon and Basil (2009) obtained good evidence for spatial learning in *Nautilus*, which in a small two-dimensional maze learned the location of a goal within three trials, and retained the memory for at least 2 weeks. These authors also found evidence that *Nautilus* tested in artificial 'reefs' retained memories of relatively complex three-dimensional spaces.

In another series of experiments in a simple maze, Crook and Basil (2013) showed not only that nautiluses can learn to navigate towards a goal but that they use visual cues, both proximate and distal, to do so. These results remind us again that we must not underestimate

the importance of vision in the life of *Nautilus*; they also pose questions about the value of spatial memory in life on a reef face.

The brain of *Nautilus* differs considerably from that of coleoids (see §2.3). In particular, it lacks the lobes that constitute the vertical lobe system, long known to be critically important in the establishment and maintenance of memories in *Octopus* (Young, 1965a, b, 1991; Hochner *et al.*, 2003; Shomrat *et al.*, 2011). It will be fascinating to investigate the functional organisation of the *Nautilus* CNS using modern techniques, to identify the regions in the brain involved in the establishment and laying down of memories in a brain quite unlike that of a coleoid.

10.5 Summary and Future Research Directions

Despite the recent surge of interest in these fascinating animals, which is yielding many exciting new findings, there is still much to be learned about their behaviour and life style. The recent, detailed longitudinal study of *N. pompilius* by Dunstan and his collaborators at Osprey Reef, Australia, has taught us, among other things, that there may be marked differences in the behaviour of different species in differing locations, so that we make generalisations about 'nautilus' at our peril.

Given this caveat, the picture of *Nautilus* emerging at the moment is of a slow-moving creature that stays close to the face of coral reefs and adjacent slopes and in so doing makes extensive vertical excursions that sample chemical trails. Its energy budget is surprisingly low (Wells & Clarke, 1996) and it is able to survive in water with very little oxygen. Without the fins and muscular mantle of a coleoid, it lacks the agility and unpredictability of movement; without ink sac and chromatophores, its behavioural response to predators is limited to withdrawal into the shell.

Nautilus is an opportunistic, largely scavenging feeder and probably locates its food using chemical rather than visual cues. Nonetheless, the eyes are large and, although poorly developed by coleoid standards, they are well developed by the standards of other molluscs and their importance should not be underestimated. Clearly, more data on the visual system are needed, such as whether nautiluses can respond to the plane of polarised light (Muntz & Wentworth, 1987). The relative contribution to behaviour of chemoreceptors on the tentacles and on the rhinophore also needs clarification. The recent experimental evidence that *Nautilus* decreases its ventilation rate in the presence of underwater vibrations, especially when the stimulus source is close (Soucier & Basil, 2008), is interesting, suggesting that mechanoreception may be more important for these animals than previously thought; however, the sensory receptors involved have yet to be identified.

There are several facets of reproductive behaviour that would repay investigation. Are there multiple matings? Is there sperm competition? What is the time course of hatching and maturation of the young?

Because of the relative ease with which *Nautilus* can be kept in aquaria, it is to be hoped that someone will address this problem: the results might throw light on the evolution of reproductive strategies in coleoid cephalopods.

Finally, it needs emphasising that *Nautilus* is so different from other cephalopods that all generalisations in this book about 'cephalopod behaviour' refer only to coleoid cephalopods. Because of this, it seems unlikely that research on *Nautilus* will add significantly to our understanding of the behaviour of present-day coleoids. On the other hand, it is beginning to help us understand the behaviour of ancestral cephalopods and how this may have affected their evolution (Wells, Well & O'Dor, 1992; Wells, 1999; Grasso & Basil, 2009).

CHAPTER ELEVEN

Synthesis: Brains, Behaviour and the Future

In this second edition, as in the first, we have tried to bring together much of what is currently known about the behaviour of cephalopods, organising the material around such familiar themes as feeding, defence, reproduction, communication and learning. In each chapter, we have provided summaries and future challenges. Here we review some of the main features of their behaviour, identify those that distinguish them from other animals, and speculate briefly on future research that may enable us to address broader questions such as why cephalopods have such large brains and diverse behaviours.

Most cephalopods 'live fast and die young': they grow extraordinarily quickly, they mature after as little as a few months and yet rarely live for more than a year or two. Their reproductive strategies must be influenced by the short life span, and only recently have we begun to unravel some of these complexities.

Despite their brief lives, cephalopods are 'brainy' animals and their behaviour is complex and diverse. They not only have good senses with which to explore their world and so select the best courses of action for survival, but they also have memory stores that can provide up-to-date information upon which to base decisions. In such short-lived animals, one might have predicted more rigid, pre-programmed behaviours, rather than flexible responses and well-developed short- and long-term memory.

The diversity of cephalopod behaviour is, of course, paralleled by the complexity of their sense organs, effectors and nervous system (Chapter 2). In no other group of molluscs is there such a well-stocked arsenal for survival, including elaborate statocysts, a lateral line analogue, well developed eyes, a fast jet-propulsion system, manipulative appendages, chromatophores, photophores, an ink sac, a poisoning and hole-drilling apparatus, and a large brain with provision for storing and retaining newly acquired information. We should also recall that most of these features are characteristic of modern coleoids (Chapter 1) rather than the ancient *Nautilus* (Chapter 10), whose behaviour is correspondingly much simpler.

On reviewing the themes discussed in this book, it becomes clear that cephalopod behaviour has two striking characteristics: versatility and plasticity. By *versatility* we mean the possibility of selecting from among several possible courses of action. Consider the different ways of hunting and capturing prey used by some cephalopods (Chapter 4). Versatility is especially characteristic of primary defence (Chapter 5), because the neurally controlled chromatophore system enables an individual to deploy different camouflage tactics quickly as it forages amidst different backgrounds. Secondary defences are highly varied, extremely fast-changing and unpredictable, and tailored to specific predators. Versatility is also apparent during reproduction (Chapter 6); for example, most loliginid squid and some octopuses can adopt more than one copulatory position, each with different spermatophore placement and sperm storage, which enables complex sperm competition mechanisms. Females exert cryptic female choice of multiple sperm sources. Males have elaborate agonistic signals and can switch conditional mating strategies instantly as behavioural contexts change. All of these attributes have consequences for fertilisation, and the complexity of cephalopod mating systems rivals almost any vertebrate. By *plasticity* we mean the ability to change the response made to a stimulus after it has (1) proved inappropriate, unpleasant or even dangerous, or (2) proved advantageous, as in adopting a new food or shelter resource. Such changes have been shown to be long-lasting for octopuses in the laboratory and, of course, constitute learning (Chapter 8): we have argued that learning is also likely to be important in the sea. Phenotypic plasticity (rapid neural polyphenism; Chapter 3) is particularly evident during secondary defences such as Deimatic and Protean behaviours (Chapter 5) as well as during reproductive signalling (e.g. sexual mimicry, Chapter 6). Sometimes the changes are so dramatic that the cephalopod can look like a diffcrent species. Versatility and plasticity are characteristic of many successful animal groups with well-developed brains: within the vertebrates, the mammals are an obvious example, and most would agree with Young (1981) that much of vertebrate success is associated with a superior brain.

At this point it is necessary to enter a caveat. Many of the statements and arguments in this book are based upon data from a few shallow-water species that have, for the most part, been studied on tropical coral reefs or temperate rock reefs or in aquaria. Cephalopod researchers have inevitably studied those species most accessible to them, or have sought out for study those that have dramatic behaviours. However, not all cephalopods are as exciting to watch as *Sepia officinalis*, *Loligo pealeii*, *Sepioteuthis sepioidea*, *Octopus cyanea* and *O. vulgaris*. For example, *Sepia elegans*, *Octopus joubini* and the Northern octopus *Eledone cirrhosa* are rather sluggish animals with more limited behavioural diversity. *Eledone cirrhosa's* behaviour is quite unlike that of *Octopus vulgaris*, although its brain appears to be morphologically very similar. Those species of slow, neutrally buoyant cranchiid squid that provide the staple food for some whales probably survive by their sheer abundance, not by their wits, their agility or their body pattern changes.

It may also be necessary to temper some of our claims for cephalopods. In many ways, their behaviour is no more remarkable than that of the many fish, birds and mammals that compete with them and prey upon them, especially if one accepts the thesis of Packard (1972) that cephalopods are 'honorary vertebrates'. It is mainly when one considers them as invertebrates, and especially as molluscs, that their behaviour seems extraordinary. The cephalopods that we know best have life styles completely unlike those of their close relations the limpets, sea slugs and clams. Yet if the behaviour of cephalopods is no more complex than that of fish or lower vertebrates, it certainly is no less so.

We now turn to the relationship between the behaviour of an animal and the nature of its brain, a topic downplayed in this book. We have noted that the cephalopod brain has been studied extensively, but in what ways exactly can brain structure be correlated with behaviour in these animals? Some answers to this question are to be found in various texts (e.g. Young, 1964a, 1966; see also Wells, 1978; Budelmann, 1995; Nixon & Young, 2003; Hochner & Shomrat, 2014) and we have alluded to them in this book. For example, we can correlate the touch-learning ability of *Octopus* with the presence in the brain of an extra set of lobes – the frontal lobes – absent from decapods; we can correlate visual learning ability with the development of the vertical lobe system; we can correlate the large size of the chromatophore lobes in *O. vulgaris* (Maddock & Young, 1987) with the complexity of their body patterns; and we can correlate the greater complexity of the coleoid optic lobe, as compared with that of *Nautilus*, with form vision (Chapter 8). However, it is harder to move from such broad generalisations to the details of neural circuitry, certainly not to the extent possible in arthropods and vertebrates. For a variety of technical reasons, there are few electrophysiological recordings from the cephalopod CNS, even in *Octopus*, so that we have little information about the ways in which the cephalopod brain analyses sensory input and organises sequences of behaviour. For example, we know nothing about what is going on in the chromatophore lobes when the animal changes body pattern. It is also regrettable that we still know little about how the brain regulates reproductive behaviour (Di Cristo, 2013), although some details of the neural pathways involved have been known for nearly 60 years (Wells & Wells, 1959). One exception is the innovative research led by Binyamin Hochner in which neural circuitry for sensing and motor control of the very flexible arms of *Octopus vulgaris* has been worked out in considerable detail (Sumbre *et al.*, 2005; Zullo *et al.*, 2009). The unusual morphology of this system, and the challenges that it presents for central versus peripheral control of eight arms, have been argued to be an example of embodied organisation, in which processes in the periphery can augment central hierarchical organisation of behaviour (Hochner, 2012, 2013).

If we accept that cephalopods have good brains, let us ask: what, apart from routine homeostatic control, do they do with them that contributes so significantly to their success? This brings us back to two outstanding behavioural features of cephalopods: learning and rapid adaptive coloration.

Why should learning be so well developed in these short-lived creatures? Is it because they are mostly soft-bodied and vulnerable, without spines, poison glands or some other active form of retaliation? Or is it because they live a difficult life in the 'behaviour space' dominated by vertebrates (Packard, 1972)? Some cephalopods certainly live in very competitive environments, such as coral reefs, replete with dangerous predators and a variety of potential prey. Field observations of *Octopus vulgaris*, *O. cyanea* and *Sepioteuthis sepioidea* (Chapter 5) suggest that they are making many decisions each day: assessing the types and approaches of predators, determining foraging paths and times, judging when and whether to compete for mates, and so forth. Such decisions must draw on information stored in the memory. It is fascinating that evidence is now emerging that life in mid-water and in the depths may be more challenging and unpredictable than previously thought (Chapter 9). This may explain why those parts of the brain that are involved in establishing and storing memories are often so well developed in forms living in these open ocean zones (Box 8.2; Maddock & Young, 1987).

Johnston (1981, 1985) has identified what he terms 'ecologically surplus abilities' in animals, whereby a capacity that has evolved for a specific purpose may be suited to cope with novel situations when these arise in the natural environment. Obviously, some of the 'arbitrary learning tasks' (Davey, 1989) imposed on octopuses by experimental psychologists in the laboratory are only distantly related to situations that these animals encounter in the sea. Nevertheless, their ability to solve tasks such as serial reversal learning (Mackintosh, 1965) may relate to a flexible feeding strategy ('win–stay but lose–shift') that depends upon learning and might be useful to a predator searching for patchily distributed prey. Field and laboratory studies of octopus foraging and spatial learning (Chapter 4) rightly emphasise that octopuses might rely on learning for predation (Ambrose, 1982; Mather & O'Dor, 1991; Forsythe & Hanlon, 1997; Boal *et al.*, 2000), especially when considering that species such as *Octopus vulgaris* and *O. cyanea* have extremely

widespread distributions that range across very diverse habitats and fauna.

Several authors have suggested that learning is irrevocably tied up with the social habit; it is clearly advantageous if individuals that live in groups can recognise their neighbours and predict their probable behaviour on the basis of past experience (Jolly, 1966; Humphrey, 1976). There is much evidence to support this in vertebrates, and, as it is now emerging that social species of some invertebrates (e.g. bees) may be better at learning than solitary ones, it is tempting to see this thesis as being universally applicable. However, it is worth recalling that octopuses are solitary creatures; they do not rear their young and the generations do not overlap. Since there is no evidence that the present-day octopods evolved from social, squid-like ancestors, one has to look elsewhere for the reasons why present-day octopuses acquired their remarkable capacity for learning and decision-making. At present, unfortunately, nothing is known about the learning abilities of cephalopods such as the shoaling squid. The Caribbean coral reef species *Sepioteuthis sepioidea* in particular is likely to show good learning abilities, including the possibility of culturally transmitted information, because young and old shoal together and the generations apparently overlap.

There seems little doubt that learning updates the information from which an animal can select appropriate courses of action, and there seems to be general agreement that learning must be important in evolution. Since present evidence suggests that cephalopods (and especially the octopuses) are at present undergoing extensive adaptive radiation (Clarke, 1988b), it is tempting to relate such success to their 'braininess'. One final word about braininess, however: there has been little rigorous experimentation into complex learning in cephalopods in the past 30 years. Once again, we need more data about their cognitive abilities, of the kind now beginning to emerge (Vitti, 2013; Darmaillacq, Dickel & Mather, 2014).

Important as learning seems to be for cephalopods, the possession of a capacity for learning does not mark them out from other animals. On the other hand, their ability to change body patterns does. There is an important distinction to be made here between animals (invertebrate and vertebrate) that change colour and pattern so that they can camouflage themselves on a variety of backgrounds. What distinguishes cephalopods from all other animals is that their brains, via the chromatophores, can rapidly generate diverse patterns on the body and that these patterns can be used for both crypsis and signalling. Chromatophores may have evolved primarily for concealment as the original shell was lost or internalised, but with neural control they have proved eminently suitable for communication.

We know little about the neural mechanisms that bring about such sophisticated adaptive coloration. The connections of the lobes involved in the control of the chromatophores are well established, and over 60 years ago Boycott (1953) drew attention to differences in the neuropils of the chromatophore lobes in some common cephalopods. After silver staining, the fibres appear more regular and 'ordered' in *Sepia* and *Octopus*, which have elaborate patterning, than in *Loligo* or *Argonauta*, which do not. We are little further forward today, and nothing is known about the way in which the optic lobes regulate the activity of the chromatophore lobes. This problem awaits the application of modern non-invasive techniques for recording from the intact, living brain. It is fascinating that – during camouflage patterning – the cuttlefish seems not to be generating a carbon copy of the visual surrounds, but rather extracting statistical estimates of the visual environment, then producing skin patterns that resemble them (Chapter 3). Thus, cuttlefish seem to pick out the essence of an image and use that information to produce a deceptive cryptic body pattern. Where and how is this visual processing occurring in the brain of cephalopods?

There are equally fascinating problems to be considered relating to signalling (Chapter 7). Cephalopods are phylogenetically remote from the chordates, yet they make sophisticated visual signals, comparable with those of the higher vertebrates. Are there general principles of active visual signalling operating in mammals, birds and cephalopods? Or are cephalopods sufficiently different from vertebrates in their signalling systems to force us to re-examine traditional concepts of signalling?

Moynihan (1975, 1985a) has drawn attention to the fact that, although each species has unique signals, quite unrelated cephalopods have similar displays. His most obvious example, already noted by Packard (1972), is

the similarity of the Deimatic Displays in a sepioid, a teuthoid and an octopod (Box 7.2). He makes a number of interesting points about the conservatism of such displays. Because the three orders to which these animals belong diverged sometime in the early Mesozoic, the body patterns and components we see today must be very ancient, and presumably have been conserved from some similar patterns and components that their common ancestors shared nearly 200 million years ago. Such conservatism could have arisen because the components and body patterns were designed for signalling to a wide variety of receivers, mainly potential predators such as other cephalopods, various kinds of fish, marine reptiles, birds and mammals. Such receivers will have always differed in their size, age or sex but probably not in their visual physiology. As Moynihan (1975) puts it, 'the more widely reflected or broadcast a signal, the more conservative it will be'. Signals destined for a more restricted audience appear to be much less conservative. Thus, sexual displays, which are necessarily intraspecific, are more diverse.

Other body patterns and components can also be considered conservative. These are what Moynihan termed 'anti-displays', but which we call cryptic body patterns. The mottles, disruptives and uniform chronic body patterns used for crypsis are all conservative, as are the flamboyant body patterns of young cephalopods, which so beautifully masquerade as floating seaweed (Hanlon *et al.*, 2011). Perhaps one reason that such body patterns are so conservative and occur in animals from the three main orders of coleoids is that the backgrounds of stones, corals, sand and algae whose appearance they match have not changed greatly over many millions of years. Moreover, as with signalling, the cryptic body patterns are exploiting key conserved features of predator vision.

This brings us to our final question: to what extent have studies of the nervous system, sense organs and behaviour of cephalopods yielded findings of fundamental importance, or of general interest, to biologists?

For cellular neuroscience, the answer is obvious. Young's discovery of the squid giant fibres in the 1930s (Young, 1939) led to the important electrophysiological experiments of Hodgkin, Huxley and their colleagues, which have provided the basis of our understanding of nervous transmission in animals. Similarly, there have been the elegant experiments of Katz, Miledi and their colleagues on the squid giant synapse, which helped establish the fundamental role of calcium in neurotransmitter release (e.g. Katz & Miledi, 1966). The important biochemical and biophysical investigations on cephalopod rhodopsins and visual transduction mechanisms by Hubbard and St George (1958), Hagins (1965), Saibil (1990) and many others (for a review, see Messenger, 1991) also have wide relevance for our understanding of photo-transduction.

On the neural systems level, there have been several findings of fundamental importance. Although the many experiments on *Octopus* learning have not unequivocally revealed the physical basis of memory (Young, 1965b; Robertson, 1994; Hochner & Shomrat, 2013), there have undoubtedly been some findings of general relevance. For example, the organisation of the vertical lobe and its neurophysiological properties (network connectivity, neuromodulation and long-term potentiation) suggest that the octopus may use some universal learning and memory neuronal mechanisms to mediate some behaviours, although there seem to be some interesting differences in the molecular mechanisms involved. Studies of octopus arms have revealed the unique organisation of motor control in highly active soft-bodied animals; these principles are different from those operating in skeletal animals and have inspired the robotics industry (Calisti *et al.*, 2011). Certainly, the presence of separate visual and tactile memory stores, the existence of short- and long-term memory systems, the distributed nature of touch memories (Wells, 1959; Young, 1983), and the operation of rules during higher-order visual processing that seem similar to those adopted by mammals such as rats (Mackintosh, 1965) all suggest that the acquisition, storage and processing of information in complex nervous systems may proceed in similar ways, irrespective of phylogeny.

There is, however, the intriguing possibility that cephalopods have evolved complex behaviours with neural ultrastructure and pathways that differ from the vertebrate lineage. This idea deserves attention, given that cephalopods are the only group to have diverged from the vertebrate line and evolved large nervous systems and complex behaviour.

There have been fewer findings of general behavioural importance, mainly because it is harder to find common ground at the level of the whole animal than at the physiological or biochemical level, but also because cephalopods are intrinsically harder to keep and study than birds, for example. Modern aquarium facilities may alter this. A noteworthy example is the development of a visual sensorimotor bioassay for cuttlefish that has elucidated fundamental principles of animal camouflage (Chapters 3, 5). Meanwhile, investigators who regard animals as models for investigating current theories should remember that most cephalopods are not as convenient to study as insects or terrestrial vertebrates. Moreover – and this is perhaps the most striking fact to emerge from the past 20 years' research – cephalopods are advanced invertebrates, as Wells (1978) emphasised long ago: almost every system that has been studied recently is extraordinarily sophisticated and unique, the result of over 200 million years of evolution.

It is worth recalling that cephalopods must solve behavioural problems common to all metazoans (detect, avoid, choose, catch, mate). They seem, at times, to express 'deep' perceptual states that are similar to those of vertebrates: aggression, pain or fear. If so (and solid data are still wanting), would it not be fascinating to understand whether these constitute ancestral states, or are the result of biological convergence? The nervous systems of molluscs are extremely varied; is there something of greater significance to discover about the evolution and molecular biology of molluscs, their behaviour, their brains? Recent progress in molecular biology and genetics, optogenetics, electrophysiology, activity indicators, and other areas provide a means to bring cephalopod brain studies of form and function into a progressive modern era. The possibility that convergent evolution of emergent properties (not mechanistic components, but rules of action and function) might reveal common high-level properties (such as partitioning of perceptual states) could be revealing and instructive.

However, if you are interested in studying the behaviour of cephalopods because they are exciting and beautiful, then you will reap your own rewards, both aesthetic and intellectual. To watch a foraging octopus, or a shoal of *Sepioteuthis* on a coral reef, is a remarkable experience; the way that the animals exhibit what appear to be caution, stealth, 'intelligence' and watchfulness will surely fascinate any biologist. Moreover, there is no shortage of challenging problems in cephalopod behaviour, as we have tried to make plain in this book. In summary, cephalopods deserve study in their own right. They are creatures that have diversified to a variety of life styles, despite phylogenetic constraints. They should be seen, not as some kind of second-class fish, nor as steps on the evolutionary road to vertebrate orthodoxy, but as highly evolved molluscs in which the original molluscan *Bauplan* has been transformed for their present mode of life no less dramatically than it has in the present-day bivalves. This transformation, which probably involved co-evolution with the vertebrates, has led to their becoming swift, agile, fish-like creatures, with excellent senses, an extraordinary skin and a large brain. The diversity and great biomass of modern coleoids testifies to their success, which, as we have argued here, depends in no small way upon their behaviour.

References

Adamo, S. A. & Hanlon, R. T. (1996). Do cuttlefish (Cephalopoda) signal their intentions to conspecifics during agonistic encounters? *Animal Behaviour*, **52**, 73–81.

Adamo, S. A. & Weichelt, K. J. (1999). Field observations of schooling in the oval squid, *Sepioteuthis lessoniana* (Lesson, 1830). *Journal of Molluscan Studies*, **65**, 377–380.

Adamo, S. A., Brown, W. M., Kinig, A. J. *et al.* (2000). Agonistic and reproductive behaviours of the cuttlefish *Sepia officinalis* in a semi-natural environment. *Journal of Molluscan Studies*, **66**, 417–419.

Adamo, S. A., Ehgoetz, K., Sangster, C. & Whitehorne, I. (2006). Signaling to the enemy? – Body pattern expression and its response to external cues during hunting in the cuttlefish *Sepia officinalis* (Cephalopoda). *Biological Bulletin*, **210**, 192–200.

Agin, V., Chichery, R. & Chichery, M. P. (2001). Effects of learning on cytochrome oxidase activity in cuttlefish brain. *Neuroreport*, **12**, 113–116.

Agin, V., Chichery, R., Chichery, M. P. *et al.* (2006). Behavioural plasticity and neural correlates in adult cuttlefish. *Vie et Milieu – Life and Environment*, **56**, 81–87.

Agin, V., Chichery, R., Maubert, E. & Chichery, M. P. (2003). Time-dependent effects of cycloheximide on long-term memory in the cuttlefish. *Pharmacology Biochemistry and Behavior*, **75**, 141–146.

Agin, V., Dickel, L., Chichery, R. & Chichery, M. P. (1998). Evidence for a specific short-term memory in the cuttlefish, *Sepia*. *Behavioural Processes*, **43**, 329–334.

Aitken, J. P., O'Dor, R. K. & Jackson, G. D. (2005). The secret life of the giant Australian cuttlefish *Sepia apama* (Cephalopoda): behaviour and energetics in nature revealed through radio acoustic positioning and telemetry (RAPT). *Journal of Experimental Marine Biology and Ecology*, **320**, 77–91.

Akkaynak, D., Allen, J. J., Mathger, L. M., Chiao, C. C. & Hanlon, R. T. (2013). Quantification of cuttlefish (*Sepia officinalis*) camouflage: a study of color and luminance using *in situ* spectrometry. *Journal of Comparative Physiology A: Neuroethology Sensory Neural and Behavioral Physiology*, **199**, 211–225.

Alcock, J. (2013). *Animal Behavior: An Evolutionary Approach*. Sunderland, MA: Sinauer Associates, Inc.

Aldred, R. G., Nixon, M. & Young, J. Z. (1978). The blind octopus, *Cirrothauma*. *Nature*, **275**, 547–549.

Aldred, R. G., Nixon, M. & Young, J. Z. (1983). *Cirrothauma murrayi* Chun, a finned octopod. *Philosophical Transactions of the Royal Society of London B*, **301**, 1–54.

Allcock, A. L., Cooke, I. R. & Strugnell, J. M. (2011). What can the mitochondrial genome reveal about higher-level phylogeny of the molluscan class Cephalopoda? *Zoological Journal of the Linnean Society*, **161**, 573–586.

Allen, A., Michels, J. & Young, J. Z. (1985). Memory and visual discrimination by squids. *Marine Behaviour and Physiology*, **11**, 271–282.

Allen, A., Michels, J. & Young, J. Z. (1986). Possible interactions between visual and tactile memories in *Octopus*. *Marine Behaviour and Physiology*, **12**, 81–97.

Allen, J. J., Bell, G. R. R., Kuzirian, A. M. & Hanlon, R. T. (2013). Cuttlefish skin papilla morphology suggests a muscular hydrostatic function for rapid changeability. *Journal of Morphology*, **274**, 645–656.

Allen, J. J., Bell, G. R. R., Kuzirian, A. M., Velankar, S. S. & Hanlon, R. T. (2014). Comparative morphology of changeable skin papillae in octopus and cuttlefish. *Journal of Morphology*, **275**, 371–390.

Allen, J. J., Mäthger, L. M., Barbosa, A. & Hanlon, R. T. (2009). Cuttlefish use visual cues to control three-dimensional skin papillae for camouflage. *Journal of Comparative Physiology A: Neuroethology Sensory Neural and Behavioral Physiology*, **195**, 547–555.

Allen, J. J., Mäthger, L. M., Barbosa, A. *et al.* (2010a). Cuttlefish dynamic camouflage: responses to substrate choice and integration of multiple visual cues. *Proceedings of the Royal Society B – Biological Sciences*, **277**, 1031–1039.

Allen, J. J., Mäthger, L. M., Buresch, K. C. *et al.* (2010b). Night vision by cuttlefish enables changeable camouflage. *Journal of Experimental Biology*, **213**, 3953–3960.

Alonso, M. K., Crespo, E. A., Garcia, N. A. *et al.* (2002). Fishery and ontogenetic driven changes in the diet of the spiny dogfish, *Squalus acanthias*, in Patagonian waters, Argentina. *Environmental Biology of Fishes*, **63**, 193–202.

Altman, J. S. (1967). The behaviour of *Octopus vulgaris* Lam. in its natural habitat: a pilot study. *Underwater Association Reports*, **1966–1976**, 77–83.

Alupay, J. S., Hadjisolomou, S. P. & Crook, R. J. (2014). Arm injury produces long-term behavioral and neural hypersensitivity in octopus. *Neuroscience Letters*, **558**, 137–142.

Alves, C., Chichery, R., Boal, J. G. & Dickel, L. (2007). Orientation in the cuttlefish *Sepia officinalis*: response versus place learning. *Animal Cognition*, **10**, 29–36.

Alves, C., Darmaillacq, A. S., Shashar, N. & Dickel, L. (2007). Field and laboratory observations of *Sepia* (*Doratosepion*) *elongata*, d'Orbigny, 1845. *Veliger*, **48**, 313–316.

Amaratunga, T. (1980). Preliminary estimates of predation by the short-finned squid (*Illex illecebrosus*) on the Scotian Shelf. *North Atlantic Fisheries Organization Scientific Council Research Document*, **80/II/31**, 1–13.

Amaratunga, T. (1983). The role of cephalopods in the marine ecosystem, in *Advances in Assessment of World Cephalopod Resources* (ed. J. F. Caddy), pp. 379–415. Rome: FAO.

Amaratunga, T. (1987). Population biology, in *Cephalopod Life Cycles,* Vol. II. *Comparative Reviews* (ed. P. R. Boyle), pp. 239–252. London: Academic Press.

Amaratunga, T., Roberge, M., Young, J. & Uozumi, Y. (1980). Summary of joint Canada/Japan research program on short-finned squid (*Illex illecebrosus*): emigration and biology. *North Atlantic Fisheries Organization Scientific Council Research Document*, **80/11/40**, No. N071, 1–20.

Ambrose, R. F. (1982). Shelter utilization by the molluscan cephalopod *Octopus bimaculatus*. *Marine Ecology Progress Series*, **7**, 67–73.

Ambrose, R. F. (1983). Midden formation by octopuses: the role of biotic and abiotic factors. *Marine Behaviour and Physiology*, **10**, 137–144.

Ambrose, R. F. (1984). Food preferences, prey availability, and the diet of *Octopus bimaculatus* Verrill. *Journal of Experimental Marine Biology and Ecology*, **77**, 29–44.

Ambrose, R. F. (1986). Effects of octopus predation on motile invertebrates in a rocky subtidal community. *Marine Ecology Progress Series*, **30**, 261–273.

Ambrose, R. F. (1988). Population dynamics of *Octopus bimaculatus*: influence of life history patterns, synchronous reproduction and recruitment. *Malacologia*, **29**, 23–39.

Ambrose, R. F. & Nelson, B. V. (1983). Predation by *Octopus vulgaris* in the Mediterranean. *Marine Ecology*, **4**, 251–261.

Anderson, J. C., Baddeley, R. J., Osorio, D. *et al.* (2003). Modular organization of adaptive colouration in flounder and cuttlefish revealed by independent component analysis. *Network-Computation in Neural Systems*, **14**, 321–333.

Anderson, J. R. (2005). *Cognitive Psychology and Its Implications*. New York: Worth.

Anderson, R. C. & Mather, J. A. (2007). The packaging problem: bivalve prey selection and prey entry techniques of the octopus *Enteroctopus dofleini*. *Journal of Comparative Psychology*, **121**, 300–305.

Anderson, R. C., Hughes, P. D., Mather, J. A. & Steele, C. W. (1999). Determination of the diet of *Octopus rubescens* Berry, 1953 (Cephalopoda: Octopodidae), through examination of its beer bottle dens in Puget Sound. *Malacologia*, **41**, 455–460.

Anderson, R. C., Mather, J. A. & Sinn, D. L. (2008). *Octopus* senescence: forgetting how to eat clams. *Festivus*, **40**, 55–57.

Anderson, R. C., Sinn, D. L. & Mather, J. A. (2008). Drilling localization on bivalve prey by *Octopus rubescens* Berry, 1953 (Cephalopoda: Octopodidae). *Veliger*, **50**, 326–328.

Anderson, R. C., Wood, J. B. & Mather, J. A. (2008). *Octopus vulgaris* in the Caribbean is a specializing generalist. *Marine Ecology Progress Series*, **371**, 199–202.

Anderson, T. J. (1997). Habitat selection and shelter use by *Octopus tetricus*. *Marine Ecology Progress Series*, **150**, 137–148.

Anderson, T. J. (1999). Morphology and biology of *Octopus maorum* Hutton 1880 in northern New Zealand. *Bulletin of Marine Science*, **65**, 676.

Andersson, M. (1982). Sexual selection, natural selection and quality advertisement. *Biological Journal of the Linnean Society*, **17**, 375–393.

Andersson, M. (1994). *Sexual Selection*. Princeton: Princeton University Press.

Andersson, M. & Simmons, L. W. (2006). Sexual selection and mate choice. *Trends in Ecology and Evolution*, **21**, 296–302.

Andre, M., Johansson, T., Delory, E. & Van Der Schaar, M. (2007). Foraging on squid: the sperm whale mid-range sonar. *Journal of the Marine Biological Association of the United Kingdom*, **87**, 59–67.

Andre, M., Sole, M., Lenoir, M. *et al.* (2011). Low-frequency sounds induce acoustic trauma in cephalopods. *Frontiers in Ecology and the Environment*, **9**, 489–493.

Antonelis, G. A., Lowry, M. S., DeMaster, D. P. & Fiscus, C. H. (1987). Assessing northern elephant seal feeding habits by stomach lavage. *Marine Mammal Science*, **3**, 308–322.

Arata, G. F. J. (1954). A note on the flying behavior of certain squids. *Nautilus*, **68**, 1–3.

Archer, J. (1988). *The Behavioural Biology of Aggression*. Cambridge: Cambridge University Press.

Aristotle (1910). *Historia Animalium. Translation by D'Arcy Wentworth Thompson*. Oxford: Clarendon Press.

Arkhipkin, A. I. & Fedulov, P. P. (1986). Diel movements of juvenile *Ilex illecebrosus* and other cephalopods in the shelf water–slope water frontal zone off the Scotian Shelf in spring. *Journal of Northwest Atlantic Fishery Science*, **7**, 15–24.

Arnold, J. M. (1962). Mating behavior and social structure in *Loligo pealii*. *Biological Bulletin*, **123**, 53–57.

Arnold, J. M. (1965). Observations on the mating behavior of the squid *Sepioteuthis sepioidea*. *Bulletin of Marine Science*, **15**, 216–222.

Arnold, J. M. (1984). Cephalopoda. Ch. 6, in *The Mollusca*, Vol. 7 (ed. A. S. Tompa, N. H. Verdonk & J. A. M. van de Biggelaar). Orlando, FL: Academic Press.

Arnold, J. M. (1985). Shell growth, trauma, and repair as an indicator of life history for *Nautilus*. *Veliger*, **27**, 386–396.

Arnold, J. M. (1987). Reproduction and embryology of *Nautilus*, in *Nautilus. The Biology and Paleobiology of a Living Fossil* (ed. W. B. Saunders & N. H. Landman), pp. 353–372. New York: Plenum.

Arnold, J. M. & Arnold, K. O. (1969). Some aspects of hole-boring predation by *Octopus vulgaris*. *American Zoologist*, **9**, 991–996.

Arnold, J. M. & Carlson, B. A. (1986). Living *Nautilus* embryos: preliminary observations. *Science*, **232**, 73–76.

Arnqvist, G. & Rowe, L. (2013). *Sexual Conflict, Monographs in Behavior and Ecology*. Princeton, NJ: Princeton University Press.

Aronson, R. B. (1982). An underwater measure of *Octopus* size. *The Veliger*, **24**, 375–377.

Aronson, R. B. (1986). Life history and den ecology of *Octopus briareus* Robson in a marine lake. *Journal of Experimental Marine Biology and Ecology*, **95**, 37–56.

Aronson, R. B. (1989). The ecology of *Octopus briareus* Robson in a Bahamian saltwater lake. *American Malacological Bulletin*, **7**, 47–56.

Aronson, R. B. (1991). Ecology, paleobiology and evolutionary constraint in the octopus. *Bulletin of Marine Science*, **49**, 245–255.

Austin, C. R., Lutwak-Mann, C. & Mann, T. (1964). Spermatophores and spermatozoa of the squid *Loligo pealii*. *Proceedings of the Royal Society of London B*, **161**, 143–152.

Azuma, A. (1981). Reasoning about flying behavior of squid. *Kagaku Asahi*, 81–85.

Azuma, A. (2006). *The Biokinetics of Flying and Swimming*. Reston, VA: American Institute of Aeronautics and Astronautics, Inc.

Baeg, G. H., Sakurai, Y. & Shimazaki, K. (1993). Maturation processes in female *Loligo bleekeri* Keferstein (Mollusca: Cephalopoda). *Veliger*, **36**, 228–235.

Baglioni, S. (1910). Zur Kenntnis der Leistungen einiger Sinnesorgane (Gesichtssinn, Tastsinn und Geruchssinn) und des Zentralnervensystems der Zephalopoden und Fische. *Zeitschrift für Biologie*, **53**, 255–286.

Baker, A. C. (1957). Underwater photographs in the study of oceanic squid. *Deep Sea Research*, **4**, 126–138.

Bakhayokho, M. (1983). Biology of the cuttlefish *Sepia officinalis* hierredda off the Senegalese coast, in *Advances in Assessment of World Cephalopod Resources. FAO Fisheries Technical Paper No. 231* (ed. J. F. Caddy), pp. 204–263: Advances in Assessment of World Cephalopod Resources.

Balda, R. P., Pepperberg, I. M. & Kamil, A. C. (1998). *Animal Cognition in Nature*: Academic Press.

Banas, P. T., Smith, D. E. & Biggs, D. C. (1982). An association between a pelagic octopod, *Argonauta* sp. Linnaeus 1758, and aggregate salps. *Fishery Bulletin*, **80**, 648–650.

Bandel, K., Reitner, J. & Sturmer, W. (1983). Coleoids from the Lower Devonian Black Slate ('Hunsruck-Schiefer') of the Hunsruck (West Germany). *Neues Jahrbuch für Geologie und Palaontologie Abhandlungen*, **165**, 397–417.

Barbato, M., Bernard, M., Borrelli, L. & Fiorito, G. (2007). Body patterns in cephalopods: 'Polyphenism' as a way of information exchange. *Pattern Recognition Letters*, **28**, 1854–1864.

Barber, V. C. & Wright, D. E. (1969). The fine structure of the sense organs of the cephalopod mollusc *Nautilus*. *Zeitschrift für Zellforschung*, **102**, 293–312.

Barbosa, A., Allen, J. J., Mäthger, L. M. & Hanlon, R. T. (2012). Cuttlefish use visual cues to determine arm postures for camouflage. *Proceedings of the Royal Society B – Biological Sciences*, **279**, 84–90.

Barbosa, A., Litman, L. & Hanlon, R. T. (2008). Changeable cuttlefish camouflage is influenced by horizontal and vertical aspects of the visual background. *Journal of Comparative Physiology A*, **194**, 405–413.

Barbosa, A., Mäthger, L. M., Buresch, K. C. et al. (2008). Cuttlefish camouflage: the effects of substrate contrast and size in evoking uniform, mottle or disruptive body patterns. *Vision Research*, **48**, 1242–1253.

Barbosa, A., Mäthger, L. M., Chubb, C. et al. (2007). Disruptive coloration in cuttlefish: a visual perception mechanism that regulates ontogenetic adjustment of skin patterning. *Journal of Experimental Biology*, **210**, 1139–1147.

Barnes, R. S. K. (1991). Reproduction, life histories and dispersal, in *Fundamentals of Aquatic Ecology* (ed. R. S. K. Barnes & K. H. Mann), pp. 145–171. Oxford: Blackwell Scientific Publications.

Barrows, E. M. (1992). *The Complete Animal Behavior Desk Reference*. Boca Raton, Florida: CRC Press.

Bartol, I. K., Mann, R. & Vecchione, M. (2002). Distribution of the euryhaline squid *Lolliguncula brevis* in Chesapeake Bay: effects of selected abiotic factors. *Marine Ecology Progress Series*, **226**, 235–247.

Bas, C. (1979). Un modelo de distribución de dos especies: *Pagellus acarne* y *Octopus vulgaris*, influidas por la pesca y las condiciónes ambientales. *Investigación Pesquera*, **43**, 141–148.

Basil, J. A., Bahctinova, I., Kuroiwa, K. et al. (2005). The function of the rhinophore and the tentacles of *Nautilus pompilius* L. (Cephalopoda, Nautiloidea) in orientation to odor. *Marine and Freshwater Behaviour and Physiology*, **38**, 209–221.

Basil, J. A., Barord, G., Crook, R. J. et al. (2011). A synthetic approach to the study of learning and memory in Chambered *Nautilus* L. (Cephalopoda, Nautiloidea). *Vie et Milieu – Life and Environment*, **61**, 231–242.

Basil, J. A., Hanlon, R. T., Sheikh, S. I. & Atema, J. (2000). Three-dimensional odor tracking by *Nautilus pompilius*. *Journal of Experimental Biology*, **203**, 1409–1414.

Basil, J. A., Lazenby, G. B., Nakanuku, L. & Hanlon, R. T. (2002). Female *Nautilus* are attracted to male conspecific odor. *Bulletin of Marine Science*, **70**, 217–225.

Bateman, A. W., Vos, M. & Anholt, B. R. (2014). When to defend: antipredator defenses and the predation sequence. *The American Naturalist*, **183**, 847–855.

Bateson, P. P. G. (1991). Levels and processes, in *The Development and Integration of Behaviour* (ed. P. P. G. Bateson), pp. 3–16. Cambridge: Cambridge University Press.

Bazzino, G., Gilly, W. F., Markaida, U., Salinas-Zavala, C. A. & Ramos-Castillejos, J. (2010). Horizontal movements, vertical-habitat utilization and diet of the jumbo squid (*Dosidicus gigas*) in the Pacific Ocean off Baja California Sur, Mexico. *Progress in Oceanography*, **86**, 59–71.

Beja, P. R. (1991). Diet of otters (*Lutra lutra*) in closely associated freshwater, brackish and marine habitats in south-west Portugal. *Journal of Zoology (London)*, **225**, 141–152.

Bell, G. R. R., Kuzirian, A. M., Senft, S. L. *et al.* (2013). Chromatophore radial muscle fibers anchor in flexible squid skin. *Invertebrate Biology*, **132**, 120–132.

Bell, G. R. R., Mäthger, L. M., Gao, M. *et al.* (2014). Diffuse white structural coloration from multilayer reflectors in a squid. *Advanced Materials*, **26**, 4352–4356.

Bello, G. (1991). Role of cephalopods in the diet of the swordfish, *Xiphias gladius*, from the eastern Mediterranean Sea. *Bulletin of Marine Science*, **49**, 312–324.

Benchley, P. (1991). *Beast*. New York: Random House.

Benoit-Bird, K. J. & Gilly, W. F. (2012). Coordinated nocturnal behavior of foraging jumbo squid *Dosidicus gigas*. *Marine Ecology Progress Series*, **455**, 211–228.

Benoit-Bird, K. J., Gilly, W. F., Au, W. W. L. & Mate, B. (2008). Controlled and *in situ* target strengths of the jumbo squid *Dosidicus gigas* and identification of potential acoustic scattering sources. *Journal of the Acoustical Society of America*, **123**, 1318–1328.

Bergmann, S., Lieb, B., Ruth, P. & Markl, J. (2006). The hemocyanin from a living fossil, the cephalopod *Nautilus pompilius*: protein structure, gene organization, and evolution. *Journal of Molecular Evolution*, **62**, 362–374.

Bergstrom, B. & Summers, W. C. (1983). *Sepietta oweniana*, in *Cephalopod Life Cycles*, Vol. 1: *Species Accounts* (ed. P. R. Boyle), pp. 75–91. London: Academic Press.

Berruti, A. & Harcus, T. (1978). Cephalopod prey of the Sooty Albatrosses *Phoebetri fusca* and *P. palpebrata* at Marion Island. *South African Journal of Antarctic Research*, **8**, 99–103.

Bert, P. (1867). Mémoire sur la physiologie de la seiche (*Sepia officinalis*, Linn.). *Mémoires de la Societe des Sciences Physiques et Naturelles de Bordeaux*, **5**, 115–138.

Biagi, V. & Bello, G. (2009). Occurrence of an egg mass of *Thysanoteuthis rhombus* (Cephalopoda: Teuthida) in the Strait of Messina (Italy), locus typicus of the species. *Bollettino Malacologia*, **45**, 35–38.

Bidder, A. M. (1962). Use of the tentacles, swimming, and buoyancy control in the pearly nautilus. *Nature*, **196**, 451–454.

Bidder, A. M. (1966). Feeding and digestion in cephalopods, in *Physiology of Mollusca*, Vol. 2 (ed. K. M. Wilbur & C. M. Yonge), pp. 97–124. New York: Academic Press.

Biedermann, G. B. & Davey, V. A. (1993). Social-learning in invertebrates. *Science*, **259**, 1627–1628.

Birkhead, T. R. (1989). The intelligent sperm? A concise review of sperm competition. *Journal of Zoology (London)*, **218**, 347–351.

Birkhead, T. R. & Moller, A. P. (1998). *Sperm Competition and Sexual Selection*. New York: Academic Press.

Birkhead, T. R. & Parker, G. A. (1997). Sperm competition and mating systems, in *Behavioural Ecology: An Evolutionary Approach* (ed. J. R. Krebs & N. B. Davies), pp. 121–145. Oxford: Blackwell Science, Ltd.

Birkhead, T. R. & Pizzari, T. (2002). Postcopulatory sexual selection. *Nature Reviews Genetics*, **3**, 262–273.

Birkhead, T. R., Hosken, D. J. & Pitnick, S. S. (2009). *Sperm Biology: An Evolutionary Perspective*: Academic Press.

Bitterman, M. E. (1975). Critical commentary, in *Invertebrate Learning* (ed. W. C. Corning, J. A. Dyal & A. O. D. Willows), pp. 139–145. New York: Plenum.

Bleckmann, H., Budelmann, B. U. & Bullock, T. H. (1991). Peripheral and central nervous responses evoked by small water movements in a cephalopod. *Journal of Comparative Physiology A*, **168**, 247–257.

Boal, J. G. (1991). Complex learning in *Octopus bimaculoides*. *American Malacological Bulletin*, **9**, 75–80.

Boal, J. G. (1996). Absence of social recognition in laboratory-reared cuttlefish, *Sepia officinalis* L. (Mollusca: Cephalopoda). *Animal Behaviour*, **52**, 529–537.

Boal, J. G. (1997). Female choice of males in cuttlefish (Mollusca: Cephalopoda). *Behaviour*, **134**, 975–988.

Boal, J. G. (2006). Social recognition: a top down view of cephalopod behavior. *Vie et Milieu – Life and Environment*, **56**, 69–79.

Boal, J. G. (2011). Behavioral research methods for octopuses and cuttlefishes. *Vie et Milieu – Life and Environment*, **61**, 203–210.

Boal, J. G. & Fenwick, J. W. (2007). Laterality in octopus eye use? *Animal Behaviour*, **74**, E1–E2.

Boal, J. G. & Golden, D. K. (1999). Distance chemoreception in the common cuttlefish, *Sepia officinalis* (Mollusca, Cephalopoda). *Journal of Experimental Marine Biology and Ecology*, **235**, 307–317.

Boal, J. G. & Gonzalez, S. A. (1998). Social behaviour of individual oval squids (Cephalopoda, Teuthoidea, Loliginidae, *Sepioteuthis lessoniana*) within a captive school. *Ethology*, **104**, 161–178.

Boal, J. G. & Marsh, S. E. (1998). Social recognition using chemical cues in cuttlefish (*Sepia officinalis* Linnaeus, 1758). *Journal of Experimental Marine Biology and Ecology*, **230**, 183–192.

Boal, J. G., Dunham, A. W., Williams, K. T. & Hanlon, R. T. (2000). Experimental evidence for spatial learning in octopuses (*Octopus bimaculoides*). *Journal of Comparative Psychology*, **114**, 246–252.

Boal, J. G., Hylton, R. A., Gonzalez, S. A. & Hanlon, R. T. (1999). Effects of crowding on the social behavior of cuttlefish (*Sepia officinalis*). *Contemporary Topics in Laboratory Animal Science*, **38**, 49–55.

Boal, J. G., Prosser, K. N., Holm, J. B. *et al.* (2010). Sexually mature cuttlefish are attracted to the eggs of conspecifics. *Journal of Chemical Ecology*, **36**, 834–836.

Boal, J. G., Shashar, N., Grable, M. M. *et al.* (2004). Behavioral evidence for intraspecific signals with achromatic and polarized light by cuttlefish (Mollusca: Cephalopoda). *Behaviour*, **141**, 837–861.

Boal, J. G., Wittenberg, K. M. & Hanlon, R. T. (2000). Observational learning does not explain improvement in predation tactics by cuttlefish (Mollusca: Cephalopoda). *Behavioural Processes*, **52**, 141–153.

Boletzky, S. von (1977). Post-hatching behaviour and mode of life in cephalopods. *Symposia of the Zoological Society of London*, **38**, 557–567.

Boletzky, S. von (1978). Nos connaissances actuelles sur le développement des octopodes. *Vie et Milieu*, **28/29**, 85–120.

Boletzky, S. von (1981). Réflexions sur les stratégies de reproduction chez les céphalopodes. *Extrait du Bulletin de la Société Zoologique de France*, **106**, 293–304.

Boletzky, S. von (1983a). *Sepiola robusta*, in *Cephalopod Life Cycles*, Vol. 1. *Species Accounts* (ed. P. R. Boyle), pp. 53–67. London: Academic Press.

Boletzky, S. von (1983b). *Sepia officinalis*, in *Cephalopod Life Cycles*, Vol. 1: *Species Accounts* (ed. P. R. Boyle), pp. 31–52. London: Academic Press.

Boletzky, S. von (1987a). Juvenile behaviour, in *Cephalopod Life Cycles*, Vol. 2: *Comparative Reviews* (ed. P. R. Boyle), pp. 45–60. London: Academic Press.

Boletzky, S. von (1987b). Fecundity variation in relation to intermittent or chronic spawning in the cuttlefish, *Sepia officinalis* L. (Mollusca, Cephalopoda). *Bulletin of Marine Science*, **40**, 382–387.

Boletzky, S. von (1988). A new record of long-continued spawning in *Sepia officinalis* (Mollusca, Cephalopoda). *Rapport Commission Internationale pour la Mer Mediterranee*, **31**, 257.

Boletzky, S. von & Boletzky, M. V. von (1970). Das Eingraben in Sand bei *Sepiola* und *Sepietta* (Mollusca, Cephalopoda). *Revue Suisse de Zoologie*, **77**, 536–548.

Boletzky, S. von & Hanlon, R. T. (1983). A review of the laboratory maintenance, rearing and culture of cephalopod molluscs. *Memoirs of the National Museum of Victoria*, **44**, 147–187.

Boletzky, S. von, Boletzky, M. V. von, Frosch, D. & Gatzi, V. (1971). Laboratory rearing of Sepiolinae (Mollusca: Cephalopoda). *Marine Biology*, **8**, 82–87.

Boletzky, S. von, Rio, M. & Roux, M. (1992). Octopod 'ballooning' response. *Nature*, **356**, 199.

Bond, A. B. (1989a). Toward a resolution of the paradox of aggressive displays: I. Optimal deceit in the communication of fighting ability. *Ethology*, **81**, 29–46.

Bond, A. B. (1989b). Toward a resolution of the paradox of aggressive displays: II. Behavioral efference and the communication of intentions. *Ethology*, **81**, 235–249.

Bone, Q. & Marshall, N. B. (1982). *Biology of Fishes*. Glasgow: Blackie.

Bone, Q., Brown, E. R. & Travers, G. (1994). On the respiratory flow in the cuttlefish *Sepia officinalis*. *Journal of Experimental Biology*, **194**, 153–165.

Bone, Q., Pulsford, A. & Chubb, A. D. (1981). Squid mantle muscle. *Journal of the Marine Biological Association of the United Kingdom*, **61**, 327–342.

Borrelli, L., Gherardi, F. & Fiorito, G. (2006). *A Catalog of Body Patterning in Cephalopoda*. Firenze: Firenze University Press.

Bott, R. (1938). Kopula und Eiablage von *Sepia officinalis* L. *Zeitschrift für Morphologie und Oekologie der Tiere*, **34**, 150–160.

Boucaud-Camou, E. & Boismery, J. (1991). The migrations of the cuttlefish (*Sepia officinalis* L) in the English Channel, in *First International Symposium on the Cuttlefish Sepia* (ed. E. Boucaud-Camou), pp. 1–358. Caen: Centre de Publications de l'Universite de Caen.

Boucaud-Camou, E. & Boucher-Rodoni, R. (1983). Feeding and digestion in cephalopods, in *The Mollusca*, Vol. 5: *Physiology*, Part 2 (ed. A. S. M. Saleuddin & K. M. Wilbur), pp. 149–187. New York: Academic Press.

Boucher-Rodoni, R., Boucaud-Camou, E. & Mangold, K. (1987). Feeding and digestion, in *Cephalopod Life Cycles*, Vol. 2: *Comparative Reviews* (ed. P. R. Boyle), pp. 85–108. London: Academic Press.

Bouligand, Y. (1961). Le dispositif d'accrochage des oeufs de *Sepia elegans* sur *Alcyonium palmatum*. *Vie et Milieu*, **12**, 589–594.

Bouth, H. F., Leite, T. S., de Lima, F. D. & Oliveira, J. E. L. (2011). Atol das Rocas: an oasis for *Octopus insularis* juveniles (Cephalopoda: Octopodidae). *Zoologia*, **28**, 45–52.

Boutilier, R. G., West, T. G., Pogson, G. H. *et al.* (1996). *Nautilus* and the art of metabolic maintenance. *Nature*, **382**, 534–536.

Boutilier, R. G., West, T. G., Webber, D. M. *et al.* (2000). The protective effects of hypoxia-induced hypometabolism in the *Nautilus*. *Journal of Comparative Physiology B – Biochemical Systemic and Environmental Physiology*, **170**, 261–268.

Bouwma, P. E. & Herrnkind, W. F. (2009). Sound production in Caribbean spiny lobster *Panulirus argus* and its role in escape during predatory attack by *Octopus briareus*. *New Zealand Journal of Marine and Freshwater Research*, **43**, 3–13.

Bower, J. R. & Ichii, T. (2005). The red flying squid (*Ommastrephes bartramii*): a review of recent research and the fishery in Japan. *Fisheries Research*, **76**, 39–55.

Bower, J. R. & Sakurai, Y. (1996). Laboratory observations on *Todarodes pacificus* (Cephalopoda:

Ommastrephidae) egg masses. *American Malacological Bulletin*, **13**, 65–71.

Bower, J. R., Seki, K., Kubodera, T., Yamamoto, J. & Nobetsu, T. (2012). Brooding in a gonatid squid off northern Japan. *Biological Bulletin*, **223**, 259–262.

Bowman, R. E., Stillwell, C. E., Michaels, W. L. & Grosslein, M. D. (2000). Food of northwest Atlantic fishes and two common species of squid. *NOAA Technical Memorandum* NMFS-NE-155, 1–135.

Boycott, B. B. (1953). The chromatophore system of cephalopods. *Proceedings of the Linnaean Society (London)*, **164**, 235–240.

Boycott, B. B. (1954). Learning in *Octopus vulgaris* and other cephalopods. *Pubblicazioni della Stazione Zoologica di Napoli*, **25**, 67–93.

Boycott, B. B. (1958). The cuttlefish – *Sepia*. *New Biology*, **25**, 98–118.

Boycott, B. B. (1960). The functioning of the statocysts of *Octopus vulgaris*. *Proceedings of the Royal Society of London B*, **152**, 78–87.

Boycott, B. B. (1961). The functional organization of the brain of the cuttlefish *Sepia officinalis*. *Proceedings of the Royal Society of London B*, **153**, 503–534.

Boycott, B. B. (1965). A comparison of living *Sepioteuthis sepioidea* and *Doryteuthis plei* with other squids, and with *Sepia officinalis*. *Journal of Zoology (London)*, **147**, 344–351.

Boycott, B. B. & Young, J. Z. (1950). The comparative study of learning. *Symposia of the Society for Experimental Biology*, **4**, 432–453.

Boycott, B. B. & Young, J. Z. (1955a). A memory system in *Octopus vulgaris* Lamarck. *Proceedings of the Royal Society B – Biological Sciences*, **143**, 449.

Boycott, B. B. & Young, J. Z. (1955b). Memories controlling attacks on food objects by *Octopus vulgaris* Lamarck. *Pubblicazioni della Stazione Zoologica di Napoli*, **27**, 232–249.

Boycott, B. B. & Young, J. Z. (1956). The subpedunculate body and nerve and other organs associated with the optic tract of cephalopods, in *Bertil Hanström. Zoological Papers in Honour of his Sixty-Fifth Birthday* (ed. K. G. Wingstrand), pp. 76–105. Lund, Sweden: Zoological Institute.

Boycott, B. B. & Young, J. Z. (1957). Effects of interference with the vertical lobe on visual discrimination in *Octopus vulgaris* Lamarck.

Proceedings of the Royal Society of London B, **146**, 439–459.

Boyd, I. L. & Arnbom, T. (1991). Diving behaviour in relation to water temperature in the southern elephant seal: foraging implications. *Polar Biology*, **11**, 259–266.

Boyle, P. R. (1980). Home occupancy by male *Octopus vulgaris* in a large seawater tank. *Animal Behaviour*, **28**, 1123–1126.

Boyle, P. R. (1983a). Ventilation rate and arousal in the octopus. *Journal of Experimental Marine Biology and Ecology*, **69**, 129–136.

Boyle, P. R. (1983b). *Cephalopod Life Cycles,* Vol. 1: *Species Accounts*. London: Academic Press.

Boyle, P. R. (1986a). Responses to water-borne chemicals by the octopus *Eledone cirrhosa* (Lamarck, 1798). *Journal of Experimental Marine Biology and Ecology*, **104**, 23–30.

Boyle, P. R. (1986b). Neural control of cephalopod behavior, in *The Mollusca*, Vol. 9: *Neurobiology and Behavior* (ed. A. O. D. Willows), pp. 1–99. Orlando: Academic Press.

Boyle, P. R. (1986c). Report on a specimen of *Architeuthis* stranded near Aberdeen, Scotland. *Journal of Molluscan Studies*, **52**, 81–82.

Boyle, P. R. (1987). *Cephalopod Life Cycles,* Vol. 2: *Comparative Reviews*. London: Academic Press.

Boyle, P. R. (1990). Prey handling and salivary secretions in octopuses, in *Trophic Relationships in the Marine Environment* (ed. M. Barnes & R. N. Gibson), pp. 541–552. Aberdeen: University Press.

Boyle, P. R. & Dubas, F. (1981). Components of body pattern displays in the octopus *Eledone cirrhosa* (Mollusca: Cephalopoda). *Marine Behaviour and Physiology*, **8**, 135–148.

Boyle, P. R. & Knobloch, D. (1981). Hole boring of crustacean prey by the octopus *Eledone cirrhosa* (Mollusca, Cephalopoda). *Journal of Zoology (London)*, **193**, 1–10.

Boyle, P. R. & Rodhouse, P. (2005). *Cephalopods. Ecology and Fisheries*: Oxford: Blackwell Science.

Boyle, P. R., Pierce, G. J. & Hastie, L. C. (1995). Flexible reproductive strategies in the squid *Loligo forbesi*. *Marine Biology*, **121**, 501–508.

Bradbury, J. W. & Vehrencamp, S. L. (2011). *Principles of Animal Communication*. Sunderland, MA: Sinauer Associates, Inc.

Bradley, E. A. & Messenger, J. B. (1977). Brightness preference in *Octopus* as a function of the background brightness. *Marine Behaviour and Physiology*, **4**, 243–251.

Bradley, E. A. & Young, J. Z. (1975). Comparison of visual and tactile learning in *Octopus* after lesions to one of two memory systems. *Journal of Neuroscience Research*, **1**, 185–205.

Braid, H. E. & Bolstad, K. S. R. (2014). Feeding ecology of the largest mastigoteuthid squid species, *Idioteuthis cordiformis* (Cephalopoda, Mastigoteuthidae). *Marine Ecology Progress Series*, **515**, 275–279.

Braithwaite, V. (2010). *Do Fish Feel Pain?* Oxford: Oxford University Press.

Briffa, M. (2013). Plastic proteans: reduced predictability in the face of predation risk in hermit crabs. *Biology Letters*, **9**, 20130592.

Briffa, M. & Hardy, I. C. W. (2013). Introduction to animal contests, in *Animal Aggression* (ed. I. C. W. Hardy & M. Briffa), pp. 1–4. Cambridge: Cambridge University Press.

Brocco, S. L. (1971). *Aspects of the Biology of the Sepiolid Squid Rossia pacifica Berry*. Victoria, British Columbia: University of Victoria.

Brooks, W. R. (1988). The influence of the location and abundance of the sea anemone *Calliactis tricolor* (Le Sueur) in protecting hermit crabs from octopus predators. *Journal of Experimental Marine Biology and Ecology*, **116**, 15–21.

Brown, C., Garwood, M. P. & Williamson, J. E. (2012). It pays to cheat: tactical deception in a cephalopod social signalling system. *Biology Letters*, **8**, 729–732.

Brown, E. R., Piscopo, S., De Stefano, R. & Giuditta, A. (2006). Brain and behavioural evidence for rest–activity cycles in *Octopus vulgaris*. *Behavioural Brain Research*, **172**, 355–359.

Brown, P. K. & Brown, P. S. (1958). Visual pigments of the octopus and cuttlefish. *Nature*, **182**, 1288–1290.

Bruun, A. F. (1943). The biology of *Spirula spirula* (L.). *Dana Reports*, **24**, 1–44.

Budelmann, B. U. (1970). Die Arbeitsweise de Statolithenorgane von *Octopus vulgaris*. *Zeitschrift für vergleichende Physiologie*, **70**, 278–312.

Budelmann, B. U. (1976). Equilibrium receptor systems in molluscs, in *Structure and Function of*

Proprioceptors in the Invertebrates (ed. P. J. Mill), pp. 529–566. London: Chapman & Hall.

Budelmann, B. U. (1990). The statocysts of squid, in *Squid as Experimental Animals* (ed. D. L. Gilbert, W. J. Adelman & J. M. Arnold), pp. 421–439. New York: Plenum Press.

Budelmann, B. U. (1992). Hearing in non-arthropod invertebrates, in *The Evolution of Hearing* (ed. D. B. Webster, R. R. Fay & A. N. Popper), pp. 141–155. New York: Springer.

Budelmann, B. U. (1994). Cephalopod sense organs, nerves and the brain: adaptations for high performance and life style. *Marine and Freshwater Behaviour and Physiology*, **25**, 13–33.

Budelmann, B. U. (1995). The cephalopod nervous system: what evolution has made of the molluscan design, in *The Nervous Systems of Invertebrates. An Evolutionary and Comparative Approach* (ed. O. Breidbach & W. Kutsch), pp. 115–138. Basel: Birkhauser.

Budelmann, B. U. (1996). Active marine predators: the sensory world of cephalopods. *Marine and Freshwater Behaviour and Physiology*, **27**, 59–75.

Budelmann, B. U. & Bleckmann, H. (1988). A lateral line analogue in cephalopods: water waves generate microphonic potentials in the epidermal head lines of *Sepia* and *Lolliguncula*. *Journal of Comparative Physiology A*, **164**, 1–5.

Budelmann, B. U. & Young, J. Z. (1984). The statocyst–oculomotor system of *Octopus vulgaris*: extraocular eye muscles, eye muscle nerves, statocyst nerves and the oculomotor centre in the central nervous system. *Philosophical Transactions of the Royal Society of London B*, **306**, 159–189.

Budelmann, B. U. & Young, J. Z. (1993). The oculomotor system of decapod cephalopods: eye muscles, eye muscle nerves, and the oculomotor neurons in the central nervous system. *Philosophical Transactions of the Royal Society of London B*, **340**, 93–125.

Budelmann, B. U., Bullock, T. H. & Williamson, R. (1995). Cephalopod brains: promising preparations for brain physiology, in *Cephalopod Neurobiology* (ed. N. J. Abbott, R. Williamson & L. Maddock). Oxford: Oxford University Press.

Budelmann, B. U., Riese, U. & Bleckmann, H. (1991). Structure, function, biological significance of the cuttlefish 'lateral lines', in *The Cuttlefish. First International Symposium on the Cuttlefish Sepia* (ed. E. Boucaud-Camou), pp. 201–209. Caen: Centre de Publications de l'Universite de Caen.

Budelmann, B. U., Sachse, M. & Staudigl, M. (1987). The angular acceleration receptor system of *Octopus vulgaris*: morphometry, ultrastructure, and neuronal and synaptic organization. *Philosophical Transactions of the Royal Society of London B: Biological Sciences*, **315**, 305–343.

Budelmann, B. U., Schipp, R. & Boletzky, S. von (1997). Cephalopoda, in *Microscopic Anatomy of Invertebrates*, Vol. 6A: *Mollusca II*, pp. 119–414. New York: Wiley-Liss, Inc.

Bullock, T. H. (1965). Mollusca: Cephalopoda, in *Structure and Function in the Nervous Systems of Invertebrates* (ed. T. H. Bullock & G. A. Horridge), pp. 1433–1515. San Francisco: Freeman.

Bullock, T. H. & Basar, E. (1988). Comparison of ongoing compound field potentials in the brains of invertebrates and vertebrates. *Brain Research Reviews*, **13**, 57–75.

Bullock, T. H. & Budelmann, B. U. (1991). Sensory evoked potentials in unanesthetized unrestrained cuttlefish: a new preparation for brain physiology in cephalopods. *Journal of Comparative Physiology A*, **168**, 141–150.

Buresch, K. C., Boal, J. G., Knowles, J. *et al.* (2003). Contact chemosensory cues in egg bundles elicit male–male agonistic conflicts in the squid *Loligo pealeii*. *Journal of Chemical Ecology*, **29**, 547–560.

Buresch, K. C., Boal, J. G., Nagle, G. T. *et al.* (2004). Experimental evidence that ovary and oviducal gland extracts influence male agonistic behavior in squids. *Biological Bulletin*, **206**, 1–3.

Buresch, K. C., Gerlach, G. & Hanlon, R. T. (2006). Multiple genetic stocks of the longfin inshore squid *Loligo pealeii* in the NW Atlantic: stocks segregate inshore in summer, but aggregate offshore in winter. *Marine Ecology Progress Series*, **310**, 263–270.

Buresch, K. C., Mäthger, L. M., Allen, J. J. *et al.* (2011). The use of background matching vs. masquerade for camouflage in cuttlefish *Sepia officinalis*. *Vision Research*, **51**, 2362–2368.

Buresch, K. C., Maxwell, M. R., Cox, M. R. & Hanlon, R. T. (2009). Temporal dynamics of mating and paternity in the squid *Loligo pealeii*. *Marine Ecology Progress Series*, **387**, 197–203.

Buresch, K. M., Hanlon, R. T., Maxwell, M. R. & Ring, S. (2001). Microsatellite DNA markers indicate a high frequency of multiple paternity within individual field-collected egg capsules of the squid *Loligo pealeii*. *Marine Ecology Progress Series*, **210**, 161–165.

Burghardt, G. M. (1999). Conceptions of play and the evolution of animal minds. *Evolutionary Cognition*, **5**, 115–123.

Burghardt, G. M. (2005). *The Genesis of Animal Play*. Cambridge, MA: MIT Press.

Bush, S. L. (2012). Economy of arm autotomy in the mesopelagic squid *Octopoteuthis deletron*. *Marine Ecology Progress Series*, **458**, 133–140.

Bush, S. L. & Robison, B. H. (2007). Ink utilization by mesopelagic squid. *Marine Biology*, **152**, 485–494.

Bush, S. L., Hoving, H. J. T., Huffard, C. L., Robison, B. H. & Zeidberg, L. D. (2012). Brooding and sperm storage by the deep-sea squid *Bathyteuthis berryi* (Cephalopoda: Decapodiformes). *Journal of the Marine Biological Association of the United Kingdom*, **92**, 1629–1636.

Bush, S. L., Robison, B. H. & Caldwell, R. L. (2009). Behaving in the dark: locomotor, chromatic, postural, and bioluminescent behaviors of the deep-sea squid *Octopoteuthis deletron* Young 1972. *Biological Bulletin, Marine Biological Laboratory, Woods Hole*, **216**, 7–22.

Butler, M. J. & Lear, J. A. (2009). Habitat-based intraguild predation by Caribbean reef octopus *Octopus briareus* on juvenile Caribbean spiny lobster *Panulirus argus*. *Marine Ecology Progress Series*, **386**, 115–122.

Butterworth, M. (1982). Shell utilization by *Octopus joubini*. Tallahassee, FL: Unpublished M.S. Thesis, Florida State University.

Byrne, R. A., Kuba, M. & Griebel, U. (2002). Lateral asymmetry of eye use in *Octopus vulgaris*. *Animal Behaviour*, **64**, 461–468.

Byrne, R. A., Kuba, M. J. & Meisel, D. V. (2004). Lateralized eye use in *Octopus vulgaris* shows antisymmetrical distribution. *Animal Behaviour*, **68**, 1107–1114.

Byrne, R. A., Kuba, M. J., Meisel, D. V., Griebel, U. & Mather, J. A. (2006a). Does *Octopus vulgaris* have preferred arms? *Journal of Comparative Psychology*, **120**, 198–204.

Byrne, R. A., Kuba, M. J., Meisel, D. V., Griebel, U. & Mather, J. A. (2006b). *Octopus* arm choice is strongly influenced by eye use. *Behavioural Brain Research*, **172**, 195–201.

Byrne, R. W. (2000). Animal cognition in nature. *Trends in Cognitive Sciences*, **4**, 73.

Byrne, R. W. & Bates, L. A. (2006). Why are animals cognitive? *Current Biology*, **16**, R445–R448.

Byzov, A. L., Orlov, O. Y. & Utina, I. A. (1962). An investigation of adaptation in the eyes of cephalopod molluscs. *Biofizika*, **7**, 318–327.

Caddy, J. F. (1983). *Advances in Assessment of World Cephalopod Resources*. FAO, United Nations, Rome: FAO Fisheries Technical Paper No. 231.

Caldwell, R. L. (2005). An observation of inking behavior protecting adult *Octopus bocki* from predation by green turtle (*Chelonia mydas*) hatchlings. *Pacific Science*, **59**, 69–72.

Caldwell, R. L. & Lamp, K. (1981). Chemically mediated recognition by the stomatopod *Gonodactylus bredini* of its competitor, the octopus *Octopus joubini*. *Marine Behaviour and Physiology*, **8**, 35–41.

Calisti, M., Giorelli, M., Levy, G. et al. (2011). An octopus-bioinspired solution to movement and manipulation for soft robots. *Bioinspiration and Biomimetics*, **6**, 036002.

Callan, H. G. (1940). Absence of a sex-hormone controlling regeneration of the hectocotylus in *Octopus vulgaris* Lam. *Pubblicazioni della Stazione Zoologica di Napoli*, **18**, 15–19.

Calow, P. (1987). Fact and theory – an overview, in *Cephalopod Life Cycles*. Vol. 2: *Comparative Reviews* (ed. P. R. Boyle), pp. 351–365. London: Academic Press.

Carere, C., Wood, J. B. & Mather, J. (2011). Species differences in captivity: where are the invertebrates? *Trends in Ecology and Evolution*, **26**, 211–211.

Cariello, L. & Zanetti, L. (1977). Alpha- and beta-cephalotoxin: two paralysing proteins from posterior salivary glands of *Octopus vulgaris*. *Comparative Biochemistry and Physiology*, **C57**, 169–173.

Carlson, B. A. (1987). Collection and aquarium maintenance of *Nautilus*, in *Nautilus. The Biology and Paleobiology of a Living Fossil* (ed. W. B. Saunders & N. H. Landman), pp. 563–578. New York: Plenum.

Carlson, B. A., Awai, M. L. & Arnold, J. M. (1992). Hatching and early growth of *Nautilus belauensis* and implications on the distribution of the genus *Nautilus*. *Proceedings of the Seventh International Coral Reef Symposium, Guam*, 1, 587–592.

Carlson, B. A., McKibben, J. N. & DeGruy, M. V. (1984). Telemetric investigation of vertical migration of *Nautilus belauensis* in Palau. *Pacific Science*, 38, 183–188.

Caro, T. (2005). *Antipredator Defenses in Birds and Mammals*. Chicago: University of Chicago Press.

Cartron, L., Darmaillacq, A. S., Jozet-Alves, C., Shashar, N. & Dickel, L. (2012). Cuttlefish rely on both polarized light and landmarks for orientation. *Animal Cognition*, 15, 591–596.

Cartron, L., Shashar, N., Dickel, L. & Darmaillacq, A. S. (2013). Effects of stimuli shape and polarization in evoking deimatic patterns in the European cuttlefish, *Sepia officinalis*, under varying turbidity conditions. *Invertebrate Neuroscience*, 13, 19–26.

Catchpole, C. K. (1980). Sexual selection and the evolution of complex songs among European warblers of the genus *Acrocephalus*. *Behaviour*, 74, 149–166.

Chamberlain, J. A. (1987). Locomotion of *Nautilus*, in *Nautilus. The Biology and Paleobiology of a Living Fossil* (ed. W. B. Saunders & N. H. Landman), pp. 489–525. New York: Plenum.

Chance, M. R. A. & Russell, W. M. S. (1959). Protean displays: a form of allaesthetic behaviour. *Proceedings of the Zoological Society of London*, 132, 65–70.

Chapman, T. (2006). Evolutionary conflicts of interest between males and females. *Current Biology*, 16, R744–R754.

Chase, R. & Wells, M. J. (1986). Chemotactic behaviour in *Octopus*. *Journal of Comparative Physiology A*, 158, 375–381.

Chedekel, M. R., Murr, B. L. & Zeise, L. (1992). Melanin standard method: empirical formula. *Pigment Cell Research*, 5, 143–147.

Cheng, M. W. & Caldwell, R. L. (2000). Sex identification and mating in the blue-ringed octopus, *Hapalochlaena lunulata*. *Animal Behaviour*, 60, 27–33.

Cherel, Y. & Hobson, K. A. (2005). Stable isotopes, beaks and predators: a new tool to study the trophic ecology of cephalopods, including giant and colossal squids. *Proceedings of the Royal Society B – Biological Sciences*, 272, 1601–1607.

Chiao, C. C. & Hanlon, R. T. (2001a). Cuttlefish camouflage: visual perception of size, contrast and number of white squares on artificial chequerboard substrata initiates disruptive coloration. *Journal of Experimental Biology*, 204, 2119–2125.

Chiao, C. C. & Hanlon, R. T. (2001b). Cuttlefish cue visually on area – not shape or aspect ratio – of light objects in the substrate to produce disruptive body patterns for camouflage. *Biological Bulletin*, 201, 269–270.

Chiao, C. C., Chubb, C., Buresch, K. C. et al. (2010). Mottle camouflage patterns in cuttlefish: quantitative characterization and visual background stimuli that evoke them. *Journal of Experimental Biology*, 213, 187–199.

Chiao, C. C., Chubb, C. & Hanlon, R. T. (2007). Interactive effects of size, contrast, intensity and configuration of background objects in evoking disruptive camouflage in cuttlefish. *Vision Research*, 47, 2223–2235.

Chiao, C. C., Kelman, E. J. & Hanlon, R. T. (2005). Disruptive body pattern of cuttlefish (*Sepia officinalis*) requires visual information regarding edges and contrast of objects in natural substrate backgrounds. *Biological Bulletin*, 208, 7–11.

Chiao, C. C., Ulmer, K. M., Siemann, L. A. et al. (2013). How visual edge features influence cuttlefish camouflage patterning. *Vision Research*, 83, 40–47.

Chiao, C.-C., Wickiser, J. K., Allen, J. J., Genter, B. & Hanlon, R. T. (2011). Hyperspectral imaging of cuttlefish camouflage indicates good color match in the eyes of fish predators. *Proceedings of the National Academy of Sciences*, 108, 9148–9153.

Chichery, M. P. & Chichery, R. (1992). Behavioral and neurohistological changes in aging *Sepia*. *Brain Research*, 574, 77–84.

Chichery, R. & Chanelet, J. (1976). Motor and behavioural responses obtained by stimulation with chronic electrodes of the optic lobe of *Sepia officinalis*. *Brain Research*, 105, 525–532.

Chiou, T. H., Mathger, L. M., Hanlon, R. T. & Cronin, T. W. (2007). Spectral and spatial properties of polarized light reflections from the arms of squid

(*Loligo pealeii*) and cuttlefish (*Sepia officinalis* L.). *Journal of Experimental Biology*, **210**, 3624–3635.

Chun, C. (1910). Die Cephalopoden. *Wissenschaftliche Ergebnisse der Deutschen Tiefsee-Expedition auf dem Dampfer 'Valdivia' 1898–1899*, **18**, 1–552.

Chun, C. (1914). Cephalopoda from the 'Michael Sars' North Atlantic Deep-Sea Expedition, 1910. *Reports on Sars North Atlantic Deep Sea Expedition*, **3**, 1–28.

Cigliano, J. (1993). Dominance hierarchy and den use in *Octopus bimaculoides*. *Animal Behaviour*, **46**, 677–684.

Cigliano, J. A. (1995). Assessment of the mating history of female pygmy octopuses and a possible sperm competition mechanism. *Animal Behaviour*, **49**, 849–851.

Clarke, A., Rodhouse, P. G. & Gore, D. J. (1994). Biochemical composition in relation to the energetics of growth and sexual maturation in the ommastrephid squid *Illex argentinus*. *Philosophical Transactions of the Royal Society of London Series B – Biological Sciences*, **344**, 201–212.

Clarke, M. R. (1962). Significance of cephalopod beaks. *Nature, London*, **193**, 560–561.

Clarke, M. R. (1966). A review of the systematics and ecology of oceanic squids. *Advances in Marine Biology*, **4**, 91–300.

Clarke, M. R. (1969). Cephalopoda collected on the SOND cruise. *Journal of the Marine Biological Association of the United Kingdom*, **49**, 961–976.

Clarke, M. R. (1977). Beaks, nets and numbers. *Symposia of the Zoological Society of London*, **38**, 89–126.

Clarke, M. R. (1980). Cephalopoda in the diet of sperm whales of the southern hemisphere and their bearing on sperm whale biology. *Discovery Reports*, **37**, 1–324.

Clarke, M. R. (1983). Cephalopod biomass – estimation from predation. *Memoirs of the National Museum of Victoria*, **44**, 95–107.

Clarke, M. R. (1985). Cephalopods in the diet of cetaceans and seals. *Rapport Commission Internationale pour la Mer Mediterranee*, **29**, 211–219.

Clarke, M. R. (1986). *Handbook for the Identification of Cephalopod Beaks*. London: Oxford University Press.

Clarke, M. R. (1988a). Evolution of buoyancy and locomotion in recent cephalopods, in *The Mollusca*, Vol. 12: *Paleontology and Neontology of Cephalopods* (ed. M. R. Clarke & E. R. Trueman), pp. 203–213. San Diego: Academic Press.

Clarke, M. R. (1988b). Evolution of recent cephalopods – a brief review, in *The Mollusca*, Vol. 12: *Paleontology and Neontology of Cephalopods* (ed. M. R. Clarke & E. R. Trueman), pp. 331–340. San Diego: Academic Press.

Clarke, M. R. (1996). Cephalopods as prey. 3. Cetaceans. *Philosophical Transactions of the Royal Society of London Series B – Biological Sciences*, **351**, 1053–1065.

Clarke, M. R. & Stevens, J. D. (1974). Cephalopods, blue sharks and migration. *Journal of the Marine Biological Association of the United Kingdom*, **54**, 949–957.

Clarke, M. R. & Trueman, E. R. (1988). *The Mollusca*, Vol. 12: *Paleontology and Neontology of Cephalopods*. San Diego, CA.: Academic Press.

Clarke, M. R., Denton, E. J. & Gilpin-Brown, J. B. (1979). Use of ammonium for buoyancy in squids. *Journal of the Marine Biological Association of the United Kingdom*, **59**, 259–276.

Cloney, R. A. & Brocco, S. L. (1983). Chromatophore organs, reflector cells, iridocytes and leucophores in cephalopods. *American Zoologist*, **23**, 581–592.

Cloney, R. A. & Florey, E. (1968). Ultrastructure of cephalopod chromatophore organs. *Zeitschrift für Zellforschung*, **89**, 250–280.

Clutton-Brock, T. H. (1988). *Reproductive Success. Studies of Individual Variation in Contrasting Breeding Systems*. Chicago and London: University of Chicago Press.

Clutton-Brock, T. H. (2009). Sexual selection in females. *Animal Behaviour*, **77**, 3–11.

Cobb, C. S., Pope, S. K. & Williamson, R. (1995). Circadian rhythms to light–dark cycles in the lesser octopus, *Eledone cirrhosa*. *Marine and Freshwater Behaviour and Physiology*, **26**, 47–57.

Coelho, M. L. (1985). Review of the influence of oceanographic factors on cephalopod distribution and life cycles. *NAFO Scientific Council Studies*, **9**, 47–57.

Coelho, M. L., Quintela, J., Bettencourt, V., Olavo, G. & Villa, H. (1994). Population structure, maturation patterns and fecundity of the squid *Loligo vulgaris* from southern Portugal. *Fisheries Research*, **21**, 87–102.

Cohen, A. C. (1976). The systematics and distribution of *Loligo* (Cephalopoda, Myopsida) in the western North Atlantic, with descriptions of two new species. *Malacologia*, **15**, 299–367.

Cole, K. S. & Gilbert, D. L. (1970). Jet propulsion of squid. *Biological Bulletin*, **138**, 245–246.

Cole, P. D. & Adamo, S. A. (2005). Cuttlefish (*Sepia officinalis*: Cephalopoda) hunting behavior and associative learning. *Animal Cognition*, **8**, 27–30.

Coleman, N. (1984). Molluscs from the diets of commercially exploited fish off the coast of Victoria, Australia. *Journal of the Malacological Society of Australia*, **6**, 143–154.

Colgan, P. W. (1978). *Quantitative Ethology*. New York: John Wiley & Sons.

Collins, M. A., Burnell, G. M. & Rodhouse, P. (1995). Reproductive strategies of male and female *Loligo forbesi* (Cephalopoda: Loliginidae). *Journal of the Marine Biological Association of the United Kingdom*, **75**, 621–634.

Colmers, W. F., Hixon, R. F., Hanlon, R. T. *et al.* (1984). 'Spinner' cephalopods: Defects of statocyst suprastructures in an invertebrate analogue of the vestibular apparatus. *Cell and Tissue Research*, **236**, 505–515.

Condit, R. & Le Boeuf, B. J. (1984). Feeding habits and feeding grounds of the northern elephant seal. *Journal of Mammalogy*, **65**, 281–290.

Cooper, K. M. & Hanlon, R. T. (1986). Correlation of iridescence with changes in iridophore platelet ultrastructure in the squid *Lolliguncula brevis*. *Journal of Experimental Biology*, **121**, 451–455.

Cooper, K. M., Hanlon, R. T. & Budelmann, B. U. (1990). Physiological color change in squid iridophores. II. Ultrastructural mechanisms in *Lolliguncula brevis*. *Cell and Tissue Research*, **259**, 15–24.

Corner, B. D. & Moore, H. T. (1980). Field observations on the reproductive behavior of *Sepia latimanus*. *Micronesica*, **16**, 235–260.

Cornwell, C. J., Messenger, J. B. & Hanlon, R. T. (1997). Chromatophores and body patterning in the squid *Alloteuthis subulata*. *Journal of the Marine Biological Association of the United Kingdom*, **77**, 1243–1246.

Cortez, T., Castro, B. G. & Guerra, A. (1998). Drilling behaviour of *Octopus mimus* Gould. *Journal of Experimental Marine Biology and Ecology*, **224**, 193–203.

Cosgrove, J. (2003). An *in situ* observation of webover hunting by the Giant Pacific Octopus, *Enteroctopus dofleini* (Wulker, 1910). *The Canadian Field-Naturalist*, **117**, 117–118.

Cott, H. B. (1940). *Adaptive Coloration in Animals*. London: Methuen.

Cousteau, J. Y. & Diolé, P. (1973). *Octopus and Squid: The Soft Intelligence*. Garden City, NY: Doubleday.

Cowdry, E. V. (1911). *The Colour Changes of* Octopus vulgaris. University of Toronto Studies, Biological Series No. 10. Toronto: The University Library, The Librarian.

Cowen, R., Gertman, R. & Wiggett, G. (1973). Camouflage patterns in *Nautilus* and their implications for cephalopod paleobiology. *Lethaia*, **6**, 201–213.

Crancher, P., King, M. G., Bennett, A. & Montgomery, R. B. (1972). Conditioning of a free operant in *Octopus cyanea* Gray. *Journal of the Experimental Analysis of Behavior*, **17**, 359–362.

Crook, R. J. & Basil, J. A. (2008). A biphasic memory curve in the chambered nautilus, *Nautilus pompilius* L. (Cephalopoda: Nautiloidea). *Journal of Experimental Biology*, **211**, 1992–1998.

Crook, R. J. & Basil, J. A. (2013). Flexible spatial orientation and navigational strategies in chambered *Nautilus*. *Ethology*, **119**, 77–85.

Crook, R. J. & Walters, E. T. (2011). Nociceptive behavior and physiology of molluscs: animal welfare implications. *Ilar Journal*, **52**, 185–195.

Crook, R. J., Dickson, K., Hanlon, R. T. & Walters, E. T. (2014). Nociceptive sensitization reduces predation risk. *Current Biology*, **24**, 1121–1125.

Crook, R. J., Hanlon, R. T. & Basil, J. A. (2009). Memory of visual and topographical features suggests spatial learning in nautilus (*Nautilus pompilius* L.). *Journal of Comparative Psychology*, **123**, 264–274.

Crook, R. J., Hanlon, R. T. & Walters, E. T. (2013). Squid have nociceptors that display widespread long-term sensitization and spontaneous activity after bodily injury. *Journal of Neuroscience*, **33**, 10021–10026.

Crook, R. J., Lewis, T., Hanlon, R. T. & Walters, E. T. (2011). Peripheral injury induces long-term sensitization of defensive responses to visual and tactile stimuli in the squid *Loligo pealeii*, Lesueur 1821. *Journal of Experimental Biology*, **214**, 3173–3185.

Crookes, W. J., Ding, L., Huang, Q. L. *et al.* (2004). Reflectins: the unusual proteins of squid reflective tissues. *Science*, **303**, 235–238.

Croxall, J. P. & Prince, P. A. (1996). Cephalopods as prey. 1. Seabirds. *Philosophical Transactions of the Royal Society of London Series B – Biological Sciences*, **351**, 1023–1043.

Cummins, S. F., Boal, J. G., Buresch, K. C. *et al.* (2011). Extreme aggression in male squid induced by a beta-MSP-like pheromone. *Current Biology*, **21**, 322–327.

Curio, E. (1976). *The Ethology of Predation*. Berlin: Springer-Verlag.

Dahlgren, H. (1916). The production of light by animals. Light production in cephalopods. *Journal of the Franklin Institute*, 525–556.

D'Aniello, A., DiCosmo, A., DiCristo, C. *et al.* (1996). Occurrence of sex steroid hormones and their binding proteins in *Octopus vulgaris* lam. *Biochemical and Biophysical Research Communications*, **227**, 782–788.

Darmaillacq, A. S. & Shashar, N. (2008). Lack of polarization optomotor response in the cuttlefish *Sepia elongata* (d'Orbigny, 1845). *Physiology and Behavior*, **94**, 616–620.

Darmaillacq, A. S., Chichery, R. & Dickel, L. (2006). Food imprinting, new evidence from the cuttlefish *Sepia officinalis*. *Biology Letters*, **2**, 345–347.

Darmaillacq, A. S., Chichery, R., Poirier, R. & Dickel, L. (2004a). Effect of early feeding experience on subsequent prey preference by cuttlefish, *Sepia officinalis*. *Developmental Psychobiology*, **45**, 239–244.

Darmaillacq, A. S., Chichery, R., Shashar, N. & Dickel, L. (2006). Early familiarization overrides innate prey preference in newly hatched *Sepia officinalis* cuttlefish. *Animal Behaviour*, **71**, 511–514.

Darmaillacq, A. S., Dickel, L., Chichery, M. P., Agin, V. & Chichery, R. (2004b). Rapid taste aversion learning in adult cuttlefish, *Sepia officinalis*. *Animal Behaviour*, **68**, 1291–1298.

Darmaillacq, A. S., Dickel, L. & Mather, J. (2014). *Cephalopod Cognition*. Cambridge: Cambridge University Press.

Darmaillacq, A. S., Lesimple, C. & Dickel, L. (2008). Embryonic visual learning in the cuttlefish, *Sepia officinalis*. *Animal Behaviour*, **76**, 131–134.

Darwin, C. (1871). *The Descent of Man and Selection in Relation to Sex*. London: Murray.

Davey, G. (1989). *Ecological Learning Theory*. London: Routledge.

Davies, N. B. (1991). Mating systems, in *Behavioural Ecology. An Evolutionary Approach* (ed. J. R. Krebs & N. B. Davies), pp. 263–294. Oxford: Blackwell Scientific Publications.

Davis, R. W., Jaquet, N., Gendron, D. *et al.* (2007). Diving behavior of sperm whales in relation to behavior of a major prey species, the jumbo squid, in the Gulf of California, Mexico. *Marine Ecology Progress Series*, **333**, 291–302.

Daw, N. W. & Pearlman, A. L. (1974). Pigment migration and adaptation in eye of squid, *Loligo pealei*. *Journal of General Physiology*, **63**, 22–36.

Dawkins, M. (1971). Perceptual changes in chicks: another look at the 'search image' concept. *Animal Behaviour*, **19**, 566–574.

Dawkins, M. S. (2007). *Observing Animal Behaviour: Design and Analysis of Quantitive Controls*. Oxford: Oxford University Press.

Dawkins, M. S. & Guilford, T. (1991). The corruption of honest signalling. *Animal Behaviour*, **41**, 865–873.

Dawkins, R. & Krebs, J. R. (1978). Animal signals: information or manipulation?, in *Behavioural Ecology: An Evolutionary Approach* (ed. J. R. Krebs & N. B. Davies), pp. 282–309. Sutherland, MA: Sinauer Associates, Inc.

de Beer, C. L. & Potts, W. M. (2013). Behavioural observations of the common octopus *Octopus vulgaris* in Baia dos Tigres, southern Angola. *African Journal of Marine Science*, **35**, 579–583.

DeLong, R. L., Kooyman, G. L., Gilmartin, W. G. & Loughlin, T. R. (1984). Hawaiian monk seal diving behavior. *Proceedings of the Third International Theriological Congress, Helsinki; Acta Zoologica Fennica*, **172**, 129–131.

DeMartini, D. G., Krogstad, D. V. & Morse, D. E. (2013). Membrane invaginations facilitate reversible water flux driving tunable iridescence in a dynamic biophotonic system. *Proceedings of the National Academy of Sciences*, **110**, 2552–2556.

Denton, E. J. (1970). On the organization of reflecting surfaces in some marine animals. *Philosophical Transactions of the Royal Society of London B*, **258**, 285–313.

Denton, E. J. (1974). On buoyancy and the lives of modern and fossil cephalopods. *Proceedings of the Royal Society of London B*, **185**, 273–299.

Denton, E. J. (1990). Light and vision at depths greater than 200 metres, in *Light and Life in the Sea* (ed. P. J. Herring, A. K. Campbell, M. Whitfield & L. Maddock), pp. 127–148. Cambridge: Cambridge University Press.

Denton, E. J. & Gilpin-Brown, J. B. (1961). Effect of light on buoyancy of cuttlefish. *Journal of the Marine Biological Association of the United Kingdom*, **41**, 343–350.

Denton, E. J. & Gilpin-Brown, J. B. (1966). On the buoyancy of the pearly nautilus. *Journal of the Marine Biological Association of the United Kingdom*, **46**, 723–759.

Denton, E. J. & Gilpin-Brown, J. B. (1973). Floatation mechanisms in modern and fossil cephalopods. *Advances in Marine Biology*, **11**, 197–268.

Denton, E. J. & Land, M. F. (1971). Mechanism of reflexion in silvery layers of fish and cephalopods. *Proceedings of the Royal Society of London B: Biological Sciences*, **178**, 43–61.

Denton, E. J. & Locket, N. A. (1989). Possible wavelength discrimination by multibank retinae in deep-sea fish. *Journal of the Marine Biological Association of the United Kingdom*, **69**, 409–435.

Denton, E. J. & Warren, F. J. (1968). Eyes of the Histioteuthidae. *Nature*, **219**, 400–401.

Denton, E. J., Gilpin-Brown, J. B. & Howarth, J. V. (1967). On the buoyancy of *Spirula spirula*. *Journal of the Marine Biological Association of the United Kingdom*, **47**, 181–191.

Deravi, L. F., Magyar, A. P., Sheehy, S. P. *et al.* (2014). The structure–function relationships of a natural nanoscale photonic device in cuttlefish chromatophores. *Journal of the Royal Society Interface*, **11**, 9.

Derby, C. D. (2007). Escape by inking and secreting: marine molluscs avoid predators through a rich array of chemicals and mechanisms. *Biological Bulletin*, **213**, 274–289.

Derby, C. D. (2014). Cephalopod ink: production, chemistry, functions and applications. *Marine Drugs*, **12**, 2700–2730.

Derby, C. D., Kicklighter, C. E., Johnson, P. M. & Zhang, X. (2007). Chemical composition of inks of diverse marine molluscs suggests convergent chemical defenses. *Journal of Chemical Ecology*, **33**, 1105–1113.

Derby, C. D., Tottempudi, M., Love-Chezem, T. & Wolfe, L. S. (2013). Ink from longfin inshore squid, *Doryteuthis pealeii*, as a chemical and visual defense against two predatory fishes, summer flounder, *Paralichthys dentatus*, and sea catfish, *Ariopsis felis*. *Biological Bulletin*, **225**, 152–160.

Dews, P. M. (1959). Some observations on an operant in the octopus. *Journal of the Experimental Analysis of Behavior*, **2**, 57–63.

Diamant, A. & Shpigel, M. (1985). Interspecific feeding associations of groupers (Teleostei: Serranidae) with octopuses and moray eels in the Gulf of Eilat (Aqaba). *Environmental Biology of Fishes*, **13**, 153–159.

Dickel, L., Chichery, M. P. & Chichery, R. (1997). Postembryonic maturation of the vertical lobe complex and early development of predatory behavior in the cuttlefish (*Sepia officinalis*). *Neurobiology of Learning and Memory*, **67**, 150–160.

Dickel, L., Chichery, M. P. & Chichery, R. (1998). Time differences in the emergence of short- and long-term memory during post-embryonic development in the cuttlefish, *Sepia*. *Behavioural Processes*, **44**, 81–86.

Dickel, L., Chichery, M. P. & Chichery, R. (2001). Increase of learning abilities and maturation of the vertical lobe complex during postembryonic development in the cuttlefish, *Sepia*. *Developmental Psychobiology*, **39**, 92–98.

Di Cosmo, A. & Di Cristo, C. (1998). Neuropeptidergic control of the optic gland of *Octopus vulgaris*: FMRF-amide and GnRH immunoreactivity. *Journal of Comparative Neurology*, **398**, 1–12.

Di Cosmo, A., Di Cristo, C. & Paolucci, M. (2001). Sex steroid hormone fluctuations and morphological changes of the reproductive system of the female of *Octopus vulgaris* throughout the annual cycle. *Journal of Experimental Zoology*, **289**, 33–47.

Di Cristo, C. (2013). Nervous control of reproduction in *Octopus vulgaris*: a new model. *Invertebrate Neuroscience*, **13**, 27–34.

Dijkgraaf, S. (1961). The statocyst of *Octopus vulgaris* as a rotation receptor. *Pubblicazioni della Stazione Zoologica di Napoli*, **32**, 64–87.

Dijkgraaf, S. (1963). Versuche über Schallwahrnehmung bei Tintenfischen. *Naturwissenschaften*, **50**, 50.

Dilly, P. N. (1963). Delayed responses in *Octopus*. *Journal of Experimental Biology*, **40**, 393–401.

Dilly, P. N. (1972). *Taonius megalops*, a squid that rolls up into a ball. *Nature*, 237, 403–404.

Dilly, P. N. & Herring, P. J. (1978). The light organ and ink sac of *Heteroteuthis dispar* (Mollusca: Cephalopoda). *Journal of Zoology (London)*, 186, 47–59.

Dilly, P. N. & Herring, P. J. (1981). Ultrastructural features of the light organs of *Histioteuthis macrohista* (Mollusca, Cephalopoda). *Journal of Zoology*, 195, 255–266.

Dilly, P. N., Nixon, M. & Packard, A. (1964). Forces exerted by *Octopus vulgaris*. *Pubblicazioni della Stazione Zoologica di Napoli*, 34, 86–97.

DiMarco, F. P. & Hanlon, R. T. (1997). Agonistic behavior in the squid *Loligo plei* (Loliginidae, Teuthoidea): fighting tactics and the effects of size and resource value. *Ethology*, 103, 89–108.

Dimitrova, M., Stobbe, N., Schaefer, H. M. & Merilaita, S. (2009). Concealed by conspicuousness: distractive prey markings and backgrounds. *Proceedings of the Royal Society B – Biological Sciences*, 276, 1905–1910.

Domingues, P., Sykes, A., Sommerfield, A. & Andrade, J. P. (2003). Effects of feeding live or frozen prey on growth, survival and the life cycle of the cuttlefish, *Sepia officinalis* (Linnaeus, 1758). *Aquaculture International*, 11, 397–410.

Donovan, D. T. (1964). Cephalopod phylogeny and classification. *Biological Reviews*, 39, 259–287.

d'Orbigny, A. (1841). *Histoire naturelle générale et particulière des Céphalopodes*. Paris.

Dorsey, E. M. (1976). Natural history and social behavior of *Octopus rubescens* Berry. Seattle: Unpublished M.S. Thesis, University of Washington.

Doubleday, Z. A., Semmens, J. M., Smolenski, A. J. & Shaw, P. W. (2009). Microsatellite DNA markers and morphometrics reveal a complex population structure in a merobenthic octopus species (*Octopus maorum*) in south-east Australia and New Zealand. *Marine Biology*, 156, 1183–1192.

Douglas, R. H., Williamson, R. & Wagner, H. J. (2005). The pupillary response of cephalopods. *Journal of Experimental Biology*, 208, 261–265.

Drazen, J. C., Goffredi, S. K., Schlining, B. & Stakes, D. S. (2003). Aggregations of egg-brooding deep-sea fish and cephalopods on the Gorda Escarpment: a reproductive hot spot. *Biological Bulletin*, 205, 1–7.

Drew, G. A. (1911). Sexual activities of the squid, *Loligo pealii* (Les.). *Journal of Morphology*, 22, 327–359.

Driver, P. M. & Humphries, D. A. (1988). *Protean Behaviour: The Biology of Unpredictability*. New York: Oxford University Press.

Drummond, H. (1981). The nature and description of behavior patterns, in *Perspectives in Ethology,* Vol. 4: *Advantages of Diversity* (ed. P. P. G. Bateson & P. H. Klopfer), pp. 1–33. New York and London: Plenum Press.

Dubas, F. & Boyle, P. R. (1985). Chromatophore motor units in *Eledone cirrhosa* (Cephalopoda: Octopoda). *Journal of Experimental Biology*, 117, 415–432.

Dubas, F., Hanlon, R. T., Ferguson, G. P. & Pinsker, H. M. (1986). Localization and stimulation of chromatophore motoneurones in the brain of the squid, *Lolliguncula brevis*. *Journal of Experimental Biology*, 121, 1–25.

Dubas, F., Leonard, R. B. & Hanlon, R. T. (1986). Chromatophore motoneurons in the brain of the squid, *Lolliguncula brevis*: an HRP study. *Brain Research*, 374, 21–29.

Dunstan, A., Alanis, O. & Marshall, J. (2010). *Nautilus pompilius* fishing and population decline in the Philippines: a comparison with an unexploited Australian *Nautilus* population. *Fisheries Research*, 106, 239–247.

Dunstan, A., Bradshaw, C. J. A. & Marshall, J. (2011). Nautilus at risk – estimating population size and demography of *Nautilus pompilius*. *PLoS One*, 6, e16716.

Dunstan, A. J., Ward, P. D. & Marshall, N. J. (2011a). *Nautilus pompilius* life history and demographics at the Osprey Reef Seamount, Coral Sea, Australia. *PLoS One*, 6, e16312.

Dunstan, A. J., Ward, P. D. & Marshall, N. J. (2011b). Vertical distribution and migration patterns of *Nautilus pompilius*. *PLoS One*, 6, e16311.

Durward, R. D., Vessey, E., O'Dor, R. K. & Amaratunga, T. (1980). Reproduction in the squid, *Illex illecebrosus*: first observations in captivity and implications for the life cycle. *International Commission for the Northwest Atlantic Fisheries Selected Papers*, 6, 7–13.

Duval, P., Chichery, M.-P. & Chichery, R. (1984). Prey capture by the cuttlefish (*Sepia officinalis* L.): an experimental study of two strategies. *Behavioural Processes*, **9**, 13–21.

Eberhard, W. G. (1996). *Female Control: Sexual Selection by Cryptic Female Choice*. New Jersey: Princeton University Press.

Eberhard, W. G. (2000). Criteria for demonstrating postcopulatory female choice. *Evolution*, **54**, 1047–1050.

Eberhard, W. G. (2009). Postcopulatory sexual selection: Darwin's omission and its consequences. *Proceedings of the National Academy of Sciences of the United States of America*, **106**, 10025–10032.

Eberhard, W. G. & Cordero, C. (2003). Sexual conflict and female choice. *Trends in Ecology & Evolution*, **18**, 438–439.

Ebisawa, S., Tsuchiya, K. & Segawa, S. (2011). Feeding behavior and oxygen consumption of *Octopus ocellatus* preying on the short-neck clam *Ruditapes philippinarum*. *Journal of Experimental Marine Biology and Ecology*, **403**, 1–8.

Edelman, D. B. & Seth, A. K. (2009). Animal consciousness: a synthetic approach. *Trends in Neurosciences*, **32**, 476–484.

Edmunds, M. (1974). *Defence in Animals. A Survey of Anti-Predator Defences*. New York: Longman Group, Ltd.

Edmunds, M. (2000). Why are there good and poor mimics? *Biological Journal of the Linnean Society*, 459–466.

Edut, S. & Eilam, D. (2004). Protean behavior under barn-owl attack: voles alternate between freezing and fleeing and spiny mice flee in alternating patterns. *Behavioural Brain Research*, **155**, 207–216.

Ehrhardt, N. M. (1991). Potential impact of a seasonal migratory jumbo squid (*Dosidicus gigas*) stock on a Gulf of California sardine (*Sardinops sagax caerulea*) population. *Bulletin of Marine Science*, **49**, 325–332.

Eibl-Eibesfeldt, I. (1975). *Ethology. The Biology of Behavior*. New York: Holt, Rinehart and Winston, Inc.

Eibl-Eibesfeldt, I. & Scheer, G. (1962). Das brutpflegeverhalten eines weiblichen *Octopus aegina* Gray. *Zeitschrift für Tierpsychologie*, **19**, 257–261.

Ellis, D. V. (1985). *Animal Behavior and its Applications*. Chelsea, MI: Lewis Publishers, Inc.

Ellis, R. (1999). *The Search for the Giant Squid*. Robert Hale, Ltd.

Elwood, R. W. & Arnott, G. (2012). Understanding how animals fight with Lloyd Morgan's canon. *Animal Behaviour*, **84**, 1095–1102.

Emery, A. M., Wilson, I. J., Craig, S., Boyle, P. R. & Noble, L. R. (2001). Assignment of paternity groups without access to parental genotypes: multiple mating and developmental plasticity in squid. *Molecular Ecology*, **10**, 1265–1278.

Emery, D. G. (1975). The histology and fine structure of the olfactory organ in the squid *Lolliguncula brevis* Blainville. *Tissue and Cell*, **7**, 357–367.

Emery, D. G. (1976). Observations on the olfactory organ of adult and juvenile *Octopus joubini*. *Tissue and Cell*, **8**, 33–46.

Emlen, S. T. & Oring, L. W. (1977). Ecology, sexual selection and the evolution of mating systems. *Science*, **197**, 215–223.

Enault, J., Zatylny-Gaudin, C., Bernay, B. *et al.* (2012). A complex set of sex pheromones identified in the cuttlefish *Sepia officinalis*. *PLoS ONE*, **7**, e46531.

Endler, J. A. (1978). A predator's view of animal color patterns. *Evolutionary Biology*, **11**, 319–364.

Endler, J. A. (1981). An overview of the relationships between mimicry and crypsis. *Biological Journal of the Linnean Society*, **16**, 25–31.

Endler, J. A. (1984). Progressive background matching in moths, and a quantitative measure of crypsis. *Biological Journal of the Linnean Society*, **22**, 187–231.

Endler, J. A. (1991). Interactions between predators and prey, in *Behavioural Ecology. An Evolutionary Approach* (ed. J. R. Krebs & N. B. Davies), pp. 169–196. Oxford: Blackwell Scientific Publications.

Endler, J. A. (1993). Some general comments on the evolution and design of animal communication systems. *Philos Trans R Soc Lond B Biol Sci*, **340**, 215–225.

Endler, J. A. (2006). Disruptive and cryptic coloration. *Proceedings of the Royal Society B – Biological Sciences*, **273**, 2425–2426.

Endler, J. A. & Mielke, P. W. (2005). Comparing entire colour patterns as birds see them. *Biological Journal of the Linnean Society*, **86**, 405–431.

Engeser, T. S. & Clarke, M. R. (1988). Cephalopod hooks, both recent and fossil, in *The Mollusca*, Vol. 12: *Paleontology and Neontology of Cephalopods* (ed. M. R. Clarke & E. R. Trueman), pp. 133–151. San Diego: Academic Press.

Englund, G. & Olsson, T. I. (1990). Fighting and assessment in the net-spinning caddis larva *Arctopsyche ladogensis*: a test of the sequential assessment game. *Animal Behaviour*, **39**, 55–62.

Enquist, M. & Leimar, O. (1983). Evolution of fighting behavior: decision rules and assessment of relative strength. *Journal of Theoretical Biology*, **102**, 387–410.

Enquist, M. & Leimar, O. (1987). Evolution of fighting behaviour: the effect of variation in resource value. *Journal of Theoretical Biology*, **127**, 187–205.

Enquist, M., Ljungberg, T. & Zandor, A. (1987). Visual assessment of fighting ability in the Cichlid fish *Nannacara anomala*. *Animal Behaviour*, **35**, 1262–1263.

Escanez, A., Riera, R., Gonzalez, A. F. & Guerra, A. (2012). On the occurrence of egg masses of the diamond-shaped squid *Thysanoteuthis rhombus* Troschel, 1857 in the subtropical eastern Atlantic (Canary Islands). A potential commercial species? *ZooKeys*, **222**, 69–76.

Eyman, M., Crispino, M., Kaplan, B. B. & Giuditta, A. (2003). Squid photoreceptor terminals synthesize calexcitin, a learning related protein. *Neuroscience Letters*, **347**, 21–24.

Fawcett, T. W. & Mowles, S. L. (2013). Assessments of fighting ability need not be cognitively complex. *Animal Behaviour*, **86**, E1–E7.

Ferguson, G. P. & Messenger, J. B. (1991). A countershading reflex in cephalopods. *Proceedings of the Royal Society of London B: Biological Sciences*, **243**, 63–67.

Ferguson, G. P., Martini, F. M. & Pinsker, H. M. (1988). Chromatophore motor fields in the squid, *Lolliguncula brevis*. *Journal of Experimental Biology*, **134**, 281–195.

Ferguson, G. P., Messenger, J. B. & Budelmann, B. U. (1994). Gravity and light influence the countershading reflexes of the cuttlefish *Sepia officinalis*. *Journal of Experimental Biology*, **191**, 247–256.

Field, J. C., Elliger, C., Baltz, K. *et al.* (2013). Foraging ecology and movement patterns of jumbo squid (*Dosidicus gigas*) in the California Current System. *Deep-Sea Research Part II – Topical Studies in Oceanography*, **95**, 37–51.

Fields, W. G. (1965). The structure, development, food relations, reproduction, and life history of the squid *Loligo opalescens* Berry. *Fishery Bulletin*, **131**, 1–108.

Finn, J. K. & Norman, M. D. (2010). The argonaut shell: gas-mediated buoyancy control in a pelagic octopus. *Proceedings of the Royal Society B – Biological Sciences*, **277**, 2967–2971.

Finn, J. K., Tregenza, T. & Norman, M. D. (2009a). Defensive tool use in a coconut-carrying octopus. *Current Biology*, **19**, R1069–R1070.

Finn, J., Tregenza, T. & Norman, M. (2009b). Preparing the perfect cuttlefish meal: complex prey handling by dolphins. *PLoS ONE*, **4**, e4217.

Fiorito, G. & Chichery, R. (1995). Lesions of the vertical lobe impair visual discrimination learning by observation in *Octopus vulgaris*. *Neuroscience Letters*, **192**, 117–120.

Fiorito, G. & Gherardi, F. (1999). Prey-handling behaviour of *Octopus vulgaris* (Mollusca, Cephalopoda) on bivalve preys. *Behavioural Processes*, **46**, 75–88.

Fiorito, G. & Scotto, P. (1992). Observational learning in *Octopus vulgaris*. *Science*, **256**, 545–547.

Fiorito, G., Agnisola, C., d'Addio, M., Valanzano, A. & Calamandrei, G. (1998). Scopolamine impairs memory recall in *Octopus vulgaris*. *Neuroscience Letters*, **253**, 87–90.

Fiorito, G., von Planta, C. & Scotto, P. (1990). Problem solving ability of *Octopus vulgaris* Lamarck (Mollusca, Cephalopoda). *Behavioral and Neural Biology*, **53**, 217–230.

Fleisher, K. J. & Case, J. F. (1995). Cephalopod predation facilitated by dinoflagellate luminescence. *Biological Bulletin*, **189**, 263–271.

Fleming, I. (1958). *Dr No*. London: Cape.

Flores, E. E. C. (1983). Visual discrimination testing in the squid *Todarodes pacificus*: experimental evidence for lack of color vision. *Memoirs of the National Museum of Victoria*, **44**, 213–227.

Flores, E. E. C., Igarashi, S. & Mikami, T. (1978). Studies on squid behavior in relation to fishing – III. On the optomotor response of squid *Todarodes pacificus* Steenstrup, to various colors. *Bulletin of the Faculty of Fisheries, Hokkaido University*, **29**, 131–140.

Florey, E. (1966). Nervous control and spontaneous activity of the chromatophores of a cephalopod,

Loligo opalescens. Comparative Biochemistry and Physiology, **18**, 305–324.

Florey, E. (1969). Ultrastructure and function of cephalopod chromatophores. *American Zoologist*, **9**, 429–442.

Florey, E. (1985). The Zoological Station at Naples and the neuron: personalities and encounters in a unique institution. *Biological Bulletin*, **168**, 137–152.

Florey, E. & Kriebel, M. E. (1969). Electrical and mechanical responses of chromatophore muscle fibers of squid, *Loligo opalescens*, to nerve stimulation and drugs. *Zeitschrift für Vergleichende Physiologie*, **65**, 98–130.

Florey, E., Dubas, F. & Hanlon, R. T. (1985). Evidence for L-glutamate as a transmitter substance of motoneurons innervating squid chromatophore muscles. *Comparative Biochemistry and Physiology C*, **82**, 259–268.

Foote, K. G., Hanlon, R. T., Iampietro, P. J. & Kvitek, R. G. (2006). Acoustic detection and quantification of benthic egg beds of the squid *Loligo opalescens* in Monterey Bay, California. *Journal of the Acoustic Society of America*, **119**, 844–856.

Ford, E. B. (1975). *Ecological Genetics*. London: Chapman & Hall.

Forsythe, J. W. & Hanlon, R. T. (1985). Aspects of egg development, post-hatching behavior, growth and reproductive biology of *Octopus burryi* Voss, 1950 (Mollusca: Cephalopoda). *Vie Milieu*, **35**, 273–282.

Forsythe, J. W. & Hanlon, R. T. (1988). Behavior, body patterning and reproductive biology of *Octopus bimaculoides* from California. *Malacologia*, **29**, 41–55.

Forsythe, J. W. & Hanlon, R. T. (1997). Foraging and associated behavior by *Octopus cyanea* Gray, 1849 on a coral atoll, French Polynesia. *Journal of Experimental Marine Biology and Ecology*, **209**, 15–31.

Forsythe, J. W. & Van Heukelem, W. F. (1987). Growth, in *Cephalopod Life Cycles*, Vol. 2: *Comparative Reviews* (ed. P. R. Boyle), pp. 135–156. London: Academic Press.

Forsythe, J. W., DeRusha, R. H. & Hanlon, R. T. (1994). Growth, reproduction and life-span of *Sepia officinalis* (Cephalopoda, Mollusca) cultured through seven consecutive generations. *Journal of Zoology*, **233**, 175–192.

Forsythe, J. W., Kangas, N. & Hanlon, R. T. (2004). Does the California market squid, *Loligo opalescens*, spawn naturally during the day or at night? A note on the successful use of ROV's to obtain basic fisheries biology data. *Fisheries Bulletin*, **102**, 389–392.

Fotheringham, N. (1974). Tropic complexity in a littoral boulderfield. *Limnology and Oceanography*, **19**, 84–91.

Fox, D. L. (1938). An illustrated note on the mating and egg-brooding habits of the two-spotted octopus. *Transactions of the San Diego Society of Natural History*, **9**, 31–34.

Foyle, T. P. & O'Dor, R. K. (1987). Predatory strategies of squid (*Illex illecebrosus*) attacking small and large fish. *Marine Behaviour and Physiology*, **13**, 155–168.

Francisco Ruiz, J., Sepulveda, R. D. & Ibanez, C. M. (2012). Behaviour of *Robsonella fontaniana* in response to a potential predator. *Latin American Journal of Aquatic Research*, **40**, 253–258.

Frank, M. G., Waldrop, R. H., Dumoulin, M., Aton, S. & Boal, J. G. (2012). A preliminary analysis of sleep-like states in the cuttlefish *Sepia officinalis*. *PLoS ONE*, **7**, e38125.

Franklin, A. M., Squires, Z. E. & Stuart-Fox, D. (2012). The energetic cost of mating in a promiscuous cephalopod. *Biology Letters*, **8**, 754–756.

Froesch, D. (1973). Projection of chromatophore nerves on the body surface of *Octopus vulgaris*. *Marine Biology*, **19**, 203–242.

Froesch, D. & Marthy, H.-J. (1975). The structure and function of the oviducal gland in octopods (Cephalopoda). *Proceedings of the Royal Society of London B*, **188**, 95–101.

Froesch, D. & Messenger, J. B. (1978). On leucophores and the chromatic unit of *Octopus vulgaris*. *Journal of Zoology (London)*, **186**, 163–173.

Fukuda, Y. (1980). Observations by SEM, in *Nautilus macromphalus in Captivity* (ed. T. Hamada, I. Obata & T. Okutani), pp. 23–33. Tokyo: Tokai University Press.

Gabe, S. H. (1975). Reproduction in the giant octopus of the North Pacific, *Octopus dofleini martini*. *Veliger*, **18**, 146–150.

Gadgil, M. & Bossert, W. H. (1970). Life historical consequences of natural selection. *The American Naturalist*, **104**, 1–24.

Garcia-Gonzalez, F. & Simmons, L. W. (2007). Shorter sperm confer higher competitive fertilization success. *Evolution*, **61**, 816–824.

Gaston, M.R. & Tublitz, N.J. (2004). Peripheral innervation patterns and central distribution of fin chromatophore motoneurons in the cuttlefish *Sepia officinalis. Journal of Experimental Biology* 207, 3089–3098.

Gauldie, R. W., West, I. F. & Forch, E. C. (1994). Statocyst, statolith, and age estimation of the giant-squid, *Architeuthis kirki. Veliger*, 37, 93–109.

Gennaro, J. F. J., Lorincz, A. E. & Brewster, H. B. (1965). The anterior gland of the octopus (*Octopus vulgaris*) and its mucous secretion. *Annals of the New York Academy of Sciences*, 118, 1021–1025.

Gerlach, G., Buresch, K. C. & Hanlon, R. T. (2012). Population structure of the squid *Doryteuthis (Loligo) pealeii* on the eastern coast of the USA: comment on Shaw *et al.* (2010). *Marine Ecology Progress Series*, 450, 281–283.

Ghiretti, F. (1959). Cephalotoxin: the crab-paralysing agent of the posterior salivary glands of cephalopods. *Nature*, 183, 1192–1193.

Ghiretti, F. (1960). Toxicity of octopus saliva against crustacea. *Annals of the New York Academy of Sciences*, 90, 726–741.

Ghiretti, F. & Cariello, L. (1977). *Gli Animali Marini Velenosi e le loro Tossine*. Padova: Piccin.

Ghoshal, A., DeMartini, D. G., Eck, E. & Morse, D. E. (2013). Optical parameters of the tunable Bragg reflectors in squid. *Journal of the Royal Society Interface*, 10, 20130386.

Gibson, R. M. (1983). Visual abilities and foraging behaviour of predatory fish. *Trends in Neuroscience*, 6, 197–199.

Gilly, W. F. & Lucero, M. T. (1992). Behavioural responses to chemical stimulation of the olfactory organ in the squid, *Loligo opalescens. Journal of Experimental Biology*, 162, 209–229.

Gilly, W. F., Elliger, C. A., Salinas, C. A. *et al.* (2006). Spawning by jumbo squid *Dosidicus gigas* in San Pedro Martir basin, Gulf of California, Mexico. *Marine Ecology Progress Series*, 313, 125–133.

Gilly, W. F., Hopkins, B. & Mackie, G. O. (1991). Development of giant motor axons and neural control of escape responses in squid embryos and hatchlings. *Biological Bulletin*, 180, 209–220.

Gilly, W. F., Markaida, U., Baxter, C. H. *et al.* (2006). Vertical and horizontal migrations by the jumbo squid *Dosidicus gigas* revealed by electronic tagging. *Marine Ecology – Progress Series*, 324, 1–17.

Gilly, W. F., Zeidberg, L. D., Booth, J. A. T. *et al.* (2012). Locomotion and behavior of Humboldt squid, *Dosidicus gigas*, in relation to natural hypoxia in the Gulf of California, Mexico. *The Journal of Experimental Biology*, 215, 3175–3190.

Glanzman, D. L. (2008). *Octopus* conditioning: a multi-armed approach to the LTP-learning question. *Current Biology*, 18, R527–R530.

Gleadall, I. G. & Shashar, N. (2004). The octopus's garden: the visual world of cephalopods, in *Complex Worlds from Simpler Nervous Systems* (ed. F. R. Prete), pp. 269–308. Cambridge, MA: MIT Press.

Gleadall, I. G., Guerrero-Kommritz, J., Hochberg, F. G. & Laptikhovsky, V. V. (2010). The inkless octopuses (Cephalopoda: Octopodidae) of the southwest Atlantic. *Zoological Science*, 27, 528–553.

Gleadall, I. G., Ohtsu, K., Gleadall, E. & Tsukahara, Y. (1993). Screening-pigment migration in the octopus retina includes control by dopaminergic efferents. *Journal of Experimental Biology*, 185, 1–16.

Godfrey-Smith, P. & Lawrence, M. (2012). Long-term high-density occupation of a site by *Octopus tetricus* and possible site modification due to foraging behavior. *Marine and Freshwater Behaviour and Physiology*, 45, 261–268.

Gonzalez-Bellido, P. T., Wardill, T. J., Buresch, K. C., Ulmer, K. M. & Hanlon, R. T. (2014). Expression of squid iridescence depends on environmental luminance and peripheral ganglion control. *Journal of Experimental Biology*, 217, 850–858.

Goss, C., Middleton, D. & Rodhouse, P. (2001). Investigations of squid stocks using acoustic survey methods. *Fisheries Research*, 54, 111–121.

Grable, M. M., Shashar, N., Gilles, N. L., Chiao, C. C. & Hanlon, R. T. (2002). Cuttlefish body patterns as a behavioral assay to determine polarization perception. *Biological Bulletin*, 203, 232–234.

Grasso, F. W. (2008). *Octopus* sucker-arm coordination in grasping and manipulation. *American Malacological Bulletin*, 24, 13–23.

Grasso, F. W. & Basil, J. A. (2009). The evolution of flexible behavioral repertoires in cephalopod molluscs. *Brain, Behavior and Evolution*, 74, 231–245.

Gratwicke, B. & Speight, M. R. (2005). The relationship between fish species richness, abundance and habitat complexity in a range of shallow tropical marine habitats. *Journal of Fish Biology*, **66**, 650–667.

Graziadei, P. (1964a). Electron microscopy of some primary receptors in the sucker of *Octopus vulgaris*. *Zeitschrift für Zellforschung*, **64**, 510–522.

Graziadei, P. (1964b). Receptors in the sucker of the cuttlefish. *Nature*, **203**, 384–386.

Graziadei, P. (1965). Sensory receptor cells and related neurons in cephalopods. *Cold Spring Harbor Symposia on Quantitative Biology*, **30**, 45–57.

Graziadei, P. (1971). The nervous system of the arms, in *The Anatomy of the Nervous System of* Octopus vulgaris (ed. J. Z. Young), pp. 45–61. Oxford: Clarendon.

Graziadei, P. P. C. & Gagne, H. T. (1976a). An unusual receptor in octopus. *Tissue and Cell*, **8**, 229–240.

Graziadei, P. P. C. & Gagne, H. T. (1976b). Sensory innervation in the rim of the octopus sucker. *Journal of Morphology*, **150**, 639–679.

Grimpe, G. (1926). Biologische Beobachtungen an *Sepia officinalis*. *Deutsche Zoologische Gesellschaft Verhandlungen*, **31**, 148–153.

Grisley, M. S. (1993). Separation and partial characterization of salivary enzymes expressed during prey handling in the octopus *Eledone cirrhosa*. *Comparative Biochemistry and Physiology*, **105B**, 183–192.

Grisley, M. S. & Boyle, P. R. (1985). A new application of serological techniques to gut content analysis. *Journal of Experimental Marine Biology and Ecology*, **90**, 1–9.

Grisley, M. S. & Boyle, P. R. (1987). Bioassay and proteolytic activity of digestive enzymes from octopus saliva. *Comparative Biochemistry and Physiology*, **88B**, 1117–1124.

Grisley, M. S. & Boyle, P. R. (1988). Recognition of food in *Octopus* digestive tract. *Journal of Experimental Marine Biology and Ecology*, **118**, 7–32.

Grisley, M. S. & Boyle, P. R. (1990). Chitinase: a novel enzyme from octopus saliva. *Comparative Biochemistry and Physiology*, **95B**, 311–316.

Grisley, M. S., Boyle, P. R. & Key, L. N. (1996). Eye puncture as a route of entry for saliva during predation on crabs by the octopus *Eledone cirrhosa*

(Lamarck). *Journal of Experimental Marine Biology and Ecology*, **202**, 225–237.

Griswold, C. A. & Prezioso, J. (1981). *In situ* observations on reproductive behavior of the long-finned squid, *Loligo pealei*. *Fishery Bulletin*, **78**, 945–947.

Groeger, G., Cotton, P. A. & Williamson, R. (2005). Ontogenetic changes in the visual acuity of *Sepia officinalis* measured using the optomotor response. *Canadian Journal of Zoology*, **83**, 274–279.

Guerra, A. (1981). Spatial distribution pattern of *Octopus vulgaris*. *Journal of Zoology (London)*, **195**, 133–146.

Guerra, A. & Nixon, M. (1987). Crab and mollusc shell drilling by *Octopus vulgaris* (Mollusca: Cephalopoda) in the Ria de Vigo (north-west Spain). *Journal of Zoology (London)*, **211**, 515–523.

Guerra, A., Gonzalez, A. F., Rocha, F. J., Sagarminaga, R. & Canadas, A. (2002a). Planktonic egg masses of the diamond-shaped squid *Thysanoteuthis rhombus* in the eastern Atlantic and the Mediterranean Sea. *Journal of Plankton Research*, **24**, 333–338.

Guerra, A., Gonzalez, A. F., Rocha, F., Segonzac, M. & Gracia, J. (2002b). Observations from submersibles of rare long-arm bathypelagic squids. *Sarsia*, **87**, 189–192.

Guerra, A., Hernandez-Urcera, J., Garci, M. E. *et al.* (2014). Dwellers in dens on sandy bottoms: ecological and behavioural traits of *Octopus vulgaris*. *Scientia Marina*, **78**, 405–414.

Guerra, A., Rocha, F., Gonzalez, A. F. & Gonzalez, J. L. (2006). First observation of sand-covering by the lesser octopus *Eledone cirrhosa*. *Iberus*, **24**, 27–31.

Guibé, M. & Dickel, L. (2011). Embryonic visual experience influences post-hatching shelter preference in cuttlefish. *Vie et Milieu – Life and Environment*, **61**, 243–246.

Guibé, M., Boal, J. G. & Dickel, L. (2010). Early exposure to odors changes later visual prey preferences in cuttlefish. *Developmental Psychobiology*, **52**, 833–837.

Guibé, M., Poirel, N., Houde, O. & Dickel, L. (2012). Food imprinting and visual generalization in embryos and newly hatched cuttlefish, *Sepia officinalis*. *Animal Behaviour*, **84**, 213–217.

Guilford, T. & Dawkins, M. S. (1991). Receiver psychology and the evolution of animal signals. *Animal Behaviour*, **42**, 1–14.

Gulland, J. A. & Garcia, S. (1984). Observed patterns in multispecies fisheries, in *Exploitation of Marine Communities* (ed. R. M. May), pp. 155–190. Berlin: Springer-Verlag.

Gutfreund, Y., Flash, T., Fiorito, G. & Hochner, B. (1998). Patterns of arm muscle activation involved in octopus reaching movements. *Journal of Neuroscience*, **18**, 5976–5987.

Gutfreund, Y., Flash, T., Yarom, Y. *et al.* (1996). Organization of octopus arm movements: a model system for studying the control of flexible arms. *Journal of Neuroscience*, **16**, 7297–7307.

Gutnick, T., Byrne, R. A., Hochner, B. & Kuba, M. (2011). *Octopus vulgaris* uses visual information to determine the location of its arm. *Current Biology*, **21**, 460–462.

Gutsal, D. K. (1989). Underwater observations on distribution and behaviour of cuttlefish *Sepia pharaonis* in the western Arabian Sea. *Biologiya Morya – Marine Biology*, **1**, 48–55.

Haddock, S. H. D., Moline, M. A. & Case, J. F. (2010). Bioluminescence in the sea. *Annual Review of Marine Science*, **2**, 443–493.

Hagins, W. A. (1965). Electrical signs of information flow in photoreceptors. *Cold Spring Harbor Symposia on Quantitative Biology*, **30**, 403–418.

Hailman, J. P. (1977). *Optical Signals. Animal Communication and Light.* Bloomington: Indiana Univ. Press.

Hall, D. N. F. (1956). Ink ejection by Cephalopoda. *Nature*, **7**, 663.

Hall, H. (1990). Mugged by a squid! *Ocean Realm*, **1990**, 6–8.

Hall, K., Fowler, A. J. & Geddes, M. C. (2007). Evidence for multiple year classes of the giant Australian cuttlefish *Sepia apama* in northern Spencer Gulf, South Australia. *Reviews in Fish Biology and Fisheries*, **17**, 367–384.

Hall, K. C. & Hanlon, R. T. (2002). Principal features of the mating system of a large spawning aggregation of the giant Australian cuttlefish *Sepia apama* (Mollusca: Cephalopoda). *Marine Biology*, **140**, 533–545.

Halliday, T. (1983). Information and communication, in *Animal Behaviour. 2: Communication* (ed. T. R. Halliday & P. J. B. Slater), pp. 43–81. Oxford: Blackwell Scientific Publications.

Hamabe, M. (1961). Experimental studies on breeding habits and development of the squid, *Ommastrephes sloani pacificus* Steenstrup I. Copulation. *Dobutsugaku Zasshi*, **70**, 378–384.

Hamabe, M. (1964). Study on the migration of squid (*Ommastrephes sloani pacificus* Steenstrup) as related to the phases of the moon. *Bulletin of the Japanese Society of Scientific Fisheries*, **30**, 209–215.

Hamabe, M. & Shimizu, T. (1957). The copulation behavior of Yariika, *Loligo bleekeri* K. *Report on the Japanese Sea Regional Fisheries Research Laboratory*, **3**, 131–136.

Hamasaki, D. I. (1968a). The electroretinogram of the intact anesthetized octopus. *Vision Research*, **8**, 247–258.

Hamasaki, D. I. (1968b). The ERG-determined spectral sensitivity of the octopus. *Vision Research*, **8**, 1013–1021.

Hand, C. (1975). Behaviour of some New Zealand sea anemones and their molluscan and crustacean hosts. *New Zealand Journal of Marine & Freshwater Research*, **9**, 509–527.

Hanlon, R. T. (1978). *Aspects of the Biology of the Squid* Loligo (Doryteuthis) plei *in Captivity*. Coral Gables: RSMAS, University of Miami.

Hanlon, R. T. (1982). The functional organization of chromatophores and iridescent cells in the body patterning of *Loligo plei* (Cephalopoda: Myopsida). *Malacologia*, **23**, 89–119.

Hanlon, R. T. (1983a). *Octopus joubini*, in *Cephalopod Life Cycles, Vol. I: Species Accounts* (ed. P. R. Boyle), pp. 293–310. London: Academic Press.

Hanlon, R. T. (1983b). *Octopus briareus*, in *Cephalopod Life Cycles, Vol. I: Species Accounts* (ed. P. R. Boyle), pp. 251–266. London: Academic Press.

Hanlon, R. T. (1988). Behavioral and body patterning characters useful in taxonomy and field identification of cephalopods. *Malacologia*, **29**, 247–264.

Hanlon, R. T. (1990). Maintenance, rearing and culture of teuthoid and sepioid squids, in *Squid as an Experimental Animal* (ed. D. L. Gilbert, W. J. Adelman & J. M. Arnold), pp. 35–62. New York: Plenum Press.

Hanlon, R. T. (2007). Cephalopod dynamic camouflage. *Current Biology*, **17**, R400–R404.

Hanlon, R. T. & Budelmann, B.-U. (1987). Why cephalopods are probably not 'deaf'. *The American Naturalist*, **129**, 312–317.

Hanlon, R. T. & Forsythe, J. W. (1985). Advances in the laboratory culture of octopuses for biomedical research. *Laboratory Animal Science*, **35**, 33–40.

Hanlon, R. T. & Forsythe, J. W. (2008). Sexual cannibalism by *Octopus cyanea* on a Pacific coral reef. *Marine and Freshwater Behaviour and Physiology*, **41**, 19–28.

Hanlon, R. T. & Hixon, R. F. (1980). Body patterning and field observations of *Octopus burryi* Voss, 1950. *Bulletin of Marine Science*, **30**, 749–755.

Hanlon, R. T. & Messenger, J. B. (1988). Adaptive coloration in young cuttlefish (*Sepia officinalis* L.): the morphology and development of body patterns and their relation to behaviour. *Philosophical Transactions of the Royal Society of London B*, **320**, 437–487.

Hanlon, R. T. & Shashar, N. (2003). Aspects of the sensory ecology of cephalopods, in *Sensory Processing in the Aquatic Environment* (ed. S. P. Collin & N. J. Marshall), pp. 266–282. Heidelberg, Germany: Springer-Verlag.

Hanlon, R. T. & Wolterding, M. R. (1989). Behavior, body patterning, growth and life history of *Octopus briareus* cultured in the laboratory. *American Malacological Bulletin*, **7**, 21–45.

Hanlon, R. T., Ament, S. A. & Gabr, H. (1999). Behavioral aspects of sperm competition in cuttlefish, *Sepia officinalis* (Sepioidea: Cephalopoda). *Marine Biology*, **134**, 719–728.

Hanlon, R. T., Benjamins, S., Beet, A. & Solow, A. (in prep). When camouflage fails: secondary defense tactics of *Octopus vulgaris* in a coral reef ecosystem. *Biological Journal of the Linnean Society*.

Hanlon, R. T., Buresch, K., Moustahfid, H. & Staudinger, M. (2013a). *Doryteuthis pealeii*, longfin inshore squid, in *Advances in Squid Biology, Ecology and Fisheries* (ed. R. Rosa, R. O'Dor & G. J. Pierce), pp. 205–240. Hauppauge, New York: Nova Science Publishers, Inc.

Hanlon, R. T., Chiao, C. C., Mäthger, L. M. & Marshall, N. J. (2013b). A fish-eye view of cuttlefish camouflage using *in-situ* spectrometry. *Biological Journal of the Linnean Society*, **109**, 535–551.

Hanlon, R. T., Chiao, C. C., Mäthger, L. M. *et al.* (2009). Cephalopod dynamic camouflage: bridging the continuum between background matching and disruptive coloration. *Philosophical Transactions of the Royal Society B – Biological Sciences*, **364**, 429–437.

Hanlon, R. T., Chiao, C. C., Mäthger, L. M. *et al.* (2011). Rapid adaptive camouflage in cephalopods, in *Animal Camouflage: Mechanisms and Functions* (ed. M. Stevens & S. Merilaita), pp. 145–163. Cambridge: Cambridge University Press.

Hanlon, R. T., Claes, M. F., Ashcraft, S. E. & Dunlap, P. V. (1997). Laboratory culture of the sepiolid squid *Euprymna scolopes*: a model system for bacteria–animal symbiosis. *Biological Bulletin*, **192**, 364–374.

Hanlon, R. T., Conroy, L. A. & Forsythe, J. W. (2008). Mimicry and foraging behaviour of two tropical sand-flat octopus species off North Sulawesi, Indonesia. *Biological Journal of the Linnean Society*, **93**, 23–38.

Hanlon, R. T., Cooper, K. M., Budelmann, B. U. & Pappas, T. C. (1990). Physiological color change in squid iridophores. I. Behavior, morphology and pharmacology in *Lolliguncula brevis*. *Cell and Tissue Research*, **259**, 3–14.

Hanlon, R. T., Forsythe, J. W. & Boletzky, S. von (1985). Field and laboratory behavior of 'macrotritopus larvae' reared to *Octopus defilippi* Verany, 1851 (Mollusca: Cephalopoda). *Vie Milieu*, **35**, 237–242.

Hanlon, R. T., Forsythe, J. W. & Joneschild, D. E. (1999). Crypsis, conspicuousness, mimicry and polyphenism as antipredator defences of foraging octopuses on Indo-Pacific coral reefs, with a method of quantifying crypsis from video tapes. *Biological Journal of the Linnean Society*, **66**, 1–22.

Hanlon, R. T., Forsythe, J. W. & Messenger, J. B. (1984). Visual discrimination training of laboratory reared octopuses. *American Malacological Bulletin*, **2**, 92.

Hanlon, R. T., Hixon, R. F., Forsythe, J. W. & Hendrix, J. P. (1979). Cephalopods attracted to experimental night lights during a saturation dive at St Croix, U.S. Virgin Islands. *Bulletin of the American Malacological Union*, **1979**, 53–58.

Hanlon, R. T., Hixon, R. F. & Hulet, W. H. (1983). Survival, growth, and behavior of the loliginid squids *Loligo plei*, *Loligo pealei*, and *Lolliguncula*

References

brevis (Mollusca: Cephalopoda) in closed sea water systems. *Biological Bulletin*, **165**, 637–685.

Hanlon, R. T., Kangas, N. & Forsythe, J. W. (2004). Egg capsule deposition and how behavioral interactions influence spawning rate in the squid *Loligo opalescens* in Monterey Bay, California. *Marine Biology*, **145**, 923–930.

Hanlon, R. T., Maxwell, M. R. & Shashar, N. (1997). Behavioral dynamics that would lead to multiple paternity within egg capsules of the squid *Loligo pealei*. *Biological Bulletin*, **193**, 212–214.

Hanlon, R. T., Maxwell, M. R., Shashar, N., Loew, E. R. & Boyle, K. L. (1999). An ethogram of body patterning behavior in the biomedically and commercially valuable squid *Loligo pealei* off Cape Cod, Massachusetts. *Biological Bulletin*, **197**, 49–62.

Hanlon, R. T., Naud, M. J., Forsythe, J. W. *et al.* (2007). Adaptable night camouflage by cuttlefish. *American Naturalist*, **169**, 543–551.

Hanlon, R. T., Naud, M. J., Shaw, P. W. & Havenhand, J. N. (2005). Behavioural ecology: transient sexual mimicry leads to fertilization. *Nature*, **430**, 212.

Hanlon, R. T., Smale, M. J. & Sauer, W. H. H. (1994). An ethogram of body patterning behaviour in the squid *Loligo vulgaris reynaudii* on spawning grounds in South Africa. *Biological Bulletin*, **187**, 363–372.

Hanlon, R. T., Smale, M. J. & Sauer, W. H. H. (2002). The mating system of the squid *Loligo vulgaris reynaudii* (Cephalopoda, Mollusca) off South Africa: fighting, guarding, sneaking, mating and egg laying behavior. *Bulletin of Marine Science*, **71**, 331–345.

Hanlon, R. T., Turk, P. E., Lee, P. G. & Yang, W. T. (1987). Laboratory rearing of the squid *Loligo pealei* to the juvenile stage: growth comparisons with fishery data. *Fishery Bulletin*, **85**, 163–167.

Hanlon, R. T., Watson, A. C. & Barbosa, A. (2010). A 'mimic octopus' in the Atlantic: flatfish mimicry and camouflage by *Macrotritopus defilippi*. *Biological Bulletin*, **218**, 15–24.

Hara, T. & Hara, R. (1972). Cephalopod retinochrome, in *Handbook of Sensory Physiology* (ed. H. J. A. Dartnall), pp. 720–746. Berlin, Heidelberg, New York: Springer.

Hara, T., Hara, R., Kishigami, A. *et al.* (1995). Rhodopsin and retinochrome in the retina of a tetrabranchiate

cephalopod, *Nautilus pompilius*. *Zoological Science*, **12**, 195–201.

Harden Jones, F. R. (1984). Could fish use inertial clues when on migration?, in *Mechanisms of Migration in Fishes* (ed. J. D. McCleave, G. P. Arnold, J. J. Dodson & W. H. Neill), pp. 67–78. New York: Plenum Press.

Hardwick, J. E. (1970). A note on the behavior of the octopod *Ocythoe tuberculata*. *California Fish and Game*, **56**, 68–70.

Hardy, A. C. (1956). *The Open Sea*. London: Collins.

Harlow, H. F. (1949). The formation of learning sets. *Psychological Review*, **56**, 51–65.

Harman, R. F., Young, R. E., Reid, S. B. *et al.* (1989). Evidence for multiple spawning in the tropical oceanic squid *Stenoteuthis oualaniensis* (Teuthoidea: Ommastrephidae). *Marine Biology*, **101**, 513–519.

Harper, D. G. C. (1991). Communication, in *Behavioural Ecology. An Evolutionary Approach* (ed. J. R. Krebs & N. B. Davies), pp. 374–397. Oxford: Blackwell Scientific Publications.

Hartline, P. H., Hurley, A. C. & Lange, G. D. (1979). Eye stabilization by statocyst mediated oculomotor reflex in *Nautilus*. *Journal of Comparative Physiology*, **132**, 117–126.

Hartwick, E. B. (1983). *Octopus dofleini*, in *Cephalopod Life Cycles, Vol. 1: Species Accounts* (ed. P. R. Boyle), pp. 277–291. London: Academic Press.

Hartwick, E. B. & Thorarinsson, G. (1978). Den associates of the giant Pacific octopus, *Octopus dofleini* (Wulker). *Ophelia*, **17**, 163–166.

Hartwick, E. B., Ambrose, R. F. & Robinson, S. M. C. (1984). Den utilization and the movements of tagged *Octopus dofleini*. *Marine Behaviour and Physiology*, **11**, 95–110.

Hartwick, E. B., Breen, P. A. & Tulloch, L. (1978). A removal experiment with *Octopus dofleini* (Wulker). *Journal of the Fisheries Research Board of Canada*, **35**, 1492–1495.

Hartwick, E. B., Thorarinsson, G. & Tulloch, L. (1978). Antipredator behavior in *Octopus dofleini* (Wulker). *Veliger*, **21**, 263–264.

Harvey, P. H. & Arnold, S. J. (1982). Female mate choice and runaway sexual selection. *Nature*, **297**, 533–534.

Harvey, P. H. & Bradbury, J. W. (1991). Sexual selection, in *Behavioural Ecology. An Evolutionary Approach* (ed. J. R. Krebs & N. B. Davies), pp. 203–233. Oxford: Blackwell Scientific Publications.

Harvey, P. H. & Pagel, M. D. (1991). *The Comparative Method in Evolutionary Biology*. Oxford: Oxford University Press.

Haven, N. (1977). The reproductive biology of *Nautilus pompilius* in the Philippines. *Marine Biology*, **42**, 177–184.

Hazlett, B. A. (1977). *Quantitative Methods in the Study of Animal Behavior*. New York: Academic Press.

Hedge, J., Bart, E. & Kersten, D. (2008). Fragment-based learning of visual object categories. *Current Biology*, **18**, 597–601.

Hendrix, J. P. J., Hulet, W. H. & Greenberg, M. J. (1981). Salininity tolerance and the responses to hypoosmotic stress of the bay squid *Lolliguncula brevis*, an euryhaline cephalopod mollusc. *Comparative Biochemistry and Physiology*, **69A**, 641–648.

Hernandez-Urcera, J., Garci, M. E., Roura, A. *et al.* (2014). Cannibalistic behavior of octopus (*Octopus vulgaris*) in the wild. *Journal of Comparative Psychology*, **128**, 427–430.

Herring, P. J. (1977). Luminescence in cephalopods and fish. *Symposia of the Zoological Society of London*, **38**, 127–159.

Herring, P. J. (1988). Luminescent organs, in *The Mollusca*, Vol. 11: *Form and Function* (ed. E. R. Trueman & M. R. Clarke), pp. 449–489. London: Academic Press.

Herring, P. J., Dilly, P. N. & Cope, C. (1985). The photophore morphology of *Selenoteuthis scintillans* Voss and other lycoteuthids (Cephalopoda: Lycoteuthidae). *Journal of Zoology (London)*, **206**, 567–589.

Herring, P. J., Dilly, P. N. & Cope, C. (1992). Different types of photophore in the oceanic squids *Octopoteuthis* and *Taningia* (Cephalopoda, Octopoteuthidae). *Journal of Zoology*, **227**, 479–491.

Heupel, M. R., Semmens, J. M. & Hobday, A. J. (2006). Automated acoustic tracking of aquatic animals: scales, design and deployment of listening station arrays. *Marine and Freshwater Research*, **57**, 1–13.

Hill, A. V. & Solandt, D. Y. (1935). Myograms from the chromatophores of *Sepia*. *Journal of Physiology (London)*, **83**, 13P–14P.

Hinde, R. A. (1970). *Animal Behaviour: A Synthesis of Ethology and Comparative Psychology*. Kogakusha: McGraw-Hill.

Hirohashi, N., Alvarez, L., Shiba, K. *et al.* (2013). Sperm from sneaker male squids exhibit chemotactic swarming to CO_2. *Current Biology*, **23**, 775–781.

Hixon, R. F. (1980). *Growth, Reproductive Biology, Distribution and Abundance of Three Species of Loliginid Squid (Myopsida, Cephalopoda) in the Northwest Gulf of Mexico. Ph.D. Dissertation*. Coral Gables, FL: University of Miami.

Hixon, R. F., Hanlon, R. T., Gillespie, S. M. & Griffin, W. L. (1980). Squid fishery in Texas: biological, economic, and market considerations. *Marine Fisheries Review*, **42**, 44–50.

Hoare, D. J. & Krause, J. (2003). Social organisation, shoal structure and information transfer. *Fish and Fisheries*, **4**, 269–279.

Hoare, D. J., Couzin, I. D., Godin, J. G. J. & Krause, J. (2004). Context-dependent group size choice in fish. *Animal Behaviour*, **67**, 155–164.

Hochachka, P. W. (1994). Oxygen efficient design of cephalopod muscle metabolism. *Marine and Freshwater Behaviour and Physiology*, **25**, 61–67.

Hochberg, F. G. & Couch, J. A. (1971). Biology of cephalopods, in *Tektite II, Scientists in the Sea, Mission 8-50*, pp. VI-221–VI-228. Washington DC: US Department of the Interior.

Hochner, B. (2012). An embodied view of octopus neurobiology. *Current Biology*, **22**, R887–R892.

Hochner, B. (2013). How nervous systems evolve in relation to their embodiment: what we can learn from octopuses and other molluscs. *Brain Behavior and Evolution*, **82**, 19–30.

Hochner, B. & Shomrat, T. (2013). The neurophysiological basis of learning and memory in advanced invertebrates: the octopus and the cuttlefish, in *Invertebrate Learning and Memory* (ed. R. Menzel & P. R. Benjamin), pp. 303–317. Dusseldorf, Germany: Elsevier/Academic Press.

Hochner, B. & Shomrat, T. (2014). The neurophysiological basis of learning and memory in an advanced invertebrate: the octopus, in *Cephalopod Cognition* (ed. A. S. Darmaillacq, L. Dickel & J. Mather), pp. 72–93. Cambridge: Cambridge University Press.

Hochner, B., Brown, E. R., Langella, M., Shomrat, T. & Fiorito, G. (2003). A learning and memory area in the octopus brain manifests a vertebrate-like long-term potentiation. *Journal of Neurophysiology*, **90**, 3547–3554.

Hochner, B., Shomrat, T. & Fiorito, G. (2006). The octopus: a model for a comparative analysis of the evolution of learning and memory mechanisms. *Biological Bulletin*, **210**, 308–317.

Hockett, C. F. (1960). The origin of speech. *Scientific American*, **203**, 89–96.

Hodgkin, A. L. (1964). *The Conduction of the Nervous Impulse*. Liverpool: Liverpool University Press.

Hofmeister, J. K., Alupay, J. S., Ross, R. & Caldwell, R. L. (2011). Observations on mating behavior and development in the lesser Pacific striped octopus, *Octopus chierchiae* (Jatta, 1889). *Integrative and Comparative Biology*, **51**, E58–E58.

Holme, N. A. (1974). The biology of *Loligo forbesi* Steenstrup (Mollusca: Cephalopoda) in the Plymouth area. *Journal of the Marine Biological Association of the United Kingdom*, **54**, 481–503.

Holmes, W. (1940). The colour changes and colour patterns of *Sepia officinalis* L. *Proceedings of the Zoological Society of London A*, **110**, 2–35.

Houck, B. A. (1982). Temporal spacing in the activity patterns of three Hawaiian shallow-water octopods. *The Nautilus*, **96**, 152–156.

House, M. R. (1988). Major features of cephalopod evolution, in *Cephalopods – Present and Past* (ed. J. Wiedmann & J. Kullmann), pp. 1–16. Stuttgart: E. Schweizerbart'sche Verlagsbuchhandlung.

Hoving, H. J. T. (2008). Reproductive biology of oceanic decapodiform cephalopods, Ph.D. Thesis, pp. 184. Haren, Netherlands: University of Groningen.

Hoving, H. J. T. & Robison, B. H. (2012). Vampire squid: detritivores in the oxygen minimum zone. *Proceedings of the Royal Society B – Biological Sciences*, **279**, 4559–4567.

Hoving, H. J. T. & Vecchione, M. (2012). Mating behavior of a deep-sea squid revealed by *in situ* videography and the study of archived specimens. *Biological Bulletin*, **223**, 263–267.

Hoving, H. J. T., Bush, S. L. & Robison, B. H. (2012). A shot in the dark: same-sex sexual behaviour in a deep-sea squid. *Biology Letters*, **8**, 287–290.

Hoving, H. J. T., Lipinski, M. R., Videler, J. J. & Bolstad, K. S. R. (2010). Sperm storage and mating in the deep-sea squid *Taningia danae* Joubin, 1931 (Oegopsida: Octopoteuthidae). *Marine Biology*, **157**, 393–400.

Hoving, H. J. T., Zeidberg, L. D., Benfield, M. C. *et al.* (2013). First *in situ* observations of the deep-sea squid *Grimalditeuthis bonplandi* reveal unique use of tentacles. *Proceedings of the Royal Society B – Biological Sciences*, **280**, 20131463.

Hu, M. Y., Yan, H. Y., Chung, W. S., Shiao, J. C. & Hwang, P. P. (2009). Acoustically evoked potentials in two cephalopods inferred using the auditory brainstem response (ABR) approach. *Comparative Biochemistry and Physiology A – Molecular & Integrative Physiology*, **153**, 278–283.

Huang, K. L. & Chiao, C. C. (2011). Can cuttlefish learn by observing others? *Journal of Shellfish Research*, **30**, 1008–1009.

Hubbard, R. & St George, R. C. C. (1958). The rhodopsin system of the squid. *Journal General Physiology*, **41**, 501–528.

Hubbard, S. J. (1960). Hearing and the octopus statocyst. *Journal of Experimental Biology*, **37**, 845–853.

Huffard, C. L. (2006). Locomotion by *Abdopus aculeatus* (Cephalopoda: Octopodidae): walking the line between primary and secondary defenses. *Journal of Experimental Biology*, **209**, 3697–3707.

Huffard, C. L. (2007). Ethogram of *Abdopus aculeatus* (D'Orbigny, 1834) (Cephalopoda: Octopodidae): can behavioural characters inform octopodid taxonomy and systematics? *Journal of Molluscan Studies*, **73**, 185–193.

Huffard, C. L. & Godfrey-Smith, P. (2010). Field observations of mating in *Octopus tetricus* Gould, 1852 and *Amphioctopus marginatus* (Taki, 1964) (Cephalopoda: Octopodidae). *Molluscan Research*, **30**, 81–86.

Huffard, C. L. & Hochberg, F. G. (2005). Description of a new species of the genus *Amphioctopus* (Mollusca: Octopodidae) from the Hawai'ian Islands. *Molluscan Research*, **25**, 113–128.

Huffard, C. L., Boneka, F. & Full, R. J. (2005). Underwater bipedal locomotion by octopuses in disguise. *Science*, **307**, 1927.

Huffard, C. L., Caldwell, R. L. & Boneka, F. (2008). Mating behavior of *Abdopus aculeatus* (d'Orbigny 1834) (Cephalopoda: Octopodidae) in the wild. *Marine Biology*, **154**, 353–362.

Huffard, C. L., Caldwell, R. L. & Boneka, F. (2010). Male–male and male–female aggression may influence mating associations in wild octopuses (*Abdopus aculeatus*). *Journal of Comparative Psychology*, **124**, 38–46.

Huffard, C. L., Caldwell, R. L., DeLoach, N. *et al.* (2008). Individually unique body color patterns in octopus (*Wunderpus photogenicus*) allow for photoidentification. *PLoS ONE*, **3**, e3732.

Huffard, C. L., Saarman, N., Hamilton, H. & Simison, W. B. (2010). The evolution of conspicuous facultative mimicry in octopuses: an example of secondary adaptation? *Biological Journal of the Linnean Society*, **101**, 68–77.

Hughes, R. N. (1980). Optimal foraging theory in the marine context. *Oceanography and Marine Biology Annual Reviews*, **18**, 423–481.

Hugo, V. (1866). *Toilers of the Sea*. Paris.

Hulet, W. H., Hanlon, R. T. & Hixon, R. F. (1980). *Lolliguncula brevis* – a new squid species for the neuroscience laboratory. *Trends in Neuroscience*, **3**, iv–v.

Humphrey, N. K. (1976). The social function of intellect, in *Growing Points in Ethology* (ed. P. P. G. Bateson & R. A. Hinde), pp. 303–317. London: Cambridge University Press.

Humphries, D. A. & Driver, P. M. (1970). Protean defence by prey animals. *Oecologia*, **5**, 285–302.

Humphries, S., Evans, J. P. & Simmons, L. W. (2008). Sperm competition: linking form to function. *BMC Evolutionary Biology*, **8**, 319.

Hunt, J. C. & Seibel, B. A. (2000). Life history of *Gonatus onyx* (Cephalopoda: Teuthoidea): ontogenetic changes in habitat, behavior and physiology. *Marine Biology*, **136**, 543–552.

Hunt, J. C., Zeidberg, L. D., Hamner, W. M. & Robison, B. H. (2000). The behaviour of *Loligo opalescens* (Mollusca: Cephalopoda) as observed by a remotely operated vehicle (ROV). *Journal of the Marine Biological Association of the United Kingdom*, **80**, 873–883.

Hurley, A. C. (1976). Feeding behavior, food consumption, growth, and respiration of the squid *Loligo opalescens* raised in the laboratory. *Fishery Bulletin*, **74**, 176–182.

Hurley, A. C. (1977). Mating behavior of the squid *Loligo opalescens*. *Marine Behaviour and Physiology*, **4**, 195–203.

Hurley, A. C. (1978). School structure of the squid *Loligo opalescens*. *Fishery Bulletin*, **76**, 433–442.

Hurley, A. C., Lange, G. D. & Hartline, P. H. (1978). Adjustable pinhole camera eye of *Nautilus*. *Journal of Experimental Zoology*, **205**, 37–43.

Hurley, G. V. & Dawe, E. G. (1980). Tagging studies on squid (*Illex illecebrosus*) in the Newfoundland area. *North Atlantic Fisheries Organization Scientific Council Research Document*, 80/II/33, #072, 1–11.

Hvorecny, L. M., Grudowski, J. L., Blakeslee, C. J. *et al.* (2007). Octopuses (*Octopus bimaculoides*) and cuttlefishes (*Sepia pharaonis, S. officinalis*) can conditionally discriminate. *Animal Cognition*, **10**, 449–459.

Ibanez, C. M. & Keyl, F. (2010). Cannibalism in cephalopods. *Reviews in Fish Biology and Fisheries*, **20**, 123–136.

Ikeda, Y., Sakurai, Y. & Shimazaki, K. (1993). Fertilizing capacity of squid (*Todarodes pacificus*) spermatozoa collected from various sperm storage sites, with special reference to the role of gelatinous substance from oviducal gland in fertilization and embryonic development. *Invertebrate Reproduction & Development*, **23**, 39–44.

Imber, M. J. (1973). The food of grey-faced petrels (*Pterodroma macroptera gouldi* (Hutton)), with special reference to diurnal migration of their prey. *Journal of Animal Ecology*, **42**, 645–662.

Iribarne, O. O. (1990). Use of shelter by the small Patagonian octopus *Octopus tehuelchus*: availability, selection and effects on fecundity. *Marine Ecology Progress Series*, **66**, 251–258.

Iribarne, O. O. (1991). Life history and distribution of the small southwestern Atlantic octopus, *Octopus tehuelchus*. *Journal of Zoology (London)*, **223**, 549–565.

Ishii, Y. & Shimada, M. (2010). The effect of learning and search images on predator–prey interactions. *Population Ecology*, **52**, 27–35.

Itami, K. (1964). The tagging of Madako (*Octopus vulgaris*) and its results. *Aquaculture*, **12**, 119–125.

Iversen, R. T. B., Perkins, P. J. & Dionne, R. D. (1963). An indication of underwater sound production by squid. *Nature*, **199**, 250–251.

Iwata, Y. (2012). Reproductive ecology in loliginid squids. *Nippon Suisan Gakkaishi*, **78**, 665–668.

Iwata, Y. & Sakurai, Y. (2007). Threshold dimorphism in ejaculate characteristics in the squid *Loligo bleekeri*. *Marine Ecology Progress Series*, **345**, 141–146.

Iwata, Y., Ito, K. & Sakurai, Y. (2008). Effect of low temperature on mating behavior of squid *Loligo bleekeri*. *Fisheries Science*, **74**, 1345–1347.

Iwata, Y., Ito, K. & Sakurai, Y. (2010). Is commercial harvesting of spawning aggregations sustainable? The reproductive status of the squid *Loligo bleekeri*. *Fisheries Research*, **102**, 286–290.

Iwata, Y., Lian, C. L. & Sakurai, Y. (2008). Development of microsatellite markers in the Japanese common squid *Todarodes pacificus* (Ommastrephidae). *Molecular Ecology Resources*, **8**, 466–468.

Iwata, Y., Munehara, H. & Sakurai, Y. (2003). Characterization of microsatellite markers in the squid, *Loligo bleekeri* (Cephalopoda: Loliginidae). *Molecular Ecology Notes*, **3**, 392–393.

Iwata, Y., Munehara, H. & Sakurai, Y. (2005). Dependence of paternity rates on alternative reproductive behaviors in the squid *Loligo bleekeri*. *Marine Ecology Progress Series*, **298**, 219–228.

Iwata, Y., Sakurai, Y. & Shaw, P. (2015). Dimorphic sperm-transfer strategies and alternative mating tactics in loliginid squid. *Journal of Molluscan Studies*, **81**, 147–151.

Iwata, Y., Shaw, P., Fujiwara, E., Shiba, K., Kakiuchi, Y. & Hirohashi, N. (2011). Why small males have big sperm: dimorphic squid sperm linked to alternative mating behaviours. *BMC Evolutionary Biology*, **11**, 236.

Izumi, M., Sweeney, A. M., DeMartini, D. *et al.* (2010). Changes in reflectin protein phosphorylation are associated with dynamic iridescence in squid. *Journal of the Royal Society Interface*, **7**, 549–560.

Jackson, G. D. (1994). Application and future potential of statolith increment analysis in squids and sepioids. *Canadian Journal of Fisheries and Aquatic Sciences*, **51**, 2612–2625.

Jackson, G. D. & Moltschaniwskyj, N. A. (2002). Spatial and temporal variation in growth rates and maturity in the Indo-Pacific squid *Sepioteuthis lessoniana* (Cephalopoda: Loliginidae). *Marine Biology*, **140**, 747–754.

Jackson, G. D., Arkhipkin, A. I., Bizikov, V. A. & Hanlon, R. T. (1993). Laboratory and field corroboration of age and growth from statoliths and gladii of the loliginid squid *Sepioteuthis lessoniana* (Mollusca: Cephalopoda), in *Recent Advances in Cephalopod Fisheries Biology* (ed. T. Okutani, R. K. O'Dor & T. Kubodera), pp. 189–199. Tokyo: Tokai University Press.

Jackson, G. D., Forsythe, J. W., Hixon, R. F. & Hanlon, R. T. (1997). Age, growth, and maturation of *Lolliguncula brevis* (Cephalopoda: Loliginidae) in the northwestern Gulf of Mexico with a comparison of length-frequency versus statolith age analysis. *Canadian Journal of Fisheries and Aquatic Sciences*, **54**, 2907–2919.

Jackson, G. D., Lu, C. C. & Dunning, M. (1991). Growth rings within the statolith microstructure of the giant squid *Architeuthis*. *Veliger*, **34**, 331–334.

Jacobson, L. (2005). Essential fish habitat source document: longfin inshore squid, *Loligo pealeii*, life history and habitat characteristics, 2nd edition. In *NOAA Technical Memorandum NMFS-NE-193*, pp. 42.

Jander, R., Daumer, K. & Waterman, T. H. (1963). Polarized light orientation by two Hawaiian decapod cephalopods. *Zeitschrift fuer vergleichende Physiologie*, **46**, 383–394.

Jantzen, T. M. & Havenhand, J. N. (2003a). Reproductive behavior in the squid *Sepioteuthis australis* from South Australia: ethogram of reproductive body patterns. *Biological Bulletin*, **204**, 290–304.

Jantzen, T. M. & Havenhand, J. N. (2003b). Reproductive behavior in the squid *Sepioteuthis australis* from South Australia: interactions on the spawning grounds. *Biological Bulletin*, **204**, 305–317.

Jatta, G. (1896). *I Cefalopodi viventi nel Golfo di Napoli. (Sistematica)*. Berlin: Verlag Von R. Friedlander & Sohn.

Jefferts, K. (1988). Zoogeography of northeastern Pacific cephalopods, in *Cephalopods – Present and Past* (ed. J. Wiedmann & J. Kullmann), pp. 317–339. Stuttgart, Germany: Schweizerbart'sche Verlagsbuchhandlung.

Jereb, P. & Roper, C. F. E. (2010). *Cephalopods of the World: An Annotated and Illustrated Catalogue of Cephalopod Species Known to Date*, Vol. 2: *Myopsid and Oegopsid Squids*. Rome, Italy: FAO.

Johnsen, S. (2000). Transparent animals. *Scientific American*, **282**, 80–89.

Johnsen, S. (2002). Cryptic and conspicuous coloration in the pelagic environment. *Proceedings of the Royal Society B – Biological Sciences*, **269**, 243–256.

Johnsen, S. (2003). Lifting the cloak of invisibility: the effects of changing optical conditions on pelagic crypsis. *Integrative and Comparative Biology*, **43**, 580–590.

Johnsen, S., Balser, E. J. & Widder, E. A. (1999). Light-emitting suckers in an octopus. *Nature*, **398**, 113–114.

Johnsen, S., Marshall, N. J. & Widder, E. A. (2011). Polarization sensitivity as a contrast enhancer in pelagic predators: lessons from *in situ* polarization imaging of transparent zooplankton. *Philosophical Transactions of the Royal Society B – Biological Sciences*, **366**, 655–670.

Johnson, W. S. & Chase, V. C. (1982). A record of cleaning symbiosis involving *Gobiosoma* sp. and a large Caribbean octopus. *Copeia*, **3**, 712–714.

Johnston, T. D. (1981). Selective costs and benefits in the evolution of learning, in *Advances in the Study of Behavior* (ed. J. S. Rosenblatt, R. A. Hinde, C. Beer & M. C. Busnel). New York: Academic Press.

Johnston, T. D. (1985). Introduction: conceptual issues in the ecological study of learning, in *Issues in the Ecological Study of Learning* (ed. T. D. Johnston & A. T. Pietrewicz). Hillsdale: Erlbaum.

Joll, L. M. (1976). Mating, egg-laying and hatching of *Octopus tetricus* (Mollusca: Cephalopoda) in the laboratory. *Marine Biology*, **36**, 327–333.

Joll, L. M. (1977). The predation of pot-caught western rock lobster (*Panulirus longipes cygnus*) by octopus. *W. Australia Department of Fish Wildlife Reports*, **29**, 1–58.

Jolly, A. (1966). Lemur social behavior and primate intelligence. *Science*, **153**, 501–506.

Jones, B. W. & Nishiguchi, M. K. (2004). Counterillumination in the Hawaiian bobtail squid, *Euprymna scolopes* Berry (Mollusca: Cephalopoda). *Marine Biology*, **144**, 1151–1155.

Jones, E. C. (1963). *Tremoctopus violaceus* uses *Physalia* tentacles as weapons. *Science*, **139**, 764–766.

Jones, K. A., Jackson, A. L. & Ruxton, G. D. (2011). Prey jitters; protean behaviour in grouped prey. *Behavioral Ecology*, **22**, 831–836.

Jordan, M., Chamberlain, J. A. & Chamberlain, R. B. (1988). Response of *Nautilus* to variation in ambient pressure. *Journal of Experimental Biology*, **137**, 175–189.

Josef, N., Amodio, P., Fiorito, G. & Shashar, N. (2012). Camouflaging in a complex environment – octopuses use specific features of their surroundings for background matching. *PLoS One*, **7**, e37579.

Jozet-Alves, C., Bertin, M. & Clayton, N. S. (2013). Evidence of episodic-like memory in cuttlefish. *Current Biology*, **23**, R1033–R1035.

Jozet-Alves, C., Moderan, J. & Dickel, L. (2008). Sex differences in spatial cognition in an invertebrate: the cuttlefish. *Proceedings of the Royal Society B – Biological Sciences*, **275**, 2049–2054.

Jozet-Alves, C., Viblanc, V. A., Romagny, S. *et al.* (2012). Visual lateralization is task and age dependent in cuttlefish, *Sepia officinalis*. *Animal Behaviour*, **83**, 1313–1318.

Kaifu, K., Akamatsu, T. & Segawa, S. (2008). Underwater sound detection by cephalopod statocyst. *Fisheries Science*, **74**, 781–786.

Kakinuma, Y., Maki, K., Tsukahara, J. & Tabata, M. (1995). The breeding behavior of *Nautilus belauensis*. *Kagoshima University Research Center for the South Pacific. Occasional Papers*, **27**, 91–105.

Karpov, K. A. & Cailliet, G. M. (1978). Feeding dynamics of *Loligo opalescens*. *California Fish and Game, Fish Bulletin*, **169**, 45–66.

Karson, M. A., Boal, J. G. & Hanlon, R. T. (2003). Experimental evidence for spatial learning in cuttlefish (*Sepia officinalis*). *Journal of Comparative Psychology*, **117**, 149–155.

Kasugai, T. (2000). Reproductive behavior of the pygmy cuttlefish *Ideosepius paradoxus* in an aquarium. *Venus*, **59**, 37–44.

Kasugai, T., Shigeno, S. & Ikeda, Y. (2004). Feeding and external digestion in the Japanese pygmy squid *Idiosepius paradoxus* (Cephalopoda: Idiosepiidae). *Journal of Molluscan Studies*, **70**, 231–236.

Katsanevakis, S. & Verriopoulos, G. (2004). Den ecology of *Octopus vulgaris* Cuvier, 1797, on soft sediment:

availability and types of shelter. *Scientia Marina*, **68**, 147–157.

Katz, B. & Miledi, R. (1966). Input–output relations of a single synapse. *Nature*, **212**, 1242–1245.

Kayes, R. J. (1974). The daily activity pattern of *Octopus vulgaris* in a natural habitat. *Marine Behaviour and Physiology*, **2**, 337–343.

Kear, A. J. (1994). Morphology and function of the mandibular muscles in some coleoid cephalopods. *Journal of the Marine Biological Association of the United Kingdom*, **74**, 801–822.

Keenleyside, M. H. A. (1979). *Diversity and Adaptation in Fish Behaviour*. Berlin: Springer.

Kelman, E. J., Baddeley, R. J., Shohet, A. J. & Osorio, D. (2007). Perception of visual texture and the expression of disruptive camouflage by the cuttlefish, *Sepia officinalis*. *Proceedings of the Royal Society B – Biological Sciences*, **274**, 1369–1375.

Kelman, E. J., Osorio, D. & Baddeley, R. J. (2008). A review of cuttlefish camouflage and object recognition and evidence for depth perception. *Journal of Experimental Biology*, **211**, 1757–1763.

Kelman, E. J., Tiptus, P. & Osorio, D. (2006). Juvenile plaice (*Pleuronectes platessa*) produce camouflage by flexibly combining two separate patterns. *Journal of Experimental Biology*, **209**, 3288–3292.

Kemp, D. J., Alcock, J. & Allen, G. R. (2006). Sequential size assessment and multicomponent decision rules mediate aerial wasp contests. *Animal Behaviour*, **71**, 279–287.

Kennedy, G. J. A. & Pitcher, T. J. (1975). Experiments on homing in shoals of the European minnow, *Phoxinus phoxinus* (L.). *Transactions of the American Fisheries Society*, **104**, 452–455.

Kennedy, J. S. (1986). Migration, behavioral and ecological, in *Migration: Mechanisms and Adaptive Significance* (ed. M. A. Rankin), pp. 5–26. Port Aransas, TX: University of Texas Marine Science Institute.

Kenyon, K. W. (1975). *The Sea Otter in the Eastern Pacific Ocean*. New York: Dover.

Kier, W. M. (1985). The musculature of squid arms and tentacles: ultrastructural evidence for functional differences. *Journal of Morphology*, **185**, 223–239.

Kier, W. M. (1987). The functional morphology of the tentacle musculature of *Nautilus pompilius*, in *Nautilus. The Biology and Paleobiology of a Living Fossil* (ed. W. B. Saunders & N. H. Landman), pp. 257–269. New York: Plenum.

Kier, W. M. (1988). The arrangement and function of molluscan muscle, in *The Mollusca*, Vol. 11: *Form and Function* (ed. E. R. Trueman & M. R. Clarke), pp. 211–252. San Diego: Academic Press.

Kier, W. M. (1991). Squid cross-striated muscle: the evolution of a specialized muscle fiber type. *Bulletin of Marine Science*, **49**, 389–403.

Kier, W. M. & Curtin, N. A. (2002). Fast muscle in squid (*Loligo pealei*): contractile properties of a specialized muscle fibre type. *Journal of Experimental Biology*, **205**, 1907–1916.

Kier, W. M. & Smith, A. M. (1990). The morphology and mechanics of octopus suckers. *Biological Bulletin*, **178**, 126–136.

Kier, W. M. & Smith, A. M. (2002). The structure and adhesive mechanism of octopus suckers. *Integrative and Comparative Biology*, **42**, 1146–1153.

Kier, W. M. & Smith, K. K. (1985). Tongues, tentacles and trunks: the biomechanics of movement in muscular-hydrostats. *Zoological Journal of the Linneaen Society*, **83**, 307–324.

Kier, W. M. & VanLeeuwen, J. L. (1997). A kinematic analysis of tentacle extension in the squid *Loligo pealei*. *Journal of Experimental Biology*, **200**, 41–53.

Kier, W. M., Messenger, J. B. & Miyan, J. A. (1985). Mechanoreceptors in the fins of the cuttlefish, *Sepia officinalis*. *Journal of Experimental Biology*, **119**, 369–373.

King, A. J. & Adamo, S. A. (2006). The ventilatory, cardiac and behavioural responses of resting cuttlefish (*Sepia officinalis* L.) to sudden visual stimuli. *Journal of Experimental Biology*, **209**, 1101–1111.

King, A. J., Adamo, S. A. & Hanlon, R. T. (2003). Squid egg mops provide sensory cues for increased agonistic behaviour between male squid. *Animal Behaviour*, **66**, 49–58.

Kito, Y., Seidou, M., Michinomae, M., Partridge, J. C. & Herring, P. J. (1992). Porphyropsin and new deep-sea visual pigment with 4-hydroxyretinal are found in some mesopelagic cephalopods in Atlantic. *Zoological Science*, **9**, 1230.

Kito, Y., Seidou, M., Michinomae, M. & Tokuyama, A. (1987). Photic environment, bioluminescence and vision of a squid *Watasenia scintillans*. *Zoological Science*, **4**, 1107.

Klages, N. (1989). Food and feeding ecology of emperor penguins in the eastern Weddell Sea. *Polar Biology*, **9**, 385–390.

Klages, N. T. W. (1996). Cephalopods as prey. 2. Seals. *Philosophical Transactions of the Royal Society of London Series B – Biological Sciences*, **351**, 1045–1052.

Klumpp, D. W. & Nichols, P. D. (1983). A study of food chain in seagrass communities. 2. Food of the rock flathead, *Platycephalus laevigatus* Cuvier, a major predator in a *Posidonia australis* seagrass bed. *Australian Journal of Marine and Freshwater Research*, **34**, 745–754.

Knight-Jones, E. W. & Morgan, E. (1966). Responses of marine animals to changes in hydrostatic pressure. *Oceanography and Marine Biology Annual Reviews*, **4**, 267–299.

Kobayashi, D. R. (1986). *Octopus* predation on hermit crabs: a test of selectivity. *Marine Behaviour and Physiology*, **12**, 125–131.

Kokko, H., Klug, H. & Jennions, M. D. (2012). Unifying cornerstones of sexual selection: operational sex ratio, Bateman gradient and the scope for competitive investment. *Ecology Letters*, **15**, 1340–1351.

Komak, S., Boal, J. G., Dickel, L. & Budelmann, B. U. (2005). Behavioural responses of juvenile cuttlefish (*Sepia officinalis*) to local water movements. *Marine and Freshwater Behaviour and Physiology*, **38**, 117–125.

Kooyman, G. L., Davis, J. P., Croxall, J. P. & Costa, D. P. (1982). Diving depths and energy requirements of king penguins. *Science*, **217**, 726–727.

Koueta, N. & Boucaud-Camou, E. (1989). Etude comparative de la sécrétion des glandes salivaires postérieures des Céphalopodes Decapodes. II. Elaboration de la sécrétion chez les calmars *Loligo vulgaris* L. et *L. forbesi* Steenstrup. *Bulletin de la Société Zoologique de France*, **114**, 47–54.

Krajewski, J. P., Bonaldo, R. M., Sazima, C. & Sazima, I. (2009). *Octopus* mimicking its follower reef fish. *Journal of Natural History* **43**, 185–190.

Krause, J. (1994). Differential fitness returns in relation to spatial position in groups. *Biological Reviews of the Cambridge Philosophical Society*, **69**, 187–206.

Krause, J., Butlin, R. K., Peuhkuri, N. & Pritchard, V. L. (2000). The social organization of fish shoals: a test of the predictive power of laboratory experiments for the field. *Biological Reviews of the Cambridge Philosophical Society*, **75**, 477–501.

Krebs, J. R. & Davies, N. B. (1993). *An Introduction to Behavioural Ecology*. Oxford: Blackwell Scientific Publications.

Krebs, J. R. & Dawkins, R. (1984). Animal rights: mind reading and manipulation, in *Behavioural Ecology: An Evolutionary Approach*, pp. 380–402. Oxford: Blackwell Scientific Publications.

Kristensen, T. K. (1983). *Gonatus fabricii*, in *Cephalopod Life Cycles*, Vol. 1 (ed. P. R. Boyle), pp. 159–173. London: Academic Press.

Kröger, B., Vinther, J. & Fuchs, D. (2011). Cephalopod origin and evolution: a congruent picture emerging from fossils, development and molecules. *Bioessays*, **33**, 602–613.

Kuba, M., Meisel, D. V., Byrne, R. A., Griebel, U. & Mather, J. A. (2003). Looking at play in *Octopus vulgaris*, in *Coleoid Cephalopods Through Time*, Vol. 3 (ed. K. Warnke, H. Keupp & S. von Boletzky), pp. 163–169: Berliner Palaobiol. Abh.

Kuba, M. J., Byrne, R. A., Meiscl, D. V. & Mather, J. A. (2006). When do octopuses play? Effects of repeated testing, object type, age, and food deprivation on object play in *Octopus vulgaris*. *Journal of Comparative Psychology*, **120**, 184–190.

Kubodera, T. & Mori, K. (2005). First-ever observations of a live giant squid in the wild. *Proceedings of the Royal Society B: Biological Sciences*, **272**, 2583–2586.

Kubodera, T., Koyama, Y. & Mori, K. (2007). Observations of wild hunting behaviour and bioluminescence of a large deep-sea, eight-armed squid, *Taningia danae*. *Proceedings of the Royal Society B – Biological Sciences*, **274**, 1029–1034.

Kubodera, T., Piatkowski, U., Okutani, T. & Clarke, M. R. (1998). Taxonomy and zoogeography of the Family Onychoteuthidae (Cephalopoda: Oegopsida). *Smithsonian Contributions to Zoology*, **586**, 277–291.

Kuipers, M. R., Pecl, G. T. & Moltschaniwskyj, N. A. (2008). Batch or trickle: understanding the multiple spawning strategy of southern calamary, *Sepioteuthis australis* (Mollusca: Cephalopoda). *Marine & Freshwater Research*, **59**, 987–997.

Kyte, M. A. & Courtney, G. W. (1978). A field observation of aggressive behavior between two North Pacific octopus, *Octopus dofleini martini*. *Veliger*, **19**, 427–428.

Laan, A., Gutnick, T., Kuba, M. J. & Laurent, G. (2014). Behavioral analysis of cuttlefish traveling waves and its implications for neural control. *Current Biology*, **24**, 1737–1742.

Land, M. F. (1972). The physics and biology of animal reflectors. *Progress in Biophysics and Molecular Biology*, **24**, 75–106.

Land, M. F. (1990). Optics of the eyes of marine animals, in *Light and Life in the Sea* (ed. P. J. Herring, A. K. Campbell, M. Whitfield & L. Maddock), pp. 149–166. Cambridge: Cambridge University Press.

Land, M. F. (1992). A note on the elongated eye of the octopus *Vitreledonella-richardi*. *Journal of the Marine Biological Association of the United Kingdom*, **72**, 89–92.

Land, M. F. & Nilsson, D.-E. (2012). *Animal Eyes*. New York: Oxford University Press.

Landman, N. H. & Cochran, J. K. (1987). Growth and longevity of *Nautilus*, in *Nautilus. The Biology and Paleobiology of a Living Fossil* (ed. W. B. Saunders & N. H. Landman), pp. 401–420. New York: Plenum.

Landman, N. H., Cochran, J. K., Cerrato, R. *et al.* (2004). Habitat and age of the giant squid (*Architeuthis sanctipauli*) inferred from isotopic analyses. *Marine Biology*, **144**, 685–691.

Landman, N. H., Cochran, J. K., Rye, D. M., Tanabe, K. & Arnold, J. M. (1994). Early-life history of *Nautilus* - evidence from isotopic analyses of aquarium-reared specimens. *Paleobiology*, **20**, 40–51.

Landman, N. H., Jones, D. S. & Davis, R. A. (2001). Hatching depth of *Nautilus pompilius* in Fiji. *The Veliger*, **44**, 333–339.

Landry, C., Garant, D., Duchesne, P. & Bernatchez, L. (2001). 'Good genes as heterozygosity': the major histocompatibility complex and mate choice in Atlantic salmon (*Salmo salar*). *Proceedings of the Royal Society B – Biological Sciences*, **268**, 1279–1285.

Lane, F. W. (1957). *Kingdom of the Octopus. The Life History of the Cephalopoda*. London: Jarrold.

Langridge, K. V. (2006). Symmetrical crypsis and asymmetrical signalling in the cuttlefish *Sepia officinalis*. *Proceedings of the Royal Society B: Biological Sciences*, **273**, 959–967.

Langridge, K. V. (2009). Cuttlefish use startle displays, but not against large predators. *Animal Behaviour*, **77**, 847–856.

Langridge, K. V., Broom, M. & Osorio, D. (2007). Selective signalling by cuttlefish to predators. *Current Biology*, **17**, R1044–R1045.

Larcombe, M. F. & Russell, B. C. (1971). Egg laying behaviour of the broad squid *Sepioteuthis bilineata* (*lessoniana*). *New Zealand Journal of Marine & Freshwater Research*, **5**, 3–11.

LaRoe, E. T. (1971). The culture and maintenance of the loliginid squids *Sepioteuthis sepioidea* and *Doryteuthis plei*. *Marine Biology*, **9**, 9–25.

Le Boeuf, B. J., Naito, Y., Huntley, A. C. & Asaga, T. (1989). Prolonged, continuous, deep diving by northern elephant seals. *Canadian Journal of Zoology*, **67**, 2514–2519.

Lee, H. (1875). *The Octopus. The 'Devil-Fish' of Fiction and of Fact*. London: Chapman and Hall.

Lee, P. G. (1992). Chemotaxis by *Octopus maya* Voss et Solis in a Y-maze. *Journal of Experimental Marine Biology and Ecology*, **153**, 53–67.

Lee, P. G., Turk, P. E., Yang, W. T. & Hanlon, R. T. (1994). Biological characteristics and biomedical applications of the squid *Sepioteuthis lessoniana* cultured through multiple generations. *Biological Bulletin*, **186**, 328–341.

Lee, Y. H., Yan, H. Y. & Chiao, C. C. (2010). Visual contrast modulates maturation of camouflage body patterning in cuttlefish (*Sepia pharaonis*). *Journal of Comparative Psychology*, **124**, 261–270.

Lee, Y.-H., Yan, H. Y. & Chiao, C.-C. (2012). Effects of early visual experience on the background preference in juvenile cuttlefish *Sepia pharaonis*. *Biology Letters*, **8**, 740–743.

Lehmann, U. (1981). *The Ammonites: Their Life and Their World*. Cambridge: Cambridge University Press.

Lehner, P. N. (1998). *Handbook of Ethological Methods* 2nd edition. Cambridge: Cambridge University Press.

Leite, T. S. & Mather, J. A. (2008). A new approach to octopuses' body pattern analysis: a framework for taxonomy and behavioral studies. *American Malacological Bulletin*, **24**, 31–41.

Lima, P. A., Nardi, G. & Brown, E. R. (2003). AMPA/ kainate and NMDA-like glutamate receptors at the chromatophore neuromuscular junction of the squid: role in synaptic transmission and skin patterning. *European Journal of Neuroscience*, **17**, 507–516.

Lima, S. L. & Dill, L. M. (1990). Behavioural decisions made under the risk of predation – a review and prospectus. *Canadian Journal of Zoology – Revue Canadienne de Zoologie*, **68**, 619–640.

Lindgren, A. R. (2010). Molecular inference of phylogenetic relationships among Decapodiformes (Mollusca: Cephalopoda) with special focus on the squid Order Oegopsida. *Molecular Phylogenetics and Evolution*, **56**, 77–90.

Lindgren, A. R., Giribet, G. & Nishiguchi, M. K. (2004). A combined approach to the phylogeny of Cephalopoda (Mollusca). *Cladistics*, **20**, 454–486.

Lindgren, A. R., Pankey, M. S., Hochberg, F. G. & Oakley, T. H. (2012). A multi-gene phylogeny of Cephalopoda supports convergent morphological evolution in association with multiple habitat shifts in the marine environment. *BMC Evolutionary Biology*, **12**, article 129.

Lipinski, M. R. (1985). Laboratory survival of *Alloteuthis subulata* (Cephalopoda: Loliginidae) from the Plymouth area. *Journal of the Marine Biological Association of the United Kingdom*, **65**, 845–855.

Lipinski, M. R. (1987). Food and feeding of *Loligo vulgaris reynaudii* from St Francis Bay, South Africa. *South Africa Journal of Marine Science*, **5**, 557–564.

Lipinski, M. R., Hampton, I., Sauer, W. H. H. & Augustyn, C. J. (1998). Daily net emigration from a spawning concentration of chokka (*Loligo vulgaris reynaudii* d'Orbigny, 1845) in Kromme Bay, South Africa squid. *ICES Journal of Marine Science*, **55**, 258–270.

Lohmann, K. J. & Willows, A. O. D. (1987). Lunar-modulated geomagnetic orientation by a marine mollusk. *Science*, **235**, 331–334.

Loi, P. K., Saunders, R. G., Young, D. C. & Tublitz, N. J. (1996). Peptidergic regulation of chromatophore function in the European cuttlefish *Sepia officinalis*. *Journal of Experimental Biology*, **199**, 1177–1187.

Long, T. M., Hanlon, R. T., Ter Maat, A. & Pinsker, H. M. (1989). Non-associative learning in the squid *Lolliguncula brevis* (Mollusca, Cephalopoda). *Marine Behaviour and Physiology*, **16**, 1–9.

Longley, W. H. (1918). Marine camoufleurs and their camouflage: the present and prospective significance of facts regarding the coloration of tropical fishes. *Annual Report of the Smithsonian Institution*, 475–485.

Lott, D. F. (1991). *Intraspecific Variation in the Social Systems of Wild Vertebrates*. Cambridge: Cambridge University Press.

Lu, C. C. & Clarke, M. R. (1975). Vertical distribution of cephalopods at 40° N, 52° N and 60° N at 20° W in the North Atlantic. *Journal of the Marine Biological Association of the United Kingdom*, **55**, 143–163.

Lucero, M. T., Farrington, H. & Gilly, W. F. (1994). Quantification of L-DOPA and dopamine in squid ink – implications for chemoreception. *Biological Bulletin*, **187**, 55–63.

Lucero, M. T., Horrigan, F. T. & Gilly, W. F. (1992). Electrical responses to chemical stimulation of squid olfactory receptor cells. *Journal of Experimental Biology*, **162**, 231–249.

Lum-Kong, A. (1993). Oogenesis, fecundity and pattern of spawning in *Loligo forbesi* (Cephalopoda: Loliginidae). *Malacological Review*, **26**, 81–88.

Lum-Kong, A., Pierce, G. J. & Yau, C. (1992). Timing of spawning and recruitment in *Loligo forbesi* (Cephalopoda: Loliginidae) in Scottish waters. *Journal of the Marine Biological Association of the United Kingdom*, **72**, 301–311.

Lutz, R. A. & Voight, J. R. (1994). Close encounter in the deep. *Nature*, **371**, 563.

Lythgoe, J. N. (1972). The adaptation of visual pigments to the photic environment, in *Handbook of Sensory Physiology,* Vol. VII/1, *Photochemistry of Vision* (ed. H. J. A. Dartnall). Berlin, Heidelberg, New York: Springer.

MacGinitie, G. E. & MacGinitie, N. (1968). *Natural History of Marine Animals*. New York: McGraw-Hill Book Company.

Machin, K. L. (2005). Avian pain: physiology and evaluation. *Compendium on Continuing Education for the Practicing Veterinarian*, **27**, 98.

Macia, S., Robinson, M. P., Craze, P., Dalton, R. & Thomas, J. D. (2004). New observations on airborne jet propulsion (flight) in squid, with a review of previous reports. *Journal of Molluscan Studies*, **70**, 297–299.

Mackintosh, N. J. (1965). Discrimination learning in the *Octopus*. *Animal Behaviour Supplement*, **1**, 129–134.

Mackintosh, N. J. (1974). *The Psychology of Animal Learning.* London: Academic Press.

Mackintosh, N. J. & Mackintosh, J. (1963). Reversal learning in *Octopus,* with and without irrelevant cues. *Quarterly Journal of Experimental Psychology,* **15,** 236–242.

Mackintosh, N. J. & Mackintosh, J. (1964a). Performance of *Octopus* over a series of reversals of a simultaneous discrimination. *Animal Behaviour,* **12,** 321–324.

Mackintosh, N. J. & Mackintosh, J. (1964b). The effect of overtraining on a nonreversal shift in *Octopus. Journal of Genetic Psychology,* **106,** 373–377.

Macy, W. K. (1982). Feeding patterns of the long-finned squid, *Loligo pealei,* in New England waters. *Biological Bulletin,* **162,** 28–38.

Macy, W. K. & Brodziak, J. K. T. (2001). Seasonal maturity and size at age of *Loligo pealeii* in waters of southern New England. *ICES Journal of Marine Science,* **58,** 852–864.

Maddock, L. & Young, J. Z. (1984). Some dimensions of the angular acceleration receptor systems of cephalopods. *Journal of the Marine Biological Association of the United Kingdom,* **64,** 55–79.

Maddock, L. & Young, J. Z. (1987). Quantitative differences among the brains of cephalopods. *Journal of Zoology (London),* **212,** 739–767.

Madsen, P. T., Wilson, M., Johnson, M. *et al.* (2007). Clicking for calamari: toothed whales can echolocate squid *Loligo pealeii. Aquatic Biology,* **1,** 141–150.

Major, P. F. (1986). Notes on a predator–prey interaction between common dolphins (*Delphinus delphis*) and short-finned squid (*Illex illecebosus*) in Lydonia submarine canyon, western North Atlantic Ocean. *Journal of Mammalogy,* **67,** 769–770.

Makino, A. & Miyazaki, T. (2010). Topographical distribution of visual cell nuclei in the retina in relation to the habitat of five species of Decapodiformes (Cephalopoda). *Journal of Molluscan Studies,* **76,** 180–185.

Maldonado, H. (1964). The control of attack by *Octopus. Zeitschrift für vergleichende Physiologie,* **47,** 656–674.

Maldonado, H. (1968). Effect of electroconvulsive shock on memory in *Octopus vulgaris. Zeitschrift für vergleichende Physiologie,* **59,** 25–37.

Maldonado, H. (1969). Further investigations on the effect of electroconvulsive shock (ECS) on memory in *Octopus vulgaris. Zeitschrift für vergleichende Physiologie,* **63,** 113–118.

Maldonado, H. (1970). The deimatic reaction in the praying mantis *Stagmatoptera biocellata. Zeitschrift für vergleichende Physiologie,* **68,** 60–71.

Mangold, K. (1983a). *Octopus vulgaris,* in *Cephalopod Life Cycles,* Vol. 1 (ed. P. R. Boyle), pp. 335–364. London: Academic Press.

Mangold, K. (1983b). Food, feeding and growth in cephalopods. *Memoirs of the National Museum of Victoria,* **44,** 81–93.

Mangold, K. (1983c). *Eledone moschata,* in *Cephalopod Life Cycles,* Vol. 1 (ed. P. R. Boyle), pp. 475. London: Academic Press.

Mangold, K. (1987). Reproduction, in *Cephalopod Life Cycles,* Vol. 2: *Comparative Reviews* (ed. P. R. Boyle), pp. 157–200. London: Academic Press.

Mangold, K. (1989). *Traité de Zoologie – Céphalopodes (Tome V, Fascicule 4).* Paris: Masson.

Mangold, K. & Boletzky, S. von (1973). New data on reproductive biology and growth of *Octopus vulgaris. Marine Biology,* **19,** 7–12.

Mangold, K., Young, R. E. & Nixon, M. (1993). Growth versus maturation in cephalopods, in *Recent Advances in Cephalopod Fisheries Biology* (ed. T. Okutani, R. K. O'Dor & T. Kubodera), pp. 697–703. Tokyo: Tokai University Press.

Mangold-Wirz, K. (1963). Biologie des Cephalopodes benthiques et nectoniques de la Mer Catalane. *Vie Milieu,* **Suppl. No. 13,** 1–285.

Maniwa, Y. (1976). Attraction of bony fish, squid and crab by sound, in *Sound Reception in Fish* (ed. A. Schuijf & A. D. Hawkins), pp. 271–283. Amsterdam: Elsevier.

Mann, T. (1984). *Spermatophores. Development, Structure, Biochemical Attributes and Role in the Transfer of Spermatozoa.* Berlin: Springer.

Mann, T., Martin, A. W. & Thiersch, J. B. (1970). Male reproductive tract, spermatophores and spermatophoric reaction in the giant octopus of the North Pacific, *Octopus dofleini martini. Proceedings of the Royal Society of London B: Biological Sciences,* **175,** 31–61.

Marian, J. E. A. R. (2012). A model to explain spermatophore implantation in cephalopods

(Mollusca: Cephalopoda) and a discussion on its evolutionary origins and significance. *Biological Journal of the Linnean Society*, **105**, 711–726

Markaida, U. (2006). Food and feeding of jumbo squid *Dosidicus gigas* in the Gulf of California and adjacent waters after the 1997–98 El Nino event. *Fisheries Research*, **79**, 16–27.

Markaida, U., Rosenthal, J. J. C. & Gilly, W. F. (2005). Tagging studies on the jumbo squid (*Dosidicus gigas*) in the Gulf of California, Mexico. *Fishery Bulletin*, **103**, 219–226.

Marler, P. & Hamilton, W. J. (1966). *Mechanisms of Animal Behavior*. New York: Wiley.

Marliave, J. B. (1981). Neustonic feeding in early larvae of *Octopus dofleini* (Wulker). *Veliger*, **23**, 350–351.

Marshall, N. B. (1954). *Aspects of Deep Sea Biology*. London: Hutchinson.

Marshall, N. J. & Messenger, J. B. (1996). Colour-blind camouflage. *Nature*, **382**, 408–409.

Marshall, N. J., Cronin, T. W. & Wehling, M. F. (2011). New directions in biological research on polarized light. *Philosophical Transactions of the Royal Society of London B: Biological Sciences*, **366**, 615–616.

Martin, A. W., Catala-Stucki, I. & Ward, P. D. (1978). The growth rate and reproductive behavior of *Nautilus macromphalus*. *Neues Jahrbuch für Geologie und Palaontologie Abhandlungen*, **156**, 207–225.

Martin, P. & Bateson, P. (2007). *Measuring Behaviour. An Introductory Guide*. Cambridge: Cambridge University Press.

Martin, R. (1965). On the structure and embryonic development of the giant fibre of the squid *Loligo vulgaris*. *Zeitschrift für Zellforschung* **67**, 77–85.

Martin, R. (1977). The giant nerve fibre system of cephalopods. Recent structural findings. *Symposia of the Zoological Society of London*, **38**, 261–275.

Martin, S. J., Grimwood, P. D. & Morris, R. G. M. (2000). Synaptic plasticity and memory: an evaluation of the hypothesis. *Annual Review of Neuroscience*, **23**, 649–711.

Mather, J. A. (1978). Mating behavior of *Octopus joubini* Robson. *Veliger*, **21**, 265–267.

Mather, J. A. (1980). Social organization and use of space by *Octopus joubini* in a semi-natural situation. *Bulletin of Marine Science*, **30**, 848–857.

Mather, J. A. (1982). Choice and competition: their effects on occupancy of shell homes by *Octopus joubini*. *Marine Behaviour and Physiology*, **8**, 285–293.

Mather, J. A. (1984). Development of behaviour in *Octopus joubini* Robson, 1929. *Vie Milieu*, **34**, 17–20.

Mather, J. A. (1985). Behavioural interactions and activity of captive *Eledone moschata*: laboratory investigations of a 'social' octopus. *Animal Behaviour*, **33**, 1138–1144.

Mather, J. A. (1986a). A female-dominated feeding hierarchy in juvenile *Sepia officinalis* in the laboratory. *Marine Behaviour and Physiology*, **12**, 233–244.

Mather, J. A. (1986b). Sand digging in *Sepia officinalis*: assessment of a cephalopod mollusc's 'fixed' behavior pattern. *Journal of Comparative Psychology*, **100**, 315–320.

Mather, J. A. (1988). Daytime activity of juvenile *Octopus vulgaris* in Bermuda. *Malacologia*, **29**, 69–76.

Mather, J. A. (1991a). Foraging, feeding, and prey remains in middens of juvenile *Octopus vulgaris* (Mollusca: Cephalopoda). *Journal of Zoology (London)*, **224**, 27–39.

Mather, J. A. (1991b). Navigation by spatial memory and use of visual landmarks in octopuses. *Journal of Comparative Physiology A*, **168**, 491–497.

Mather, J. A. (1994). Home choice and modification by juvenile *Octopus vulgaris* (Mollusca, Cephalopoda) – specialized intelligence and tool use. *Journal of Zoology*, **233**, 359–368.

Mather, J. A. (1998). How do octopuses use their arms? *Journal of Comparative Psychology*, **112**, 306–316.

Mather, J. A. (2008a). To boldly go where no mollusc has gone before: personality, play, thinking, and consciousness in cephalopods. *American Malacological Bulletin*, **24**, 51–58.

Mather, J. A. (2008b). Cephalopod consciousness: behavioural evidence. *Consciousness and cognition*, **17**, 37–48.

Mather, J. A. (2010). Vigilance and antipredator responses of Caribbean reef squid. *Marine and Freshwater Behaviour and Physiology*, **43**, 357–370.

Mather, J. A. & Anderson, R. C. (1993). Personalities of octopuses (*Octopus rubescens*). *Journal of Comparative Psychology*, **107**, 336–340.

Mather, J. A. & Anderson, R. C. (1999). Exploration, play, and habituation in octopuses (*Octopus dofleini*). *Journal of Comparative Psychology*, 113, 333–338.

Mather, J. A. & Kuba, M. J. (2013). The cephalopod specialties: complex nervous system, learning, and cognition. *Canadian Journal of Zoology – Revue Canadienne de Zoologie*, 91, 431–449.

Mather, J. A. & Mather, D. L. (2004). Apparent movement in a visual display: the 'passing cloud' of *Octopus cyanea* (Mollusca: Cephalopoda). *Journal of Zoology*, 263, 89–94.

Mather, J. A. & O'Dor, R. K. (1984). Spatial organization of schools of the squid *Illex illecebrosus*. *Marine Behaviour and Physiology*, 10, 259–271.

Mather, J. A. & O'Dor, R. K. (1991). Foraging strategies and predation risk shape the natural history of juvenile *Octopus vulgaris*. *Bulletin of Marine Science*, 49, 256–269.

Mather, J. A., Anderson, R. C. & Wood, J. B. (2010). *Octopus: The Ocean's Intelligent Invertebrate*. Portland, OR: Timber Press, Inc.

Mather, J. A., Griebel, U. & Byrne, R. A. (2010). Squid dances: an ethogram of postures and actions of *Sepioteuthis sepioidea* squid with a muscular hydrostatic system. *Marine and Freshwater Behaviour and Physiology*, 43, 45–61.

Mäthger, L. M. (2003). The response of squid and fish to changes in the angular distribution of light. *Journal of the Marine Biological Association of the United Kingdom*, 83, 849–856.

Mäthger, L. M. & Denton, E. J. (2001). Reflective properties of iridophores and fluorescent 'eyespots' in the loliginid squid *Alloteuthis subulata* and *Loligo vulgaris*. *Journal of Experimental Biology*, 204, 2103–2118.

Mäthger, L. M. & Hanlon, R. T. (2006). Anatomical basis for camouflaged polarized light communication in squid. *Biology Letters*, 2, 494–496.

Mäthger, L. M. & Hanlon, R. T. (2007). Malleable skin coloration in cephalopods: selective reflectance, transmission and absorbance of light by chromatophores and iridophores. *Cell and Tissue Research*, 329, 179–186.

Mäthger, L. M., Barbosa, A., Miner, S. & Hanlon, R. T. (2006). Color blindness and contrast perception in cuttlefish (*Sepia officinalis*) determined by a visual sensorimotor assay. *Vision Research*, 46, 1746–1753.

Mäthger, L. M., Bell, G. R. R., Kuzirian, A. M., Allen, J. J. & Hanlon, R. T. (2012). How does the blue-ringed octopus (*Hapalochlaena lunulata*) flash its blue rings? *Journal of Experimental Biology*, 215, 3752–3757.

Mäthger, L. M., Chiao, C. C., Barbosa, A. *et al.* (2007). Disruptive coloration elicited on controlled natural substrates in cuttlefish, *Sepia officinalis*. *Journal of Experimental Biology*, 210, 2657–2666.

Mäthger, L. M., Chiao, C.-C., Barbosa, A. & Hanlon, R. T. (2008). Color matching on natural substrates in cuttlefish, *Sepia officinalis*. *Journal of Comparative Physiology A*, 194, 577–585.

Mäthger, L. M., Collins, T. F. T. & Lima, P. A. (2004). The role of muscarinic receptors and intracellular Ca^{2+} in the spectral reflectivity changes of squid iridophores. *Journal of Experimental Biology* 207, 1759–1769.

Mäthger, L. M., Denton, E. J., Marshall, N. J. & Hanlon, R. T. (2009). Mechanisms and behavioural functions of structural coloration in cephalopods. *Journal of the Royal Society Interface*, 6 Suppl. 2, S149–S163.

Mäthger, L. M., Roberts, S. B. & Hanlon, R. T. (2010). Evidence for distributed light sensing in the skin of cuttlefish, *Sepia officinalis*. *Biology Letters*, 6, 600–603.

Mäthger, L. M., Senft, S. L., Gao, M. *et al.* (2013). Bright white scattering from protein spheres in color changing, flexible cuttlefish skin. *Advanced Functional Materials*, 23, 3980–3989.

Mäthger, L. M., Shashar, N. & Hanlon, R. T. (2009). Do cephalopods communicate using polarized light reflections from their skin? *J Exp Biol*, 212, 2133–2140.

Matsui, S., Seidou, M., Horiuchi, S., Uchiyama, I. & Kito, Y. (1988). Adaptations of a deep-sea cephalopod to the photic environment. Evidence for three visual pigments. *Journal of General Physiology*, 92, 55–66.

Matteson, R. S., Benoit-Bird, K. J. & Gilly, W. F. (2009). Humboldt squid distribution in three-dimensional space as measured by acoustics in the Gulf of California. *Journal of the Acoustical Society of America*, 125.

Mattiello, T., Fiore, G., Brown, E. R., d'Ischia, M. & Palumbo, A. (2010). Nitric oxide mediates the glutamate-dependent pathway for neurotransmission in *Sepia officinalis* chromatophore organs. *Journal of Biological Chemistry*, 285, 24154–24163.

Maturana, H. R. & Sperling, S. (1963). Unidirectional response to angular acceleration recorded from the middle cristal nerve in the statocyst of *Octopus vulgaris*. *Nature*, **197**, 815–816.

Matzner, H., Gutfreund, Y. & Hochner, B. (2000). Neuromuscular system of the flexible arm of the octopus: physiological characterization. *Journal of Neurophysiology*, **83**, 1315–1328.

Mauris, E. (1989). Colour patterns and body postures related to prey capture in *Sepiola affinis* (Mollusca: Cephalopoda). *Marine Behaviour and Physiology*, **14**, 189–200.

Mauro, A. (1977). Extra-ocular photoreceptors in cephalopods. *Symposia of the Zoological Society of London*, **38**, 287–308.

Maxwell, G. (1965). *Ring of Bright Water*. London: Longmans, Green.

Maxwell, M. R. & Hanlon, R. T. (2000). Female reproductive output in the squid *Loligo pealeii*: multiple egg clutches and implications for a spawning strategy. *Marine Ecology Progress Series*, **199**, 159–170.

Maynard, D. M. (1967). Organization of central ganglia, in *Invertebrate Nervous Systems* (ed. C. G. A. Wiersma). Chicago: University Press.

Maynard Smith, J. (1970). The causes of polymorphism. *Symposia of the Zoological Society of London*, **26**, 371–383.

Maynard Smith, J. (1974). The theory of games and the evolution of animal conflicts. *Journal of Theoretical Biology*, **47**, 209–221.

Maynard Smith, J. (1982). Do animals convey information about their intentions? *Journal of Theoretical Biology*, **97**, 1–5.

Maynard Smith, J. & Harper, D. (2003). *Animal Signals*. Oxford: Oxford University Press.

McClean, R. (1983). Gastropod shells: a dynamic resource that helps shape benthic community structure. *Journal of Experimental Marine Biology and Ecology*, **69**, 151–174.

McCleneghan, K. & Ames, J. A. (1976). A unique method of prey capture by a sea otter, *Enhydra lutris*. *Journal of Mammalogy*, **57**, 410–412.

McFall-Ngai, M. J. (1990). Crypsis in the pelagic environment. *American Zoologist*, **30**, 175–188.

McFarland, D. (1981). *The Oxford Companion to Animal Behavior*. Oxford: Oxford University Press.

McGowan, J. A. (1954). Observations on the sexual behavior and spawning of the squid, *Loligo opalescens*, at LaJolla, CA. *California Fish and Game*, **40**, 47–54.

Meisel, D. V., Byrne, R. A., Kuba, M. *et al.* (2006). Contrasting activity patterns of two related octopus species, *Octopus macropus* and *Octopus vulgaris*. *Journal of Comparative Psychology*, **120**, 191–197.

Meisel, D. V., Byrne, R. A., Mather, J. A. & Kuba, M. (2011). Behavioral sleep in *Octopus vulgaris*. *Vie et Milieu – Life and Environment*, **61**, 185–190.

Mellin, C., Parrott, L., Andrefouet, S. *et al.* (2012). Multi-scale marine biodiversity patterns inferred efficiently from habitat image processing. *Ecological Applications*, **22**, 792–803.

Melo, Y. & Sauer, W. H. H. (2007). Determining the daily spawning cycle of the chokka squid, *Loligo reynaudii* off the South African coast. *Reviews in Fish Biology and Fisheries*, **17**, 247–257.

Melo, Y. C. & Sauer, W. H. H. (1999). Confirmation of serial spawning in the chokka squid *Loligo vulgaris reynaudii* off the coast of South Africa. *Marine Biology*, **135**, 307–313.

Merilaita, S., Schaefer, H. M. & Dimitrova, M. (2013). What is camouflage through distractive markings? *Behavioral Ecology*, **24**, e1271–e1272.

Merskey, H. M. & Bogduk, N. (1994). *Classification of Chronic Pain*. Seattle: IASP Press.

Messenger, J. B. (1963). Behaviour of young *Octopus briareus* Robson. *Nature*, **197**, 1186–1187.

Messenger, J. B. (1968). The visual attack of the cuttlefish, *Sepia officinalis*. *Animal Behaviour*, **16**, 342–357.

Messenger, J. B. (1970). Optomotor responses and nystagmus in intact, blinded and statocystless cuttlefish (*Sepia officinalis* L.). *Journal of Experimental Biology*, **53**, 789–796.

Messenger, J. B. (1971). Two-stage recovery of a response in *Sepia*. *Nature*, **232**, 202–203.

Messenger, J. B. (1973a). Learning performance and brain structure: a study in development. *Brain Research*, **58**, 519–523.

Messenger, J. B. (1973b). Learning in the cuttlefish, *Sepia*. *Animal Behaviour*, **21**, 801–826.

Messenger, J. B. (1974). Reflecting elements in cephalopod skin and their importance for camouflage. *Journal of Zoology (London)*, **174**, 387–395.

Messenger, J. B. (1977a). Evidence that *Octopus* is colour blind. *Journal of Experimental Biology*, **70**, 49–55.

Messenger, J. B. (1977b). Prey-capture and learning in the cuttlefish, *Sepia*. *Symposia of the Zoological Society of London*, **38**, 347–376.

Messenger, J. B. (1979a). The eyes and skin of *Octopus*: compensating for sensory deficiencies. *Endeavour*, **3**, 92–98.

Messenger, J. B. (1979b). *Nerves, Brain and Behaviour*. London: Arnold.

Messenger, J. B. (1979c). The nervous system of *Loligo*. IV. The peduncle and olfactory lobes. *Philosophical Transactions of the Royal Society of London B*, **285**, 275–309.

Messenger, J. B. (1981). Comparative physiology of vision in Molluscs, in *Handbook of Sensory Physiology*, Vol. VII/6C, *Comparative Physiology and Evolution of Vision in Invertebrates* (ed. H. Autrum), pp. 93–200. Berlin, Heidelberg, New York: Springer-Verlag.

Messenger, J. B. (1983). Multimodal convergence and the regulation of motor programs in cephalopods, in *Multimodal Convergences in Sensory Systems*. *Fortschritte der Zoologie 28* (ed. E. Horn), pp. 77–98. Stuttgart: Gustav Fischer.

Messenger, J. B. (1991). Photoreception and vision in molluscs, in *Evolution of the Eye and Visual System* (ed. J. R. Cronly-Dillon & R. L. Gregory), pp. 364–367. London: Macmillan Press.

Messenger, J. B. (1996). Neurotransmitters of cephalopods. *Invertebrate Neuroscience*, **2**, 95–114.

Messenger, J. B. (2001). Cephalopod chromatophores: neurobiology and natural history. *Biological Reviews*, **76**, 473–528.

Messenger, J. B. & Sanders, G. D. (1972). Visual preference and two-cue discrimination learning in *Octopus*. *Animal Behaviour*, **20**, 580–585.

Messenger, J. B. & Young, J. Z. (1999). The radular apparatus of cephalopods. *Philosophical Transactions of the Royal Society of London B*, **354**, 161–182.

Messenger, J. B., Cornwell, C. & Reed, C. (1997). L-Glutamate and serotonin are endogenous in squid chromatophore nerves. *Journal of Experimental Biology*, 200, 3043–3054.

Messenger, J. B., Wilson, A. P. & Hedge, A. (1973). Some evidence for colour-blindness in *Octopus*. *Journal of Experimental Biology*, 59, 77–94.

Michinomae, M., Masuda, H., Seidou, M. & Kito, Y. (1994). Structural basis for wavelength discrimination in the banked retina of the firefly squid *Watasenia scintillans*. *Journal of Experimental Biology*, **193**, 1–12.

Mikami, S. & Okutani, T. (1977). Preliminary observations on maneuvering, feeding, copulating and spawning behaviors of *Nautilus macromphalus* in captivity. *Japanese Journal of Malacology (Venus)*, **36**, 29–41.

Milinski, M. & Parker, G. A. (1991). Competition for resources, in *Behavioural Ecology. An Evolutionary Approach* (ed. J. R. Krebs & N. B. Davies), pp. 137–168. Oxford: Blackwell Scientific Publications.

Minton, J. W., Walsh, L. S., Lee, P. G. & Forsythe, J. W. (2001). First multi-generation culture of the tropical cuttlefish *Sepia pharaonis* Ehrenberg, 1831. *Aquaculture International*, **9**, 379–392.

Mirow, S. (1972). Skin color in the squids *Loligo pealii* and *Loligo opalescens*. II. Iridophores. *Zeitschrift für Zellforschung*, **125**, 176–190.

Miske, V. & Kirchhauser, J. (2006). First record of brooding and early life cycle stages in *Wunderpus photogenicus* Hochberg, Norman and Finn, 2006 (Cephalopoda: Octopodidae). *Molluscan Research*, **26**, 169–171.

Miyan, J. & Messenger, J. B. (1995). Intracellular recordings from the chromatophore lobes of *Octopus*, in *Cephalopod Neurobiology* (ed. N. J. Abbott, R. Williamson & L. Maddock), pp. 415–429. Oxford: Oxford University Press.

Mizerez, A., Weaver, J. C., Pedersen, P. B., *et al.* (2009). Microstructural and biochemical characterization of the nanoporous sucker rings from *Dosidicus gigas*. *Advanced Materials*, **20**, 1–6.

Mobley, A. S., Michel, W. C. & Lucero, M. T. (2008). Odorant responsiveness of squid olfactory receptor neurons. *Anatomical Record – Advances in Integrative Anatomy and Evolutionary Biology*, **291**, 763–774.

Moiseev, S. I. (1991). Observation of the vertical distribution and behavior of nektonic squids using manned submersibles. *Bulletin of Marine Science*, **49**, 446–456.

Moltschaniwskyj, N. A. (1995). Multiple spawning in the tropical squid *Photololigo* sp.: what is the cost in somatic growth? *Marine Biology*, **124**, 127–135.

Moltschaniwskyj, N. A. & Pecl, G. T. (2007). Spawning aggregations of squid (*Sepioteuthis australis*) populations: a continuum of 'microcohorts'. *Reviews in Fish Biology and Fisheries*, **17**, 183–195.

Moltschaniwskyj, N. A. & Steer, M. A. (2004). Spatial and seasonal variation in reproductive characteristics and spawning of southern calamary (*Sepioteuthis australis*): spreading the mortality risk. *ICES Journal of Marine Science*, **61**, 921–927.

Moody, M. F. & Parriss, J. R. (1960). The visual system of *Octopus*. Discrimination of polarized light by *Octopus*. *Nature*, **186**, 839–840.

Moody, M. F. & Parriss, J. R. (1961). The discrimination of polarized light by *Octopus*: a behavioural and morphological study. *Zeitschrift für vergleichende Physiologie*, **44**, 268–291.

Mooney, T. A., Hanlon, R. T., Christensen-Dalsgaard, J. et al. (2010). Sound detection by the longfin squid (*Loligo pealeii*) studied with auditory evoked potentials: sensitivity to low-frequency particle motion and not pressure. *Journal of Experimental Biology*, **213**, 3748–3759.

Morejohn, G. V., Harvey, J. T. & Krasnow, L. T. (1978). The importance of *Loligo opalescens* in the food web of marine vertebrates in Monterey Bay, California, in *Department of Fish and Game, Fish Bulletin 169* (ed. C. W. Recksiek & H. W. Frey), pp. 67–98. Los Angeles: Department of Fish and Game.

Moroshita, T. (1974). Participation in digestion by the proteolytic enzymes of the posterior salivary gland in *Octopus* – III. Some properties of purified enzymes from the posterior salivary gland. *Bulletin of the Japanese Society of Scientific Fisheries*, **40**, 927–936.

Moynihan, M. (1975). Conservatism of displays and comparable stereotyped patterns among cephalopods, in *Function and Evolution in Behaviour. Essays in Honor of Professor Niko Tinbergen, F.R.S.* (ed. G. Baerends, C. Beer & A. Manning), pp. 276–291. New York: Oxford University Press.

Moynihan, M. (1983a). Notes on the behavior of *Euprymna scolopes* (Cephalopoda: Sepiolidae). *Behaviour*, **85**, 25–41.

Moynihan, M. (1983b). Notes on the behavior of *Idiosepius pygmaeus* (Cephalopoda; Idiosepiidae). *Behaviour*, **85**, 42–57.

Moynihan, M. (1985a). *Communication and Noncommunication by Cephalopods*. Bloomington: Indiana University Press.

Moynihan, M. (1985b). Why are cephalopods deaf? *The American Naturalist*, **125**, 465–469.

Moynihan, M. & Rodaniche, A. F. (1982). The behavior and natural history of the Caribbean reef squid *Sepioteuthis sepioidea*. With a consideration of social, signal and defensive patterns for difficult and dangerous environments. *Advances in Ethology*, **25**, 1–150.

Moynihan, M. & Rodaniche, F. (1977). Communication, crypsis, and mimicry among cephalopods, in *How Animals Communicate* (ed. T. A. Sebeok), pp. 293–302. Bloomington: Indiana University Press.

Muñoz, J. L. P., Patino, M. A. L., Hermosilla, C. et al. (2011). Melatonin in octopus (*Octopus vulgaris*): tissue distribution, daily changes and relation with serotonin and its acid metabolite. *Journal of Comparative Physiology A –Neuroethology Sensory Neural and Behavioral Physiology*, **197**, 789–797.

Muntz, W. R. A. (1961). Interocular transfer in *Octopus vulgaris*. *Journal of Comparative and Physiological Psychology*, **54**, 49–55.

Muntz, W. R. A. (1962). Stimulus generalisation following monocular training in *Octopus*. *Journal of Comparative and Physiological Psychology*, **55**, 535–540.

Muntz, W. R. A. (1976). On yellow lenses in mesopelagic animals. *Journal of the Marine Biological Association of the United Kingdom*, **56**, 963–976.

Muntz, W. R. A. (1977). Pupillary response of cephalopods. *Symposia of the Zoological Society of London*, **38**, 277–285.

Muntz, W. R. A. (1986). The spectral sensitivity of *Nautilus pompilius*. *Journal of Experimental Biology*, **126**, 513–517.

Muntz, W. R. A. (1987). Visual behavior and visual sensitivity of *Nautilus pompilius*, in *Nautilus: The Biology and Paleobiology of a Living Fossil* (ed. W. B. Saunders & N. H. Landman), pp. 231–244. New York: Plenum Press.

Muntz, W. R. A. (1991). Anatomical and behavioural studies on vision in *Nautilus* and *Octopus*. *American Malacological Bulletin*, **9**, 69–74.

Muntz, W. R. A. (1994a). Spatial summation in the phototactic behavior of *Nautilus pompilius*. *Marine Behaviour and Physiology*, **24**, 183–187.

Muntz, W. R. A. (1994b). Effects of light on the efficacy of traps for *Nautilus pompilius*. *Marine Behaviour and Physiology*, **24**, 189–193.

Muntz, W. R. A. & Gwyther, J. (1988a). Visual discrimination of distance by octopuses. *Journal of Experimental Biology*, **140**, 345–353.

Muntz, W. R. A. & Gwyther, J. (1988b). Visual acuity in *Octopus pallidus* and *Octopus australis*. *Journal of Experimental Biology* **134**, 119–129.

Muntz, W. R. A. & Johnson, M. S. (1978). Rhodopsins of oceanic decapods. *Vision Research*, **18**, 601–602.

Muntz, W. R. A. & Raj, U. (1984). On the visual system of *Nautilus pompilius*. *Journal of Experimental Biology*, **109**, 253–263.

Muntz, W. R. A. & Wentworth, S. L. (1987). An anatomical study of the retina of *Nautilus pompilius*. *Biological Bulletin*, **173**, 387–397.

Muntz, W. R. A. & Wentworth, S. L. (1995). Structure of the adhesive surface of the digital tentacles of *Nautilus pompilius*. *Journal of the Marine Biological Association of the United Kingdom*, **75**, 747–750.

Munz, F. W. (1958). Photosensitive pigments from the retinae of certain deep-sea fishes. *Journal of Physiology (London)*, **140**, 220–225.

Munz, F. W. (1964). The visual pigments of epipelagic and rocky shore fishes. *Vision Research*, **4**, 441–454.

Muramatsu, K., Yamamoto, J., Abe, T. *et al.* (2013). Oceanic squid do fly. *Marine Biology*, **160**, 1171–1175.

Murata, M. (1990). Oceanic resources of squids. *Marine Behaviour and Physiology B*, **18**, 19–71.

Murata, M., Ishii, M. & Osako, M. (1982). Some information on copulation of the oceanic squid *Onychoteuthis borealijaponica* Okada. *Bulletin of the Japanese Society of Scientific Fisheries*, **48**, 351–354.

Myrberg, A. A. (1973). Underwater television – a tool for the marine biologist. *Bulletin of Marine Science*, **23**, 824–836.

Nabhitabhata, J. (1998). Distinctive behaviour of Thai pygmy squid, *Idiosepius thailandicus* Chotiyaputta, Okutani & Chaitiamvong, 1991. *Phuket Marine Biological Center Special Publication*, **18**, 25–40.

Nabhitabhata, J. & Suwanamala, J. (2008). Reproductive behaviour and cross-mating of two closely related pygmy squids *Idiosepius biserialis* and *Idiosepius thailandicus* (Cephalopoda: Idiosepiidae). *Journal of the Marine Biological Association of the United Kingdom*, **88**, 987–993.

Naef, A. (1923). Die Cephalopoden. Systematik. *Fauna e Flora del Golfo di Napoli*, **35**, 1–863.

Naef, A. (1928). Die Cephalopoden. Embryologie. *Fauna e Flora del Golfo di Napoli*, **35**, 1–357.

Nagasawa, K., Takayanagi, S. & Takami, T. (1993). Cephalopod tagging and marking in Japan: a review, in *Recent Advances in Cephalopod Fisheries Biology* (ed. T. Okutani, R. K. O'Dor & T. Kubodera), pp. 313–329. Tokyo: Tokai University Press.

Nakamura, Y. (1991). Tracking of the mature female of flying squid, *Ommastrephes bartrami*, by an ultrasonic transmitter. *Bulletin of the Hokkaido National Fisheries Research Institute*, **55**, 205–207.

Narvarte, M., Gonzalez, R. A., Storero, L. & Fernandez, M. (2013). Effects of competition and egg predation on shelter use by *Octopus tehuelchus* females. *Marine Ecology Progress Series*, **482**, 141–151.

Natsukari, Y. (1970). Egg-laying behavior, embryonic development and hatched larva of the pygmy cuttlefish, *Idiosepius pygmaeus paradoxus* Ortmann. *Bulletin of the Faculty of Fisheries of Nagasaki University*, **30**, 15–29.

Natsukari, Y. & Tashiro, M. (1991). Neritic squid resources and cuttlefish resources in Japan. *Marine Behaviour and Physiology B*, **18**, 149–226.

Naud, M. J. & Havenhand, J. N. (2006). Sperm motility and longevity in the giant cuttlefish, *Sepia apama* (Mollusca: Cephalopoda). *Marine Biology*, **148**, 559–566.

Naud, M. J., Hanlon, R. T., Hall, K. C., Shaw, P. W. & Havenhand, J. N. (2004). Behavioral and genetic assessment of mating success in a natural spawning aggregation of the giant cuttlefish (*Sepia apama*) in southern Australia. *Animal Behaviour*, **67**, 1043–1050.

Naud, M. J., Shaw, P. W., Hanlon, R. T. & Havenhand, J. N. (2005). Evidence for biased use of sperm sources in wild female giant cuttlefish (*Sepia apama*). *Proceedings of the Royal Society B: Biological Sciences*, **272**, 1047–1051.

Neill, S. R. S. J. (1971). Notes on squid and cuttlefish; keeping, handling and colour-patterns. *Pubblicazioni della Stazione Zoologica di Napoli*, **39**, 64–69.

Neill, S. R. S. J. & Cullen, J. M. (1974). Experiments on whether schooling by their prey affects the hunting behaviour of cephalopods and fish predators. *Journal of Zoology (London)*, **172**, 549–569.

Nesher, N., Levy, G., Grasso, F. W. & Hochner, B. (2014). Self-recognition mechanism between skin and suckers prevents octopus arms from intefering with each other. *Current Biology*, **24**, 1271–1275.

Nesis, K. N. (1965). Distribution and feeding of young squids *Gonatus fabricii* (Licht.) in the Labrador Sea and the Norwegian Sea. *Oceanology (Washington)*, **5**, 102–108.

Nesis, K. N. (1970). The biology of the giant squid of Peru and Chile, *Dosidicus gigas*. *Oceanology (Washington)*, **10**, 108–118.

Nesis, K. N. (1982). *Cephalopods of the World. Squids, Cuttlefishes, Octopuses, and Allies,* translated from Russian by B. F. Levitov (ed. L.A. Burgess) Neptune City, NJ: TFH Publications.

Nesis, K. N. (1983). *Dosidicus gigas*, in *Cephalopod Life Cycles*, Vol. 1 (ed. P. R. Boyle), pp. 475. London: Academic Press.

Nesis, K. N. (1995). Mating, spawning and death in oceanic cephalopods: a review. *Ruthenica*, **6**, 23–64.

Nesis, K. N. & Nikitina, I. V. (1981). Macrotritopus, a planktonic larva of the benthic octopus, *Octopus defilippi*: identification and distribution. *Zoologicheskifi Zhurnal*, **60**, 835–847.

Neumeister, H. & Budelmann, B. U. (1997). Structure and function of the *Nautilus* statocyst. *Philosophical Transactions of the Royal Society of London Series B – Biological Sciences*, **352**, 1565–1588.

Nicol, S. & O'Dor, R. K. (1985). Predatory behaviour of squid (*Illex illecebrosus*) feeding on surface swarms of euphausiids. *Canadian Journal of Zoology*, **63**, 15–17.

Nigmatullin, C. M. & Ostapenko, A. A. (1976). Feeding of *Octopus vulgaris* Lam. from the northwest African coast. *Shellfish & Benthos Committee. International Council for the Exploration of the Sea, C.M. 1976/K*, **6**, 1–15.

Nigmatullin, C. M., Nesis, K. N. & Arkhipkin, A. I. (2001). A review of the biology of the jumbo squid *Dosidicus gigas* (Cephalopoda: Ommastrephidae). *Fisheries Research*, **54**, 9–19.

Nilsson, D. E., Warrant, E. J., Johnsen, S., Hanlon, R. & Shashar, N. (2012). A unique advantage for giant eyes in giant squid. *Current Biology*, **22**, 683–688.

Nilsson, D.-E., Warrant, E. J., Johnsen, S., Hanlon, R. T. & Shashar, N. (2013). The giant eyes of giant squid are indeed unexpectedly large, but not if used for spotting sperm whales. *BMC Evolutionary Biology*, **13**, Article 187.

Nishimura, M. (1961). Frequency characteristics of sea noise and fish sound. *Technical Report of Fishing Boat, Tokyo*, **15**, 111–118.

Nixon, M. (1979). Hole-boring in shells by *Octopus vulgaris* Cuvier in the Mediterranean. *Malacologia*, **18**, 431–443.

Nixon, M. (1980). The salivary papilla of *Octopus* as an accessory radula for drilling shells. *Journal of Zoology (London)*, **190**, 53–57.

Nixon, M. (1984). Is there external digestion by *Octopus*? *Journal of Zoology (London)*, **202**, 441–447.

Nixon, M. (1987). Cephalopod diets, in *Cephalopod Life Cycles,* Vol. 2: *Comparative Reviews* (ed. P. R. Boyle), pp. 201–219. London: Academic Press.

Nixon, M. & Budelmann, B. U. (1984). Scale-worms – occasional food of *Octopus*. *Journal of Molluscan Studies*, **50**, 39–42.

Nixon, M. & Dilly, P. N. (1977). Sucker surfaces and prey capture. *Symposia of the Zoological Society of London*, **38**, 447–511.

Nixon, M. & Maconnachie, E. (1988). Drilling by *Octopus vulgaris* (Mollusca: Cephalopoda) in the Mediterranean. *Journal of Zoology (London)*, **216**, 687–716.

Nixon, M. & Messenger, J. B. (1977). *The Biology of Cephalopods*. London: Academic Press.

Nixon, M. & Young, J. Z. (2003). *The Brains and Lives of Cephalopods*. Oxford: Clarendon.

Nixon, M., Maconnachie, E. & Howell, P. G. T. (1980). The effects on shells of drilling by *Octopus*. *Journal of Zoology (London)*, **191**, 75–88.

Noakes, D. L. G. & Baylis, J. R. (1990). Behavior, in *Methods for Fish Biology* (ed. C. B. Schreck & P. B. Moyle), pp. 555–583. Bethesda, MD: American Fisheries Society.

Norman, M. D. (1991). *Octopus cyanea* Gray, 1849 (Mollusca, Cephalopoda) in Australian waters – description, distribution and taxonomy. *Bulletin of Marine Science*, **49**, 20–38.

Norman, M. D. (1992a). *Ameloctopus litoralis*, gen. et sp. nov. (Cephalopoda: Octopodidae), a new shallow-water octopus from tropical Australian waters. *Invertebrate Taxonomy*, **6**, 567–582.

Norman, M. D. (1992b). Ocellate octopuses (Cephalopoda: Octopodidae) of the Great Barrier Reef, Australia: description of two new species and redescription of *Octopus polyzenia* Gray, 1949. *Memoirs of the Museum of Victoria*, **53**, 309–344.

Norman, M. D. (2000). *Cephalopods: A World Guide*. Hackenheim: ConchBooks.

Norman, M. D. & Lu, C. C. (1997). Sex in giant squid. *Nature*, **389**, 683–684.

Norman, M. D., Finn, J. & Tregenza, T. (1999). Female impersonation as an alternative reproductive strategy in giant cuttlefish. *Proceedings of the Royal Society B: Biological Sciences*, **266**, 1347–1349.

Norman, M. D., Finn, J. & Tregenza, T. (2001). Dynamic mimicry in an Indo-Malayan octopus. *Proceedings of the Royal Society of London B: Biological Sciences*, **268**, 1755–1758.

Norman, M. D., Paul, D., Finn, J. & Tregenza, T. (2002). First encounter with a live male blanket octopus: the world's most sexually size-dimorphic large animal. *New Zealand Journal of Marine and Freshwater Research*, **36**, 733–736.

Norris, K. S. & Mohl, B. (1983). Can odontocetes debilitate prey with sound? *The American Naturalist*, **122**, 85–104.

Novicki, A., Messenger, J. B., Budelmann, B. U., Terrell, M. L. & Kadekaro, M. (1992). [14C] Deoxyglucose labelling of functional activity in the cephalopod central nervous system. *Proceedings of the Royal Society of London B*, **249**, 77–82.

Nyholm, S. V. & Nishiguchi, M. K. (2008). The evolutionary ecology of a sepiolid squid–*Vibrio* association: from cell to environment. *Vie et Milieu – Life and Environment*, **58**, 175–184.

O'Brien, W. J., Browman, H. I. & Evans, B. I. (1990). Search strategies of foraging animals. *American Scientist*, **78**, 152–160.

O'Dor, R. (2002). Telemetered cephalopod energetics: swimming, soaring, and blimping. *Integrative and Comparative Biology*, **42**, 1065–1070.

O'Dor, R., Balch, N. E. & Amaratunga, T. (1982). Laboratory observations of midwater spawning by *Illex illecebrosus*. *NAFO Scientific Council Studies*, **9**, 69–133.

O'Dor, R., Stewart, J., Gilly, W. *et al.* (2013). Squid rocket science: how squid launch into air. *Deep-Sea Research Part II: Tropical Studies in Oceanography*, **95**, 113–118.

O'Dor, R. K. (1983). *Illex illecebrosus*, in *Cephalopod Life Cycles*, Vol. 1 (ed. P. R. Boyle), pp. 175–199. London: Academic Press.

O'Dor, R. K. (1988). The energetic limits on squid distributions. *Malacologia*, **29**, 113–119.

O'Dor, R. K. (1998). Can understanding squid life-history strategies and recruitment improve management? *South African Journal of Marine Science – Suid-Afrikaanse Tydskrif vir Seewetenskap*, **20**, 193–206.

O'Dor, R. K. (2012). The incredible flying squid. *New Scientist*, **214**, 39–41.

O'Dor, R. K. & Balch, N. (1985). Properties of *Illex illecebrosus* egg masses potentially influencing larval oceanographic distribution. *NAFO Scientific Council Studies*, **9**, 69–76.

O'Dor, R. K. & Macalaster, E. G. (1983). *Bathypolypus arcticus*, in *Cephalopod Life Cycles*, Vol. 1 (ed. P. R. Boyle), pp. 475. London: Academic Press.

O'Dor, R. K. & Webber, D. M. (1986). The constraints on cephalopods: why squid aren't fish. *Canadian Journal of Zoology*, **64**, 1591–1605.

O'Dor, R. K. & Webber, D. M. (1991). Invertebrate athletes: trade-offs between transport efficiency and power density in cephalopod evolution. *Journal of Experimental Biology*, **160**, 93–112.

O'Dor, R. K. & Wells, M. J. (1978). Reproduction versus somatic growth: hormonal control in *Octopus vulgaris*. *Journal of Experimental Biology*, **77**, 15–31.

O'Dor, R. K. & Wells, M. J. (1987). Energy and nutrient flow, in *Cephalopod Life Cycles*, Vol. 2: *Comparative*

Reviews (ed. P. R. Boyle), pp. 109–133. London: Academic Press.

O'Dor, R. K., Adamo, S., Aitken, J. P. *et al.* (2002). Currents as environmental constraints on the behavior, energetics and distribution of squid and cuttlefish. *Bulletin of Marine Science*, **71**, 601–617.

O'Dor, R. K., Andrade, Y., Webber, D. M. *et al.* (1998). Applications and performance of Radio-Acoustic Positioning and Telemetry (RAPT) systems. *Hydrobiologia*, **372**, 1–8.

O'Dor, R. K., Carey, F. G., Webber, D. M. & Voegeli, F. M. (1991). Behavior and energenetics of Azorean squid, *Loligo forbesi*, in *Biotelemetry XI, Proceedings of the Eleventh International Symposium on Biotelemetry* (ed. A. Uchiyama & C. J. Amlaner Jr), pp. 191–195. Tokyo: Waseda University Press.

O'Dor, R. K., Forsythe, J., Webber, D. M., Wells, J. & Wells, M. J. (1993). Activity levels of *Nautilus*. *Nature*, **362**, 626–627.

O'Dor, R. K., Helm, P. & Balch, N. (1985). Can rhynchoteuthions suspension feed? (Mollusca: Cephalopoda). *Vie et Milieu*, **35**, 267–271.

O'Dor, R. K., Hoar, J. A., Webber, D. M. *et al.* (1995). Squid (*Loligo forbesi*) performance and metabolic rates in nature. *Marine and Freshwater Behaviour and Physiology*, **25**, 163–177.

O'Dor, R. K., Stewart, J., Gilly, W. *et al.* (2013). Squid rocket science: how squid launch into air. *Deep-Sea Research Part II: Tropical Studies in Oceanography*, **95**, 113–118.

O'Dor, R. K., Vessey, E. & Amaratunga, T. (1980). Factors affecting fecundity and larval distribution in the squid, *Illex illecebrosus*. *North Atlantic Fisheries Organization Scientific Council Research Document*, **2**, 1–9.

O'Dor, R. K., Wells, J. & Wells, M. J. (1990). Speed, jet pressure and oxygen consumption relationships in free-swimming *Nautilus*. *Journal of Experimental Biology*, **154**, 383–396.

Okutani, T. (1960). *Argonauta boettgeri* preys on *Cavolinia tridentata*. *Venus*, **21**, 39–41.

Okutani, T. (1974). Epipelagic decapod cephalopods collected by micronekton tows during the EASTROPAC Expeditions, 1967–1968 (Systematic Part). *Bulletin of the Tokai Regional Fisheries Research Laboratory*, **80**, 29–118.

Okutani, T. (1983). *Todarodes pacificus*, in *Cephalopod Life Cycles*, Vol. 1 (ed. P. R. Boyle), pp. 475. London: Academic Press.

Okutani, T. & Osuga, K. (1986). A peculiar nesting behavior of *Ocythoe tuberculata* in the test of a gigantic salp, *Tethys vagina*. *Venus: Japanese Journal of Malacology*, **45**, 67–69.

Oliveira, R. F., Taborsky, M. & Brockmann, H. J. (ed.) (2008). *Alternative Reproductive Tactics: An Integrative Approach*. Cambridge: Cambridge University Press.

Olofsson, M., Eriksson, S., Jakobsson, S. & Wiklund, C. (2012). Deimatic display in the European swallowtail butterfly as a secondary defence against attacks from great tits. *PLoS One*, **7**.

Ord, T. J., Peters, R. A., Evans, C. S. & Taylor, A. J. (2002). Digital video playback and visual communication in lizards. *Animal Behaviour*, **63**, 879–890.

Orelli, M. v. (1962). Die Ubertragung der spermatophore von *Octopus vulgaris* and *Eledone* (Cephalopoda). *Revue Suisse de Zoologie*, **69**, 193–202.

Orlov, O. Y. & Byzov, A. L. (1961). Colorimetric research on the vision of molluscs (Cephalopoda). *Doklady Akademii NaukSSSR*, **139**, 723–725.

Orlov, O. Y. & Byzov, A. L. (1962). Vision in cephalopod molluscs. *Priroda Moskva*, **3**, 115–118.

Ormond, R. F. G. (1980). Aggressive mimicry and other interspecific feeding associations among Red Sea coral reef predators. *Journal of Zoology*, **191**, 247–262.

O'Shea, S. & Bolstad, K. S. (2004). First records of egg masses of *Nototodarus gouldi* McCoy, 1888 (Mollusca: Cephalopoda: Ommastrephidae), with comments on egg-mass susceptibility to damage by fisheries trawl. *New Zealand Journal of Zoology*, **31**, 161–166.

Otis, T. S. & Gilly, W. F. (1990). Jet-propelled escape in the squid *Loligo opalescens*: concerted control by giant and non-giant motor axon pathways. *Proceedings of the National Academy of Sciences USA*, **87**, 2911–2915.

Packard, A. (1961). Sucker display of *Octopus*. *Nature*, **190**, 736–737.

Packard, A. (1963). The behaviour of *Octopus vulgaris*. *Bulletin de l'Institut Océanographique (Monaco)*, **No. 1 D**, 35–49.

Packard, A. (1969a). Jet propulsion and giant fibre response of *Loligo*. *Nature*, **221**, 875.

Packard, A. (1969b). Visual acuity and eye growth in *Octopus vulgaris* (Lamarck). *Monitore Zoologico Italiano*, **3**, 19–32.

Packard, A. (1972). Cephalopods and fish: the limits of convergence. *Biological Reviews*, **47**, 241–307.

Packard, A. (1974). Chromatophore fields in skin of *Octopus*. *Journal of Physiology-London*, **238**, P38–P40.

Packard, A. (1982). Morphogenesis of chromatophore patterns in cephalopods: are morphological and physiological 'units' the same? *Malacologia*, **23**, 193–201.

Packard, A. (1985). Sizes and distribution of chromatophores during post-embryonic development in cephalopods. *Vie et Milieu*, **35**, 285–298.

Packard, A. (1988a). Visual tactics and evolutionary strategies, in *Cephalopods – Present and Past* (ed. J. Wiedmann & J. Kullmann), pp. 89–103. Stuttgart, Germany: Schweizerbart'sche Verlagsbuchhandlung.

Packard, A. (1988b). The skin of cephalopods (Coleoids): general and special adaptations, in *The Mollusca*, Vol. 11: *Form and Function* (ed. E. R. Trueman & M. R. Clarke), pp. 37–67. San Diego: Academic Press.

Packard, A. (1995). Organization of cephalopod chromatophore systems: a neuromuscular image-generator, in *Cephalopod Neurobiology* (ed. N. J. Abbott, R. Williamson & L. Maddock), pp. 331–368. Oxford: Oxford University Press.

Packard, A. & Brancato, D. (1993). Some responses of octopus chromatophores to light. *Journal of Physiology London* 459, P429–P429.

Packard, A. & Hochberg, F. G. (1977). Skin patterning in *Octopus* and other genera. *Symposia of the Zoological Society of London*, **38**, 191–231.

Packard, A. & Sanders, G. (1969). What the octopus shows to the world. *Endeavour*, **28**, 92–99.

Packard, A. & Sanders, G. D. (1971). Body patterns of *Octopus vulgaris* and maturation of the response to disturbance. *Animal Behaviour*, **19**, 780–790.

Packard, A. & Wurtz, M. (1994). An *Octopus, Ocythoe*, with a swimbladder and triple jets. *Philosophical Transactions of the Royal Society of London Series B – Biological Sciences*, **344**, 261–275.

Packard, A., Bone, Q. & Hignette, M. (1980). Breathing and swimming movements in a captive *Nautilus*. *Journal of the Marine Biological Association of the United Kingdom*, **60**, 313–327.

Packard, A., Karlsen, H. E. & Sand, O. (1990). Low frequency hearing in cephalopods. *Journal of Comparative Physiology A*, **166**, 501–505.

Palmer, B. W. & O'Dor, R. K. (1978). Changes in vertical migration patterns of captive *Illex illecebrosus* in varying light regimes and salinity gradients. *Fisheries and Marine Service Technical Report*, No. **833**, 23.1–23.12.

Palmer, M. E., Calvé, M. R. & Adamo, S. A. (2006). Response of female cuttlefish *Sepia officinalis* (Cephalopoda) to mirrors and conspecifics: evidence for signaling in female cuttlefish. *Animal Cognition*, **9**, 151–155.

Papini, M. R. & Bitterman, M. E. (1991). Appetitive conditioning in *Octopus cyanea*. *Journal of Comparative Psychology*, **105**, 107–114.

Parker, G. A. (1970). Sperm competition and its evolutionary consequences in the insects. *Biological Reviews*, **45**, 525–567.

Parker, G. A. (1990). Sperm competition games: sneaks and extra-pair copulations. *Proceedings of the Royal Society of London B*, **242**, 127–133.

Parker, G. A. & Pizzari, T. (2010). Sperm competition and ejaculate economics. *Biological Reviews*, **85**, 897–934.

Parker, G. A., Simmons, L. W. & Kirk, H. (1990). Analyzing sperm competition data – simple models for predicting mechanisms. *Behavioral Ecology and Sociobiology*, **27**, 55–65.

Parry, M. (2000). A description of the nuchal organ, a possible photoreceptor, in *Euprymna scolopes* and other cephalopods. *Journal of Zoology*, **252**, 163–177.

Parry, M. (2006). Feeding behavior of two ommastrephid squids *Ommastrephes bartramii* and *Sthenoteuthis oualaniensis* off Hawaii. *Marine Ecology Progress Series*, **318**, 229–235.

Partridge, B. L. & Pitcher, T. (1980). The sensory basis of fish schools: relative roles of lateral line and vision. *Journal of Comparative Physiology A*, **135**, 315–325.

Pascual, E. (1978). Crecimiento y alimentación de tres generaciónes de *Sepia officinalis* en cultivo. *Investigación Pesquera*, **42**, 421–442.

Passarella, K. C. & Hopkins, T. L. (1991). Species composition and food habits of the micronektonic cephalopod assemblage in the Eastern Gulf of Mexico. *Bulletin of Marine Science*, **49**, 638–659.

Payne, N. L., Gillanders, B. M. & Semmens, J. (2011). Breeding durations as estimators of adult sex ratios and population size. *Oecologia*, **165**, 341–347.

Payne, N. L., Gillanders, B. M., Seymour, R. S. *et al.* (2010a). Accelerometry estimates field metabolic rate in giant Australian cuttlefish *Sepia apama* during breeding. *Journal of Animal Ecology*, 80, 422–430.

Payne, N. L., Gillanders, B. M., Webber, D. M. & Semmens, J. M. (2010b). Interpreting diel activity patterns from acoustic telemetry: the need for controls. *Marine Ecology Progress Series*, **419**, 295–301.

Payne, R. J. H. (1998). Gradually escalating fights and displays: the cumulative assessment model. *Animal Behaviour*, **56**, 651–662.

Payne, R. J. H. & Pagel, M. (1996). Escalation and time costs in displays of endurance. *Journal of Theoretical Biology*, **183**, 185–193.

Payne, R. J. H. & Pagel, M. (1997). Why do animals repeat displays? *Animal Behaviour*, **54**, 109–119.

Pearce, J. M. (2008). *Animal Learning and Cognition*. Hove, UK: Psychology Press.

Pham, C. K., Carreira, G. P., Porteiro, F. M. *et al.* (2009). First description of spawning in a deep water loliginid squid, *Loligo forbesi* (Cephalopoda: Myopsida). *Journal of the Marine Biological Association of the United Kingdom*, **89**, 171–177.

Philips, M. & Austad, S. N. (1992). Animal communication and social evolution, in *Interpretation and Explanation in the Study of Animal Behaviour*, Vol. 1: *Interpretation, Intentionality, and Communication* (ed. M. Behoff & D. Jamieson), pp. 254–268. Boulder, CO: Westview Press.

Pickford, G. E. & McConnaughey, B. H. (1949). The *Octopus bimaculatus* problem: a study in sibling species. *Bulletin of the Bingham Oceanographic Collection*, **12**, 1–66.

Pieron, H. (1911). Contribution à la pyschologie du poulpe. *Bulletin de l'Institut General Pyschologique*, **11**, 111–119.

Pietrewicz, A. T. & Kamil, A. C. (1981). Search images and the detection of cryptic prey: an operant approach, in *Foraging Behavior: Ecological, Ethological, and Psychological Approaches* (ed. A. C. Kamil & T. D. Sargent), pp. 311–331. New York, NY: Garland STPM Press.

Pignatelli, V., Temple, S. E., Chiou, T. H. *et al.* (2011). Behavioural relevance of polarization sensitivity as a target detection mechanism in cephalopods and fishes. *Philosophical Transactions of the Royal Society B – Biological Sciences*, **366**, 734–741.

Pimm, S. L., Lawton, J. H. & Cohen, J. E. (1991). Food web patterns and their consequences. *Nature*, **350**, 669–674.

Pitcher, T. J. (1983). Heuristic definitions of fish shoaling behaviour. *Animal Behaviour*, **31**, 611–612.

Pitcher, T. J. (1986). Functions of shoaling behavior in teleosts, in *The Behaviour of Teleost Fishes* (ed. T. J. Pitcher), pp. 294–337. Croon Helm: London.

Pitcher, T. J. (1993). *The Behaviour of Teleost Fishes*. London: Chapman & Hall.

Pitcher, T. J. & Parrish, J. K. (1993). Functions of shoaling behaviour in teleosts, in *Behaviour of Teleost Fishes* (ed. T. J. Pitcher), pp. 363–440. London: Chapman & Hall.

Pitcher, T. J. & Partridge, B. L. (1979). Fish school density and volume. *Marine Biology*, **54**, 383–394.

Pitcher, T. J. & Wyche, C. J. (1983). Predator-avoidance behaviours of sand-eel schools: why schools seldom split, in *Predators and Prey in Fishes* (ed. D. L. G. Noakes, D. G. Linquist, G. S. Helfman & J. A. Ward), pp. 193–204. The Hague, Netherlands: Dr. W. Junk Publ.

Pitcher, T. J., Partridge, B. L. & Wardle, C. S. (1976). A blind fish can school. *Science*, **194**, 963–965.

Pliny (Plinius, G. S.) (AD 77; this edition 1963). *Natural History*. London: Heinemann.

Ploger, B. J. & Yasukawa, K. (2003). *Exploring Animal Behavior in Laboratory and Field: An Hypothesis-Testing Approach to the Development, Causation, Function, and Evolution of Animal Behavior*: Elsevier.

Poirier, R., Chichery, R. & Dickel, L. (2004). Effects of rearing conditions on sand digging efficiency in juvenile cuttlefish. *Behavioural Processes*, **67**, 273–279.

Poirier, R., Chichery, R. & Dickel, L. (2005). Early experience and postembryonic maturation of body patterns in cuttlefish (*Sepia officinalis*). *Journal of Comparative Psychology*, **119**, 230–237.

Polese, G., Bertapelle, C. & Di Cosmo, A. (2015). Role of olfaction in *Octopus vulgaris* reproduction. *General and Comparative Endocrinology*, **210**, 55–62.

Polimanti, O. (1910). Les cephalopodes ont-ils une memorie? *Archives de Psychologie Geneve*, **10**, 84–87.

Porteiro, F. M., Martins, H. R. & Hanlon, R. T. (1990). Some observations on the behavior of adult squids, *Loligo forbesi*, in captivity. *Journal of the Marine Biological Association of the United Kingdom*, **70**, 459–472.

Pörtner, H. O. (2002). Environmental and functional limits to muscular exercise and body size in marine invertebrate athletes. *Comparative Biochemistry and Physiology A – Molecular and Integrative Physiology*, **133**, 303–321.

Preuss, T. & Budelmann, B. U. (1995a). Proprioceptive hair-cells on the neck of the squid *Lolliguncula brevis* – a sense organ in cephalopods for the control of head-to-body position. *Philosophical Transactions of the Royal Society of London Series B – Biological Sciences*, **349**, 153–178.

Preuss, T. & Budelmann, B. U. (1995b). A dorsal light reflex in a squid. *Journal of Experimental Biology*, **198**, 1157–1159.

Preuss, T. & Gilly, W. F. (2000). Role of prey-capture experience in the development of the escape response in the squid *Loligo opalescens*: a physiological correlate in an identified neuron. *Journal of Experimental Biology*, **203**, 559–565.

Pronk, R., Wilson, D. R. & Harcourt, R. (2010). Video playback demonstrates episodic personality in the gloomy octopus. *Journal of Experimental Biology*, **213**, 1035–1041.

Prota, G. J., Ortonne, P., Voulot, C. *et al.* (1981). Occurrence and properties of tyrosinase in the ejected ink of cephalopods. *Comparative Biochemistry and Physiology*, **15**, 453–466.

Pumphrey, R. J. & Young, J. Z. (1938). The rates of conduction of nerve fibres of various diameters in cephalopods. *Journal of Experimental Biology*, **15**, 453–466.

Quetglas, A., Alemany, F., Carbonell, A., Merella, P. & Sanchez, P. (1999). Diet of the European flying squid *Todarodes sagittatus* (Cephalopoda: Ommastrephidae) in the Balearic Sea (western Mediterranean). *Journal of the Marine Biological Association of the United Kingdom*, **79**, 479–486.

Quinteiro, J., Baibai, T., Oukhattar, L. *et al.* (2011). Multiple paternity in the common octopus *Octopus vulgaris* (Cuvier, 1797), as revealed by microsatellite DNA analysis. *Molluscan Research*, **31**, 15–20.

Racovitza, E. G. (1894). Sur l'accouplement de quelques Céphalopodes *Sepiola rondeletti* (Leach), *Rossia macrosoma* (d. Ch.) et *Octopus vulgaris* (Lam.) [About the mating behavior of the cephalopods . . .]. *Comptes Rendus Hebdomadaires des Seances de l'Academie des Sciences, Paris, Series D*, **118**, 722–724.

Radakov, D. V. (1973). *Schooling in the Ecology of Fish*, Israel Programme for Scientific Translations.

Ramachandran, V. S., Tyler, C. W., Gregory, R. L., *et al.* (1996). Rapid adaptive camouflage in tropical flounders. *Nature*, **379**, 815–818.

Randall, J. E. (1967). Food habits of reef fishes of West Indies. *Studies in Tropical Oceanography*, **5**, 665–847.

Reynolds, J. D. (1996). Animal breeding systems. *Trends in Ecology and Evolution*, **11**, 68–72.

Richard, A. (1971). Action qualitative de la lumiere dans le determinisme du cycle sexuel chez le Cephalopode *Sepia officinalis* L. *Comptes Rendus Hebdomadaires des Seances de l'Academie des Sciences, Paris, Series D*, **272**, 106–109.

Ricklefs, R. E. & Miller, G. L. (1999). *Ecology*. New York: W.H. Freeman & Co.

Rigby, P. R. & Sakurai, Y. (2005). Multidimensional tracking of giant Pacific octopuses in northern Japan reveals unexpected foraging behaviour. *Marine Technology Society Journal*, **39**, 64–67.

Roberts, M. J., Barange, M., Lipinski, M. R. & Prowse, M. R. (2002). Direct hydroacoustic observations of chokka squid *Loligo vulgaris reynaudii* spawning activity in deep water. *South African Journal of Marine Science – Suid-Afrikaanse Tydskrif vir Seewetenskap*, **24**, 387–393.

Roberts, M. J., Downey, N. J. & Sauer, W. H. (2012). The relative importance of shallow and deep shelf spawning habitats for the South African chokka squid (*Loligo reynaudii*). *ICES Journal of Marine Science*, **69**, 563–571.

Robertson, J. D. (1994). Cytochalasin-D blocks touch learning in *Octopus vulgaris*. *Proceedings of the Royal Society of London Series B – Biological Sciences*, **258**, 61–66.

Robertson, J. D. & Lee, P. (1990). An electron microscopic and behavioral study of tactile learning and memory in *Octopus vulgaris. Progress in Cell Research*, **1**, 287–306.

Robertson, J. D., Bonaventura, J. & Kohm, A. (1995). Nitric oxide synthase inhibition blocks octopus touch learning without producing sensory or motor dysfunction. *Proceedings of the Royal Society B – Biological Sciences*, **261**, 167–172.

Robertson, J. D., Bonaventura, J. & Kohm, A. P. (1994). Nitric oxide is required for tactile learning in *Octopus vulgaris. Proceedings of the Royal Society of London B – Biological Sciences*, **256**, 269–273.

Robins, C. R., Bailey, R. M., Bond, C. E. *et al.* (1991). *Common and scientific names of fishes from the United States and Canada*, American Fisheries Society Special Publication. Bethesda, MD: American Fisheries Society.

Robinson, M. H. (1969). Defenses against visually hunting predators, in *Evolutionary Biology* (ed. T. Dobzhansky, M. K. Hecht & W. C. Steere), pp. 225–259. New York, NY: Appleton-Century-Crofts Publ.

Robison, B. H. & Young, R. E. (1981). Bioluminescence in pelagic octopods. *Pacific Science*, **35**, 39–44.

Robison, B. H., Reisenbichler, K. R., Hunt, J. C. & Haddock, S. H. D. (2003). Light production by the arm tips of the deep-sea cephalopod *Vampyroteuthis infernalis. Biological Bulletin*, **205**, 102–109.

Rocha, F. & Guerra, A. (1996). Signs of an extended and intermittent terminal spawning in the squids *Loligo vulgaris* Lamarck and *Loligo forbesi* Steenstrup (Cephalopoda: Loliginidae). *Journal of Experimental Marine Biology and Ecology*, **207**, 177–189.

Rocha, F., Gonzalez, A. F., Segonzac, M. & Guerra, A. (2002). Behavioural observations of the cephalopod *Vulcanoctopus hydrothermalis. Cahiers de Biologie Marine*, **43**, 299–302.

Rocha, F., Guerra, A. & Gonzalez, A. F. (2001). A review of reproductive strategies in cephalopods. *Biological Reviews*, **76**, 291–304.

Rocha, L., Ross, R. & Kopp, G. (2012). Opportunistic mimicry by a Jawfish. *Coral Reefs*, **31**, 285–285.

Rodaniche, A. F. (1984). Iteroparity in the lesser Pacific striped octopus *Octopus chierchiae. Bulletin of Marine Science*, **35**, 99–104.

Rodaniche, A. F. (1991). Notes on the behavior of the larger Pacific striped octopus, an undescribed species

of the genus *Octopus. Bulletin of Marine Science*, **49**, 667.

Rodhouse, P. G., Prince, P. A., Clarke, M. R. & Murray, A. W. A. (1990). Cephalopod prey of the grey-headed albatross *Diomedea chrysostoma. Marine Biology*, **104**, 353–362.

Rodhouse, P. G., Swinfen, R. C. & Murray, A. W. A. (1988). Life cycle, demography and reproductive investment in the myopsid squid *Alloteuthis subulata. Marine Ecology Progress Series*, **45**, 245–253.

Rodrigues, M., Garci, M. E., Troncoso, J. S. & Guerra, A. (2010). Burying behaviour in the bobtail squid *Sepiola atlantica* (Cephalopoda: Sepiolidae). *Italian Journal of Zoology*, **77**, 247–251.

Rodrigues, M., Garci, M. E., Troncoso, J. S. & Guerra, A. (2011). Spawning strategy in Atlantic bobtail squid *Sepiola atlantica* (Cephalopoda: Sepiolidae). *Helgoland Marine Research*, **65**, 43–49.

Roeleveld, M. A. & Lipinski, M. R. (1991). The giant squid *Architeuthis* in southern African waters. *Journal of Zoology (London)*, **224**, 431–477.

Roffe, T. (1975). Spectral perception in *Octopus*: a behavioral study. *Vision Research*, **15**, 353–356.

Romagny, S., Darmaillacq, A. S., Guibé, M., Bellanger, C. & Dickel, L. (2012). Feel, smell and see in an egg: emergence of perception and learning in an immature invertebrate, the cuttlefish embryo. *Journal of Experimental Biology*, **215**, 4125–4130.

Romanini, M. G. (1952). Osservazioni sulla ialuronidasi delle ghiandole salivari anteriori e posteriori degli Octopodi. *Pubblicazioni della Stazione Zoologica di Napoli*, **23**, 251–270.

Roper, C. F. E. & Boss, K. J. (1982). The giant squid. *Scientific American*, **246**, 96–105.

Roper, C. F. E. & Brundage, W. L. (1972). Cirrate octopods with associated deep-sea organisms: new biological data based on deep benthic photographs (Cephalopoda). *Smithsonian Contributions to Zoology*, **121**, 1–46.

Roper, C. F. E. & Hochberg, F. G. (1988). Behavior and systematics of cephalopods from Lizard Island, Australia, based on color and body patterns. *Malacologia*, **29**, 153–193.

Roper, C. F. E. & Vecchione, M. (1993). A geographic and taxonomic review of *Taningia danae* Joubin, 1931 (Cephalopoda: Octopoteuthidae), with new records and observations on bioluminescence, in *Recent Advances in Fisheries Biology* (ed. T. Okutani, R. K. O'Dor & T. Kubodera), pp. 441–456. Tokyo: Tokai University Press.

Roper, C. F. E. & Vecchione, M. (1996). *In situ* observations on *Brachioteuthis beanii* Verrill: paired behavior, probably mating (Cephalopoda, Oegopsida). *American Malacological Bulletin*, **13**, 55–60.

Roper, C. F. E. & Young, R. E. (1975). Vertical distribution of pelagic cephalopods. *Smithsonian Contributions to Zoology*, **209**, 51.

Roper, C. F. E., Sweeney, M. J. & Nauen, C. E. (1984). FAO Species Catalogue. Cephalopods of the World. An annotated and illustrated catalogue of species of interest to fisheries. *FAO Fisheries Synopsis*, **3**, 1–277.

Roper, C. F. E., Young, R. E. & Voss, G. L. (1969). An illustrated key to the families of the order Teuthoidea (Cephalopoda). *Smithsonian Contributions to Zoology*, **13**, 1–32.

Rosa, R. & Seibel, B. A. (2010a). Slow pace of life of the Antarctic colossal squid. *Journal of the Marine Biological Association of the United Kingdom*, **90**, 1375–1378.

Rosa, R. & Seibel, B. A. (2010b). Voyage of the argonauts in the pelagic realm: physiological and behavioural ecology of the rare Paper Nautilus, *Argonauta nouryi*. *ICES Journal of Marine Science: Journal du Conseil*, **67**, 1494–1500.

Rosa, R., O'Dor, R. & Pierce, G. J. (2013). *Advances in Squid Biology, Ecology and Fisheries. Part 1: Myopsid Squids*. Hauppauge, NY: Nova Publishers.

Ross, D. M. (1971). Protection of hermit crabs (*Dardanus* spp.) from octopus by commensal sea anemones (*Calliactis* spp.). *Nature*, **230**, 401–402.

Ross, D. M. & Boletzky, S. von (1979). The association between the pagurid *Dardanus arrosor* and the actinian *Calliactis parasitica*. Recovery of activity in 'inactive' *D. arrosor* in the presence of cephalopods. *Marine Behaviour and Physiology*, **6**, 175–184.

Rowell, C. H. F. & Wells, M. J. (1961). Retinal orientation and the discrimination of polarized light by octopuses. *Journal of Experimental Biology*, **38**, 827–831.

Rowland, H. M. (2009). From Abbott Thayer to the present day: what have we learned about the function of countershading? *Philosophical Transactions of the Royal Society B – Biological Sciences*, **364**, 519–527.

Royan, A., Muir, A. P. & Downie, J. R. (2010). Variability in escape trajectory in the Trinidadian stream frog and two treefrogs at different life-history stages. *Canadian Journal of Zoology – Revue Canadienne de Zoologie*, **88**, 922–934.

Ruth, P., Schmidtberg, H., Westermann, B. & Schipp, R. (2002). The sensory epithelium of the tentacles and the rhinophore of *Nautilus pompilius* L. (Cephalopoda, Nautiloidea). *Journal of Morphology*, **251**, 239–255.

Ruxton, G. D., Sherratt, T. N. & Speed, M. P. (2004). *Avoiding Attack: The Evolutionary Ecology of Crypsis, Warning Signals, and Mimicry*. Oxford: Oxford University Press.

Saibil, H. R. (1990). Cell and molecular biology of photoreceptors. *Seminars in the Neurosciences*, **2**, 15–23.

Saibil, H. R. & Hewat, E. (1987). Ordered transmembrane and extracellular structure in squid photoreceptor microvilli. *Journal of Cell Biology*, **105**, 19–28.

Saidel, W. M., Lettvin, J. Y. & MacNichol, E. F. J. (1983). Processing of polarized light by squid photoreceptors. *Nature*, **304**, 534–536.

Saidel, W. M., Shashar, N., Schmolesky, M. T. & Hanlon, R. T. (2005). Discriminative responses of squid (*Loligo pealeii*) photoreceptors to polarized light. *Comparative Biochemistry and Physiology A*, **142**, 340–346.

Sakamoto, W., Naito, Y., Huntley, A. C. & Le Boeuf, B. J. (1989). Daily gross energy requirements of a female northern elephant seal *Mirounga angustirostris* at sea. *Nippon Suisan Gakkaishi*, **55**, 2057–2063.

Sakurai, Y., Bower, J. R. & Ikeda, Y. (2003). Reproductive characteristics of the ommastrephid squid *Todarodes pacificus*, in *Modern Approaches to Assess Maturity and Fecundity of Warm- and Cold-Water Fish and Squids*, Vol. 12 (ed. O. S. Kjesbu, J. R. Hunter & P. R. Witthames), pp. 105–115. Bergen, Norway: Institute of Marine Research.

Sakurai, Y., Kidokoro, H., Yamashita, N. *et al.* (2013). *Todarodes pacificus*, Japanese common squid, in *Advances in Squid Biology, Ecology and Fisheries. Part II* (ed. R. Rosa, G. J. Pierce & R. O'Dor), pp. 249–271.

Samson, J. E., Mooney, T. A., Gussekloo, S. W. S. & Hanlon, R. T. (2014). Graded behavioral responses and habituation to sound in the common cuttlefish *Sepia officinalis*. *Journal of Experimental Biology*, **217**, 4347–4355.

Sanchez, P. (2003). Cephalopods from off the Pacific coast of Mexico: biological aspects of the most abundant species. *Scientia Marina*, **67**, 81–90.

Sanders, F. K. & Young, J. Z. (1940). Learning and other functions of the higher nervous centres of *Sepia*. *Journal of Neurophysiology*, **3**, 501–526.

Sanders, G. D. (1970). Long-term memory of a tactile discrimination in *Octopus vulgaris* and the effect of vertical lobe removal. *Brain Research*, **20**, 59–73.

Sanders, G. D. (1975). The Cephalopods, in *Invertebrate Learning*, Vol. 3: *Cephalopods and Echinoderms* (ed. W. C. Corning, J. A. Dyal & A. O. D. Willows), pp. 1–101. New York, NY: Plenum Press.

Sato, N., Kasugai, T., Ikeda, Y. & Munehara, H. (2010). Structure of the seminal receptacle and sperm storage in the Japanese pygmy squid. *Journal of Zoology*, **282**, 151–156.

Sato, N., Kasugai, T. & Munehara, H. (2013). Sperm transfer or spermatangia removal: postcopulatory behaviour of picking up spermatangium by female Japanese pygmy squid. *Marine Biology*, **160**, 553–561.

Sato, N., Kasugai, T. & Munehara, H. (2014). Female pygmy squid cryptically favour small males and fast copulation as observed by removal of spermatangia. *Evolutionary Biology*, **41**, 221–228.

Sato, N., Yoshida, M.-A., Fujiwara, E. & Kasugai, T. (2013). High-speed camera observations of copulatory behaviour in *Idiosepius paradoxus*: function of the dimorphic hectocotyli. *Journal of Molluscan Studies*, **79**, 183–186.

Sauer, W. H. H. (1995). The impact of fishing on chokka squid *Loligo vulgaris reynaudii* concentrations on inshore spawning grounds in the South-Eastern Cape, South Africa. *South African Journal of Marine Science – Suid-Afrikaanse Tydskrif vir Seewetenskap*, **16**, 185–193.

Sauer, W. H. H. & Smale, M. J. (1993). Spawning behaviour of *Loligo vulgaris reynaudii* in shallow coastal waters of the South Eastern Cape, South Africa, in *Recent Advances in Fisheries Biology* (ed. T. Okutani, R. K. O'Dor & T. Kubodera), pp. 489–498. Tokyo: Tokai University Press.

Sauer, W. H. H., Roberts, M. J., Lipinski, M. R. *et al.* (1997). Choreography of the squid's 'nuptial dance'. *Biological Bulletin*, **192**, 203–207.

Sauer, W. H. H., Smale, M. J. & Lipinski, M. R. (1992). The location of spawning grounds, spawning and schooling behaviour of the squid *Loligo vulgaris reynaudii* (Cephalopoda: Myopsida) off the Eastern Cape Coast, South Africa. *Marine Biology*, **114**, 97–107.

Saunders, W. B. (1983). Natural rates of growth and longevity of *Nautilus belauensis*. *Paleobiology*, **9**, 280–288.

Saunders, W. B. (1984). The role and status of *Nautilus* in its natural habitat: evidence from deep-water remote camera photosequences. *Paleobiology*, **10**, 469–486.

Saunders, W. B. (1985). Studies of living *Nautilus* in Palau. *National Geographic Society Research Reports*, **18**, 669–682.

Saunders, W. B. & Landman, N. H. (2009). *Nautilus: The Biology and Paleobiology of a Living Fossil*. New York: Plenum Press.

Saunders, W. B. & Spinosa, C. (1979). *Nautilus* movement and distribution in Palau, Western Caroline Islands. *Science*, **204**, 1199–1201.

Saunders, W. B. & Ward, P. D. (1987). Ecology, distribution and population characteristics of *Nautilus*, in *Nautilus: The Biology and Paleobiology of a Living Fossil* (ed. W. B. Saunders & N. H. Landman), pp. 137–162. New York: Plenum Press.

Saunders, W. B., Knight, R. L. & Bond, P. N. (1991). *Octopus* predation on *Nautilus*: evidence from Papua New Guinea. *Bulletin of Marine Science*, **49**, 280–287.

Saunders, W. B., Spinosa, C. & Davis, L. E. (1987). Predation on *Nautilus*, in *Nautilus: The Biology and Paleobiology of a Living Fossil* (ed. W. B. Saunders & N. H. Landman), pp. 201–212. New York: Plenum Press.

Schäfer, W. (1936). Bau, Entwicklung und Farbenentstehung bei den Flitterzellen von *Sepia officinalis*. *Zeitschrift für Zellforschung und Mikroskopische Anatomie*, **27**, 222–245.

Schäfer, W. (1956). Die Schutzwirkung der Tintenfisch-Tinte. *Natur und Volk*, **86**, 24–26.

Scheel, D. & Anderson, R. (2012). Variability in the diet specialization of *Enteroctopus dofleini* (Cephalopoda: Octopodidae) in the eastern Pacific examined from midden contents. *American Malacological Bulletin*, **30**, 267–279.

Scheel, D. & Bisson, L. (2012). Movement patterns of giant Pacific octopuses, *Enteroctopus dofleini* (Walker, 1910). *Journal of Experimental Marine Biology and Ecology*, **416–417**, 21–31.

Schiller, P. H. (1949). Delayed detour response in the octopus. *Journal of Comparative and Physiological Psychology*, **42**, 220–225.

Schmitt, R. J. (1982). Consequences of dissimilar defenses against predation in a subtidal marine community. *Ecology*, **63**, 1588–1601.

Schmitt, R. J. (1987). Indirect interactions between prey: apparent competition, predator aggregation, and habitat segregation. *Ecology*, **68**, 1887–1897.

Schnell, A. K. (2014). Signalling, mating and conflict resolution in the Giant Australian cuttlefish, *Sepia apama*. Ph.D. Thesis. Sydney, Australia: Macquarie University.

Schnell, A. K., Smith, C. L., Hanlon, R. T. & Harcourt, R. (2015). Giant Australian cuttlefish use mutual assessment to resolve male–male contests. *Animal Behaviour*, **107**, 31–40.

Schoener, T. W. (1974). Resource partitioning in ecological communities. *Science*, **185**, 27–39.

Scott, W. B. & Tibbo, S. N. (1968). Food and feeding habits of swordfish, *Xiphias gladius*, in the western North Atlantic. *Journal of the Fisheries Research Board of Canada*, **25**, 903–919.

Segawa, S. (1987). Life history of the oval squid, *Sepioteuthis lessoniana*, in Kominato and adjacent waters central Honshu, Japan. *Journal of the Tokyo University of Fisheries*, **74**, 67–105.

Segawa, S., Izuka, T., Tamashiro, T. & Okutani, T. (1993). A note on mating and egg deposition by *Sepioteuthis lessoniana* in Ishigaki Island, Okinawa, Southwestern Japan. *Venus: Japanese Journal of Malacology*, **52**, 101–108.

Segawa, S., Yang, W. T., Marthy, H. J. & Hanlon, R. T. (1988). Illustrated embryonic stages of the eastern Atlantic squid *Loligo forbesi*. *Veliger*, **30**, 230–243.

Seibel, B. A. (2011). Critical oxygen levels and metabolic suppression in oceanic oxygen minimum zones. *Journal of Experimental Biology*, **214**, 326–336.

Seibel, B. A., Hochberg, F. G. & Carlini, D. B. (2000). Life history of *Gonatus onyx* (Cephalopoda: Teuthoidea): deep-sea spawning and post-spawning egg care. *Marine Biology*, **137**, 519–526.

Seibel, B. A., Robison, B. H. & Haddock, S. H. D. (2005). Post-spawning egg care by a squid. *Nature*, **438**, 929.

Seidou, M., Sugahara, M., Uchiyama, H. *et al.* (1990). On the three visual pigments in the retina of the firefly squid, *Watasenia scintillans*. *Journal of Comparative Physiology A*, **166**, 769–773.

Sereni, E. & Young, J. Z. (1932). Nervous degeneration and regeneration in cephalopods. *Pubblicazioni della Stazione Zoologica di Napoli*, **12**, 173–208.

Seyfarth, R. M. & Cheney, D. L. (2003). Signalers and receivers in animal communication. *Annual Review of Psychology*, **54**, 145–173.

Shashar, N. & Cronin, T. W. (1996). Polarization contrast vision in octopus. *Journal of Experimental Biology*, **199**, 999–1004.

Shashar, N. & Hanlon, R. T. (1997). Squids (*Loligo pealei* and *Euprymna scolopes*) can exhibit polarized light patterns produced by their skin. *Biological Bulletin*, **193**, 207–208.

Shashar, N. & Hanlon, R. T. (2013). Spawning behavior dynamics at communal egg beds in the squid *Doryteuthis* (*Loligo*) *pealeii*. *Journal of Experimental Marine Biology and Ecology*, **447**, 65–74.

Shashar, N., Hagan, R., Boal, J. G. & Hanlon, R. T. (2000). Cuttlefish use polarization sensitivity in predation on silvery fish. *Vision Research*, **40**, 71–75.

Shashar, N., Hanlon, R. T. & Petz, A. D. (1998). Polarization vision helps detect transparent prey. *Nature*, **393**, 222–223.

Shashar, N., Harosi, F. I., Banaszak, A. T. & Hanlon, R. T. (1998). UV radiation blocking compounds in the eye of the cuttlefish *Sepia officinalis*. *Biological Bulletin*, **195**, 187–188.

Shashar, N., Johnsen, S., Lerner, A. *et al.* (2011). Underwater linear polarization: physical limitations to biological functions. *Philosophical Transactions of the Royal Society B – Biological Sciences*, **366**, 649–654.

Shashar, N., Milbury, C. A. & Hanlon, R. T. (2002). Polarization vision in cephalopods: neuroanatomical and behavioral features that illustrate aspects of form and function. *Marine and Freshwater Behaviour and Physiology*, **35**, 57–68.

Shashar, N., Rutledge, P. S. & Cronin, T. W. (1996). Polarization vision in cuttlefish – a concealed communication channel? *Journal of Experimental Biology*, **199**, 2077–2084.

Shaw, E. (1978). Schooling fishes. *American Scientist*, **66**, 166–175.

Shaw, P. W. (1997). Polymorphic microsatellite markers in a cephalopod: the veined squid *Loligo forbesi*. *Molecular Ecology*, **6**, 297–298.

Shaw, P. W. (2002). Past, present and future applications of DNA-based markers in cephalopod biology: workshop report. *Bulletin of Marine Science*, **71**, 67–78.

Shaw, P. W. (2003). Polymorphic microsatellite DNA markers for the assessment of genetic diversity and paternity testing in the giant cuttlefish, *Sepia apama* (Cephalopoda). *Conservation Genetics*, **4**, 533–535.

Shaw, P. W. & Boyle, P. R. (1997). Multiple paternity within the brood of single females of *Loligo forbesi* (Cephalopoda: Loliginidae), demonstrated with microsatellite DNA markers. *Marine Ecology Progress Series*, **160**, 279–282.

Shaw, P. W. & Sauer, W. H. H. (2004). Multiple paternity and complex fertilisation dynamics in the squid *Loligo vulgaris reynaudii*. *Marine Ecology Progress Series*, **270**, 173–179.

Shaw, P. W., Arkhipkin, A. I., Adcock, G. J. et al. (2004). DNA markers indicate that distinct spawning cohorts and aggregations of Patagonian squid, *Loligo gahi*, do not represent genetically discrete subpopulations. *Marine Biology*, **144**, 961–970.

Shaw, P. W., Hendrickson, L., McKeown, N. J. et al. (2010). Discrete spawning aggregations of loliginid squid do not represent genetically distinct populations. *Marine Ecology Progress Series*, **408**, 117–127.

Shaw, P. W., Hendrickson, L., McKeown, N. J. et al. (2012). Population structure of the squid *Doryteuthis* (*Loligo*) *pealeii* on the eastern coast of the USA: reply to Gerlach et al. (2012). *Marine Ecology Progress Series*, 450, 285–287.

Shchetinnikov, A. S. (1992). Feeding spectrum of squid *Sthenoteuthis oualaniensis* (Oegopsida) in the Eastern Pacific. *Journal of the Marine Biological Association of the United Kingdom*, **72**, 849–860.

Shears, J. (1988). The use of a sand-coat in relation to feeding and diel activity in the sepiolid squid *Euprymna scolopes*. *Malacologia*, **29**, 121–133.

Shettleworth, S. J. (2010). *Cognition, Evolution and Behaviour*. Oxford: Oxford University Press.

Sheumack, D. D., Howden, M. E. H., Spence, I. & Quinn, R. J. (1978). Maculotoxin: a neurotoxin for the venom glands of the octopus *Hapalochlaena maculosa* identified as tetrodotoxin. *Science*, **199**, 188–189.

Shigeno, S., Sasaki, T., Moritaki, T. et al. (2008). Evolution of the cephalopod head complex by assembly of multiple molluscan body parts: evidence from *Nautilus* embryonic development. *Journal of Morphology*, **269**, 1–17.

Shohet, A., Baddeley, O., Anderson, J. & Osorio, D. (2007). Cuttlefish camouflage: a quantitative study of patterning. *Biological Journal of the Linnean Society*, **92**, 335–345.

Shohet, A. J., Baddeley, R. J., Anderson, J. C., Kelman, E. J. & Osorio, D. (2006). Cuttlefish response to visual orientation of substrates, water flow and a model of motion camouflage. *Journal of Experimental Biology*, **209**, 4717–4723.

Shomrat, T., Graindorge, N., Bellanger, C., Fiorito, G., Loewenstein, Y. & Hochner, B. (2011). Alternative sites of synaptic plasticity in two homologous 'fan-out fan-in' learning and memory networks. *Current Biology*, **21**, 1773 1782.

Shomrat, T., Zarrella, I., Fiorito, G. & Hochner, B. (2008). The octopus vertical lobe modulates short-term learning rate and uses LTP to acquire long-term memory. *Current Biology*, **18**, 337–342.

Shuster, S. M. & Wade, M. J. (2003). *Mating Systems and Strategies*. Princeton, NJ: Princeton University Press.

Sifner, S. K. & Vrgoc, N. (2004). Population structure, maturation and reproduction of the European squid, *Loligo vulgaris*, in the Central Adriatic Sea. *Fisheries Research*, **69**, 239–249.

Simmons, L. W. (2005). The evolution of polyandry: sperm competition, sperm selection, and offspring viability. *Annual Review of Ecology Evolution and Systematics*, **36**, 125–146.

Singley, C. T. (1982). Histochemistry and fine structure of the ectodermal epithelium of the sepiolid squid *Euprymna scolopes*. *Malacologia*, **23**, 177–192.

Singley, C. T. (1983). *Euprymna scolopes*, in *Cephalopod Life Cycles*, Vol. 1 (ed. P. R. Boyle), pp. 69–74. London: Academic Press.

Sinn, D. L. & Moltschaniwskyj, N. A. (2005). Personality traits in dumpling squid (*Euprymna tasmanica*):

context-specific traits and their correlation with biological characteristics. *Journal of Comparative Psychology*, **119**, 99–110.

Sinn, D. L., Apiolaza, L. A. & Moltschaniwskyj, N. A. (2006). Heritability and fitness-related consequences of squid personality traits. *Journal of Evolutionary Biology*, **19**, 1437–1447.

Sinn, D. L., Gosling, S. D. & Moltschaniwskyj, N. A. (2008). Development of shy/bold behaviour in squid: context-specific phenotypes associated with developmental plasticity. *Animal Behaviour*, **75**, 433–442.

Sinn, D. L., Moltschaniwskyj, N., Wapstra, E. & Dall, S. (2010). Are behavioral syndromes invariant? Spatiotemporal variation in shy/bold behavior in squid. *Behavioral Ecology and Sociobiology*, **64**, 693–702.

Sinn, D. L., Perrin, N. A., Mather, J. A. & Anderson, R. C. (2001). Early temperamental traits in an octopus (*Octopus bimaculoides*). *Journal of Comparative Psychology*, **115**, 351–364.

Skelhorn, J., Rowland, H. M. & Ruxton, G. D. (2010). The evolution and ecology of masquerade. *Biological Journal of the Linnean Society*, **99**, 1–8.

Skelhorn, J., Rowland, H. M., Speed, M. P. & Ruxton, G. D. (2010). Masquerade: camouflage without crypsis. *Science*, **327**, 51.

Slater, P. J. B. (1983). The study of communication, in *Communication* (ed. T. R. Halliday & P. J. B. Slater). Oxford: Blackwell Scientific Publications.

Smale, M. J. (1996). Cephalopods as prey. 4. Fishes. *Philosophical Transactions of the Royal Society of London Series B – Biological Sciences*, **351**, 1067–1081.

Smale, M. J., Sauer, W. H. H. & Hanlon, R. T. (1995). Attempted ambush predation on spawning squids *Loligo vulgaris reynaudii* by benthic pyjama sharks, *Poroderma africanum*, off South Africa. *Journal of the Marine Biological Association of the United Kingdom*, **75**, 739–742.

Smith, R. L. (1984). *Sperm Competition and the Evolution of Animal Mating Systems*. Orlando: Academic Press.

Smith, W. J. (1997). The behavior of communicating, after twenty years, in *Communication* (ed. D. H. Owings, M. D. Beecher & N. S. Thompson). New York: Plenum Press.

Sneddon, L. U. (2009). Pain perception in fish: indicators and endpoints. *Ilar Journal*, **50**, 338–342.

Soeda, J. (1965). Migration of the 'Surume' squid *Ommastrephes sloani pacificus* (Steenstrup), in the coastal waters of Japan. *Fisheries Research Board of Canada*, **533**, 1–38.

Soucier, C. P. & Basil, J. A. (2008). Chambered nautilus (*Nautilus pompilius pompilius*) responds to underwater vibrations. *Americal Malacological Bulletin*, **24**, 3–11.

Squires, H. J. (1957). Squid, *Illex illecebrosus* (Lesueur), in the Newfoundland fishing area. *Journal of the Fisheries Research Board of Canada*, **14**, 693–728.

Squires, Z. E., Norman, M. D. & Stuart-Fox, D. (2013). Mating behaviour and general spawning patterns of the southern dumpling squid *Euprymna tasmanica* (Sepiolidae): a laboratory study. *Journal of Molluscan Studies*, **79**, 263–269.

Squires, Z. E., Wong, B. B. M., Norman, M. D. & Stuart-Fox, D. (2012). Multiple fitness benefits of polyandry in a cephalopod. *PLoS One*, **7**, e37074.

Staaf, D. J., Camarillo-Coop, S., Haddock, S. H. D. *et al.* (2008). Natural egg mass deposition by the Humboldt squid (*Dosidicus gigas*) in the Gulf of California and characteristics of hatchlings and paralarvae. *Journal of the Marine Biological Association of the United Kingdom*, **88**, 759–770.

Staples, J. F., Webber, D. M. & Boutilier, R. G. (2003). Environmental hypoxia does not constrain the diurnal depth distribution of free-swimming *Nautilus pompilius*. *Physiological and Biochemical Zoology*, **76**, 644–651.

Stark, K. E., Jackson, G. D. & Lyle, J. M. (2005). Tracking arrow squid movements with an automated acoustic telemetry system. *Marine Ecology Progress Series*, **299**, 167–177.

Staudinger, M. D. (2006). Seasonal and size-based predation on two species of squid by four fish predators on the Northwest Atlantic continental shelf. *Fishery Bulletin*, **104**, 605–615.

Staudinger, M. D. & Juanes, F. (2010a). Feeding tactics of a behaviorally plastic predator, summer flounder (*Paralichthys dentatus*). *Journal of Sea Research*, **64**, 68–75.

Staudinger, M. D. & Juanes, F. (2010b). Size-dependent susceptibility of longfin inshore squid (*Loligo pealeii*) to attack and capture by two predators. *Journal of Experimental Marine Biology and Ecology*, **393**, 106–113.

Staudinger, M. D., Buresch, K. C., Mathger, L. M. *et al.* (2013a). Defensive responses of cuttlefish to different teleost predators. *Biological Bulletin*, **225**, 161–174.

Staudinger, M. D., Hanlon, R. T. & Juanes, F. (2011). Primary and secondary defenses of squid to cruising and ambush fish predators: variable tactics and their survival value. *Animal Behaviour*, **81**, 585–594.

Staudinger, M. D., Juanes, F., Salmon, B. & Teffer, A. K. (2013b). The distribution, diversity, and importance of cephalopods in top predator diets from offshore habitats of the Northwest Atlantic Ocean. *Deep-Sea Research Part II –Topical Studies in Oceanography*, **95**, 182–192.

Stearns, S. C. (1992). *The Evolution of Life Histories*. Oxford: Oxford University Press.

Stella, M. P. & Kier, W. M. (2004). The morphology and mechanics of octopus arms: inspiration for novel robotics. *Integrative and Comparative Biology*, **44**, 645–645.

Stephens, P. R. & Young, J. Z. (1982). The statocyst of the squid *Loligo*. *Journal of Zoology (London)*, **197**, 241–266.

Stevens, M. (2005). The role of eyespots as anti-predator mechanisms, principally demonstrated in the Lepidoptera. *Biological Reviews*, **80**, 573–588.

Stevens, M. (2013). *Sensory Ecology, Behaviour and Evolution*. Oxford: Oxford University Press.

Stevens, M. & Cuthill, I. C. (2006). Disruptive coloration, crypsis and edge detection in early visual processing. *Proceedings of the Royal Society B – Biological Sciences* **273**, 2141–2147.

Stevens, M. & Merilaita, S. (2009a). Animal camouflage: current issues and new perspectives. *Philosophical Transactions of the Royal Society of London B: Biological Sciences*, **364**, 423–557.

Stevens, M. & Merilaita, S. (2009b). Defining disruptive coloration and distinguishing its functions. *Philosophical Transactions of the Royal Society of London B: Biological Sciences*, **364**, 481–488.

Stevens, M. & Merilaita, S. (ed.) (2011a). *Animal Camouflage: Mechanisms and Function*. Cambridge: Cambridge University Press.

Stevens, M. & Merilaita, S. (2011b). Animal camouflage: function and mechanisms, in *Animal Camouflage: Mechanisms and Function* (ed. M. Stevens & S. Merilaita), pp. 1–16. Cambridge: Cambridge University Press.

Stevens, M. & Ruxton, G. D. (2014). Do animal eyespots really mimic eyes? *Current Zoology*, **60**, 26–36.

Stevens, M., Cuthill, I. C., Windsor, A. M. M. & Walker, H. J. (2006). Disruptive contrast in animal camouflage. *Proceedings of the Royal Society B: Biological Sciences*, **273**, 2433–2438.

Stevens, M., Graham, J., Winney, I. S. & Cantor, A. (2008). Testing Thayer's hypothesis: can camouflage work by distraction? *Biology Letters*, **4**, 648–650.

Stevens, M., Marshall, K. L. A., Troscianko, J. *et al.* (2013). Revealed by conspicuousness: distractive markings reduce camouflage. *Behavioral Ecology*, **24**, 213–222.

Stevens, M., Stubbins, C. L. & Hardman, C. J. (2008). The anti-predator function of 'eyespots' on camouflaged and conspicuous prey. *Behavioral Ecology and Sociobiology*, **62**, 1787–1793.

Stevens, M., Winney, I. S., Cantor, A. & Graham, J. (2009). Outline and surface disruption in animal camouflage. *Proceedings of the Royal Society B – Biological Sciences*, **276**, 781–786.

Stevenson, J. A. (1934). On the behaviour of the long-finned squid (*Loligo pealii* (Lesueur)). *Canadian Field Naturalist*, **48**, 4–7.

Stewart, J. S., Field, J. C., Markaida, U. & Gilly, W. F. (2013). Behavioral ecology of jumbo squid (*Dosidicus gigas*) in relation to oxygen minimum zones. *Deep-Sea Research Part II – Topical Studies in Oceanography*, **95**, 197–208.

Stewart, J. S., Hazen, E. L., Foley, D. G., Bograd, S. J. & Gilly, W. F. (2012). Marine predator migration during range expansion: Humboldt squid *Dosidicus gigas* in the northern California Current System. *Marine Ecology Progress Series*, **471**, 135–150.

Stillwell, C. E. & Kohler, N. E. (1985). Food and feeding ecology of the swordfish *Xiphias gladius* in the western north-Atlantic Ocean with estimates of daily ration. *Marine Ecology Progress Series*, **22**, 239–247.

Strand, S. (1988). Following behavior: interspecific foraging associations among Gulf of California reef fishes. *Copeia*, **1988**, 351–357.

Strugnell, J., Jackson, J., Drummond, A. J. & Cooper, A. (2006). Divergence time estimates for major cephalopod groups: evidence from multiple genes. *Cladistics*, **22**, 89–96.

Strugnell, J., Norman, M., Jackson, J., Drummond, A. J. & Cooper, A. (2005). Molecular phylogeny of coleoid cephalopods (Mollusca: Cephalopoda) using a multigene approach; the effect of data partitioning on resolving phylogenies in a Bayesian framework. *Molecular Phylogenetics and Evolution*, **37**, 426–441.

Stürmer, W. (1985). A small coleoid cephalopod with soft parts from the Lower Devonian discovered using radiography. *Nature*, **318**, 53–55.

Suboski, M. D., Muir, D. & Hall, D. (1993). Social learning in invertebrates. *Science*, **259**, 1628–1629.

Sugawara, K., Katagiri, Y. & Tomita, T. (1971). Polarized light respones from octopus single reticula cells. *Journal of the Faculty of Sciences, Hokkaido University*, **17**, 581–586.

Sugimoto, C., Yanagisawa, R., Nakajima, R. & Ikeda, Y. (2013). Observations of schooling behaviour in the oval squid *Sepioteuthis lessoniana* in coastal waters of Okinawa Island. *Marine Biodiversity Records*, **6**, e34.

Sumbre, G., Fiorito, G., Flash, T. & Hochner, B. (2005). Neurobiology: motor control of flexible octopus arms. *Nature*, **433**, 595–596.

Sumbre, G., Fiorito, G., Flash, T. & Hochner, B. (2006). Octopuses use a human-like strategy to control precise point-to-point arm movements. *Current Biology*, **16**, 767–772.

Sumbre, G., Gutfreund, Y., Fiorito, G., Flash, T. & Hochner, B. (2001). Control of octopus arm extension by a peripheral motor program. *Science*, **293**, 1845–1848.

Summers, W. C. (1983). *Loligo pealei*, in *Cephalopod Life Cycles*, Vol. I: *Species Accounts* (ed. P. R. Boyle), New York: Academic Press, pp. 115–142.

Summers, W. C. (1985). Comparative life history adaptations of some myopsid and sepiolid squids. *NAFO Scientific Council Studies*, **9**, 139–142.

Sundermann, G. (1983). The fine structure of epidermal lines on arms and head of postembryonic *Sepia officinalis* and *Loligo vulgaris* (Mollusca, Cephalopoda). *Cell and Tissue Research*, **232**, 669–677.

Sundermann, G. (1990). Development and hatching state of ectodermal vesicle organs in the head of *Sepia officinalis*, *Loligo vulgaris* and *Loligo forbesi*

(Cephalopoda, Decabrachia). *Zoomorphology*, **109**, 343–352.

Sutherland, N. S. (1957a). Visual discrimination of orientation by *Octopus*. *British Journal of Psychology*, **48**, 55–71.

Sutherland, N. S. (1957b). Visual discrimination of orientation and shape by the octopus. *Nature*, **179**, 11–13.

Sutherland, N. S. (1958). Visual discrimination of shape by *Octopus*. Squares and triangles. *Quarterly Journal of Experimental Psychology*, **10**, 40–47.

Sutherland, N. S. (1959). A test of a theory of shape discrimination in *Octopus vulgaris* Lamarck. *Journal of Comparative and Physiological Psychology*, **52**, 135–141.

Sutherland, N. S. (1960). Theories of shape discrimination in *Octopus*. *Nature*, **186**, 840–844.

Sutherland, N. S. (1962). Visual discrimination of shape by *Octopus*: squares and crosses. *Journal of Comparative and Physiological Psychology*, **55**, 939–943.

Sutherland, N. S. (1963). Shape discrimination and receptive fields. *Nature*, **197**, 118–122.

Sutherland, N. S. (1968). Outlines of a theory of visual pattern recognition in animals and man. *Proceedings of the Royal Society of London B*, **171**, 297–317.

Sutherland, N. S. (1969). Shape discrimination in rat, octopus and goldfish: a comparative study. *Journal of Comparative and Physiological Psychology*, **67**, 160–176.

Sutherland, N. S. & Carr, A. E. (1963). The visual discrimination of shape by *Octopus*: the effects of stimulus size. *Quarterly Journal of Experimental Psychology*, **15**, 225–235.

Sutherland, N. S. & Mackintosh, N. J. (1971). *Mechanisms of Animal Discrimination Learning*. New York: Academic Press.

Sutherland, N. S. & Muntz, W. R. A. (1959). Simultaneous discrimination training and preferred directions of motion in visual discrimination of shape in *Octopus vulgaris* Lamarck. *Pubblicazioni della Stazione Zoologica di Napoli*, **31**, 109–126.

Sutherland, N. S., Mackintosh, N. J. & Mackintosh, J. (1963). Simultaneous discrimination training in *Octopus* and transfer of a discrimination along a continuum. *Journal of Comparative and Physiological Psychology*, **56**, 150–156.

Sutherland, N. S., Mackintosh, N. J. & Mackintosh, J. (1965). Shape and size discrimination in *Octopus*: the effects of pretraining along different dimensions. *Journal of Genetic Psychology*, **107**, 1–10.

Sutherland, R. L., Mäthger, L. M., Hanlon, R. T., Urbas, A. M. & Stone, M. O. (2008a). Cephalopod coloration model. I. Squid chromatophores and iridophores. *Journal of the Optical Society of America A*, **25**, 588–599.

Sutherland, R. L., Mäthger, L. M., Hanlon, R. T., Urbas, A. M. & Stone, M. O. (2008b). Cephalopod coloration model. II. Multiple layer skin effects. *Journal of the Optical Society of America A*, **25**, 2044–2054.

Suzuki, M., Kimura, T., Ogawa, H., Hotta, K. & Oka, K. (2011). Chromatophore activity during natural pattern expression by the squid *Sepioteuthis lessoniana*: contributions of miniature oscillation. *PLoS ONE*, **6**, e18244.

Sweeney, A. M., Haddock, S. H. D. & Johnsen, S. (2007). Comparative visual acuity of coleoid cephalopods. *Integrative and Comparative Biology*, **47**, 808–814.

Sweeney, M. J., Roper, C. F. E., Mangold, K. M., Clarke, M. R. & Boletzky, S. von (1992). Larval and juvenile cephalopods: a manual for their identification. *Smithsonian Contributions to Zoology*, **513**, 1–282.

Sykes, A. V., Pereira, D., Rodriguez, C., Lorenzo, A. & Andrade, J. P. (2013). Effects of increased tank bottom areas on cuttlefish (*Sepia officinalis*, L.) reproduction performance. *Aquaculture Research*, **44**, 1017–1028.

Takami, T. & Suzu-Uchi, T. (1993). Southward migration of the Japanese common squid (*Todarodes pacificus*) from northern Japanese waters, in *Recent Advances in Cephalopod Fisheries Biology* (ed. T. Okutani, R. K. O'Dor & T. Kubodera), pp. 537–543. Tokyo: Tokai University Press.

Taki, I. (1941). On keeping octopods in an aquarium for physiological experiments, with remarks on some operative techniques. *Venus: Japanese Journal of Malacology*, **10**, 140–156.

Talbot, C. M. & Marshall, J. (2010a). Polarization sensitivity in two species of cuttlefish – *Sepia plangon* (Gray 1849) and *Sepia mestus* (Gray 1849) – demonstrated with polarized optomotor stimuli. *Journal of Experimental Biology*, **213**, 3364–3370.

Talbot, C. M. & Marshall, J. (2010b). Polarization sensitivity and retinal topography of the striped pyjama squid (*Sepioloidea lineolata* – Quoy/Gaimard 1832). *Journal of Experimental Biology*, **213**, 3371–3377.

Talbot, C. M. & Marshall, J. N. (2011). The retinal topography of three species of coleoid cephalopod: significance for perception of polarized light. *Philosophical Transactions of the Royal Society B – Biological Sciences*, **366**, 724–733.

Tansley, K. (1965). *Vision in Vertebrates*. London: Chapman & Hall.

Tao, A. R., DeMartini, D. G., Izumi, M. *et al.* (2010). The role of protein assembly in dynamically tunable bio-optical tissues. *Biomaterials*, **31**, 793–801.

Tardent, P. (1962). Keeping *Loligo vulgaris* L. in the Naples aquarium. *1st Congres International d'Aquariologie*, A, 41–46.

Tasaki, K. & Karita, K. (1966). Intraretinal discrimination of horizontal and vertical planes of polarized light by *Octopus*. *Nature*, **209**, 934.

Taylor, M. A. (1986). Stunning whales and deaf squids. *Nature*, **323**, 298–299.

Taylor, P. B. & Chen, L. C. (1969). The predator–prey relationship between the octopus (*Octopus bimaculatus*) and the California scorpionfish (*Scorpaena guttata*). *Pacific Science*, **23**, 311–316.

Teichert, C. (1988). Main features of cephalopod evolution, in *The Mollusca*, Vol. 12: *Paleontology and Neontology of Cephalopods*. (ed. M. R. Clarke & E. R. Trueman), pp. 11–79. San Diego, CA: Academic Press.

Temple, S. E., Pignatelli, V., Cook, T. *et al.* (2012). High-resolution polarisation vision in a cuttlefish. *Current Biology*, **22**, R121–R122.

Terrace, H. S., Petitto, L. A., Sanders, R. J. & Bever, T. G. (1979). Can an ape create a sentence? *Science*, **200**, 891–902.

Thayer, G. H. (1909). *Concealing-Coloration in the Animal Kingdom. An Exposition of the Laws of Disguise through Color and Pattern: Being a Summary of Abbott H. Thayer's Discoveries*. New York: The Macmillan Company.

Thomas, R. F. (1977). Systematics, distribution, and biology of cephalopods of the genus *Tremoctopus* (Octopoda: Tremoctopodidae). *Bulletin of Marine Science*, **27**, 353–392.

Thompson, J. T. & Voight, J. R. (2003). Erectile tissue in an invertebrate animal: the *Octopus* copulatory organ. *Journal of Zoology*, **261**, 101–108.

Thorpe, W. H. (1963). *Learning and Instinct in Animals*. London: Methuen.

Tinbergen, L. (1939). Zur Fortpflanzungsethologie von *Sepia officinalis* L. *Archives Néerlandaises de Zoologie*, **3**, 323–364.

Tinbergen, L. (1960). The natural control of insects in pinewoods. I. Factors influencing the intensity of predation by songbirds. *Archives Néerlandaises Zoologie*, **13**, 265–343.

Tinbergen, N. (1951). *The Study of Instinct*. Oxford: Clarendon Press.

Tinbergen, N. (1959). Comparative studies of the behaviour of gulls (Laridae): a progress report. *Behaviour*, **15**, 1–70.

Tinbergen, N. (1963). On aims and methods of ethology. *Zeitschrift für Tierpsychologie*, **20**, 410–433.

Toll, R. B. (1983). The lycoteuthid genus *Oregoniateuthis* Voss, 1956, a synonym of *Lycoteuthis* Pfeffer, 1900 (Cephalopoda: Teuthoidea). *Proceedings of the Biological Society of Washington*, **96**, 365–369.

Toll, R. B. & Hess, S. C. (1981). Cephalopods in the diet of the swordfish, *Xiphias gladius*, from the Florida Straits. *Fishery Bulletin*, **79**, 765–774.

Tollrian, R. & Harvell, C. D. (ed.) (1999). *The Ecology and Evolution of Inducible Defenses*. Princeton, NJ: Princeton University Press.

Tong, D., Rozas, N. S., Oakley, T. H. *et al.* (2009). Evidence for light perception in a bioluminescent organ. *Proceedings of the National Academy of Sciences USA*, **106**, 9386–9841.

Tranter, D. J. & Augustine, O. (1973). Observations on the life history of the blue-ringed octopus *Hapalochlaena maculosa*. *Marine Biology*, **18**, 115–128.

Tricarico, E., Borrelli, L., Gherardi, F. & Fiorito, G. (2011). I know my neighbour: individual recognition in *Octopus vulgaris*. *PLoS ONE*, **6**.

Troscianko, T., Benton, C. P., Lovell, P. G., Tolhurst, D. J. & Pizlo, Z. (2009). Camouflage and visual perception. *Philosophical Transactions of the Royal Society B – Biological Sciences*, **364**, 449–461.

Trueman, E. R. & Packard, A. (1968). Motor performances of some cephalopods. *Journal of Experimental Biology*, **49**, 495–507.

Tsuchiya, K. & Uzu, T. (1997). Sneaker male in octopus. *Venus: Japanese Journal of Malacology*, **56**, 177–181.

Tublitz, N. J., Gaston, M. R. & Loi, P. K. (2006). Neural regulation of a complex behavior: body patterning in cephalopod molluscs. *Integrative and Comparative Biology*, **46**, 880–889.

Tucker, J. K. & Mapes, R. H. (1978). Possible predation on *Nautilus pompilius*. *Veliger*, **21**, 95–98.

Tyrie, E. K., Hanlon, R. T., Siemann, L. A. & Uyarra, M. C. (2015). Coral reef flounders, *Bothus lunatus*, choose substrates on which they can achieve camouflage with their limited body pattern repertoire. *Biological Journal of the Linnean Society*, **114**, 629–638.

Uchiyama, K. & Tanabe, K. (1999). Hatching of *Nautilus macromphalus* in the Toba aquarium, Japan, in *Advancing Research on Living and Fossil Cephalopods* (ed. F. Olóriz & F. J. Rodríguez-Tovar), pp. 11–16. New York: Kluwer Academic.

Ueda, K. (1985). Studies on the growth, maturation and migration of the shiriyake-ika, *Sepiella japonica* Sasaki. *Bulletin of the Nansei Regional Fisheries Research Laboratory*, **19**, 1–42.

Uexküll, J. von (1905). *Leitfaden in das Studium des Experimentale Biologie der Wassertiere*. Wiesbaden: J.F. Bergmann.

Ulmer, K. M., Buresch, K. C., Kossodo, M. M. *et al.* (2013). Vertical visual features have a strong influence on cuttlefish camouflage. *The Biological Bulletin*, **224**, 110–118.

Vallin, A., Dimitrova, M., Kodandaramaiah, U. & Merilaita, S. (2011). Deflective effect and the effect of prey detectability on anti-predator function of eyespots. *Behavioral Ecology and Sociobiology*, **65**, 1629–1636.

van Camp, L. M., Donnellan, S. C., Dyer, A. R. & Fairweather, P. G. (2004). Multiple paternity in field- and captive-laid egg strands of *Sepioteuthis australis* (Cephalopoda: Loliginidae). *Marine and Freshwater Research*, **55**, 819–823.

van Camp, L. M., Fairweather, P. G., Steer, M. A., Donnellan, S. C. & Havenhand, J. N. (2005). Linking male and female morphology to reproductive success in captive southern calamary (*Sepioteuthis australis*). *Marine and Freshwater Research*, **56**, 933–941.

Van Heukelem, W. F. (1966). Some aspects of the ecology and ethology of *Octopus cyanea* Gray. M.S. Thesis, Honolulu, HI: University of Hawaii.

Van Heukelem, W. F. (1977). Laboratory maintenance, breeding, rearing and biomedical research potential of the Yucatan octopus (*Octopus maya*). *Laboratory Animal Science*, **27**, 852–859.

Van Heukelem, W. F. (1983a). *Octopus cyanea*, in *Cephalopod Life Cycles,* Vol. 1. (ed. P. R. Boyle), pp. 267–276. London: Academic Press.

Van Heukelem, W. F. (1983b). *Octopus maya*, in *Cephalopod Life Cycles,* Vol. 1 (ed. P. R. Boyle), pp. 311–323. London: Academic Press.

van Staaden, M. J., Searcy, W. A. & Hanlon, R. T. (2011). Signaling aggression, in *Advances in Genetics* (ed. R. Huber, D. L. Bannasch & P. Brennan), pp. 23–49. Elsevier, Inc.

Vecchione, M. (1987). Juvenile ecology, in *Cephalopod Life Cycles,* Vol. 2: *Comparative Reviews* (ed. P. R. Boyle), pp. 61–84. London: Academic Press.

Vecchione, M. (1988). *In-situ* observations on a large squid-spawning bed in the eastern Gulf of Mexico. *Malacologia*, **29**, 135–141.

Vecchione, M. & Roper, C. F. E. (1991). Cephalopods observed from submersibles in the western North Atlantic. *Bulletin of Marine Science*, **49**, 433–445.

Vecchione, M. & Young, R. E. (2006). The squid family Magnapinnidae (Mollusca: Cephalopoda) in the Atlantic Ocean, with a description of a new species. *Proceedings of the Biological Society of Washington*, **119**, 365–372.

Vecchione, M., Roper, C. F. E., Widder, E. A. & Frank, T. M. (2002). *In situ* observations on three species of large-finned deep-sea squids. *Bulletin of Marine Science*, **71**, 893–901.

Vecchione, M., Young, R. E., Guerra, A. *et al.* (2001). Worldwide observations of remarkable deep-sea squids. *Science*, **294**, 2505–2506.

Vermeij, G. J. (1987). *Evolution and Escalation: An Ecological History of Life*. Princeton: Princeton University Press.

Verne, J. (1869). *Twenty Thousand Leagues under the Sea*. Paris.

Verrill, A. E. (1880–81). The cephalopods of the northeastern coast of America. Part II. The smaller cephalopods, including the 'squids' and the octopi, with other allied forms. *Transactions of the Connecticut Academy of Sciences*, **5**, 259–446.

Verwoerd, D. J. (1987). Observations on the food and status of the Cape Clawless Otter *Aonyx capensis* at Betty's Bay, South Africa. *South African Journal of Zoology*, **22**, 33–39.

Villanueva, R. (1992). Deep-sea cephalopods of the north-western Mediterranean: indications of up-slope ontogenetic migration in two bathybenthic species. *Journal of Zoology (London)*, **227**, 267–276.

Villanueva, R. (1994). Decapod crab zoeae as food for rearing cephalopod paralarvae. *Aquaculture*, **128**, 143–152.

Villanueva, R. & Guerra, A. (1991). Food and prey detection in two deep-sea cephalopods: *Opisthoteuthis agassizi* and *O. vossi* (Octopoda: Cirrata). *Bulletin of Marine Science*, **49**, 288–299.

Vincent, T. L. S., Scheel, D. & Hough, K. R. (1998). Some aspects of diet and foraging behavior of *Octopus dofleini* (Wülker, 1910) in its northernmost range. *Marine Ecology*, **19**, 13–29.

Vitti, J. J. (2013). Cephalopod cognition in an evolutionary context: implications for ethology. *Biosemiotics*, **6**, 393–401.

Voight, J. R. (1991a). Ligula length and courtship in *Octopus digueti*: a potential mechanism of mate choice. *Evolution*, **45**, 1726–1730.

Voight, J. R. (1991b). Enlarged suckers as an indicator of male maturity in *Octopus*. *Bulletin of Marine Science*, **49**, 98–106.

Voight, J. R. (1992). Movement, injuries and growth of members of a natural population of the Pacific pygmy octopus, *Octopus digueti*. *Journal of Zoology (London)*, **228**, 247–263.

Voight, J. R. (1996). The hectocotylus and other reproductive structures of *Berryteuthis magister* (Teuthoidea: Gonatidae). *Veliger*, **39**, 117–124.

Voight, J. R. (2000). A deep-sea octopus (*Graneledone* cf. *boreopacifica*) as a shell-crushing hydrothermal vent predator. *Journal of Zoology*, **252**, 335–341.

Voight, J. R. (2001). The relationship between sperm reservoir and spermatophore length in benthic octopuses (Cephalopoda: Octopodidae). *Journal of the Marine Biological Association of the United Kingdom*, **81**, 983–986.

Voight, J. R. (2005). Hydrothermal vent octopuses of *Vulcanoctopus hydrothermalis*, feed on bathypelagic amphipods of *Halice hesmonectes*. *Journal of the*

Marine Biological Association of the United Kingdom, **85**, 985–988.

Voight, J. R. (2008). Observations of deep-sea octopodid behavior from undersea vehicles. *American Malacological Bulletin*, **24**, 43–50.

Voight, J. R. & Feldheim, K. A. 2009 Microsatellite inheritance and multiple paternity in the deep-sea octopus *Graneledone boreopacifica* (Mollusca: Cephalopoda). *Invertebrate Biology* 128, 26–30.

Voight, J. R., Portner, H. O. & O'Dor, R. K. (1994). A review of ammonia-mediated buoyancy in squids (Cephalopoda: Teuthoidea). *Marine and Freshwater Behaviour and Physiology*, **25**, 193–203.

von Byern, J. & Klepal, W. (2006). Adhesive mechanisms in cephalopods: a review. *Biofouling*, **22**, 329–338.

von Byern, J., Scott, R., Griffiths, C. *et al.* (2011). Characterization of the adhesive areas in *Sepia tuberculata* (Mollusca, Cephalopoda). *Journal of Morphology*, **272**, 1245–1258.

von Byern, J., Wani, R., Schwaha, T., Grunwald, I. & Cyran, N. (2012). Old and sticky – adhesive mechanisms in the living fossil *Nautilus pompilius* (Mollusca, Cephalopoda). *Zoology*, **115**, 1–11.

Voss, G. L. (1956). A review of the cephalopods of the Gulf of Mexico. *Bulletin of Marine Science*, **6**, 85–178.

Voss, N. A. (1980). A generic revision of the Cranchiidae (Cephalopoda, Oegopsida). *Bulletin of Marine Science*, **30**, 365–412.

Vovk, A. N. (1974). Feeding habits of the North American squid *Loligo pealei* Lesueur. *Fisheries and Marine Service, Translation Series*, **3304**, 1–14.

Vovk, A. N. (1977). The position of the longfin squid *Loligo pealei* Les. in the ecosystem. *Fisheries and Marine Service, Translation Series*, **3977**, 1–13.

Vovk, A. N. & Khvichiya, L. A. (1980). On feeding of long-finned squid (*Loligo pealei*) juveniles in Subareas 5 and 6. *North Atlantic Fisheries Organization Scientific Council Research Document*, **80/VI/50, N087**, 1–9.

Wada, T., Takegaki, T., Mori, T. & Natsukari, Y. (2005a). Alternative male mating behaviors dependent on relative body size in captive oval squid *Sepioteuthis lessoniana* (Cephalopoda, Loliginidae). *Zoological Science*, **22**, 645–651.

Wada, T., Takegaki, T., Mori, T. & Natsukari, Y. (2005b). Sperm displacement behavior of the cuttlefish *Sepia esculenta* (Cephalopoda: Sepiidae). *Journal of Ethology*, **23**, 85–92.

Wada, T., Takegaki, T., Mori, T. & Natsukari, Y. (2006). Reproductive behavior of the Japanese spineless cuttlefish *Sepiella japonica*. *Venus*, **65**, 221–228.

Wada, T., Takegaki, T., Mori, T. & Natsukari, Y. (2010). Sperm removal, ejaculation and their behavioural interaction in male cuttlefish in response to female mating history. *Animal Behaviour*, **79**, 613–619.

Walderon, M. D., Nolt, K. J., Haas, R. E. *et al.* (2011). Distance chemoreception and the detection of conspecifics in *Octopus bimaculoides*. *Journal of Molluscan Studies*, **77**, 309–311.

Walker, J. J., Longo, N. & Bitterman, M. E. (1970). The octopus in the laboratory. Handling, maintenance, and training. *Behavior Research Methods and Instrumentation*, **2**, 15–18.

Waller, R. A. & Wicklund, R. I. (1968). Observations from a research submersible – mating and spawning of the squid, *Doryteuthis plei*. *Bioscience*, **18**, 110–111.

Walton, S. A., Korn, D. & Klug, C. (2010). Size distribution of the Late Devonian ammonoid *Prolobites*: indication for possible mass spawning events. *Swiss Journal of Geosciences*, **103**, 475–494.

Ward, A. J. W., Herbert-Read, J. E., Sumpter, D. J. T. & Krause, J. (2011). Fast and accurate decisions through collective vigilance in fish shoals. *Proceedings of the National Academy of Sciences of the United States of America*, **108**, 2312–2315.

Ward, D. V. & Wainwright, S. A. (1972). Locomotory aspects of squid mantle structure. *Journal of Zoology (London)*, **167**, 437–449.

Ward, P. D. (1987). *The Natural History of Nautilus*. London: Allen & Unwin.

Ward, P. D. & Bandel, K. (1987). Life history strategies in fossil cephalopods, in *Cephalopod Life Cycles*, Vol. 2: *Comparative Reviews* (ed. P. R. Boyle), pp. 329–350. London: Academic Press.

Ward, P. D. & Martin, A. W. (1980). Depth distribution of *Nautilus pompilius* in Fiji and *Nautilus macromphalus* in New Caledonia. *The Veliger*, **22**, 259–264.

Ward, P. D. & Saunders, W. B. (1997). *Allonautilus*: a new genus of living nautiloid cephalopod and its bearing on phylogeny of the Nautilida. *Journal of Paleontology*, **71**, 1054–1064.

Ward, P. D. & Wicksten, M. K. (1980). Food sources and feeding behavior of *Nautilus macromphalus*. *Veliger*, **23**, 119–124.

Ward, P. D., Carlson, B., Weekly, M. & Brumbaugh, B. (1984). Remote telemetry of daily vertical and horizontal movement of *Nautilus* in Palau. *Nature*, **309**, 248–252.

Wardill, T. J., Gonzalez-Bellido, P. T., Crook, R. J. & Hanlon, R. T. (2012). Neural control of tuneable skin iridescence in squid. *Proceedings of the Royal Society B – Biological Sciences*, **279**, 4243–4252.

Warnke, K. (1994). Some aspects of social interactions during feeding in *Sepia officinalis* (Mollusca) hatched and reared in the laboratory. *Vie et Milieu*, **44**, 125–131.

Warnke, K. M., Kaiser, R. & Hasselmann, M. (2012). First observations of a snail-like body pattern in juvenile *Sepia bandensis* (Cephalopoda: Sepiidae). A note. *Neues Jahrbuch für Geologie und Palaontologie-Abhandlungen*, **266**, 51–57.

Warnke, K. M., Meyer, A., Ebner, B. & Lieb, B. (2011). Assessing divergence time of Spirulida and Sepiida (Cephalopoda) based on hemocyanin sequences. *Molecular Phylogenetics and Evolution*, **58**, 390–394.

Warrant, E. (2004). Vision in the dimmest habitats on Earth. *Journal of Comparative Physiology A – Neuroethology Sensory Neural and Behavioral Physiology*, **190**, 765–789.

Warrant, E. J. (2007). Visual ecology: hiding in the dark. *Current Biology*, **17**, R209–R211.

Warren, L. R., Scheier, M. F. & Riley, D. A. (1974). Colour changes of *Octopus rubescens* during attacks on unconditioned and conditioned stimuli. *Animal Behaviour*, **22**, 211–219.

Wegener, B. J., Stuart-Fox, D. M., Norman, M. D. & Wong, B. B. M. (2013a). Strategic male mate choice minimizes ejaculate consumption. *Behavioral Ecology*, **24**, 668–671.

Wegener, B. J., Stuart-Fox, D., Norman, M. D. & Wong, B. B. M. (2013b). Spermatophore consumption in a cephalopod. *Biology Letters*, **9**, DOI: 10.1098/rsbl.2013.0192.

Weihs, D. & Moser, H. G. (1981). Stalked eyes as an adaptation towards more efficient foraging in marine fish larvae. *Bulletin of Marine Science*, **31**, 31–36.

Wells, M. J. (1958). Factors affecting reactions to *Mysis* by newly hatched *Sepia*. *Behaviour*, **13**, 96–111.

Wells, M. J. (1959a). A touch learning centre in *Octopus*. *Journal of Experimental Biology*, **36**, 590–612.

Wells, M. J. (1959b). Functional evidence for neurone fields representing the individual arms within the central nervous system of *Octopus*. *Journal of Experimental Biology*, **36**, 501–511.

Wells, M. J. (1960). Proprioception and visual discrimination of orientation in *Octopus*. *Journal of Experimental Biology*, **37**, 489–499.

Wells, M. J. (1961). Weight discrimination by *Octopus*. *Journal of Experimental Biology*, **38**, 127–133.

Wells, M. J. (1962a). *Brain and Behaviour in Cephalopods*. London: Heinemann.

Wells, M. J. (1962b). Early learning in *Sepia*. *Symposia of the Zoological Society of London*, **8**, 149–169.

Wells, M. J. (1963). Taste by touch: some experiments with *Octopus*. *Journal of Experimental Biology*, **40**, 187–193.

Wells, M. J. (1964). Detour experiments with octopuses. *Journal of Experimental Biology*, **41**, 621.

Wells, M. J. (1966). Learning in the *Octopus*. *Symposia of the Society for Experimental Biology*, **20**, 477–507.

Wells, M. J. (1967a). Sensitization and the evolution of associative learning, in *Symposium on Neurobiology of Invertebrates, 1967*, pp. 391–411. Budapest: Hungarian Academy of Sciences.

Wells, M. J. (1967b). Short-term learning and interocular transfer in detour experiments with octopuses. *Journal of Experimental Biology*, 47, 393–408.

Wells, M. J. (1978). *Octopus: Physiology and Behaviour of an Advanced Invertebrate*. London: Chapman and Hall.

Wells, M. J. (1990). Oxygen extraction and jet propulsion in cephalopods. *Canadian Journal of Zoology*, **68**, 815–824.

Wells, M. J. (1999). Why the ammonites snuffed it. *Marine and Freshwater Behaviour and Physiology*, **32**, 103–111.

Wells, M. J. & Clarke, A. (1996). Energetics: the costs of living and reproducing for an individual cephalopod. *Philosophical Transactions of the Royal Society B: Biological Sciences*, **351**.

Wells, M. J. & O'Dor, R. K. (1991). Jet propulsion and the evolution of the cephalopods. *Bulletin of Marine Science*, **49**, 419–432.

Wells, M. J. & Wells, J. (1956). Tactile discrimination and the behaviour of blind *Octopus*. *Pubblicazioni della Stazione Zoologica di Napoli*, **28**, 94–126.

Wells, M. J. & Wells, J. (1957a). The function of the brain of *Octopus* in tactile discrimination. *Journal of Experimental Biology*, **34**, 131–142.

Wells, M. J. & Wells, J. (1957b). Repeated presentation experiments and the function of the vertical lobe in *Octopus*. *Journal of Experimental Biology*, **34**, 469–477.

Wells, M. J. & Wells, J. (1957c). The effect of lesions to the vertical and optic lobes on tactile discrimination in *Octopus*. *Journal of Experimental Biology*, **34**, 378–393.

Wells, M. J. & Wells, J. (1958). The effect of vertical lobe removal on the performance of octopuses in retention tests. *Journal of Experimental Biology*, **35**, 337–348.

Wells, M. J. & Wells, J. (1959). Hormonal control of sexual maturity in *Octopus*. *Journal of Experimental Biology*, **36**, 1–33.

Wells, M. J. & Wells, J. (1970). Observations on the feeding, growth rate and habits of newly settled *Octopus cyanea*. *Journal of Zoology (London)*, **161**, 65–74.

Wells, M. J. & Wells, J. (1972). Sexual displays and mating of *Octopus vulgaris* Cuvier and *O. cyanea* Gray and attempts to alter the performance by manipulating the glandular condition of the animals. *Animal Behaviour*, **20**, 293–308.

Wells, M. J. & Wells, J. (1977). Cephalopoda: Octopoda, in *Reproduction of Marine Invertebrates,* Vol. 4: *Molluscs: Gastropods and Cephalopods* (ed. A. C. Giese & J. S. Pearse), pp. 291–336. New York, NY: Academic Press.

Wells, M. J. & Young, J. Z. (1968). Learning with delayed rewards in *Octopus*. *Zeitschrift für vergleichende Physiologie*, **61**, 103–128.

Wells, M. J. & Young, J. Z. (1970). Stimulus generalisation in the tactile system of *Octopus*. *Journal of Neurobiology*, **2**, 31–46.

Wells, M. J., Freeman, N. H. & Ashburner, M. (1965). Some experiments on the chemotactile sense of octopuses. *Journal of Experimental Biology*, **43**, 553–563.

Wells, M. J., Wells, J. & O'Dor, R. K. (1992). Life at low oxygen tensions: the behaviour and physiology of *Nautilus pompilius* and the biology of extinct forms. *Journal of the Marine Biological Association of the United Kingdom*, **72**, 313–328.

Westermann, B. & Beuerlein, K. (2005). Y-maze experiments on the chemotactic behaviour of the tetrabranchiate cephalopod *Nautilus pompilius* (Mollusca). *Marine Biology*, **147**, 145–151.

Westermann, B. & Schipp, R. (1998). Cytological and enzyme-histochemical investigations on the digestive organs of *Nautilus pompilius* (Cephalopoda, Tetrabranchiata). *Cell and Tissue Research*, **293**, 327–336.

Westermann, B., Beck-Schildwachter, I., Beuerlein, K., Kaleta, E. F. & Schipp, R. (2004). Shell growth and chamber formation of aquarium-reared *Nautilus pompilius* (Mollusca, Cephalopoda) by X-ray analysis. *Journal of Experimental Zoology*, **301A**, 930–937.

Westermann, B., Ruth, P., Litzlbauer, H. D. *et al.* (2002). The digestive tract of *Nautilus pompilius* (Cephalopoda, Tetrabranchiata): an X-ray analytical and computational tomography study on the living animal. *Journal of Experimental Biology*, **205**, 1617–1624.

Whitehead, M. R. (1990). *Language and Literacy in the Early Years*. London: Chapman.

Wickler, W. (1968). *Mimicry*. London: Weidenfeld and Nicolson.

Wickstead, J. (1956). An unusual method of capturing prey by a cuttlefish. *Nature*, **178**, 929.

Wilbur, K. M. (1983–1988). *Physiology of the Mollusca*. San Diego: Academic Press.

Wiley, R. H. (1983). The evolution of communication: information and manipulation, in *Communication* (ed. T. R. Halliday & P. J. B. Slater), pp. 82–113. Oxford: Blackwell.

Willey, A. (1899). General account of a zoological expedition to the South Seas during the years 1894–1897. *Proceedings of the Zoological Society of London*, **1899**, 7–9.

Williams, B. L., Lovenburg, V., Huffard, C. L. & Caldwell, R. L. (2011). Chemical defense in pelagic octopus paralarvae: tetrodotoxin alone does not protect individual paralarvae of the greater blue-ringed octopus (*Hapalochlaena lunulata*) from common reef predators. *Chemoecology*, **21**, 131–141.

Williams, S. B., Pizarro, O., How, M. *et al.* (2009). Surveying nocturnal cuttlefish camouflage behaviour using an AUV. *2009 IEEE International Conference on Robotics and Automation*, 214–219.

Williamson, G. R. (1965). Underwater observations of the squid *Illex illecebrosus* Lesueur in Newfoundland waters. *Canadian Field-Naturalist*, 79, 239–247.

Williamson, R. (1988). Vibration sensitivity in the statocyst of the northern octopus, *Eledone cirrosa*. *Journal of Experimental Biology*, 134, 451–454.

Williamson, R. (1991). Factors affecting the sensory response characteristics of the cephalopod statocyst and their relevance in predicting swimming performance. *Biological Bulletin*, 180, 221–227.

Williamson, R. & Budelmann, B. U. (1991). Convergent inputs to octopus oculomotor neurones demonstrated in a brain slice preparation. *Neuroscience Letters*, 121, 215–218.

Wilson, E. O. (1975). *Sociobiology*. Boston: Harvard University Press.

Wilson, M., Hanlon, R. T., Tyack, P. L. & Madsen, P. T. (2007). Intense ultrasonic clicks from echolocating toothed whales do not elicit anti-predator responses or debilitate the squid *Loligo pealeii*. *Biology Letters*, 3, 225–227.

Winkelmann, I., Campos, P. F., Strugnell, J. *et al.* (2013). Mitochondrial genome diversity and population structure of the giant squid *Architeuthis*: genetics sheds new light on one of the most enigmatic marine species. *Proceedings of the Royal Society B – Biological Sciences*, 280.

Wirz, K. (1954). Études quantitatives sur le système nerveux des Céphalopodes. *Comptes Rendus de L'Académie des Sciences, Paris*, 238, 1353–1355.

Wirz, K. (1959). Étude biométrique du système nerveux des Céphalopodes. *Bulletin Biologique de la France et de la Belgique*, 93, 78–117.

Wodinsky, J. (1969). Penetration of the shell and feeding on gastropods by *Octopus*. *American Zoologist*, 9, 997–1010.

Wodinsky, J. (1973). Ventilation rate and copulation in *Octopus vulgaris*. *Marine Biology*, 20, 154–164.

Wodinsky, J. (1978). Feeding behaviour of broody female *Octopus vulgaris*. *Animal Behaviour*, 26, 803–813.

Wodinsky, J. (2008). Reversal and transfer of spermatophores by *Octopus vulgaris* and *O. hummelinki*. *Marine Biology*, 155, 91–103.

Wolken, J. J. (1958). Retinal structure. Mollusc Cephalopods: *Octopus, Sepia*. *J. Biophysics and Biochem. Cytol.*, 4.

Wood, F. G. (1963). Observations on the behavior of *Octopus*. *Proceedings of the International Congress of Zoology*, 16, 73.

Wood, J. B., Maynard, A. E., Lawlor, A. G. *et al.* (2010). Caribbean reef squid, *Sepioteuthis sepioidea*, use ink as a defense against predatory French grunts, *Haemulon flavolineatum*. *Journal of Experimental Marine Biology and Ecology*, 388, 20–27.

Wood, J. B., Pennoyer, K. E. & Derby, C. D. (2008). Ink is a conspecific alarm cue in the Caribbean reef squid, *Sepioteuthis sepioidea*. *Journal of Experimental Marine Biology and Ecology*, 367, 11–16.

Woodhams, P. L. & Messenger, J. B. (1974). A note on the ultrastructure of the octopus olfactory organ. *Cell and Tissue Research*, 152, 253–258.

Woods, J. (1965). Octopus-watching off Capri. *Animals*, 7, 324–327.

Wooton, R. J. (1990). *Ecology of Teleost Fishes*. London: Chapman and Hall.

Worms, J. (1983). World fisheries for cephalopods: a synoptic overview, in *Advances in Assessment of World Cephalopod Resources* (ed. J. F. Caddy), pp. 1–20. Rome: FAO.

Wormuth, J. H. (1976). *The Biogeography and Numerical Taxonomy of the Oegopsid Squid Family Ommastrephidae in the Pacific Ocean*. Berkeley: University of California Press.

Würsig, B. (1986). Delphinid foraging strategies, in *Dolphin Cognition and Behavior: A Comparative Approach* (ed. R. J. Schusterman, J. A. Thomas & F. G. Wood), pp. 347–359. Hillsdale: Lawrence Erlbaum Associates.

Xavier, J. C. & Cherel, Y. (2009). *Cephalopod Beak Guide for the Southern Ocean*. Cambridge: British Antarctic Survey.

Xavier, J. C., Cherel, Y., Assis, C. A., Sendao, J. & Borges, T. C. (2010). Feeding ecology of conger eels (*Conger conger*) in north-east Atlantic waters. *Journal of the Marine Biological Association of the United Kingdom*, 90, 493–501.

Yamamoto, M. (1985). Ontogeny of the visual system in the cuttlefish, *Sepiella japonica*. 1. Morphological differentiation of the visual cell. *Journal of Comparative Neurology*, **232**, 347–361.

Yang, W. T., Hanlon, R. T., Krejci, M. E., Hixon, R. F. & Hulet, W. H. (1983). Laboratory rearing of *Loligo opalescens*, the market squid of California. *Aquaculture*, **31**, 77–88.

Yang, W. T., Hanlon, R. T., Lee, P. G. & Turk, P. E. (1989). Design and function of closed seawater systems for culturing loliginid squids. *Aquacultural Engineering*, **8**, 47–65.

Yang, W. T., Hixon, R. F., Turk, P. E. *et al.* (1986). Growth, behavior, and sexual maturation of the market squid, *Loligo opalescens*, cultured through the life cycle. *Fishery Bulletin*, **84**, 771–798.

Yano, K. & Tanaka, S. (1984). Some biological aspects of the deep sea squaloid shark *Centroscymus* from Suruga Bay, Japan. *Bulletin of the Japanese Society of Scientific Fisheries*, **50**, 249.

Yarnall, J. L. (1969). Aspects of the behaviour of *Octopus cyanea* Gray. *Animal Behaviour*, **17**, 747–754.

Young, J. Z. (1939). Fused neurons and synaptic contacts in the giant nerve fibres of cephalopods. *Philosophical Transactions of the Royal Society of London B*, **229**, 465–503.

Young, J. Z. (1950). *The Life of Vertebrates*. Oxford: Oxford University Press.

Young, J. Z. (1958). Effect of removal of various amounts of the vertical lobes on visual discrimination by *Octopus*. *Proceedings of the Royal Society B – Biological Sciences*, **149**, 441–462.

Young, J. Z. (1960a). Observations on *Argonauta* and especially its method of feeding. *Proceedings of the Zoological Society of London*, **133**, 471–479.

Young, J. Z. (1960b). The statocysts of *Octopus vulgaris*. *Proceedings of the Royal Society B: Biological Sciences*, **152**, 3–29.

Young, J. Z. (1960c). Unit processes in the formation of representations in the memory of *Octopus*. *Proceedings of the Royal Society B: Biological Sciences*, **153**, 1–17.

Young, J. Z. (1960d). The failures of discrimination learning following the removal of the vertical lobes in *Octopus*. *Proceedings of the Royal Society B: Biological Sciences*, **153**, 18–46.

Young, J. Z. (1962a). Courtship and mating by a coral reef octopus (*Octopus horridus*). *Proceedings of the Zoological Society of London*, **138**, 157–162.

Young, J. Z. (1962b). Why do we have two brains?, in *Interhemispheric Relations and Cerebral Dominance* (ed. V. B. Mountcastle), pp. 7–24. Baltimore: Johns Hopkins Press.

Young, J. Z. (1963). Light- and dark-adaptation in the eyes of some cephalopods. *Proceedings of the Zoological Society of London*, **140**, 255–272.

Young, J. Z. (1964a). *A Model of the Brain*. Oxford: Clarendon Press.

Young, J. Z. (1964b). Paired centres for the control of attack by *Octopus*. *Proceedings of the Royal Society B: Biological Sciences*, **159**, 565–588.

Young, J. Z. (1965a). The central nervous system of *Nautilus*. *Philosophical Transactions of the Royal Society of London B*, **249**, 1–25.

Young, J. Z. (1965b). The organization of a memory system. *Proceedings of the Royal Society B: Biological Sciences*, **163**, 285–320.

Young, J. Z. (1966). *The Memory System of the Brain*. Oxford: Oxford University Press.

Young, J. Z. (1970). Stalked eyes of *Bathothauma* (Mollusca, Cephalopoda). *Journal of Zoology*, **162**, 437–447.

Young, J. Z. (1971). *The Anatomy of the Nervous System of Octopus vulgaris*. Oxford: Clarendon Press.

Young, J. Z. (1974). The central nervous system of *Loligo*. I. The optic lobe. *Philosophical Transactions of the Royal Society of London B*, **267**, 263–302.

Young, J. Z. (1976). The nervous system of *Loligo*. II. Suboesophageal centres. *Philosophical Transactions of the Royal Society of London B*, **274**, 101–167.

Young, J. Z. (1977a). Brain, behaviour and evolution of cephalopods. *Symposia of the Zoological Society of London*, **38**, 377–434.

Young, J. Z. (1977b). The nervous system of *Loligo*. III. Higher motor centres: the basal supraoesophageal lobes. *Philosophical Transactions of the Royal Society of London B*, **276**, 351–398.

Young, J. Z. (1978). *Programs of the Brain*. Oxford: Clarendon Press.

Young, J. Z. (1979). The nervous system of *Loligo*. V. The vertical lobe complex. *Philosophical Transactions of the Royal Society of London B*, **285**, 311–354.

Young, J. Z. (1981). *The Life of Vertebrates*. Oxford: Clarendon Press.

Young, J. Z. (1983). The distributed tactile memory system of *Octopus*. *Proceedings of the Royal Society B: Biological Sciences*, **218**, 135–176.

Young, J. Z. (1984). The statocysts of cranchiid squids (Cephalopoda). *Journal of Zoology (London)*, **203**, 1–21.

Young, J. Z. (1985). Cephalopods and neuroscience. *Biological Bulletin*, **168**, 153–158.

Young, J. Z. (1988). Evolution of the cephalopod brain, in *The Mollusca*, Vol.12: *Paleontology and Neontology of Cephalopods* (ed. M. R. Clarke & E. R. Trueman), pp. 215–228. San Diego: Academic Press.

Young, J. Z. (1989). The angular acceleration receptor system of diverse cephalopods. *Philosophical Transactions of the Royal Society of London B*, **325**, 189–238.

Young, J. Z. (1991). Computation in the learning system of cephalopods. *Biological Bulletin*, **180**, 200–208.

Young, M., Kvitek, R. G., Iampietro, P. J., Hanlon, R. T. & Malliet, R. (2011). Seafloor mapping and landscape ecology analyses used to monitor variations in spawning site preference and benthic egg mop abundance for the California market squid (*Doryteuthis opalescens*). *Journal of Experimental Marine Biology and Ecology*, **407**, 226–233.

Young, R. E. (1972a). Brooding in a bathypelagic octopus. *Pacific Science*, **26**, 400–404.

Young, R. E. (1972b). The systematics and areal distribution of pelagic cephalopods from the seas off southern California. *Smithsonian Contributions to Zoology*, **97**, 1–159.

Young, R. E. (1973). Information feedback from photophores and ventral countershading in mid-water squid. *Pacific Science*, **27**, 1–7.

Young, R. E. (1975a). Transitory eye shapes and the vertical distribution of two mid-water squids. *Pacific Science*, **29**, 243–255.

Young, R. E. (1975b). Function of the dimorphic eyes in the midwater squid *Histioteuthis dofleini*. *Pacific Science*, **29**, 211–218.

Young, R. E. (1977). Ventral bioluminescent countershading in midwater cephalopods. *Symposia of the Zoological Society of London*, **38**, 161–190.

Young, R. E. (1978). Vertical distribution and photosensitive vesicles of pelagic cephalopods from Hawaiian waters. *Fishery Bulletin*, **76**, 583–615.

Young, R. E. (1983). Oceanic bioluminescence: an overview of general functions. *Bulletin of Marine Science*, **33**, 829–845.

Young, R. E. (1995). Aspects of the natural history of pelagic cephalopods of the Hawaiian mesopelagic-boundary region. *Pacific Science*, **49**, 143–155.

Young, R. E. & Arnold, J. M. (1982). The functional morphology of a ventral photophore from the mesopelagic squid, *Abralia trigonura*. *Malacologia*, **23**, 135–163.

Young, R. E. & Bennett, T. M. (1988). Photophore structure and evolution within the Enoploteuthinae (Cephalopoda), in *The Mollusca*, Vol. 12: *Paleontology and Neontology of Cephalopods* (ed. M. R. Clarke & E. R. Trueman), pp. 241–251. San Diego, CA.: Academic Press.

Young, R. E. & Harman, R. F. (1988). 'Larva,' 'paralarva' and 'subadult' in cephalopod terminology. *Malacologia*, **29**, 201–207.

Young, R. E. & Roper, C. F. E. (1976). Bioluminescent countershading in midwater animals: evidence from living squid. *Science*, **191**, 1046–1048.

Young, R. E. & Roper, C. F. E. (1977). Intensity regulation of bioluminescence during countershading in living midwater animals. *Fishery Bulletin*, **75**, 239–252.

Young, R. E., Kampa, E. M., Maynard, S. D., Mencher, F. M. & Roper, C. F. E. (1980). Counterillumination and the upper depth limits of midwater animals. *Deep-Sea Research Part A – Oceanographic Research Papers*, **27**, 671–691.

Young, R. E., Roper, C. F. E. & Walters, J. F. (1979). Eyes and extraocular photoreceptors in midwater cephalopods and fishes: their roles in detecting downwelling light for counterillumination. *Marine Biology*, **51**, 371–380.

Young, R. E., Seapy, R. R., Mangold, K. & Hochberg, F. G. (1982). Luminescent flashing in the midwater squids *Pterygioteuthis microlampas* and *P. giardi*. *Marine Biology*, **69**, 299–308.

Young, R. E., Vecchione, M. & Mangold, K. (2008). Cephalopoda Cuvier 1797: octopods, squids, cuttlefish, nautiluses. *The Tree of Life Project*, Version 21, April 2008.

Zahavi, A. (1975). Mate selection – a selection for handicap. *Journal of Theoretical Biology*, **53**, 205–214.

Zahavi, A. (1980). Ritualization and the evolution of movement signals. *Behaviour*, **72**, 77–81.

Zahavi, A. (1987). The theory of signal selection and some of its implications, in *International Symposium on Biological Evolution* (ed. V. P. Delfino). Bari, Italy: Adriatica Editrice.

Zakharov, Y. D., Shigeta, Y., Smyshlyaeva, O. P., Popov, A. M. & Ignatiev, A. V. (2006). Relationship between delta C-13 and delta O-19 values of the recent *Nautilus* and brachiopod shells in the wild and the problem of reconstruction of fossil cephalopod habitat. *Geosciences Journal*, **10**, 331–345.

Zann, L. P. (1984). The rhythmic activity of *Nautilus pompilius*, with notes on its ecology and behavior in Fiji. *Veliger*, **27**, 19–28.

Zatylny, C., Gagnon, J., Boucaud-Camou, E. & Henry, J. (2000). ILME: A waterborne pheromonal peptide released by the eggs of *Sepia officinalis*. *Biochemical and Biophysical Research Communications*, **275**, 217–222.

Zeidberg, L. D. (2009). First observations of 'sneaker mating' in the California market squid, *Doryteuthis opalescens* (Cephalopoda: Myopsida). *Marine Biodiversity Records*, **2**, e6.

Zeidberg, L. D., Butler, J. L., Ramon, D. *et al.* (2012). Estimation of spawning habitats of market squid (*Doryteuthis opalescens*) from field surveys of eggs off central and southern California. *Marine Ecology – An Evolutionary Perspective*, **33**, 326–336.

Zeidberg, L. D., Hamner, W., Moorehead, K. & Kristof, E. (2004). Egg masses of *Loligo opalescens* (Cephalopoda: Myopsida) in Monterey Bay, California following the El Nino event of 1997–1998. *Bulletin of Marine Science*, **74**, 129–141.

Zonana, H. V. (1961). Fine structure of squid retina. *Bulletin of Johns Hopkins Hospital*, **109**, 185–205.

Zullo, L., Sumbre, G., Agnisola, C., Flash, T. & Hochner, B. (2009). Nonsomatotopic organization of the higher motor centers in *Octopus*. *Current Biology*, **19**, 1632–1636.

Zylinski, S. & Johnsen, S. (2011). Mesopelagic cephalopods switch between transparency and pigmentation to optimize camouflage in the deep. *Current Biology*, **21**, 1937–1941.

Zylinski, S. & Osorio, D. (2011). What can camouflage tell us about non-human visual perception? A case study of multiple cue use in cuttlefish (*Sepia officinalis*), in *Animal Camouflage: Mechanisms and Function* (ed. M. Stevens & S. Merilaita), pp. 164–185. Cambridge: Cambridge University Press.

Zylinski, S., Darmaillacq, A. S. & Shashar, N. (2012). Visual interpolation for contour completion by the European cuttlefish (*Sepia officinalis*) and its use in dynamic camouflage. *Proceedings of the Royal Society B – Biological Sciences*, **279**, 2386–2390.

Zylinski, S., How, M. J., Osorio, D., Hanlon, R. T. & Marshall, N. J. (2011). To be seen or to hide: visual characteristics of body patterns for camouflage and communication in the Australian giant cuttlefish *Sepia apama*. *American Naturalist*, **177**, 681–690.

Zylinski, S., Osorio, D. & Shohet, A. (2009a). Cuttlefish camouflage: context-dependent body pattern use during motion. *Proceedings of the Royal Society B: Biological Sciences*, **276**, 3963–3969.

Zylinski, S., Osorio, D. & Shohet, A. (2009b). Edge detection and texture classification by cuttlefish. *Journal of Vision*, **9**, 1–10.

Zylinski, S., Osorio, D. & Shohet, A. J. (2009c). Perception of edges and visual texture in the camouflage of the common cuttlefish, *Sepia officinalis*. *Philosophical Transactions of the Royal Society B: Biological Sciences*, **364**, 439–448.

Scientific Names

Systematists have an irritating habit of every now and then changing the names of animals. This is not our fault. However, it can lead to confusion: the giant Pacific octopus that many of us used to know as *Octopus dofleini* is now called *Enteroctopus dofleini*. A squid often referred to in this book as *Loligo pleii* has recently become *Doryteuthis pleii*.

In response to this problem we have decided, throughout this book, to use the generic name of an animal *as it appears in the original published article*.

Index of Organisms

Bold typeface indicates figures.

Subject Index

Bold typeface indicates figures; italic indicates boxes or tables.

cannibalism, *90*, 94–96, 181, 183, 252, 279

catch rate, 79

cephalotoxin, 39

chemical trails, 278, 283

chemoreception, 31–32, 44, 78, 135, 210, 274–275

chemotaxis, 31

chitin, 39
  in beak, 39

chitinous, 13

chromatophore
  organs, 36
  pigmentation, 109

chromatophores, 35
  colour class, 45–46, **51**
  control, **58**
  density, 46
  lobes, *see* brains
  radial muscles, 33, 35–36, **37**, 46, 59, 67

cilia, 92

circadian rhythms, 257 *see* rhythms

cirri, 39, 74

clandestine escape, 129, 133, **134**

closed-loop visual control, 79

CNS (central nervous system), 40, 67, 76, 102, 221, 225, 283, 286

cognition, 221, 236, 241

collar, 34

colonies, 243

colour
  blindness, 22–23
  change, 45, 58
  matching, 44, 101

colour vision, 20, 22, 24, 169, 264

colours, 25, 36, 45–46, 57–58, 70, 102, 119, 169, 261
  pigmentary, 45, *46*, 58
  structural, 36, *46*

communication channels, 206, 209–210, 218

communities, 268

compensatory movements, 18, 276

competition, 9, **11**, 249, *see also* sperm
    competition
  for resources, 247, 270

components
  chromatic, *207*
  locomotor, 45, 53, 72, *207*
  postural, 53, 99, 171, *207*
  textural, 53, *207*, 223

components (of body patterns), 58, 218
  chromatic, 45, 70, 207

concealment, 34, **57**, 67, 73, 97, 107, 112, 116, 120, **224**, 261, 288 *see* crypsis

conditioned stimulus, 80

conditioning
  classical, 228, 240
  instrumental (operant), 30, 228

contest duration, 190, **190**, 191, 194

copulation
  extra-pair, 187, *188*, 189–190

coral reefs, 71, 73, 110, 116, 118, 121, 130, 144, 165, 177, 189, 246–249, 268, 277, 283, 286–287

counter-illumination, ventral, 12, 38, 109, 261

countershading, 26, 72, **101**, 107–108, 138, 262, 273

countershading reflex, 25, 59, 107, **108**

courting parties, 165, 169

courtship, 70, 72, 148, 150, 152, 155, 159, 162, 164–165, 168, 171, 177, 180–181, 184, 187–189, **192**, 196, 204–206, 210–211, 216–217, 244, 258, 280, *see also* vertical migration

crepuscular activity, 119

crista–cupula system, 25

crop, 77, 276, 279

crypsis, **103**, **109**

crypsis (cryptic coloration), 67, 99, 138, 184
  as primary defence, 97, **99**, 146

cryptic behaviour, 97, 119, 121

cue-additivity, 237

cuttlebone, 106, 144, 157, 257

dark-adaptation, 19

deceptive resemblance, 70, 112, **114**, **117**, 130

decision-making, 143, 146

deep-sea cephalopods, 12, 14, 177, 179, 205, 259, 271

defence
  primary, 70, 97
  secondary, 68, 70–71, 99, 121–122, 124, 135, 138–139, 141, **141**, 144, 146–147, 206, 246, 262, 285–286, 333

defence sequences, 133, **140**

defences
  secondary, 121, **134**, 135, 139, **140**, 143

defensive postures, 131, 141

deflective marks, **131**, 131

deimatic (= dymantic)
  behaviour, 122, 124–126, 129, 135, 141, 146, 181
  displays, 68, 70, **99**, 124–126, 133, **134**, 135, 138, 141, 146, 207, *209*, 215, 217, 219, 223–224, 288

delayed rewards, 233

dens, 86–87, 89, 91, **119**, 119, 121, 183, 185, 230–231, 237, 243, 248–249, 257–258, 270

depth distribution, 12, 256, 266, 278

depth perception, 17

dermis, 36, 45, 50, 72

development
  escape response, 221
  prey capture, 225, 240

Devonian, 9

diel rhythm, *see* rhythms, diel

diet, 14, 92
  ontogenetic change, 14, 92

digestion, 13, 77
  external, 77, 319, 331

digging behaviour, 221, 279

dilution effect, 132

directive marks, 83

discrimination
  size, 238

discrimination learning, 227–229
  chemotactile, 39
  tactile, 30, 227, 229
  visual, **229**, 236

dishabituation, 227, **227**

displays, *see also* deimatic displays,
    flamboyant displays
  agonistic, *72*, 153, 161, 169, 187, **193**, **207**, 215
  ritualised, 207
  sexual, 288
  sucker, 151, 181, *188*
  unilateral, 125, 168, 198
  visual, 39, 126, 183

disruptive coloration, 70, 103, 107, 146

distant touch, 77

diversity, **1**, 9, 13, 45, 58, 70, 72, 116, 118, 142, 146, 177, 263, 285–286, 290

diving, SCUBA, 11–12, 89, 176

dominance relationships, 243

dopamine, 32, 38–39, 130, 210

downwelling light, 23, 107–108, **109**, 109, 261–262, 271

dymantic, 125, *see also* deimatic

ECS (electro-convulsive shock), 233

effectors, 16, **17**, 33, 36, 40, 44, 92, 285, *see also* arm appendages, buccal mass, chromatophores, ink sac, muscles, photophores, reflecting cells

elective group size, 270